Have you been to our website?

For code downloads, print and e-book bundles, extensive samples from all books, special deals, and our blog, please visit us at:

www.rheinwerk-computing.com

Rheinwerk Computing

The Rheinwerk Computing series offers new and established professionals comprehensive guidance to enrich their skillsets and enhance their career prospects. Our publications are written by the leading experts in their fields. Each book is detailed and hands-on to help readers develop essential, practical skills that they can apply to their daily work.

Explore more of the Rheinwerk Computing library!

Joachim Steinwendner, Roland Schwaiger
Programming Neural Networks with Python
2025, 457 pages, paperback and e-book
www.rheinwerk-computing.com/6059

Johannes Ernesti, Peter Kaiser
Python 3: The Comprehensive Guide
2022, 1078 pages, paperback and e-book
www.rheinwerk-computing.com/5566

Veit Steinkamp
Python for Engineering and Scientific Computing
2024, 511 pages, paperback and e-book
www.rheinwerk-computing.com/5852

Bert Gollnick
Generative AI with Python: The Developer's Guide to Pretrained LLMs, Vector Databases, Retrieval-Augmented Generation, and Agentic Systems
2025, 392 pages, paperback and e-book
www.rheinwerk-computing.com/6057

Metin Karatas
Developing AI Applications: An Introduction
2024, 402 pages, paperback and e-book
www.rheinwerk-computing.com/5899

www.rheinwerk-computing.com

Mohammad Nauman

Keras 3

The Comprehensive Guide to Deep Learning
with the Keras API and Python

Editor Rachel Gibson
Acquisitions Editor Hareem Shafi
Copyeditor Julie McNamee
Cover Design Graham Geary
Photo Credits Shutterstock: 1305456070/© optimarc; iStockphoto: 2153711283/© shulz
Layout Design Vera Brauner
Production Kelly O'Callaghan
Typesetting III-satz, Germany
Printed and bound in Canada, on paper from sustainable sources

ISBN 978-1-4932-2739-6
1st edition 2026

© 2026 by:
Rheinwerk Publishing, Inc.
2 Heritage Drive, Suite 305
Quincy, MA 02171
USA
info@rheinwerk-publishing.com
+1.781.228.5070

Represented in the E.U. by:
Rheinwerk Verlag GmbH
Rheinwerkallee 4
53227 Bonn
Germany
service@rheinwerk-verlag.de
+49 (0) 228 42150-0

Library of Congress Cataloging-in-Publication Control Number: 2025034926

All rights reserved. Neither this publication nor any part of it may be copied or reproduced in any form or by any means or translated into another language, without the prior consent of Rheinwerk Publishing.

Rheinwerk Publishing makes no warranties or representations with respect to the content hereof and specifically disclaims any implied warranties of merchantability or fitness for any particular purpose. Rheinwerk Publishing assumes no responsibility for any errors that may appear in this publication.

"Rheinwerk Publishing", "Rheinwerk Computing", and the Rheinwerk Publishing and Rheinwerk Computing logos are registered trademarks of Rheinwerk Verlag GmbH, Bonn, Germany.

All products mentioned in this book are registered or unregistered trademarks of their respective companies.

No part of this book may be used or reproduced in any manner for the purpose of training artificial intelligence technologies or systems. In accordance with Article 4(3) of the Digital Single Market Directive 2019/790, Rheinwerk Publishing, Inc. expressly reserves this work from text and data mining.

Contents at a Glance

1	Introduction	17
2	Introduction to the Core of Machine Learning	33
3	Fundamentals of Gradient Descent	79
4	Classification Through Gradient Descent	117
5	Deep Dive into Keras	167
6	Regularization Techniques	223
7	Convolutional Neural Networks	265
8	Exploring the Keras Functional API	315
9	Understanding Transformers	375
10	Reinforcement Learning: The Secret Sauce	429
11	Autoencoders and Generative AI	495
12	Advanced Generative AI: Stable Diffusion	553
13	Recap of Key Concepts	605

Contents

1 Introduction — 17

- **1.1 Overview of Deep Learning** — 18
  - 1.1.1 The Success of Deep Learning — 18
  - 1.1.2 The Two Pillars Supporting These Breakthroughs — 21
  - 1.1.3 Why Now? — 22
- **1.2 Why Keras** — 23
- **1.3 The Structure of This Book** — 25
  - 1.3.1 Text Boxes — 27
- **1.4 How to Use This Book** — 28
  - 1.4.1 Why Keras Installation Comes Later — 29
  - 1.4.2 Suggested Reading Strategy — 30

2 Introduction to the Core of Machine Learning — 33

- **2.1 What Is Machine Learning?** — 35
  - 2.1.1 Helping the Machine Recognize Digits — 35
  - 2.1.2 The Outdated Method—Rule-Based Learning — 36
  - 2.1.3 Case Study in Machine Learning Code — 38
  - 2.1.4 Structuring the Learning Process — 41
  - 2.1.5 Analyzing the Machine's Learning Process — 46
  - 2.1.6 Lessons Learned from the Code Case Study — 48
- **2.2 Types of Machine Learning** — 49
  - 2.2.1 Supervised Learning — 50
  - 2.2.2 Unsupervised Learning — 59
- **2.3 The Magic Sauce: Reinforcement Learning** — 65
  - 2.3.1 The Building Blocks of Reinforcement Learning — 65
  - 2.3.2 Key Applications for Reinforcement Learning — 67
  - 2.3.3 Challenges in Implementing Reinforcement Learning — 68
- **2.4 Basics of Neural Networks** — 69
  - 2.4.1 Core Components of Neural Networks — 70
  - 2.4.2 The Unintuitive Process of Learning — 72
  - 2.4.3 Clearing Up Some Misconceptions — 73

2.5	Setting Up Your Environment		73
2.6	Summary		78

3   Fundamentals of Gradient Descent   79

3.1	**Understanding Gradient Descent**		80
	3.1.1	The Basic Setup	80
	3.1.2	Formalizing the Process: Training Phase	82
	3.1.3	From Training to Deployment: Making It Work in the Real World	84
	3.1.4	The Process of Learning	85
	3.1.5	Finding the Best Parameters: Minimizing the Loss	89
	3.1.6	Lifting the Assumptions of a Single Input Feature	94
	3.1.7	Higher Dimensions in Gradient Descent	99
3.2	**Types of Gradient Descent: Batch, Stochastic, Mini-Batch**		101
	3.2.1	Contour Plots for Visualization of Gradient Descent	102
	3.2.2	Improving the Efficiency of Batch Gradient Descent	103
	3.2.3	A Paradox in the Use of Stochastic Gradient Descent	105
3.3	**Learning Rate and Optimization**		107
3.4	**Implementing Gradient Descent in Code**		110
	3.4.1	Gradient Descent from Scratch	110
	3.4.2	Gradient Descent Using Keras	113
3.5	**Summary**		116

4   Classification Through Gradient Descent   117

4.1	**Classification Basics**		118
	4.1.1	Classification Problem Setup	119
	4.1.2	First Attempt Using Gradient Descent	121
	4.1.3	Second Attempt: Fixing the Issues in the First Attempt	122
	4.1.4	Third Attempt: Fixing the Loss Function	128
	4.1.5	Squishing Functions and Decision Boundaries	131
	4.1.6	Learning Process Summary	134
4.2	**Nonlinear Relationships and Neural Networks**		136
	4.2.1	Feature Transformations	137
	4.2.2	The Kernel Trick, Logistic Regression, and All of Machine Learning	140

4.3	**Binary vs. Multi-Class Classification**	147
	4.3.1 The One-vs-All Approach	147
	4.3.2 The Softmax Classifier	152
4.4	**Loss Functions: Cross-Entropy**	155
	4.4.1 Categorical Cross-Entropy	156
	4.4.2 Pitfall to Avoid When Using Cross Entropies	157
	4.4.3 Sparse Categorical Cross-Entropy	158
4.5	**Building a Classifier with Gradient Descent**	161
	4.5.1 Layers in Code	161
	4.5.2 Choice of Loss Function	163
	4.5.3 Parameter Counts	164
	4.5.4 The Density of Keras Code	165
4.6	**Summary**	166

5 Deep Dive into Keras   167

5.1	**Introduction to Keras Framework**	168
	5.1.1 The Philosophy Behind Keras: Making AI Human-Friendly	169
	5.1.2 Evolution Through Adaptability	169
	5.1.3 Keras 3.0: The Multi-Engine Framework	171
	5.1.4 Key Strengths of Keras	172
	5.1.5 Keras in the Real World	173
5.2	**Setting Up Keras**	174
	5.2.1 Setting Up Python	175
	5.2.2 TensorFlow Installation and Points to Keep in Mind	177
	5.2.3 Setting Up CUDA for GPU Acceleration	180
	5.2.4 Installing Keras	185
	5.2.5 Using a GPU in Google Collaboratory	186
5.3	**Building Your First Model**	188
	5.3.1 Why NumPy Matters for Machine Learning	188
	5.3.2 Symbolic Computation: The Magic Behind Neural Networks	200
5.4	**Implementing Core Concepts in Keras: Gradient Descent and Classification**	205
	5.4.1 The Building Blocks of a Cat-Dog Classifier	205
	5.4.2 Getting and Fixing the Data	206
	5.4.3 Performance Optimization and Model Specification	210

		5.4.4	Evaluation Metrics	212
		5.4.5	Guiding the Training Through Callbacks and Checkpoints	216
		5.4.6	Evaluating the Model	218
	5.5	Summary		222

6 Regularization Techniques — 223

	6.1	An Overview of Overfitting and Underfitting: Do You Need More Data?	224
		6.1.1 From Lines to Curves: Adding Polynomial Features	225
		6.1.2 Using Increasingly Complex Models	228
		6.1.3 The Balance of Complexity	229
		6.1.4 Regularization Term	234
		6.1.5 Adjusting the Complexity Knob	237
		6.1.6 Do I Need More Data	240
		6.1.7 Reporting the Final Results: Validation Set	241
	6.2	Dropout: Concept and Implementation	243
		6.2.1 The Problem: Co-Adaptation and Overfitting	243
		6.2.2 The Road to Memorization	244
		6.2.3 The Ensemble Intuition Behind Dropout	245
		6.2.4 Dropout Mechanics	246
		6.2.5 Finding the Sweet Spot: Dropout Rates in Practice	247
		6.2.6 Implementing Dropout in Pure Python	248
		6.2.7 Common Pitfalls and Debugging Tips	250
	6.3	Other Regularization Methods: L1 and L2 Regularization	251
		6.3.1 L1 Regularization (Lasso)	252
		6.3.2 Elastic Net: Combining L1 and L2	252
		6.3.3 When Not to Use Regularization	253
		6.3.4 Practical Considerations: Dropout vs. L1/L2 in Neural Networks	254
	6.4	Applying Regularization in Keras	254
		6.4.1 Implementing L2 Regularization in Keras	255
		6.4.2 Dropout in Keras	255
		6.4.3 Beyond Basic Dropout: Specialized Variants	257
		6.4.4 Finding the Perfect Dropout Rate: A Systematic Approach	259
	6.5	Summary	264

7 Convolutional Neural Networks — 265

7.1 Introduction to Convolutional Neural Networks — 266
7.1.1 The Limitation of Fully Connected Networks — 266
7.1.2 The Learning Standstill — 269
7.1.3 Solving the Vanishing Gradient Problem — 270
7.1.4 Dense vs. Sparse Connections — 272
7.1.5 From Convolution to Neural Networks: The Conv2D Layer — 279

7.2 Convolutional Layers, Pooling Layers and Fully Connected Layers — 287
7.2.1 Core Implementation of a Convolutional Layer — 288
7.2.2 The Hidden Superpowers of Convolutional Layers — 291
7.2.3 Pooling Layers: The Image Simplifiers — 293
7.2.4 Global Pooling — 295
7.2.5 Bringing It All Together: Fully Connected Layers in CNNs — 299

7.3 Implementing CNNs with Keras — 301
7.3.1 Conv2D vs. Conv1D — 301
7.3.2 The Opposite of Convolution: Deconvolution Layers — 301

7.4 The "Shapes" Problem — 303

7.5 Case Study: Image Classification — 307

7.6 Summary — 313

8 Exploring the Keras Functional API — 315

8.1 Overview of Keras Functional API — 316
8.1.1 The Information Bottleneck — 317
8.1.2 Networks as Directed Acyclic Graphs — 318
8.1.3 The Functional Programming Heritage — 319
8.1.4 Key Advantages of the Functional API — 320

8.2 Building Complex Models with the Functional API — 323
8.2.1 Overview of the Functional API Syntax — 323
8.2.2 Creating Models with the Functional API — 324
8.2.3 Best Practices for Complex Models — 327
8.2.4 Handling Multiple Inputs and Outputs — 328
8.2.5 Example: Building an Image Captioning Model — 331
8.2.6 Residual Connections — 332
8.2.7 Branching Architectures — 336

8.3 Use Cases and Examples — 340
8.3.1 Image Classification with ResNet — 341

		8.3.2 Siamese Networks for Similarity Learning	346
		8.3.3 U-Net for Image Segmentation	354
8.4	**Using Transfer Learning to Customize Models for Your Organization**		364
		8.4.1 The Why of Transfer Learning	365
		8.4.2 Leveraging Pretrained Models from Keras	366
		8.4.3 The Process of Transfer Learning	368
		8.4.4 Reloading an Existing Model	368
8.5	**Summary**		373

9 Understanding Transformers   375

9.1 The Theory Behind Transformers   376
- 9.1.1 A Simple Time Series Example   377
- 9.1.2 From Numbers to Words: The Challenge of Text Data   382
- 9.1.3 GloVe: Learning the Language of Vectors   383
- 9.1.4 A Gentle Introduction to Attention   388
- 9.1.5 Why Transformers Revolutionized Natural Language Processing and Beyond   391

9.2 Components: Attention Mechanism, Encoder, Decoder   393
- 9.2.1 The Conversation Between Words   394
- 9.2.2 Why Position Information Matters   398
- 9.2.3 Encoder Structure: The Information Processing Powerhouse   401
- 9.2.4 Decoder Structure: Creating New Sequences from Understanding   403

9.3 Implementing Transformers in Keras   406
- 9.3.1 The Encoder Block   407
- 9.3.2 The Decoder Block   413
- 9.3.3 The Transformer: Putting the Encoder and Decoder Together   415

9.4 Case Study: Large Language Model Chatbot   418
- 9.4.1 Structure of Modern Keras Transformer Models   418
- 9.4.2 Working with Pretrained Models from Kaggle Hub   422

9.5 Summary   427

10 Reinforcement Learning: The Secret Sauce   429

10.1 Introduction to Reinforcement Learning   430
- 10.1.1 The Problem of Learning by Doing   430
- 10.1.2 Brief History and Major Breakthroughs   433

	10.1.3	Real-World Applications	434
	10.1.4	Challenges Unique to Reinforcement Learning	436
10.2	**Key Concepts: Agents, Environments, Rewards**		**438**
	10.2.1	Structure of the Reinforcement Learning Framework	438
	10.2.2	Environment Design and State Representation	440
	10.2.3	Understanding Agents and Policy Functions	442
	10.2.4	Reward Engineering and Signal Design	443
	10.2.5	The Exploration vs. Exploitation Dilemma	445
10.3	**Popular Algorithms: Q-Learning, Policy Gradients, and Deep Q-Networks**		**447**
	10.3.1	The Markov Decision Processes	447
	10.3.2	Value Functions and Q-Tables	449
	10.3.3	Building the Q-Table	451
	10.3.4	Q-Learning Algorithm: A Worked Example	453
	10.3.5	Q-Learning and Associated Issues	458
	10.3.6	The Limits of Tabular Q-Learning	460
10.4	**Implementing Reinforcement Learning Models in Keras**		**464**
	10.4.1	Our First Reinforcement Learning Environment	465
	10.4.2	Implementing the Deep Q-Network Algorithm with Keras	473
	10.4.3	Experience Replay and Target Networks: The Foundations of Stable Deep Reinforcement Learning	482
10.5	**Reinforcement Learning in Large Language Models**		**486**
	10.5.1	The Fundamental Challenge: Moving Beyond Prediction	487
	10.5.2	Challenges and Limitations	489
	10.5.3	Future Directions and Emerging Approaches	491
10.6	**Summary**		**493**

11 Autoencoders and Generative AI   495

11.1	**Introduction to Autoencoders**		**496**
	11.1.1	What Are Autoencoders?	497
	11.1.2	Autoencoder Architecture Deep Dive	500
	11.1.3	Building Your First Autoencoder in Keras	505
	11.1.4	Types of Autoencoders	514
11.2	**Variational Autoencoders**		**519**
	11.2.1	Navigating the Space with Uncertainty	520
	11.2.2	Mathematical Framework of Variational Autoencoders	521
	11.2.3	Variational Autoencoder Implementation in Keras	525

Contents

11.3 Generative Adversarial Networks 535
    11.3.1 The Adversarial Game 535
    11.3.2 Generative Adversarial Network Architecture 537
    11.3.3 Other Variations 541
    11.3.4 Generative Adversarial Network Implementation in Keras 543
    11.3.5 Implementation Challenges 551

11.4 Summary 552

12 Advanced Generative AI: Stable Diffusion   553

12.1 Theory Behind Stable Diffusion 554
    12.1.1 From Previous Generative Models to Diffusion 555
    12.1.2 Diffusion Process Fundamentals 557
    12.1.3 Reverse Diffusion: Learning to Denoise 558
    12.1.4 Connections to Physical Processes 559
    12.1.5 Denoising Diffusion Probabilistic Models 559
    12.1.6 Latent Diffusion and Stable Diffusion Architecture 562
    12.1.7 Cross-Attention: The Bridge Between Text and Images 564

12.2 How Stable Diffusion Uses Core Concepts 565
    12.2.1 Efficient Diffusion Through Learned Representations 566
    12.2.2 Advanced Attention Mechanisms 567
    12.2.3 Training Strategies and Optimization 570

12.3 Implementing Stable Diffusion Models 572
    12.3.1 Environment and Data Prep 573
    12.3.2 Setting Up an Evaluation Measure 574
    12.3.3 Model Description and Time-Step Encodings 578
    12.3.4 Diffusion and Reverse Diffusion 582
    12.3.5 The Generation Engine 584
    12.3.6 Following Progress in the Training Process 589

12.4 Case Study: Image Generation 593
    12.4.1 Loading Pretrained Models from Keras Hub 594
    12.4.2 Loading Models Through Keras Hub 595
    12.4.3 Using Stable Diffusion Models 596
    12.4.4 Beyond Image Generation to More Complex Workflows 599

12.5 Summary 603

13 Recap of Key Concepts — 605

13.1 Future Trends in Deep Learning — 606
- 13.1.1 Advanced Architecture Trajectories — 607
- 13.1.2 Reinforcement Learning Frontiers — 608
- 13.1.3 Generative AI Revolution — 609

13.2 Tips for Staying Updated with Advancements — 611
- 13.2.1 Technical Skills Maintenance — 611
- 13.2.2 Following Tutorials and Keras Codebase — 612
- 13.2.3 Research Consumption Strategy — 613
- 13.2.4 Community Engagement — 614

13.3 Following the Latest Research — 615
- 13.3.1 Technical Deep Dives — 616
- 13.3.2 Practical Research Integration — 617
- 13.3.3 Parting Words — 618

The Author — 619
Index — 621

Chapter 1
Introduction

In this chapter, we'll explore some success stories of deep learning and discuss why Keras is the ideal framework for learning this impressive technology, as well as what you can expect from this book's comprehensive journey and how to use it most effectively to build your expertise.

Welcome to your deep dive into the world of deep learning and modern artificial intelligence (AI). Whether you're a curious beginner taking your first steps into machine learning or an experienced programmer looking to master Keras, this book is designed to guide you through the concepts that power today's most remarkable AI systems. From core algorithms that are deceptively simple to the modern large language models (LLMs) and photorealistic image generation giants, the techniques you'll learn here form the foundation of modern AI. In this opening chapter, we won't dive into the technical depths of neural networks or start building models just yet. Instead, we'll focus on understanding what makes this book unique and how it can help you master state-of-the-art machine learning techniques. The concepts we'll explore throughout these pages represent the same fundamental principles that drive breakthrough technologies such as ChatGPT, image generation systems, and autonomous vehicles.

Every great journey begins with understanding the path ahead, and that's exactly what we'll accomplish in this chapter. We'll explore why deep learning has become so powerful, why Keras is the perfect tool for learning these concepts, and how this book is structured to take you from foundational principles to advanced applications. This foundation will prove invaluable as we progress through increasingly sophisticated topics, ensuring you're well-equipped to tackle each new concept with confidence. With that groundwork in mind, let's begin by exploring what deep learning really means and why it has become the driving force behind today's AI revolution.

> **Objective of This Chapter**
> While we'll begin the technical details of machine learning fundamentals in the next chapter, this introductory chapter serves the important purpose of helping you understand how to use this book most effectively.

1.1 Overview of Deep Learning

Picture a helicopter hovering steadily in turbulent wind, its rotors adjusting hundreds of times per second to maintain perfect balance. Now imagine that helicopter has no human pilot. Instead, it's controlled by AI that learned to fly by crashing thousands of times in simulation, each failure teaching it something new about the delicate dance of rotor physics and wind dynamics. This is perfectly possible today in research labs around the world, where AI systems are mastering tasks that once seemed impossibly complex for machines.

We're living through what many consider the most remarkable period in the history of AI. Every few months brings news of another breakthrough that would have seemed like magic just a decade ago. Machines are now solving scientific problems that have puzzled researchers for decades, creating art that moves us emotionally and making decisions in split seconds that save lives. These achievements signal a fundamental shift in what we can expect from computer systems in the future. What makes this moment particularly exciting is that these aren't isolated tricks or narrow solutions to specific problems. Instead, they emerge from the same underlying principles and techniques that you'll learn throughout this book. The algorithms powering a self-driving car share fundamental similarities with those generating photorealistic images or discovering new *precision medicines*. Understanding these will enable you to see not just how these systems work but how to build them yourself.

In this section, we'll discuss some of the important success stories of deep learning in the recent past, study briefly what made these successes possible, and discuss the importance of why this happened at this particular time in history.

1.1.1 The Success of Deep Learning

The breadth of applications where AI now excels was unimaginable just a few years ago. Let's explore some of the most remarkable achievements that showcase what's possible when we harness the power of modern machine learning. These examples will hopefully motivate you to build systems that solve problems in many more domains.

The New Frontier of Autonomous Systems

Consider the challenge of *autonomous driving*—a problem that requires processing visual information faster than any human while making life-or-death decisions in milliseconds. Modern self-driving cars use deep learning that can identify pedestrians in shadows, predict the behavior of other drivers, and navigate complex intersections where multiple vehicles, cyclists, and pedestrians interact simultaneously. These systems process input from cameras, radar, and lidar sensors, combining all of this information into a coherent understanding of their environment.

What makes autonomous vehicles particularly impressive is their ability to handle edge cases—unexpected situations that engineers never explicitly programmed them to encounter. Examples include a child's ball rolling into the street, construction zones with unusual traffic patterns, or emergency vehicles requiring right of way.

AI systems learn to recognize these scenarios and respond appropriately through exposure to millions of miles of driving data and countless simulated scenarios. Beyond cars, we're seeing *autonomous systems* master even more challenging environments. Research helicopters now perform acrobatic maneuvers that push the boundaries of what's physically possible, executing rolls and loops that would challenge the most skilled human pilots. These systems learn through trial and error in simulation, gradually building an intuitive understanding of aerodynamics and control that allows them to perform feats that seem to defy gravity.

Intelligence That Masters Games and Strategy

The world watched in amazement when *AlphaGo* defeated Lee Sedol, one of the world's strongest Go players, in a match that many experts thought would take decades for AI to win. Go, with its 10^{170} possible board configurations, represents a complexity that dwarfs chess. Yet AlphaGo didn't just win—it played moves that initially seemed like mistakes but revealed deep strategic insights that human masters had never considered. Even more remarkable was AlphaGo's successor, *AlphaZero*, which learned to play chess, Go, and shogi from scratch without any human guidance. Given only the rules of each game, it discovered strategies and principles that took human civilization centuries to develop, often finding novel approaches that overturned conventional wisdom. In chess, AlphaZero developed an aggressive, dynamic style that prioritized piece activity over material, leading to games that grand masters described as beautiful and instructive.

These game-playing systems demonstrate something profound about machine learning: the ability to discover knowledge rather than simply storing and retrieving information. They develop intuition about position evaluation, long-term planning, and strategic principles that mirror the way human experts think about complex decisions.

Scientific Discovery and Understanding

Building on the same concepts as the two systems just mentioned, *AlphaFold*, developed by DeepMind, solved the protein folding problem—predicting how amino acid sequences fold into 3D structures that determine their biological function. This problem had challenged scientists for more than 50 years, yet AlphaFold achieved accuracy levels that approached experimental methods while requiring just minutes rather than months or years of laboratory work. My own research work in *protein function prediction* through deep learning predates AlphaFold by a couple of years. I was able to beat expert-designed systems even though I had no training in bioinformatics. I just knew how to write a deep learning system that could learn from the given data.

The implications of this data-driven approach extend far beyond academic curiosity. Understanding protein structure is crucial for drug discovery, disease research, and biotechnology development. AlphaFold has already accelerated research into malaria, Parkinson's disease, and COVID-19 treatments. The system essentially compressed decades of potential research time into computational predictions, opening new avenues for medical breakthroughs.

Similarly, in materials science, AI systems are discovering new compounds with desired properties, from superconductors to batteries with improved energy density. These systems can explore vast spaces of possible molecular configurations, identifying promising candidates for synthesis and testing. What once required years of trial-and-error experimentation can now be guided by intelligent predictions about material properties.

The Creative Revolution

On the more abstract side, the boundary between human and machine creativity has blurred dramatically with recent advances in *generative AI*. Systems can now produce photorealistic images from simple text descriptions and generate videos that seamlessly blend reality with imagination. Creating systems that can generate photorealistic images from a given text prompt is, in fact, the objective of the final chapter in this book. *Neural radiance fields* (NeRFs) can reconstruct complete 3D scenes from just a handful of photographs, allowing virtual cameras to move through spaces with realistic lighting and shadows.

These aren't simple cut-and-paste operations or basic filters applied to existing content. Modern generative systems learn the underlying patterns and structures that make images, text, or audio compelling, and then use this understanding to create entirely new content. The technology has progressed to the point where AI-generated content regularly appears in movies, advertisements, and art galleries. Fashion designers use AI to explore new patterns and styles, architects visualize buildings that exist only as concepts, and filmmakers create impossible scenes without expensive special effects. The creative process itself is being augmented and enhanced by systems that can generate hundreds of variations on an idea in seconds.

Language Understanding and Generation

Large language models (LLMs) represent perhaps the most visible breakthrough in recent AI development. These systems can engage in conversations that feel natural and helpful, write code in dozens of programming languages, translate between languages with impressive accuracy, and explain complex concepts in ways adapted to different audiences. They demonstrate a form of language understanding that goes beyond simple pattern matching to something that resembles genuine comprehension.

What makes these systems particularly remarkable is their *emergent capabilities*—abilities that weren't explicitly programmed but arise from the complex interactions within their parts. They can perform mathematical reasoning, creative writing, code debugging, and analogical thinking. When asked to solve a problem they've never encountered, they can often break it down into components and work through solutions step-by-step. The applications continue expanding as researchers discover new ways to leverage these language capabilities. The technology is transforming how we interact with computers, moving from rigid command structures to natural conversation.

Robotics and Physical Intelligence

What's more, this progress isn't limited to software. Modern *robotics* combines computer vision, motor control, and decision-making in ways that enable machines to operate in unstructured, real-world, brick-and-mortar environments. Robots can now navigate cluttered warehouses, manipulate delicate objects, and adapt to unexpected obstacles or changes in their environment. Robots created at Boston Dynamics demonstrate parkour moves, dance routines, and recovery behaviors that showcase remarkable physical intelligence.

Industrial robots equipped with AI can learn new tasks through demonstration rather than explicit programming. A human worker might show the robot how to assemble a component, and the system learns to generalize this knowledge to variations in part positions, orientations, and even slightly different components. This flexibility makes automation practical for smaller production runs and more varied manufacturing environments. Service robots are beginning to operate in hospitals, hotels, and homes, performing tasks such as cleaning, delivery, and basic assistance. These systems must navigate around people, understand social cues, and operate safely in environments designed for humans. In all of these contexts, the AI capabilities required include not just movement and manipulation but understanding of human behavior and social context.

1.1.2 The Two Pillars Supporting These Breakthroughs

While these applications might seem diverse and unrelated, they all rest on two fundamental approaches to machine learning: *deep learning* and *reinforcement learning*. Understanding this connection is crucial because it reveals why learning these core techniques gives you access to such a broad range of capabilities.

Deep learning excels at finding patterns in complex data. Whether processing images to identify objects, analyzing text to understand meaning, or examining protein sequences to predict structure, deep learning systems can automatically discover relevant features and relationships without explicit programming. The "deep" refers to the many layers of processing that transform raw input into increasingly sophisticated representations.

Reinforcement learning, on the other hand, enables systems to learn through interaction and feedback. Rather than learning from fixed examples, these systems explore their environment, try different actions, and gradually improve their behavior based on the results they achieve. This approach is essential for systems that must make sequential decisions in dynamic environments. Many of the most impressive AI breakthroughs combine both approaches. AlphaGo uses deep learning to evaluate board positions and reinforcement learning to develop strategy through self-play. Autonomous vehicles use deep learning to process sensor data and reinforcement learning to improve driving policies. Language models use deep learning for text understanding and generation, followed by reinforcement learning to align their responses with human preferences.

This convergence reflects fundamental aspects of intelligence itself. The ability to learn from experience is essential for navigating a complex world. By mastering deep learning and reinforcement learning using the techniques in this book, you'll understand the foundation underlying virtually all modern AI applications.

1.1.3 Why Now?

These breakthroughs aren't the result of a single invention or discovery. Instead, they emerge from the convergence of several factors that have aligned in recent years. *Computational power* has increased dramatically with specialized processors designed for machine learning workloads. The internet has generated vast datasets that provide the training material these systems need. Algorithmic advances have made it possible to train larger and more capable models than ever before. Perhaps most importantly, these technologies have reached a level of maturity where they can move from research laboratories into real-world applications. The gap between what works in controlled experimental conditions and what functions reliably in the messy real world has narrowed considerably. This transition from research curiosity to practical tool represents a fundamental shift in the technology landscape.

The systems we've explored represent just the beginning of what's possible. As techniques improve and computational resources continue to expand, there will definitely be more remarkable applications. Researchers are working on AI systems that can conduct scientific experiments, design new materials at the atomic level, and solve complex optimization problems that affect everything from traffic flow to energy distribution.

Building systems like these requires more than theoretical knowledge. Practical tools are essential for experimentation, iteration, and efficient implementation of ideas. The choice of framework can make the difference between a smooth learning experience and frustrating struggles with unnecessary complexity. This brings us to a crucial question: Which tool should you use to transform your understanding into working systems?

1.2 Why Keras

If you've picked up this book, you're probably already familiar with Keras and its widespread popularity in the machine learning community. Perhaps you've seen it mentioned in research papers, used in online tutorials, or recommended by colleagues who work with neural networks. That familiarity serves you well because Keras represents the perfect choice for learning and applying deep learning concepts.

Keras has earned its reputation through several key strengths that make it uniquely suited for both beginners and experts:

- First and foremost is its remarkable *usability*. Where other frameworks might require dozens of lines of boilerplate code just to set up a basic neural network, Keras lets you define complex architectures with intuitive, readable syntax that closely mirrors how you think about the problem. The code reads almost like a natural description of your model's structure.
- Beyond ease of use, Keras delivers highly efficient performance that scales from laptop experiments to production systems running on massive cloud infrastructures. The framework maintains an extremely up-to-date knowledge base, incorporating the latest research advances often within weeks of their publication.
- Perhaps most impressively, Keras demonstrates remarkable *backwards compatibility*—code written years ago typically runs with only minor modifications, which represents no small feat in this rapidly evolving field where frameworks often break existing code with each major update. This also means that the concepts you learn in this book will stay with you for a long time, which is a major benefit in this fast-paced domain.
- One of Keras' most valuable features is how it abstracts away underlying implementation details that aren't relevant to high-level model design. When you're building a neural network, you don't need to worry about the intricate optimizations happening during matrix multiplications, memory management strategies, or the dozens of other low-level considerations that could easily distract from your core objective. Keras handles these complexities behind the scenes, letting you focus on the architecture and logic of your models.
- This abstraction doesn't come at the cost of flexibility, however. When you do need to dive deeper—perhaps when creating novel architectures that push beyond established patterns—Keras gracefully exposes the underlying mechanisms with remarkable ease. This *progressive disclosure* represents the hallmark of truly great software design. You can work at the level of abstraction that matches your current needs,

whether that's rapidly prototyping a standard model or implementing cutting-edge research that requires custom components.

These strengths and the balance between usability and flexibility has made Keras the go-to framework for learning about deep learning across universities, research institutions, and industry teams worldwide. Students appreciate its gentle learning curve, while researchers value its flexibility and rapid iteration capabilities. The framework has become something of a lingua franca in the deep learning community, making code sharing and collaboration more seamless than ever before.

Keras code is remarkably dense and concise. You can create extremely complex models, train them on large datasets, and evaluate their performance using only tens of lines of code, not hundreds. Some smaller pipelines require fewer than 30 lines to implement complete end-to-end solutions. This conciseness represents both Keras's greatest strength and its potential challenge for newcomers. The density of Keras code means that each line often encapsulates sophisticated deep learning concepts and operations. While this efficiency is powerful, it demands genuine understanding of the underlying principles. Using Keras blindly—changing parameter values or adding layers without comprehending their purpose—leads to a frustrating cycle of trial and error. The framework's flexibility ensures you won't encounter immediate errors, but you'll find yourself making changes without clear direction, unable to diagnose problems or make meaningful progress toward your goals.

This challenge highlights why simply learning Keras syntax isn't enough. Understanding how to use the framework effectively requires deep comprehension of the machine learning concepts it implements. You need to know not just which functions to call but why those functions exist, what they accomplish, and how they interact with other components of your model. Without this foundation, you'll find yourself stuck in unproductive loops, adjusting function argument values randomly and hoping for improvement.

Throughout this book, we'll address this challenge head-on by explaining Keras syntax as well as the fundamental concepts that give that syntax meaning. You'll learn why certain architectural choices make sense, how different loss functions affect training dynamics, and what various optimization strategies accomplish. This approach ensures that when your models don't perform as expected, you'll have the knowledge to diagnose problems systematically rather than resorting to random experimentation. Our goal is to make you fluent in both the Keras framework and the deep learning principles it embodies. This dual fluency will serve you well whether you're adapting existing models to new problems, implementing research papers, or developing entirely novel approaches. You'll gain the confidence that comes from genuine understanding, enabling you to make informed decisions about model architecture, training strategies, and debugging approaches.

With this foundation established and our tool of choice justified, let's see how this book is structured to guide you through the complete journey from fundamental concepts to advanced applications.

1.3 The Structure of This Book

Understanding deep learning is similar to building a house. First, it requires a solid foundation. If even one corner of the foundation is weak, it will be impossible to construct a strong house. Once the foundation is strong and secure, you can add the walls, and you need those walls in place before you can put on the roof. Along the same lines, each chapter in this book represents a crucial step in constructing your understanding, with every concept building naturally on what came before. By the time you reach the final chapter, you'll have built something remarkable: a deep, intuitive understanding of the same techniques that power today's most advanced AI systems. The chapters in this book follow a carefully crafted progression that mirrors how these concepts actually work together in practice.

We start with a practical case study that illustrates the strengths of Keras but then immediately jump to the mathematical bedrock that supports everything else. We then gradually add layers of sophistication until you're working with cutting-edge architectures. This approach ensures you never feel lost or overwhelmed—each new idea will feel like a natural extension of what you already understand.

Here's what this progression looks like in terms of the chapter contents:

- **Chapter 2: Introduction to the Core of Machine Learning** serves as your entry point into the world of AI. Here, we'll explore what machine learning really means and how it differs from traditional programming. You'll discover the three main types of learning—supervised learning, unsupervised learning, and the particularly exciting world of reinforcement learning. We'll also introduce neural networks at a conceptual level and get your coding environment set up. Consider this chapter your orientation session, where we establish the vocabulary and basic concepts that will guide everything that follows.

- **Chapter 3: Fundamentals of Gradient Descent** dives into the engine that makes machine learning possible. This technique enables our models to learn from mistakes and improve performance over time. We'll explore different varieties of gradient descent and implement them in code, giving you hands-on experience with the optimization process that drives all neural network training.

- **Chapter 4: Classification Through Gradient Descent** puts your newfound understanding to work on one of machine learning's most important tasks: classification. You'll learn how to build systems that can categorize data into different groups, moving from simple binary decisions to complex multi-class problems. We'll introduce loss functions such as cross-entropy, which help our models understand what "good" and "bad" predictions look like. By the end of this chapter, you'll have built your first real classifier using the gradient descent principles from Chapter 3. This chapter will also discuss some pitfalls that can arise when using gradient descent for solving new problems. You'll learn the method to identify these issues, which will be a critical skill when working with the latest and greatest models.

- **Chapter 5: Deep Dive into Keras** introduces you to the powerful framework that will become your primary tool for building neural networks. You'll get a deep understanding of the underlying principles and methods that enable Keras to do what it does best. You'll also see how Keras elegantly implements the gradient descent and classification concepts you've already learned. This chapter bridges the gap between understanding principles and applying them practically.

- **Chapter 6: Regularization Techniques** addresses one of machine learning's greatest challenges: ensuring your models work well on new, unseen data. We'll explore what happens when models become too specialized on their training data, as well as introduce techniques such as dropout that help create more robust systems. You'll learn when you might need more data versus when you need better techniques, giving you the judgment to diagnose and fix common training problems.

- **Chapter 7: Convolutional Neural Networks (CNNs)** opens the door to computer vision by introducing specialized neural architectures designed for processing images. You'll learn about convolutional layers, pooling operations, and how to handle the infamous "shape mismatch problem" that often confuses even the more expert deep learning practitioners. We'll cap off with a complete image classification project that brings together everything you've learned so far.

- **Chapter 8: Exploring the Functional API of Keras** expands your toolkit by showing how to build more sophisticated model architectures. While earlier chapters focused on sequential models that flow data in a straight line, the functional API lets you create complex networks with multiple inputs, outputs, and intricate internal connections. You'll also explore transfer learning, a powerful technique that lets you adapt pretrained models to your specific needs. This chapter is where we start diving deeper into the nitty-gritty of the more modern, more complex models.

- **Chapter 9: Understanding Transformers** introduces you to the architecture that revolutionized natural language processing (NLP) and powers systems such as GPT and ChatGPT. Transformers use an attention mechanism that lets them focus on relevant parts of their input, much like how you might scan a paragraph and pay special attention to certain key phrases. You'll implement transformers in Keras and build your own language model chatbot, seeing firsthand how these systems process and generate text.

- **Chapter 10: Reinforcement Learning: The Secret Sauce** explores the learning paradigm that enables AI systems to master games, control robots, and make sequential decisions in complex environments. Unlike supervised learning, where we show the system correct answers, reinforcement learning lets agents discover successful strategies through trial and error. You'll implement popular algorithms such as Q-learning and policy gradients, and then discover how reinforcement learning enhances LLMs through techniques such as human feedback training.

- **Chapter 11: Autoencoders and Generative AI** shifts focus from recognition to creation, introducing you to systems that can generate new content. Autoencoders learn to compress and reconstruct data, developing internal representations that capture essential features. We'll explore Variational Autoencoders (VAEs) that can generate new examples and Generative Adversarial Networks (GANs), where two neural networks compete against each other, driving each other to improve. These techniques form the foundation of modern generative AI.

- **Chapter 12: Advanced Generative AI: Stable Diffusion** builds on our generative exploration with one of the most impressive recent breakthroughs in AI. Stable Diffusion represents the cutting edge of image generation, capable of creating photorealistic images from text descriptions. You'll understand the theory behind diffusion models and see how they use many of the core concepts you've learned throughout the book. We'll implement diffusion models and explore their applications in image generation and enhancement.

- **Chapter 13: Recap of Key Concepts** brings our journey full circle, connecting all the pieces you've learned into a coherent whole. We'll explore emerging trends in deep learning, discuss how to stay current with rapidly evolving research, and provide guidance for continued learning. You'll leave with not just technical knowledge but the confidence and framework to tackle new developments as they emerge.

This progression is a carefully designed learning experience that transforms you from a newcomer curious about AI and Keras into someone who truly understands how these remarkable systems work. Each chapter provides both the conceptual understanding and practical skills you need, while the overall arc ensures you see how everything connects. By the time you complete all of this, you'll have the same foundational knowledge that researchers use to push the boundaries of what's possible with AI.

The path ahead is challenging but rewarding. Every concept you master opens doors to new possibilities, and every project you complete builds confidence in your growing expertise. Most importantly, you'll develop the deep understanding needed to adapt these techniques to your own problems and contribute to the ongoing AI revolution.

1.3.1 Text Boxes

Throughout the book, we've also provided several elements that will help you access useful information:

1 Introduction

> **[+] Tips and Tricks**
> Boxes with this symbol provide you with recommendations as to how you can simplify your work.

> **[»] Notes**
> Boxes marked with this symbol contain additional information or important contents that you should keep in mind.

> **[!] Warnings**
> Boxes with this symbol contain details worth considering. Moreover, it warns you of common errors or problems that might occur.

1.4 How to Use This Book

As we've established, Keras code is remarkably dense and concise. The framework has been designed to eliminate bloat and boilerplate code, allowing you to express complex machine learning concepts with minimal syntax. This elegant efficiency comes with an important implication: understanding Keras deeply really means understanding the underlying machine learning concepts thoroughly. There's no separation between knowing the framework and knowing the field. This fundamental connection explains why any Keras book that focuses solely on syntax and function names will never truly serve you well. Learning to call functions without understanding their purpose leaves you helpless when those functions don't produce the results you expect. You'll find yourself copying code patterns without the knowledge needed to adapt them to your specific problems or diagnose issues when they arise.

This book has one clear objective: to help you use Keras effectively and efficiently. Achieving this goal requires developing a deep understanding of the core principles that underlie machine learning, deep learning, and all the related concepts that make these systems work.

Consider regularization as an example of why this depth is important. In Keras, applying regularization might look as simple as adding a single argument to a function call shown in Listing 1.1.

```
Dense(128, activation='relu',
      kernel_regularizer=regularizers.l2(0.01))
```

Listing 1.1 Keras Code Snippet with Regularization Parameter

That single function call encapsulates a sophisticated technique for preventing memorization by the machine, but determining the right value for that 0.01 parameter requires understanding contour plots, loss landscapes, and how different regularization strengths interact with your model's learning dynamics. Without this foundational knowledge, you're reduced to guessing parameter values and hoping for the best, which usually doesn't work out in the complex landscape of this field. This is why we dedicate several sections to regularization concepts in Chapter 6, ensuring you understand not just how to apply these techniques but when and why they work and even when to not use them at all!

This philosophy shapes our entire approach to learning Keras. We won't start by explaining every aspect of the framework immediately. Instead, while we begin using Keras right from the next chapter, we won't explain every part of the code at first. Attempting to cover all details simultaneously would be counterproductive and overwhelming. Our strategy follows a more natural learning progression. We'll show you fully working code examples—starting with a complete, functioning system at the very beginning of the next chapter. We'll explain some parts of this code while acknowledging that other parts remain mysterious for now. Then, we'll step back to explore the mathematical foundations and underlying concepts that give meaning to what you've seen. Armed with this deeper understanding, we'll return to the previous code, this time explaining additional components that were previously opaque. We'll also progressively add more features of Keras as we go along, adding them based on need rather than because they are next in line.

This iterative approach ensures you see tangible progress in terms of working code while simultaneously building the conceptual foundation that will serve you throughout your career. You'll develop the ability not just to read other people's code but to modify existing systems and write completely new ones from scratch when your projects demand novel solutions.

1.4.1 Why Keras Installation Comes Later

You might notice from the table of contents that we don't dive deep into Keras installation and setup until Chapter 5. This timing is deliberate and reflects how learning actually happens in practice. By weaving Keras into our narrative early, you gain hands-on exposure to the framework without needing complete mastery of every installation detail up front. This approach mirrors real-world learning, where tools and concepts evolve together rather than in isolation.

You'll see Keras in action and work with code early on. However, delaying the comprehensive Keras installation and deep dive until after we've covered foundational machine learning concepts ensures clarity while maintaining momentum through your early learning experience.

1 Introduction

We can start working with Keras immediately in the next chapter without requiring local installation. Online platforms and cloud-based notebooks let you run code and experiment with concepts before committing to any particular setup. We'll provide instructions on these in Chapter 2 to bring you up to speed. After we've established the foundational machine learning concepts through Chapter 4, we'll provide everything you need for a complete Keras environment in Chapter 5. This will serve you through the advanced concepts in Chapters 6 and beyond.

1.4.2 Suggested Reading Strategy

The most effective way to work with this book is to read it from start to finish. This isn't a reference manual where you can jump around freely. While this back-and-forth is possible, the text of this book is one cohesive story that builds systematically toward deep understanding. Each chapter assumes knowledge from previous chapters, and skipping sections will leave gaps that become problematic later. Still, we've heavily cross-referenced relevant portions of previous chapters in later ones so that you can easily go back to refresh important concepts.

Even topics you think you already know deserve at least a light reading. For example, the types of machine learning covered in Chapter 2 are often presented as purely theoretical concepts in other resources. Here, they've been crafted specifically to impact your learning in later chapters. The way we introduce supervised, unsupervised, and reinforcement learning sets up crucial distinctions that become essential when you're choosing architectures and training strategies in advanced chapters.

More importantly, whenever you encounter code snippets, always run the code yourself. We're making all the code in this book available through the book resources page, and it's highly recommended that you type the examples yourself rather than simply copying and pasting. The act of typing forces you to engage with each line, often revealing details you might otherwise miss. If typing every example becomes impractical, running the provided code is still far better than just reading it passively.

> **Book Resources Page: Source Code Listings and Full Color Images**
> You can access all source code listings and full color versions of images and diagrams used throughout the book on the book's web page at *http://rheinwerk-computing.com/6142* as well as on the resources page here: *https://recluze.net/keras-book*. Resources are arranged by chapters, and we'll provide specifics of which code listing to use as they are discussed throughout the text.

In summary, your learning process should follow this pattern: read through chapters in order (skipping only when absolutely necessary), run code whenever it's discussed, and experiment with different parameter values to see how changes affect results. This experimentation is crucial. Seeing what happens when you modify values builds intuition that no amount of theoretical explanation can provide. Don't worry if some pieces of code remain unexplained initially—this is intentional, not an oversight. By the end of Chapter 6, we'll have covered enough foundational concepts that every component can be explained thoroughly. From that point forward, we won't skip explanations, and you'll have the knowledge base needed to understand sophisticated implementations.

This patient, systematic approach ensures that you develop both the practical skills needed to implement solutions and the deep understanding required to innovate beyond existing patterns. The journey requires commitment, but the destination—true fluency in both Keras and the machine learning concepts it embodies—will serve you throughout your career in this rapidly evolving field.

With all of that out of the way, let's start writing some Keras code!

Chapter 2
Introduction to the Core of Machine Learning

At its core, machine learning is about finding patterns in data and using them to make decisions. In this chapter, we'll explore the fundamental ways machines can learn—through direct instruction with examples, by discovering patterns on their own, or through trial and error. We'll see how these approaches mirror learning in the natural world, and we'll begin to understand the powerful systems that make this learning possible. By the end, you'll be ready to start experimenting with these ideas yourself.

Computers have evolved from simple instruction-followers to systems that can learn from data and improve their performance over time. This shift represents the core of machine learning—creating programs that discover patterns and develop solutions without being explicitly programmed for every scenario. This is the world of machine learning, and in this chapter, we're taking our first steps into this fascinating landscape. You might have heard buzzwords such as AI, deep learning, or neural networks thrown around in tech conversations. Perhaps you've wondered how your phone can recognize your face even in poor lighting, how your email service filters out spam with remarkable accuracy, or how self-driving cars can distinguish between pedestrians, road signs, and other vehicles in real time. Maybe you've been amazed by how modern language translation services can capture nuances and context that simple word-for-word translation could never achieve. These are only a few examples of machine learning in action, and by the end of this chapter, you'll understand the fundamental principles that make them (and many more) possible.

We're going to approach this journey a bit differently than traditional textbooks. Instead of diving straight into formal definitions and mathematical notation, we'll start with real-world examples that show why traditional programming falls short and how machine learning fills that gap. Think of it as learning to cook by actually making dishes, rather than memorizing recipes.

2 Introduction to the Core of Machine Learning

> **[+] How to Make the Most of This Chapter**
>
> While this chapter may appear theoretical, it's crucial to devote time and thought to these concepts. The understanding you develop here will prove invaluable when implementing the concepts in Keras. The framework's concise syntax means that adjusting a single parameter can dramatically impact performance. This chapter equips you with the analytical framework needed to make informed design decisions when confronting complex challenges in later implementations.

First, as a case study, you'll see why teaching a computer to recognize handwritten numbers becomes surprisingly challenging when we try to write explicit rules but becomes elegantly solvable when we let the computer learn from examples. This case study will serve as our gateway into understanding the core principles of machine learning, illuminating the path that leads from data to intelligence. We'll then explore the different flavors of machine learning: supervised, unsupervised, and semi-supervised learning. But more importantly, we'll discuss the common misconceptions that often trip up newcomers to the field. You'll learn why "more data" isn't always the answer and why some problems that seem perfect for machine learning might be better solved with traditional methods.

One of the most exciting areas we'll explore is reinforcement learning, which you can think of as teaching through experience rather than examples. This is how computers learn to play chess better than grand masters or control robotic systems with superhuman precision. It's often overlooked in introductory texts, but understanding reinforcement learning early will give you a more complete picture of what's possible in machine learning.

We'll then introduce neural networks, showing how they serve as a bridge between traditional machine learning and reinforcement learning. Don't worry if you've heard they're complex—we'll break them down into digestible concepts and expand on them piece by piece to simplify them.

Finally, we'll get practical. You'll set up your own machine learning environment, installing the tools you'll need for the rest of this book. We've streamlined this process to get you up and running quickly, focusing on writing code rather than fighting with installations. By the end of this chapter, you'll have a solid foundation in machine learning concepts, hands-on experience with basic tools, and most importantly, the confidence to tackle more advanced topics in the chapters ahead. So, let's begin our journey into the heart of machine learning using a case study.

2.1 What Is Machine Learning?

Imagine walking into a room filled with little sticky notes with handwritten numbers on them. Each digit, from 0 to 9, scrawled in different styles—some neat and precise, others barely legible. Now, picture being asked to teach a computer to recognize these numbers, just as effortlessly as you do. This is the famous *MNIST dataset*, a collection of handwritten digits that has become the "Hello World" of machine learning. Created by researchers Yann LeCun, Corinna Cortes, and Christopher Burges through a careful modification of an earlier dataset from the National Institute of Standards and Technology (NIST), MNIST (the Modified National Institute of Standards and Technology database) became the first large-scale, standardized testing ground for machine learning algorithms. Think of MNIST as the primer that sparked a revolution. In the late 1990s, when researchers were still debating whether computers could reliably recognize handwriting, MNIST provided a perfect challenge: 60,000+ images, each a carefully scanned and size-normalized handwritten digit. It gave the entire machine learning community a standardized test so that everyone could work on the same problem and compare their results directly.

Even today, in an era of massive datasets and powerful computers, MNIST remains invaluable. It's seemingly basic but contains all the fundamental challenges of *computer vision*. The dataset is small enough to experiment with quickly (you can train a model in minutes, not days), yet complex enough to demonstrate real-world challenges. Each digit shows variations in style, thickness, and angle, presenting the same kind of difficulties that modern vision systems face with more complex images. This perfect balance between simplicity and complexity makes MNIST an ideal laboratory for testing new ideas and teaching machine learning concepts.

Starting with MNIST ensures that you'll learn the core principles behind how machines actually learn while also laying the foundations for the more complex models to come in future chapters of this book. So, let's begin by learning how to recognize MNIST digits.

2.1.1 Helping the Machine Recognize Digits

But how do you teach a computer to recognize these digits? The traditional way is to write explicit *rules*: "if there's a closed loop, it might be a 0, 6, 8, or 9" or "if there's a straight line with a hook at the top, it's probably a 7." However, anyone who's ever tried to read someone else's handwriting knows that rules like these quickly fall apart. There are simply too many variations, too many exceptions, and too many ways to write each number. Consider how a hastily written 7 might look nearly identical to a 1—both could appear as a single vertical stroke with a slight angle. Or think about the subtle difference between a 5 and a 6—in many handwriting styles, both start with a curved top, and if the writer doesn't clearly close the loop of the 6, these digits become almost indistinguishable. Even more confounding are the regional variations: some people cross their 7s horizontally, while others don't; some write their 4s with an open top, while others close it completely. The variations are endless, and what seems obvious to human eyes often defies simple computational rules. Figure 2.1 shows some examples from the MNIST dataset.

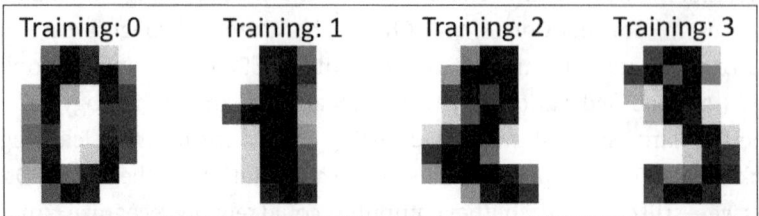

Figure 2.1 Sample Digits in the MNIST Dataset

Let me explain these images, keeping in mind their importance in understanding machine learning's core concepts. What you're seeing are four handwritten digits from the MNIST dataset but shown in a way that reveals their digital nature. Each image is a grid of pixels—tiny squares that, when put together, form the complete picture. While the original images are 8×8 pixels (giving us 64 pixels per digit), they're greatly magnified here so we can see the individual pixels clearly. That means you have to squint a little bit to recognize the digits. Go ahead and squint so that you can see the 0, 1, 2 and 3 in the figure.

Each pixel's shade represents how dark that particular point was in the original handwritten digit. The darkest pixels (shown in black) indicate where the pen made the strongest contact with the paper, while the lighter shades of gray show where the pen pressure was lighter or where the ink spread slightly. Think of it like a detailed pressure map of someone writing these numbers.

2.1.2 The Outdated Method—Rule-Based Learning

To truly appreciate the complexity of the task here, let's look at how early attempts at AI worked through another fascinating example: *ELIZA*, one of the first chatbots ever created. While today's chatbots use complex systems, ELIZA used very complex *pattern-matching* rules written in LISP—one of the oldest programming languages that was used extensively for AI back in the day—that looked something like that shown in Listing 2.1.

```
(rule "HELLO"
  ((* "Hello" * "I'm" *name))
  ("Hi there" *name "I'm Eliza"))
```

Listing 2.1 A Simple Rule for Greetings in ELIZA

This rule is a template with blanks to fill in—like a sophisticated game of Mad Libs. The asterisks (*) are wildcards that can match any text, while *name captures whatever the person typed after "I'm". So, if someone types "Hello, I'm Sarah", ELIZA would match this pattern and respond with "Hi there Sarah, I'm Eliza".

This rule-based approach is elegant in its simplicity, but it also shows us why modern machine learning became necessary. ELIZA needed explicit rules for every possible conversation pattern. What if someone said "Hey" instead of "Hello"? What if they used "my name is" instead of "I'm"? Each variation would need its own rule as shown in Listing 2.2. This gets complicated pretty quickly when you start to think about global languages, slang, and typing mistakes people make.

```
(rule "HEY"
  ((* "Hey" * "I'm" *name))
  ("Hi there" *name "I'm Eliza"))

(rule "NAME-IS"
  ((* "my name is" *name))
  ("Hi there" *name "I'm Eliza"))
```

Listing 2.2 More Complex Rules in ELIZA

In our MNIST problem, we have similar or even more complex issues. Look at the digits and try to imagine writing rules to identify them. The task becomes surprisingly complex for several reasons:

- **Grayscale complexity**
  It's not just black and white but multiple shades of gray. How do we write a rule that says, "this level of gray counts as part of the digit, but that slightly lighter shade does not"? It's almost like trying to define exactly where dusk ends and night begins.

- **Style variations**
  Notice how each digit has its own character, that is, its own "handwriting personality." The zero has uneven thickness, the one is slightly tilted, the two has a distinctive curve, and the three has its own unique style. It's similar to how everyone's handwriting is unique—some people loop their twos, and others make them angular.

- **Imperfections and noise**
  Look at the edges of these digits—they're not clean, crisp lines. There are pixels that fade from dark to light, spots where the ink might have bled slightly, and areas where the stroke isn't perfectly continuous. How do we write rules that account for all of these natural variations?

- **Pixel patterns**
  Even if we tried to define rules based on pixel positions, we'd quickly find ourselves in trouble. A slight shift in how someone writes the number could move all the pixels by a few positions, completely breaking our rigid rules.

This is precisely why traditional programming approaches fall short here. We simply can't write explicit rules like "if pixel (15,20) is black and pixel (16,20) is gray, then this is part of a zero." The variations are too numerous, the patterns too subtle, and the distinctions too nuanced for explicit programming rules to capture effectively. Instead of trying to write rules about pixel values and positions, machine learning approaches this problem differently—by learning from examples. Rather than telling the computer "here's how to recognize a three," we show it thousands of examples of threes and let it discover the patterns on its own. This approach can handle all the variations, imperfections, and subtleties that make explicit rule-writing impossible. Let's try to do that using some code next.

Before we get started though, note that as you journey through this book, you'll encounter snippets of code that demonstrate key concepts. Don't worry if some of the code examples in the first few chapters look complex or unfamiliar—they're not meant to be fully understood right away. Think of them as windows into the practical world of machine learning, giving you glimpses of how theoretical concepts transform into working solutions. Instead, these code snippets serve a deeper purpose: they're here to help you develop an intuitive feel for machine learning. Just as a chef might learn about flavors by tasting rather than just reading recipes, you'll gain insights by experimenting with these small pieces of code. Each example is carefully chosen to illuminate a specific concept, letting you see machine learning principles in action.

For instance, when we show you how to load and process the MNIST dataset, the code might include terms or functions that seem mysterious now. That's perfectly normal! The goal isn't to understand every line immediately, but to start recognizing patterns and building familiarity with how machine learning code works.

As we progress through the book, we'll dive into the internals of Keras—the powerful tool that makes much of this code possible. That's when these initial code examples will start to click into place like puzzle pieces finally finding their spot in the bigger picture. You'll begin to understand not just what the code does, but why it's written that way. For now, focus on the big picture these code examples paint. Run them—we'll show you how to do that in a bit—experiment with them, and observe their behavior. Your understanding will grow naturally as we build your knowledge foundation piece by piece.

2.1.3 Case Study in Machine Learning Code

With that at the back of your mind, let's look at our example code for MNIST. Listing 2.3 shows the importing of some essential *libraries*—think of these as specialized toolkits that will help us explore the world of machine learning. Just as a carpenter carefully selects their tools before starting a project, we'll import three powerful libraries that form the foundation of our work.

> **Accessing Fully Operational Code**
>
> As mentioned in Chapter 1, all code used in this book will be available in the resources. You can access the complete code for this section on the resources page at *https://recluze.net/keras-book* in the **02-01-mnist-basic** notebook. Section 2.5 will guide you on how to actually run these code snippets. These resources will be available on the book's official web page at *http://rheinwerk-computing.com/6142* as well.

First up is *matplotlib*, accessed through its plotting interface plt. Think of matplotlib as your digital artist's toolkit. It helps us visualize data in various ways—from simple line graphs to complex heatmaps. When we're working with image data such as handwritten digits, matplotlib lets us actually see what our computer is working with, turning arrays of numbers into visible images that make sense to our human eyes.

Next, we have *NumPy* (imported as np), which is like the mathematical brain of our operation. Imagine trying to juggle thousands of numbers in your head. It's not an easy feat, but NumPy makes this effortless. It's designed to handle large, complex arrays of numbers efficiently, performing calculations at lightning speed. In machine learning, we're constantly working with large collections of numbers (think pixels in images), and NumPy helps us manage these with grace.

The third import, *sklearn* (specifically its *datasets* module), gives you access to a well-organized library of data. Scikit-learn, as it's fully known, is a treasure trove of machine learning tools and example datasets. In this case, we're only interested in its collection of digit images that we'll use for learning. We use the load_digits function to get the whole MNIST dataset. Later on in the book, we'll come back to sklearn and its various facilities—it's one of my favorite libraries, and I can't get enough of it

```
import matplotlib.pyplot as plt
import numpy as np
from sklearn import datasets
digits = datasets.load_digits()

print(digits.data.shape)
```

Listing 2.3 Loading the MNIST Dataset

Finally, we peek at the structure of our data using the last line: digits.data.shape. This command reveals the *dimensions* of our dataset—how many digits we have and how many pixels make up each digit. It's similar to asking how many pages are in our book, and how detailed each page is. Understanding this structure is crucial because it tells us the scope of our learning task and helps us plan our approach. At the moment, the shape is a simple (1797, 64). This means that we have 1,797 handwritten images, and each image is composed of 64 pixels. This is a little difficult to wrap our heads around and visualize. Let's help our libraries make sense of it.

Listing 2.4 shows the code for creating the visualizations shown in Figure 2.1. Even though this visualization is just for us and the machine doesn't need it (the machine works with the underlying numbers only), it's a great idea to practice with such visualizations. They will help you get closer to the way machines are working with the data and help you debug your systems later. So, let's see what each line of this code does to display our digits. Line 1 creates our display area, setting up a figure with four subplot spaces arranged in a single row. The figure will be 15 units wide and 3 units tall. We store these subplot spaces in the axes variable (the underscore is used because we don't need the first returned value).

Line 2 starts a loop that processes three things simultaneously: each subplot space (ax), the corresponding digit image, and its label (the actual number it represents). The zip function helps us iterate through all three sequences at once.

Lines 3–5 work with each individual subplot:

- Line 3 removes the x- and y-axis markings for cleaner visualization.
- Line 4 displays the actual digit image in grayscale, keeping the pixels sharp and clear.
- Line 5 adds a title above the image showing which number this digit represents.

When executed, this code will show us four different handwritten digits from our dataset, each one properly labeled and displayed in grayscale. The zip function in line 2, while basic syntax in Python, is a very versatile tool used often to structure data into more complex forms. It's highly recommended that you practice with it a bit and get used to what it does.

```
_, axes = plt.subplots(nrows=1, ncols=4, figsize=(15, 3))
for ax, image, label in zip(axes, digits.images, digits.target):
    ax.set_axis_off()
    ax.imshow(image, cmap=plt.cm.gray_r, interpolation="nearest")
    ax.set_title("Training: %i" % label)
```

Listing 2.4 Visualizing the Handwritten Digits Using Matplotlib

We can now see the images and understand that these are pixel values. Let's study what the machine sees though. The code in Listing 2.5 allows us to do just this. On line 1, we're getting the size of our dataset. By counting the number of images, we learn exactly how many handwritten digits we have to work with. This count is crucial because it tells us the scope of our learning task, that is, how many examples our computer will have to learn from.

The digits object contains various pieces of information, but we're particularly interested in the raw pixel values. By extracting just the data portion in line 2, we're streamlining our work. Lines 3 and 4 give us our first glimpse into the actual structure of our digit data. When we print the raw data in line 3, we see one-dimensional (1D) arrays of numbers that represent pixel values. You can see them at the end of the listing where the output is included. At first glance, these numbers might seem overwhelming or meaningless. That's perfectly fine—the goal here isn't to understand each number, but to start developing a sense for how digital images are represented as data. This exposure will prove invaluable later when we work with more complex datasets.

The *shape* printed in line 4 tells us something more concrete about our data's organization: we have 1,797 different digit images, and each image consists of 64 pixels (arranged in an 8×8 grid). In the pixel values, 0 means a completely white pixel and 255 means a completely black pixel with intermediate values capturing shades of gray. This seemingly simple piece of information is actually a crucial blueprint of our data's structure. When we move on to more complex models where we can't easily visualize the data, understanding these structural patterns becomes essential for debugging and optimization.

This early exposure to raw data structures, even when they seem abstract or puzzling, is building your intuition for machine learning. We'll keep repeating this pattern of looking at things that only marginally make sense until we get to the nitty-gritty of Keras where we'll explore these in great detail and explain everything.

```
# Study the data and it's shape
n_samples = len(digits.images)
data = digits.data
print(data)
print("Shape: ", data.shape)

# Following is the output produced by this code:
[[ 0.  0.  5. ...  0.  0.  0.]
 [ 0.  0.  0. ... 10.  0.  0.]
 ...
 [ 0.  0.  2. ... 12.  0.  0.]
 [ 0.  0. 10. ... 12.  1.  0.]]
Shape:  (1797, 64)
```

Listing 2.5 Studying the Data to Get an Intuitive Grasp

2.1.4 Structuring the Learning Process

Now that we and the machine are on the same page with the understanding of the data, let's go ahead and enable the machine to actually *learn*. Again, this is going to be abstract at the moment. In the next chapter, we'll begin to define the process cleanly, but for now, let's explain the code given in Listing 2.6 in plain terms. While we won't dive into all the technical details just yet, understanding the basic structure will set the foundation for deeper exploration later.

Lines 1–2 bring in our essential tools from the Keras library. Keras is a specially designed workshop that makes building machine learning systems more approachable. While the exact names and organization of these tools may change as the field evolves (we'll show you how to stay current later), the core concepts remain the same.

Lines 4–7 are where we construct our learning system. The word `Sequential` in line 5 tells us we're building a system where information flows in one direction, from input to output, like an assembly line. Each *layer* in this assembly line has a specific role in transforming our *input* (digit images) into our desired *output* (digit classifications).

Line 6 introduces our first `Dense` layer with 128 units. Pause here for a moment—we're essentially asking our machine to discover 128 important "things" (or features) about each digit image. We don't specify what these features should be; the machine will figure that out on its own. Some units might learn to recognize curves, others might focus on angles, and others might detect more abstract patterns we humans haven't even thought about. The point is that we're not defining what is to be learned. The input shape is (64,) because the number of pixel values in each image is 64.

> **Is That a Typo?**
>
> This (64,) isn't a typing mistake. If we wrote just (64) without a trailing comma, that would be a number; however, (64,) with the comma after 64 makes it a *tuple* with just one element. This tip is going to be very useful when we work with Keras input shapes. Keep this in mind to avoid many debugging headaches later.

Finally, line 7 creates another `Dense` layer, but this time with 10 units. The reason for the specific number 10 is that we want our machine to categorize each image into one of 10 possible buckets or classes (0–9). This layer will help our machine make its final decision about which digit it's looking at. The *softmax* function in this line ensures that our machine's confidence in each possibility adds up to 100%—it might be 95% sure an image is a 7, 4% sure it's a 1, and have tiny amounts of confidence distributed among the other possibilities.

Understanding this structure is crucial because it represents a fundamental pattern in machine learning—taking raw input (in our case, 64 pixels), finding meaningful patterns (our 128-unit layer), and using those patterns to make decisions (our 10-unit output layer). As we progress, you'll see this same pattern repeated and expanded in increasingly sophisticated ways.

```
import keras
from keras.layers import Dense

# Define the model
model = keras.Sequential([
```

```
    Dense(128, activation='relu', input_shape=(64,)),
    Dense(10, activation='softmax')  # 10 classes (digits 0-9)
])
```

Listing 2.6 Making Our Learning Black Box

Next, let's see how we get our machine to actually learn from our digit images. This is where we transition from setting up the structure of our machine to enabling it to discover patterns on its own.

Line 1 in Listing 2.7 introduces the *compile* step, where we prepare our machine for learning. We're effectively giving it a learning strategy—a methodology for how it should approach the problem of recognizing digits. We'll look in detail into the most important of these learning strategies in the next chapter. For now, think of this as setting up the rules of the game—similar to how you might teach a child to play darts. You don't tell them exactly how to move their arm or precisely what angle to use, but you do give them a strategy: "Start by aiming for the center," "Pay attention to whether you're hitting too high or too low," "Make small adjustments based on where your last throw landed." The child learns through practice, gradually refining their technique based on the results of each throw.

In our machine, the compile step establishes similar *learning* guidelines. We're not telling the machine "thick loops mean it's a zero" or "a vertical line with a hat means it's a seven." Instead, we're giving it a systematic way to learn from its mistakes: how big of adjustments to make when it's wrong, how to measure how far off it was, and how to balance between dramatic changes and subtle refinements. Just as the dart player learns to make increasingly precise throws by observing their misses, our machine will learn to make increasingly accurate predictions by analyzing its errors.

In line 2, we prepare our *target* values—the correct answers for each digit image. When we examined our dataset earlier, we had the images themselves (the input) and labels telling us what digit each image represents (the output). These labels are crucial because they're what the machine will use to check its work and improve its understanding.

Line 3 contains the part where actual learning takes place—the *fit* operation. During this process, the machine begins examining our digit images and their corresponding labels. It starts making predictions about what digit each image represents, compares these predictions to the correct answers, and adjusts its internal understanding based on where it made mistakes.

What's particularly intriguing about this approach is how different it is from our earlier look at ELIZA's rule-based system given in Listing 2.1 and Listing 2.2. In ELIZA, we had to explicitly write rules like "if you see this pattern, respond this way." But here, we're not writing any rules about what makes a 7 look like a 7. Instead, we're letting the machine discover these patterns on its own within those 128 learning units we specified earlier. This is the true power of machine learning—we don't program the rules, we program

2 Introduction to the Core of Machine Learning

the ability to *learn rules*. Those 128 things our machine learns about each image aren't features we chose or defined; they're patterns the machine discovers through its training process. Some might be obvious features such as loops or straight lines, while others might be more abstract patterns that even we humans might not have thought to look for.

```
model.compile(optimizer='adam',
              loss='sparse_categorical_crossentropy',
              metrics=['accuracy'])

# Reshape the target variable to be a 2D array for Keras
y = digits.target.reshape(-1, 1) # Learn!
model.fit(data, digits.target, epochs=10)
```

Listing 2.7 Learning from Data

Now that our machine has gone through its learning process, it's time to see what it has actually learned. Think of this as the moment after a study session when we check whether real understanding has taken place. Lines 1–2 in Listing 2.8 show our machine performing a self-assessment. Just as a student might work through practice problems and check their answers, our machine looks at the digit images and makes *predictions*, comparing them against the correct answers we provided. This *evaluation* gives us metrics about how well the machine is performing overall.

At this point, you might be thinking that since we've just shown the machine these same images during training, maybe it's just memorizing everything, like a student who remembers exactly where each answer appears in their textbook rather than understanding the underlying concepts. This observation gets at the heart of machine learning. The difference between memorization and true learning is crucial, and we'll explore this distinction in detail later when we discuss concepts such as *overfitting* and generalization. For now, keep this skepticism in mind—it's exactly the kind of critical thinking that makes for good machine learning practitioners.

Lines 3–4 represent something slightly different—here we're doing our own independent evaluation. Rather than asking the machine to grade itself, we're asking it to make predictions that we can then analyze in detail. The `predicted` variable now contains all the machine's guesses about what digit each image represents. This is like collecting a student's answers without marking them yet, allowing us to perform our own detailed analysis of where they succeeded and where they might have struggled.

This two-pronged approach to evaluation—both self-assessment and independent analysis—gives us a more complete picture of what our machine has learned. The machine's self-evaluation provides quick performance *metrics*, while our separate prediction collection allows us to dig deeper into specific cases where the machine might be consistently making certain types of mistakes or showing interesting patterns in its predictions.

```
# Evaluate the model
_, _ = model.evaluate(data, digits.target, verbose=0)
# Predict the value of the digit on the test
predicted = model.predict(data)
```

Listing 2.8 The Machine Makes Predictions

The code in Listing 2.9 helps us visualize both what we show the machine and what it thinks it's seeing. Lines 1–2 set up our plot where we'll display our results. In line 3, we're selecting the digit from our dataset to examine. This could be any number in our dataset, and changing this value lets us explore different examples to see how our machine performs with various digits.

Lines 4–5 handle the display of our chosen digit. Here, where we reshape the 64-element array back into an 8×8 grid so that we can actually see the digit. The machine learns from the flat list of 64 numbers, but we humans prefer to see it as a proper digit image. Finally, line 7 looks up what digit this actually is in our original dataset, and line 8 examines what our machine thought this digit was by looking at its predictions. The argmax function here finds which of the 10 possible digits (0–9) our machine thought was most likely.

```
fig, ax = plt.subplots(figsize=(7, 5))

num_to_show = 25
image = data[num_to_show].reshape(8, 8)
_ = ax.imshow(image, cmap=plt.cm.gray_r, interpolation="nearest")

actual_digit = digits.target[num_to_show]
predicted_digit = np.argmax(predicted[num_to_show])
```

Listing 2.9 Visualizing and Evaluating Predictions

> **Argmax for Finding Predictions**
>
> The argmax function is similar to but not the same as the max function. argmax finds the position of the maximum value in a list as opposed to the maximum value itself. For instance, max([1.1, 2.2, 8.4, 5.6]) will return 8.4, but argmax([1.1, 2.2, 8.4, 5.6]) will return 2—the position in the list of the maximum value. It's a simple concept that makes our code very compact.

2.1.5 Analyzing the Machine's Learning Process

Let's go ahead and look at some of the predictions made by the machine. The first one in Figure 2.2. is a simple example. Looking at the pixel layout in this image, we can see why our machine learning model had no trouble identifying this as the digit 5. It's a relatively clean and well-formed example of the digit, making it an ideal case for our model to recognize. The black pixels form a distinctive pattern that's characteristic of a 5—we can trace the clear top horizontal line, the curve down the right side, and the loop at the bottom. The grayscale values are also quite distinct, with strong contrasts between the digit and the background, making the features of the number stand out clearly. Do keep in mind that we don't really know what patterns the machine has actually learned. We just think this is what the machine might have learned.

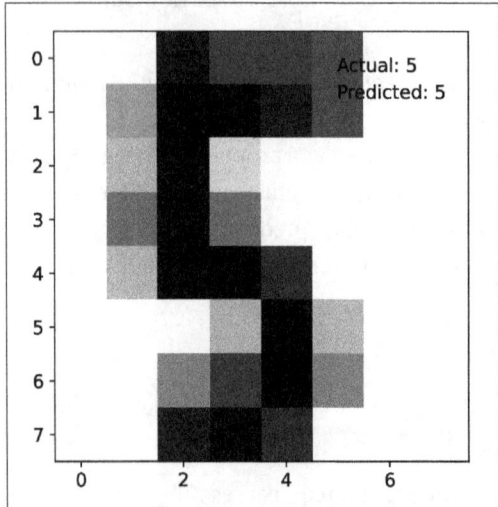

Figure 2.2 A Simple Example of a Digit Correctly Identified

Let's look at another example in Figure 2.3 that's slightly more difficult—at least for humans.

Looking at this image, we can see a slightly more challenging example of the digit 2, but our model is still performing nicely. Despite the increased complexity, the machine correctly identified it, which tells us something interesting about how well our model has learned to recognize key features of numbers.

This particular 2 has some characteristics that make it more complex than simpler examples. The top curve is less defined, and there's some variability in the pixel intensities along its path. The diagonal stroke that typically forms the bottom of a 2 shows varying degrees of darkness, suggesting either the original writing pressure varied or the digitization process captured some nuance in the pen stroke. What's particularly impressive is how our model handled these variations. The pixels show a range of grayscale values—you can see the gradient from dark to light, especially in the middle section of the digit. In a simpler example, we might see more binary black-and-white values, but the model had to work with more subtle distinctions here.

Despite these complexities, our model confidently and correctly identified this as a 2. This suggests the model learned not just the basic shape of a 2 but also how to handle variations in writing style and intensity. It's like the difference between recognizing someone's face in perfect lighting versus being able to identify them in varying light conditions—our model seems to have developed this more robust kind of understanding.

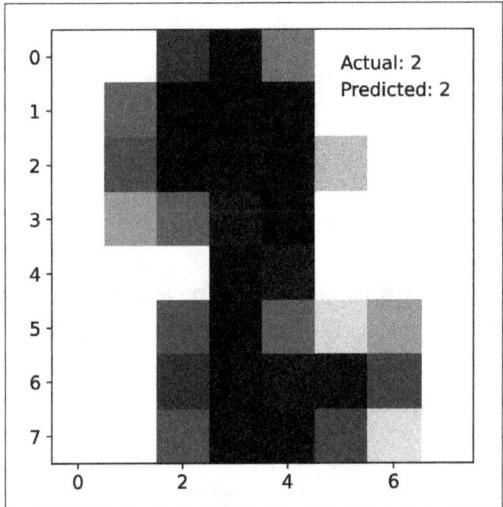

Figure 2.3 An Intermediate Example with Some Complexity

Finally, let's look at a much more complex example in Figure 2.4. This is a great example that really highlights the surprising capabilities of machine learning! Here, we have a digit that, to human eyes like mine (and possibly yours), barely resembles a traditional 9. If someone showed this to us without context, we might struggle to identify it. However, both the original writer and our machine learning model confidently identified it as a 9.

This reveals something remarkable about pattern recognition in machine learning. Our model isn't constrained by the same preconceptions humans have about how numbers should look. It's learned to recognize features and patterns that we might not even be consciously aware of. While we might be looking for the classic loop-and-tail structure of a 9, our model has learned a broader set of characteristics that can identify a 9 in various forms. The pixel intensity pattern here is particularly interesting. You can see a sort of curved structure at the top and a diagonal element trailing down—perhaps these are the key features that both the original writer and our model associate with a 9, even though they're arranged in what looks to us like an unusual configuration.

2 Introduction to the Core of Machine Learning

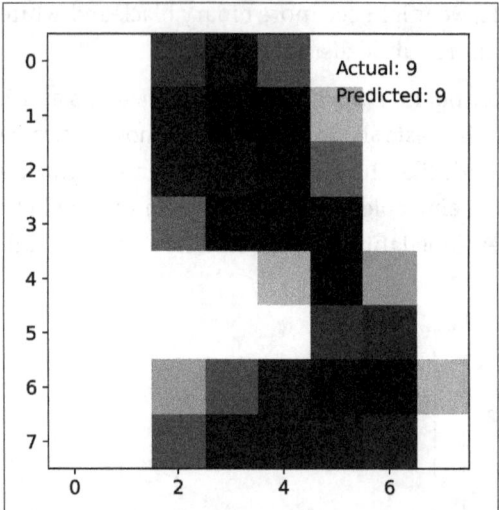

Figure 2.4 A Difficult-to-Identify Example Still Correctly Recognized

This example also reminds us of an important principle in machine learning: sometimes the hardest cases for humans aren't necessarily the hardest cases for machines, and vice versa. It's a humbling reminder that artificial and human intelligence can complement each other by bringing their own unique strengths to pattern recognition tasks.

2.1.6 Lessons Learned from the Code Case Study

Through this example with handwritten digits, we've uncovered something remarkable about machine learning. What started as a simple question—"Can a computer recognize handwritten numbers?"—led us to discover fundamental principles that shape how machines learn. Remember how we began by considering the traditional programming approach with ELIZA's explicit rules? We saw how quickly that approach becomes overwhelming when dealing with the countless variations in handwriting. Then we witnessed a very elegant alternative: by showing our machine thousands of examples and letting it discover patterns on its own, it learned to recognize digits we could barely decipher ourselves.

Our exploration revealed several key insights. First, we saw how data becomes the teacher, replacing rigid rules with learned patterns. Those 128 things our network learned about each image weren't features we prescribed; instead, they were patterns the machine discovered through exposure to many examples. Some of these patterns were obvious, such as loops and lines, while others were subtle features that even human eyes might not consciously recognize. The process we worked through—preparing data, building a model, training it, and evaluating its performance—represents a fundamental *workflow* in machine learning. These steps, though implemented differently, appear across various machine learning applications, from speech recognition to medical diagnosis. We'll flesh them out in much more detail in the next chapter.

But what we've explored here is just one type of machine learning. In this approach, we provided our machine with both the input (digit images) and the desired output (correct digit labels). It's like learning with a teacher who shows you both the questions and the answers.

In the next section, we'll discover that machines can learn in other ways too. Sometimes, they learn without being told the correct answers, and they find patterns and structures on their own. Other times, they learn through trial and error, like a child exploring a new game. Each approach has its own strengths and ideal applications, expanding the possibilities of what machines can learn and how they can help us solve increasingly complex problems.

2.2 Types of Machine Learning

Before we dive into how machines learn, let's reflect on what we mean by learning in general, using the most natural example we have: a child's development. Children learn to interact with the world in many different ways, each suited to different types of knowledge and skills. These learning processes mirror the varied approaches used in machine learning. Take a look at Figure 2.5 for different sources of learning. Sometimes, learning is explicit and straightforward. A parent points to an apple and says, "This is red." Or a teacher marks math homework as correct or incorrect. These are examples of direct instruction, where the "right answer" is unambiguous and feedback is immediate.

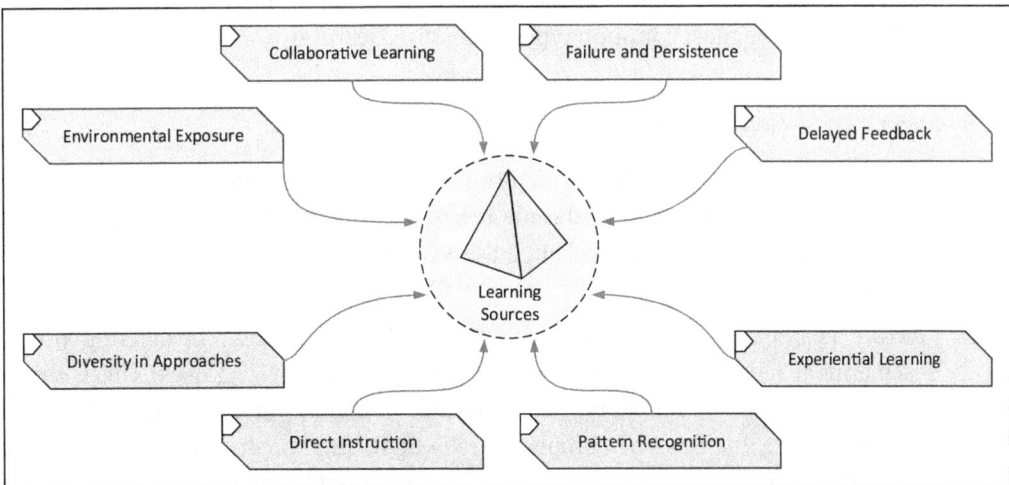

Figure 2.5 Common Sources of Learning

2 Introduction to the Core of Machine Learning

But much of childhood learning is more nuanced. Consider how children learn language—not through grammar books, but by absorbing patterns from conversations. They make mistakes such as saying "goed" instead of "went," showing how they internalize rules through observation and trial and error. Or think about how they learn to balance on a bicycle. They can't explain the physics, yet through repeated failure and persistence, they figure it out. This kind of learning—intuitive, experiential, and often hard to verbalize—is especially relevant when we later explore reinforcement learning.

Social learning adds another layer. Children also learn by watching others—picking up table manners, understanding social cues, and figuring out playground games not just through instruction but through participation and imitation.

What's remarkable is how these types of learning often work together. A child baking cookies with a parent might receive direct instructions, observe patterns, experiment, and develop intuition—all at once. Similarly, machine learning involves a combination of methods, each suited to different problems and data. Understanding this diversity in human learning helps us appreciate the range of techniques used in machine learning.

When a machine learns with clear, labeled examples—like a child being taught that "this is red"—we call this *supervised learning*. When it must discover patterns and structure on its own—like a child figuring out how to group similar toys together—we call this *unsupervised learning*. Sometimes, we provide a mix of *labeled* and *unlabeled* data, leading to what we call *semi-supervised learning*. And in some fascinating cases, we let machines learn through trial and error and rewards—just as a child learns to ride a bike or play a game—which we call *reinforcement learning*.

In the following sections, we'll explore each of these learning paradigms in detail. We'll see how they mirror different aspects of human learning while developing their own unique characteristics that make them powerful tools in the world of AI. Let's begin by talking about the most common type—supervised learning.

2.2.1 Supervised Learning

Supervised learning is the most straightforward type of learning, where there's a clear relationship between teacher and student. Earlier, we talked about a parent pointing to an apple and saying, "This is red," but let's see how this type of learning actually works in practice.

Picture a parent and child sitting with a picture book of animals. The parent points to each animal and provides its name: "This is a giraffe. See its long neck? This is a penguin—notice how it has flippers instead of wings?" What's interesting here isn't just the simple labeling. It's how the parent naturally highlights the distinctive features that make each animal unique. Over time, the child learns not just to memorize specific pictures but to recognize the key characteristics that define each animal. When they see a new picture of a giraffe from a different angle, they can still identify it because they've learned what makes a giraffe, a giraffe.

This process becomes even more critical when the stakes are higher. Consider a cooking instructor teaching students about mushroom foraging. They can't possibly show examples of every mushroom their students might encounter. Instead, they teach by highlighting crucial patterns: "See these distinctive gills? This particular color? The ring around the stem?" They'll show many examples of both safe and dangerous mushrooms, repeatedly emphasizing the critical features that distinguish them. The goal isn't just to memorize specific mushrooms but to develop a reliable system for identifying safe ones in new situations.

In addition, think about a master craftsperson teaching an apprentice to select quality wood. The teaching goes beyond simple yes/no judgments. The master might knock on different pieces to demonstrate how good wood sounds different from poor-quality timber. They might show how to examine grain patterns, demonstrate the feel of different moisture levels, or explain how to spot signs of disease or infestation. Each piece of wood becomes a lesson, with the master carefully pointing out the subtle indicators that an untrained eye might miss.

In all of these cases, notice how the learning follows a clear pattern. First, there's an expert who knows the *correct answers*. Second, they provide examples where both the input (the animal picture, mushroom, or piece of wood) and the desired output (the animal's name, safe/dangerous *classification*, or quality assessment) are clearly defined. Finally, and most importantly, the goal isn't just to memorize specific cases but to learn the underlying patterns that will help identify new, unseen examples correctly.

This type of guided learning with clear right and wrong answers forms the foundation of supervised learning in machine learning. Let's see what makes this type of learning work so well. At its heart, this approach relies on having examples where we know the correct answer. Think back to our mushroom foraging class. Each time the instructor shows a mushroom, they provide two crucial pieces of information: the mushroom itself (with all of its characteristics) and the definitive answer about whether it's safe to eat. This pairing of example and answer is critical—it's not enough to just see lots of mushrooms or to just hear about which ones are safe. The learning happens when you connect the two.

The role of the teacher—whether it's our mushroom instructor, the parent with the picture book, or the master craftsperson—goes beyond just providing answers. They're curating a *learning experience*. They choose examples that highlight important differences, they ensure all important variations are covered, and they often deliberately show tricky cases that might trip up a learner. When teaching about safe mushrooms, a good instructor will make sure to show examples of dangerous mushrooms that look similar to safe ones.

But here's where things get interesting: the true measure of learning isn't how well you remember the examples you've seen—it's how well you handle new situations. A child who has only memorized the specific animals in their picture book hasn't truly learned what a giraffe is. Real learning happens when they can identify a giraffe they've never seen before, perhaps from a different angle or partially hidden behind trees. This ability to handle new, unseen examples is what we're really after.

This brings us to a crucial point: the number of examples matters a lot. Consider trying to learn to identify safe mushrooms from just three examples. Even if those examples are perfect, they simply can't capture enough variation to build reliable judgment. You need to see many examples of both safe and dangerous mushrooms in different lighting conditions, at different stages of growth, and from different angles. Each new example adds to your understanding, helping you build a more complete picture of what makes a mushroom safe or dangerous.

Variation in the Number of Examples

Beginners usually make this mistake of thinking that more data is always good. While that's true in many cases, simply getting more of the same type of data isn't that useful. It's the *variety* of data that really helps machine learn. Be sure to keep this in mind when we get to engineering issues related to machine learning.

The same applies to our woodworking example. The apprentice can't develop reliable judgment by examining just a few pieces of wood. They need to see hundreds of examples: different species, different cuts, different conditions. Each piece of wood, along with the master's assessment, adds another data point to their understanding. Over time, patterns emerge, and what once seemed like subtle differences become obvious distinctions.

Real-World Stakes

Let's see how this type of learning applies in real-world situations, where the stakes can range from minor inconvenience to life-and-death decisions. Each application shows both the power and responsibility that comes with automated decision-making.

Consider email *spam filtering*—perhaps the most ubiquitous example of supervised learning that most of us interact with daily. While a misclassified email might seem trivial, the global impact is enormous. Businesses lose an estimated $20.5 billion annually to spam-related productivity drops and cybersecurity issues. Every day, email systems must process billions of messages, learning to distinguish between legitimate business communications and potentially harmful spam. The system learns from millions of examples where humans have previously marked emails as either spam or legitimate. What makes this task particularly challenging is that spammers constantly adapt their tactics, meaning the system must continuously learn from new examples to stay effective.

Moving to higher stakes, medical imaging systems represent a crucial application where accuracy can be a matter of life and death. These systems assist radiologists in identifying potential tumors in X-rays, MRIs, and other medical images. The human impact here is profound—early detection can significantly improve the chances of an agreeable outcome. The cost of errors runs both ways: a *false positive* (telling a patient they have cancer when they don't) leads to unnecessary stress and expensive follow-up procedures, while a *false negative* (telling a patient they're healthy when they actually have a malignant tumor) could mean missing a critical early-stage tumor. Training these systems requires thousands of carefully labeled medical images, each requiring expert radiologist time to annotate—a resource that's both expensive and scarce.

Banks use supervised learning to assess loan applications, a task where fairness and accuracy have significant financial and social implications. These systems learn from historical data about which loans were repaid and which weren't. While this automation can process applications more quickly and consistently than human reviewers, it comes with serious responsibilities. A single incorrect decision can affect someone's ability to buy a home or start a business. Moreover, if the historical data contains biases, the system might perpetuate or even amplify these biases, which is a crucial ethical consideration that banks must actively address.

Voice assistants represent a more everyday application but one that showcases the complexity of modern supervised learning. These systems must learn to understand spoken commands in different accents, with background noise, and in various contexts. The cost of errors here might seem small—having to repeat a command is merely annoying—but multiply that by millions of users, and the economic impact becomes significant. Companies invest heavily in collecting diverse voice samples to train these systems, ensuring they work reliably for users regardless of accent, age, or speech pattern.

Perhaps one of the most visible applications is in *autonomous vehicles*, where supervised learning helps cars identify traffic signs, pedestrians, and other vehicles. The stakes here are immediately apparent as mistakes could lead to accidents with severe consequences. Training these systems requires enormous datasets of labeled images and video taken under every conceivable condition, such as different weather, lighting, and traffic situations. The cost of collecting and labeling this data is substantial, but it's a necessary investment, given the potential human cost of errors. A system that fails to recognize a stop sign even 0.1% of the time is unacceptable for deployment on real roads.

What's particularly interesting about these examples is how they demonstrate different aspects of supervised learning's challenges and responsibilities. In email filtering, the challenge is adapting to ever-changing patterns. In medical imaging, it's achieving extremely high accuracy with limited training data. In loan assessment, it's ensuring fairness and transparency. In voice recognition, it's handling infinite variations of human speech. And, in autonomous vehicles, it's maintaining reliability under any possible condition.

Key Aspects of the Learning Process

Now that we've seen where supervised learning is used in the real world, let's peek under the hood to understand how this learning actually happens. The process might surprise you with its similarity to how we learn through trial and error. When a machine first starts learning, it begins much like a student taking their first guess on a new topic—often quite poorly. Consider our email system looking at its first email. Having no real understanding yet, it might make completely *random guesses* about whether messages are spam or not. Or think of a medical imaging system seeing its first X-ray—its initial attempts at identifying abnormalities would be no better than random chance.

Every time the system makes a guess, it can compare its answer to the correct one provided by human experts. An email system might guess that a message about a family reunion is spam, but when it sees that users consistently mark similar messages as legitimate, it begins to adjust its thinking. This comparison between the guess and the correct answer is crucial—it's like a teacher marking a test and showing exactly where the student went wrong. The real magic happens in how the system adjusts based on these mistakes. Unlike a human student who might need to figure out why they got something wrong, our machine follows a more systematic approach. When it makes a mistake, it doesn't just try to memorize that specific example. Instead, it makes small adjustments to its general understanding. If it mistakenly flags emails containing the word "family" as spam, it gradually learns to weight that word differently in its decisions.

This learning process is intentionally gradual. Think back about learning to ride a bicycle—you don't suddenly go from not being able to balance to being an expert cyclist. Instead, you make small adjustments with each attempt. Maybe you learn to lean slightly left when you start tilting right, or to pedal a bit faster when you feel unsteady. Our machine learning systems follow a similar principle. Each example leads to small adjustments, and these small changes add up over time to create significant improvements.

> **Gradual Learning Is Good Learning**
> The point about gradual learning is very important. You'll see in the next chapter that it has an impact on whether our machine can actually learn. Consider what might happen if we learn something too fast.

The power of this approach lies in its ability to handle many examples quickly. While a human might get overwhelmed trying to remember thousands of specific cases, our machine can process millions of examples, making tiny adjustments for each one. This is why the email spam filter can handle the enormous variety of possible spam messages, or why a medical imaging system can recognize subtle patterns that might not be obvious to the human eye.

Consider how this plays out in our autonomous vehicle example. The first time the system sees a stop sign in the rain, at night, partially obscured by a tree branch, it might miss it entirely. But after seeing thousands of stop signs in different conditions, it starts to develop a robust understanding. Each mistake leads to an adjustment, and each adjustment makes the system a bit more reliable. The system isn't memorizing specific stop signs it has seen—it's learning the fundamental patterns that make a stop sign recognizable under any conditions.

What's particularly interesting is how this process mirrors human learning while also differing from it in important ways. Like humans, these systems learn through experience and improve with practice. But unlike humans, they can do the following:

- Process thousands of examples in seconds.
- Make precise, mathematical adjustments to their understanding.
- Maintain consistent performance without getting tired or distracted.
- Apply their learning with the same thoroughness to every new example.

This combination of human-like learning patterns with machine-like precision and scale is what makes supervised learning so powerful in practice. But it also highlights some of the key challenges we face in implementing these systems effectively—challenges we'll explore next.

Common Challenges and Limitations

While supervised learning is powerful, it comes with its own set of challenges that anyone working with these systems needs to understand. Let's study these limitations through our real-world examples as they reveal important practical concerns that affect how we implement these systems.

The first and most fundamental challenge is the need for many *labeled examples*—often many more than you might initially expect. Think back to our medical imaging system. To reliably identify potential tumors, it needs to see thousands of different examples of both healthy tissue and various types of tumors. Each of these images needs to be carefully labeled by experienced radiologists. It's trying to teach someone to identify mushrooms, but instead of needing dozens of examples, you need thousands or even millions to achieve reliable performance.

This leads us to our second challenge: the cost and time required to get these labels. In our email spam example, we can easily get labels because users naturally mark spam as they encounter it. However, consider autonomous vehicles—every frame of video needs to be meticulously labeled to identify cars, pedestrians, signs, and other objects. This labeling process is expensive and time-consuming. A single hour of video might take days to label properly, with teams of people carefully marking every relevant object in every frame. It can take millions of dollars and huge amounts of human effort to create these labeled datasets. As an example, see the growth of labeled versus unlabeled *protein data* in Figure 2.6 (you can find a full color version of this chart and all other images in this book online at *http://rheinwerk-computing.com/6142* or on the resources page here: *https://recluze.net/keras-book*). The GenBank and EMBL data is *unlabeled data* about protein sequences, whereas the PDB line shows data that has received labels from either humans or automated processes. It's not necessary for you to understand what the data actually shows about proteins. It's sufficient to note that the unlabeled data grows at a much faster pace. This gap has been increasing exponentially, and it doesn't seem that we'll be able to bridge it in the foreseeable future. This is true almost universally for all sorts of domains, and the problem is getting even worse. The amount of data generated in all fields of life is exponentially increasing due to the permeating use of technology, and the human effort required simply can't keep up.

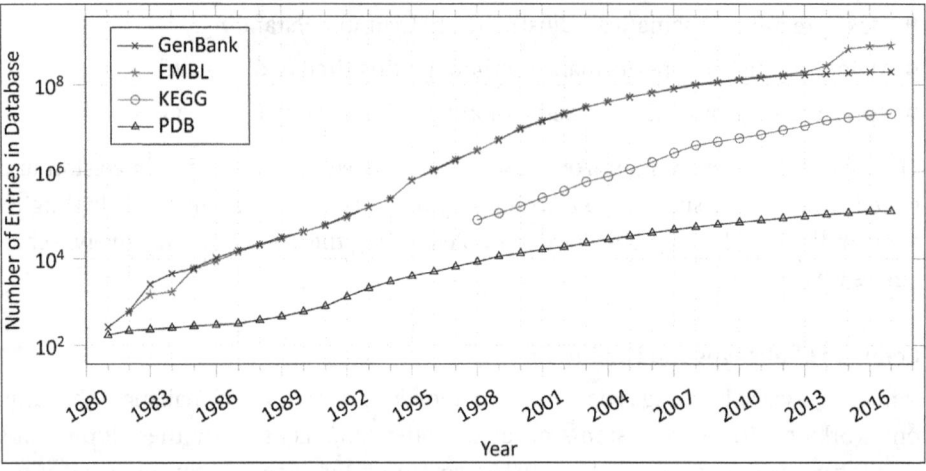

Figure 2.6 Growth of Labeled versus Unlabeled Data

But even when we're willing to invest the time and money, we face another subtle but critical challenge: What happens when some of our labels are wrong? Consider the impact of a tired radiologist accidentally *mislabeling* some medical images or a distracted worker incorrectly marking objects in autonomous vehicle training data. These incorrect labels are like a teacher giving wrong answers—they can confuse our learning system and lead to decreased performance. What makes this particularly challenging is that it's often hard to identify which labels are incorrect. In our mushroom example, even expert foragers occasionally disagree about certain specimens.

Perhaps the most interesting challenge emerges when our systems face completely new situations—scenarios they haven't encountered in their training data. Consider our autonomous vehicle encountering a new type of construction sign it's never seen before or our medical imaging system facing a rare type of abnormality that wasn't present in its training data. This is similar to our mushroom forager encountering an entirely new species—without prior examples to learn from, making reliable decisions becomes much more difficult.

These limitations can interact in complex ways. For example, the need for lots of labeled data becomes even more challenging when we're trying to prepare our systems for rare situations. How do we get enough labeled examples of rare medical conditions? How do we ensure our autonomous vehicles can handle unusual traffic scenarios they might only encounter once in a million miles of driving?

> **Rare Situations in the Age of Big Data**
> Having limited data is a critical challenge in modern computing ecosystems where machines rely on massive datasets. Domains with scarce data present significant obstacles for systems designed to learn from abundant inputs. To address these edge cases effectively, we must develop approaches that can function reliably even with limited examples. Creating models that perform well on rare or underrepresented data remains essential for building truly robust and generalizable AI systems.

Particularly important is how these challenges often force us to make practical *trade-offs*. Do we spend more money to get more labeled data, or do we try to make better use of the data we have? Do we accept that our system might be less reliable in rare situations, or do we invest heavily in finding and labeling examples of these edge cases? There's rarely a perfect answer, and different applications might require different approaches. This doesn't mean supervised learning isn't valuable—far from it. However, understanding these limitations helps us make better decisions about when and how to apply these systems. It also helps explain why we sometimes need to combine supervised learning with other approaches, something we'll explore in later chapters. Most importantly, it reminds us that while these systems can be remarkably powerful, they're not magic—they're tools that need to be used thoughtfully and with a clear awareness of their limitations.

Types of Target Answers

So far, we've looked at various examples of supervised learning, but you might have noticed that the type of answer we're looking for isn't always the same. Let's discuss these different types of answers our systems might need to provide, as understanding them helps us better grasp what's possible with supervised learning. Based on these answer types, we can categorize machine learning into subtypes, as shown in Figure 2.7.

The simplest type of answer is a straightforward yes or no decision. Our email spam filter is a perfect example—each message is either spam or not spam, with no middle ground. In machine learning, we call this *binary classification* because we're classifying inputs into one of two possible categories. Similar binary decisions appear in many critical applications: Is this transaction fraudulent? Should we approve this loan? Does this medical test indicate a particular condition? While these decisions might be based on a complex analysis of many factors, the final answer is always one of two possibilities.

2 Introduction to the Core of Machine Learning

However, this simplicity in the output can be deceptive because the reasoning behind that binary classification often involves weighing many subtle factors.

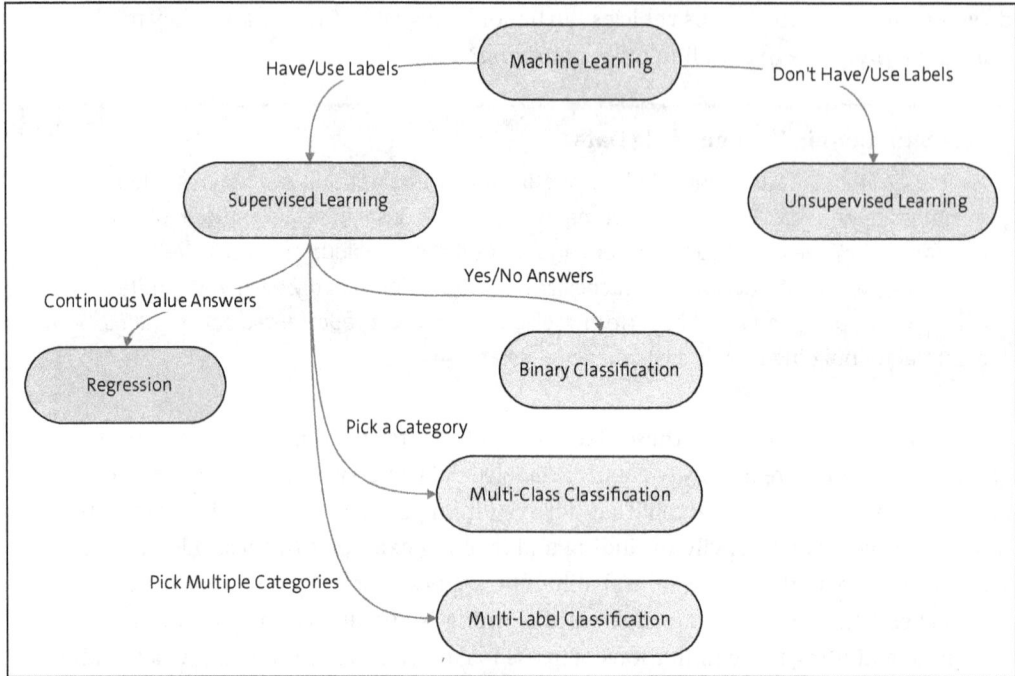

Figure 2.7 Types of Machine Learning Based on Answers

Moving up in complexity, we often want to choose from several categories—what we call *multi-class classification*. Remember our earlier example of identifying animals in photos? This isn't a simple yes/no question; instead, the system needs to decide which animal from many possibilities is in the image. A system might need to distinguish between dozens or even hundreds of different species. Similarly, in a medical diagnosis, a system might need to categorize an X-ray into one of many possible conditions, or in manufacturing, a quality control system might need to classify defects into different types. Each possible answer is distinct, and there's usually no meaningful "in-between" state—an animal is either a giraffe or it isn't.

Sometimes, what we want isn't a classification at all but a number. In machine learning, we call this *regression*, which is predicting a continuous numerical value. Think about predicting house prices—we're not asking for a yes/no decision or a category but rather a specific numerical value. This type of answer brings its own interesting challenges. When predicting a house's price, being off by $1,000 on a house that costs $5 million might be acceptable, while being off by the same amount on a $50,000 house would be a significant error. Other regression examples include predicting a customer's likely purchase amount, estimating the remaining lifetime of a machine part, or forecasting tomorrow's temperature. In these cases, our answers exist on a continuous spectrum rather than in discrete categories.

The most complex situations involve outputs that combine multiple elements or have intricate structures—sometimes combining classification and regression in the same problem. Consider an autonomous vehicle analyzing a street scene—it needs to identify multiple objects (multi-class classification for cars, pedestrians, signs), their locations (regression for precise coordinates), their movements, and how they relate to each other. Or think about a system that generates detailed descriptions of medical images, not just identifying abnormalities but describing their size, location, and characteristics. These complex outputs might combine numbers, categories, and multiple classifications all at once.

These different types of answers require different approaches to learning:

- For binary classification, the system needs to learn a clear boundary between the two possibilities.
- For multi-class classification, the system needs to learn the distinctive features that separate each category from all others.
- For *multi-label classification*, we still predict categories, but we don't have to pick just one. Categories in this case won't be mutually exclusive.
- For regression, the system needs to understand how different factors influence the final numerical prediction.
- For complex outputs, the system needs to learn not just individual elements but how they fit together.

Understanding these different types of answers helps us appreciate both the flexibility and the limitations of supervised learning. It also helps explain why some problems are harder than others as binary classification is generally easier than multi-class classification, which in turn is often simpler than complex problems combining multiple types of predictions. But regardless of whether we're doing classification or regression, the core principle remains the same: we learn from examples where we know the correct answer, whether that answer is a simple binary choice or a rich, detailed description.

But what happens when we don't have the correct answers? What if we need to learn from data without any labels at all? This brings us to our next topic: unsupervised learning.

2.2.2 Unsupervised Learning

Unsupervised learning is the type of learning where there are no right or wrong answers, only patterns waiting to be discovered. While this might sound strange after our discussion of supervised learning, it's actually something we encounter every day.

2 Introduction to the Core of Machine Learning

Watch a young child playing with a collection of toys. Without any instruction, they might naturally start sorting them into groups. Some children might organize by color, putting all the red toys together. Others might group by size, or by type (all the cars in one pile, all the stuffed animals in another). See Figure 2.8 for an example of grouping different objects based on different criteria. What's important to note is that none of these arrangements is "correct" or "incorrect"—they're just different ways of recognizing patterns. The child isn't following rules they were taught; they're discovering natural relationships in their environment.

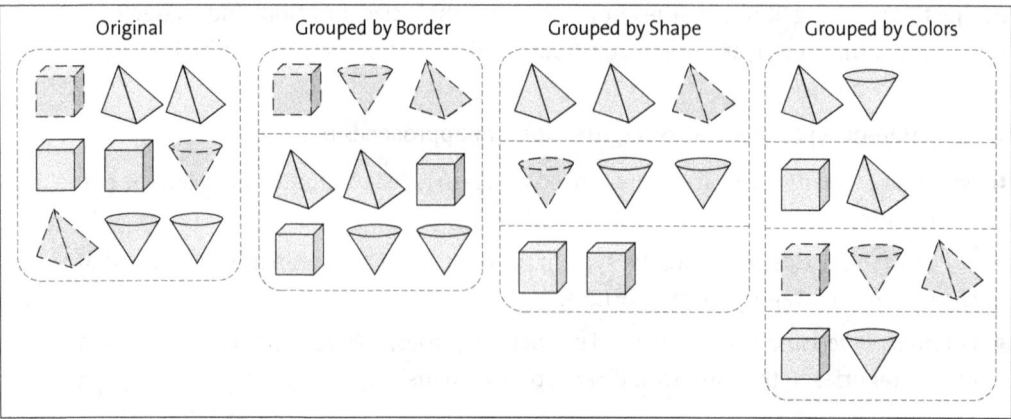

Figure 2.8 Unguided Unsupervised Learning Might Produce Unneeded Results

This natural pattern-finding behavior appears in many contexts. Consider how a grocery store organizes its products. While some groupings are obvious (all produce together), others are less clear-cut. Should tomatoes go with vegetables or fruits? Should gluten-free products be in their own section or distributed among their respective categories? Different stores make different choices because there's no single "right" way to organize—it depends on what patterns and relationships you consider most important.

Perhaps the most poetic example of human pattern-finding is how we see constellations in the night sky. The stars aren't actually connected or grouped in any physical way—they might be vastly different distances from Earth. Yet our brains naturally find patterns in these random points of light, creating pictures and stories. Different cultures have found different patterns in the same stars, showing how pattern recognition can be both natural and culturally influenced.

These examples highlight a key aspect of this type of learning: the patterns are discovered rather than taught. Unlike in supervised learning, where we had clear examples of right and wrong answers, here we're looking for natural structure in the data itself. This approach opens up fascinating possibilities for discovering patterns we might not have known to look for—but it also brings its own unique challenges depending on the unique practical application.

Real-World Applications of Unsupervised Learning

Let's explore how this pattern-finding approach translates into practical applications, starting with familiar business scenarios and moving to more complex scientific and technical uses.

Marketing teams have long sought to understand their customers better, but manually grouping thousands or millions of customers is impossible. This is the concept of customer *segmentation* through unsupervised learning. By analyzing purchase histories, browsing behavior, and demographic data, these systems discover natural *clusters* of customers. One group might be price-sensitive bargain hunters who mainly shop during sales, while another might be early adopters who always buy the latest products. These discovered patterns help businesses tailor their marketing strategies, optimize inventory, and even design new products for specific customer segments. This is the process of *clustering*.

In manufacturing, the challenge often isn't finding similar things but finding what doesn't fit the pattern. *Anomaly detection* systems monitor production lines, analyzing data from sensors to identify potential problems before they become critical. Consider a factory producing computer chips: the system might notice that one machine's temperature patterns are slightly different from all others, flagging it for maintenance before it fails. This predictive approach can save millions in prevented downtime and reduce waste.

Document clustering tackles the massive scale of digital information. When you use a search engine, it needs to understand how billions of documents relate to each other. Unsupervised learning helps organize this vast collection by finding documents that cover similar topics, even if they use different words. This is why searching for "car maintenance" might also bring up relevant articles about "automobile repair" the system has learned these topics are related without being explicitly told.

Recommendation systems combine multiple types of patterns to suggest products, content, or connections. While Netflix and Amazon's recommendations might seem like magic, they're actually finding patterns in what different users watch or buy, what items are often consumed together, and how preferences change over time. These systems don't just match you with things similar to what you've liked before. Instead, they discover complex patterns of user behavior that help predict what you might enjoy next.

In biology, unsupervised learning has opened new frontiers in understanding genes. Scientists use these techniques to analyze vast datasets of genetic information, discovering groups of genes that work together or patterns in how genes are expressed in different conditions. These discoveries have led to breakthroughs in understanding diseases, developing treatments, and even predicting how different patients might respond to specific medications. This field of *precision medicine* has huge potential for the future.

Perhaps one of the most critical applications is in cybersecurity, where *network traffic analysis* helps protect systems from attacks. By learning the normal patterns of network activity, these systems can identify suspicious behavior that doesn't fit the usual pattern. What makes this particularly challenging is that "normal" traffic patterns change constantly, and attackers are always developing new techniques. The system must continuously update its understanding of what constitutes normal behavior while being sensitive enough to flag genuine anomalies.

What's particularly interesting about these applications is how they highlight different aspects of pattern finding:

- Customer segmentation shows how patterns can reveal *hidden structures* in seemingly chaotic data.
- Anomaly detection demonstrates how understanding normal patterns helps identify what's unusual.
- Document clustering illustrates how patterns can exist at massive scales.
- Recommendation systems show how multiple patterns can interact in complex ways.
- Genetic analysis reveals how patterns can lead to scientific discoveries.
- Network analysis shows how patterns can evolve over time.

In each case, the power of unsupervised learning comes from its ability to discover patterns we might not have known to look for.

> [!]
> **Unsupervised Learning Is a Double-edged Sword**
> Unlike supervised learning, where we start with known categories or outcomes, these applications let the data reveal its own structure, which often leads to insights that human experts might have missed. It might also lead to results that aren't of interest to us though. That's a double-edged sword we need to be wary of.

The Learning Process and Challenges

Let's peek behind the curtain to understand how unsupervised learning actually works and why it can be both powerful and challenging. Remember our child organizing toys? Their process of grouping similar items together reveals many of the key principles—and difficulties—that our machine learning systems face. This brings us to one of the core concepts: measuring *similarity*. Try to think of applying the concept of similarity on two books., We might say two books are similar if they share the same author. But what if one book has multiple authors? What if we care about both author and genre? This is where the concept of distance comes in—not physical distance, but how different two items are across multiple characteristics. Just as you might say two colors are "close" to each other on the color spectrum, our systems need ways to calculate how close or far apart items are in terms of their features.

Consider our earlier example of Figure 2.8. If we specify the "distance" between two shapes to be the difference in the number of sides they have, we can possibly get a grouping that is based on shape. Depending on the situation, different types of distances can be crafted carefully to arrive at the intended solution.

This leads to our first major challenge: there's no clear "right answer" to validate against. Unlike in supervised learning, where we could check if our email was correctly identified as spam, here we're discovering patterns that might be equally valid in different ways. It's like asking whether books should be organized by genre or author—the answer depends on what you're trying to achieve. This is known as the *clustering problem*, and it's one of the most challenging aspects of unsupervised learning.

Then, there's the question of *outliers*—items that don't fit neatly into any group. In customer segmentation, there might be customers with unique buying patterns that don't fit any major group. Should we create special categories for these outliers, force them into existing groups, or treat them as noise in the data?

This connects to a broader challenge: *validating results* without ground truth. In supervised learning, we could measure our success by how often we got the right answer. But what's the right answer when you're discovering patterns? We might look at whether the patterns are consistent, whether they're stable over time, or whether they lead to useful insights—but these are often subjective measures. These challenges don't make unsupervised learning any less valuable—in fact, they're part of what makes it so powerful. By grappling with these questions, we often discover insights we wouldn't have found if we were limited to predefined categories. The key is understanding these limitations and challenges so we can use unsupervised learning effectively, choosing the right approaches for our specific needs and being thoughtful about how we interpret the results.

Flavors of Unsupervised Learning

Unsupervised learning is a collection of approaches, each designed to uncover different kinds of hidden patterns in data. Some look for groups, others for relationships, and still others for things that just don't fit in.

The most common type is clustering, where we group similar items together. This is what we're doing when we divide customers into segments or group similar products. The simplest form, called *k-means clustering*, creates distinct groups where each item belongs to exactly one cluster—simple, straightforward, and surprisingly powerful.

Sometimes, we need something more sophisticated than simple groups leading to what is termed as *hierarchical clustering*. Instead of just creating separate groups, it builds a tree-like structure of relationships, like branches splitting into smaller branches, and then into twigs. Think of biological classification where an animal belongs to a species, which belongs to a genus, which belongs to a family, and so on up the tree. When you're dealing with complex relationships in data, such as organizing documents by topics and subtopics, this hierarchical approach really shines.

Anomaly detection takes a different approach entirely. Rather than looking for similarities, it hunts for things that stand out. First, it learns what "normal" looks like. Then, it flags anything that doesn't fit this pattern. This might sound simple, but it's incredibly powerful in practice. It's how credit card companies spot fraudulent transactions, manufacturers identify defective products, and security systems detect network intrusions.

Association rule learning digs deeper, uncovering hidden relationships between different elements in data. When an online store suggests "customers who bought X also bought Y," that's association rule learning at work. It's understanding how different things connect and relate to each other in meaningful ways. These connections can reveal surprising patterns that even human experts might miss.

Then there's *dimensionality reduction*—a fancy term for something we do naturally. When you describe a complex object like a car, you might focus on just a few key features: color, size, and style. You're reducing many dimensions of information into a simpler description. *Principal Component Analysis* (PCA) does something similar, finding the most important "directions" in data while ignoring the less important ones. This makes complex data easier to work with while keeping the patterns that matter most.

[+]
> **Dimensionality Reduction as a Means to a Goal**
> We'll be using dimensionality reduction to make sense of some very complex models in later chapters. Keep this basic principle in mind: we can't always understand high dimensional data easily, so we reduce the number of factors we must keep in mind to grasp a complex concept.

Perhaps most fascinating is learning *intermediate representations*, which has become crucial in modern deep learning. Imagine trying to understand a foreign language by first translating it into something simpler than your native language—a set of universal concepts that bridge the gap. That's what systems such as *autoencoders* do with data, learning to transform it into forms that make complex patterns easier to recognize. This approach has revolutionized how we handle tasks such as *image recognition* and *natural language processing (NLP)*. We'll return to these concepts in Chapter 9 onward in a lot of detail.

In practice, these approaches often work together like parts of an engine. A system might use dimensionality reduction to simplify the data and then apply clustering to find groups while simultaneously checking for anomalies. Each type of unsupervised learning brings its own strengths to the ensemble. This is the true strength of unsupervised learning—using it in a much larger model setting. This is also how we'll be using this type of learning. So, the plan is to spend some chapters of this book focusing on supervised learning and then return to unsupervised learning as part of a much larger supervised learning model. Of course, there is another type of learning that sits somewhere in the middle of supervised and unsupervised models—semi-supervised learning more commonly known as reinforcement learning.

2.3 The Magic Sauce: Reinforcement Learning

Let's explore the third major type of machine learning through a thought-provoking example about learning to drive. Consider a new driver taking their first long journey. They start driving, and everything goes smoothly for two hours—maintaining proper speed, staying in their lane, and keeping safe distances. Then, fatigue sets in. They begin to doze off, and when they suddenly come to, they immediately hit the brakes hard. Despite this reaction, they still end up hitting the car in front of them. If we were to collect data about accidents, we'd notice something curious: in many cases, the last action before a collision is hitting the brakes. If we were to analyze this data too simplistically, we might conclude that braking causes accidents! Of course, this is clearly wrong—the brakes were a desperate attempt to prevent the accident, not its cause. The real problem started much earlier, with the decision to continue driving while tired.

This example illustrates one of the most challenging problems in learning: How do we correctly attribute outcomes to the actions that truly caused them? The accident wasn't caused by the final brake action but by a series of decisions made much earlier. This is what we call the *credit assignment problem*—figuring out which actions in a long sequence were actually responsible for the final outcome, whether good or bad.

This is the core challenge of reinforcement learning. Unlike supervised learning, where we immediately know if an answer is right or wrong, or unsupervised learning, where we find patterns, reinforcement learning deals with situations where the consequences of our actions might not become apparent until much later. It's like playing a turn-based game—the move you make on the first turn might seem fine at the time but could lead to your defeat 20 moves later. How do we learn which early moves were actually responsible for the eventual loss?

This type of learning is fundamentally different from what we've seen before. In reinforcement learning, our machine must learn through trial and error, dealing with *delayed feedback* about its actions. The challenge isn't just learning what to do but understanding which actions in a long sequence were truly important for the final outcome.

2.3.1 The Building Blocks of Reinforcement Learning

Let's break down reinforcement learning into its fundamental pieces. While we'll explore this topic in much more detail later in the book, understanding these basic elements will help us see how this type of learning differs from what we've discussed so

2 Introduction to the Core of Machine Learning

far. At its heart, reinforcement learning involves two main players: an *agent* and its *environment*. The agent is the learner—like our driver in the previous example, or a robot learning to walk. The environment is everything the agent interacts with—the car and the road, or the robot's physical surroundings. This interaction is at the core of reinforcement learning.

The agent exists in what we call a state, that is, its current situation. For our driver, the state might include their position on the road, their speed, and how alert they are. The agent can take various *actions*—such as pressing the gas pedal, turning the steering wheel, or applying the brakes. Each action might change the state, leading to a new situation.

After each action, the environment provides feedback through *rewards* or penalties. This is how the agent learns what's good and what's bad. A reward might be positive (successfully navigating a turn), negative (hitting another car), or neutral. Sometimes, these rewards are immediate, such as the instant feedback of bumping into a wall. Other times, they're delayed, such as realizing much later that an early game move led to victory.

To visualize this in a simplistic environment, take a look at the mini world shown in Figure 2.9. This is a simple world where an agent (shown as a cone with a ball on top) can move between different blocks or states, each marked with coordinates, for example, (0,0). The agent can move up, down, left, or right to neighboring blocks. There's something valuable to reach—a gem-like reward in the top-right corner at (0,4). However, there are also dangerous areas to avoid, marked with crosses in the bottom-right at (2,3) and (2,4). Some blocks are inaccessible, shown by diagonal lines. The agent's challenge is to find a path from its starting position at (1,0) to the gem while avoiding the dangerous areas. The agent needs a strategy for doing this.

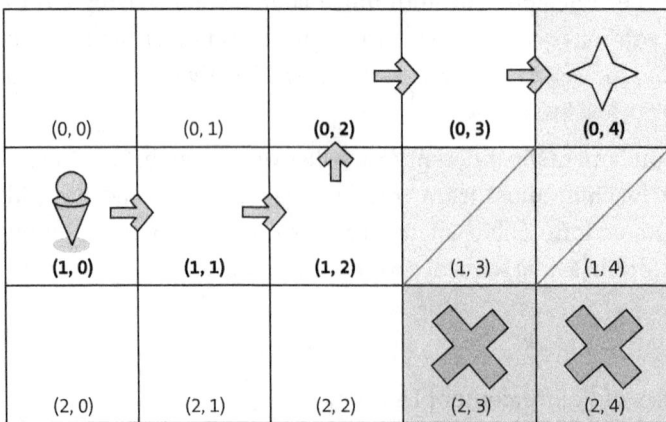

Figure 2.9 State Changes, Rewards, and Actions in Reinforcement Learning

This brings us to one of the most important concepts in reinforcement learning: the *policy*. A policy is essentially the agent's strategy—what action it should take in each situation. Initially, the agent might make random choices, but over time, it learns which actions tend to lead to good outcomes in different states. It's like developing good habits where experience teaches you what works best in different situations.

Let's see how this plays out in a real example. Consider a robot learning to walk:

- The agent is the robot's control system.
- The environment is the physical world with its surfaces and gravity.
- States include the position of each joint and the robot's balance.
- Actions are different motor movements.
- Rewards might be positive for successfully reaching a target object across the room and negative for failing to reach it.
- The policy is the learned rules for which movements work best in different positions.

What makes this challenging is how the robot must deal with *delayed rewards*. A slight imbalance might not cause problems immediately but could lead to a fall several steps later. The robot needs to learn which actions truly contributed to success or failure—just like our driver example where the real cause of the accident wasn't the final brake action but the earlier decision to drive while tired. We'll return to reinforcement learning in much more detail in Chapter 10 after we've built a solid foundation with supervised learning and gained hands-on experience with Keras. For now, the key is understanding that this type of learning is about an agent learning through trial and error, dealing with delayed consequences, and gradually developing better strategies for achieving its goals.

This might all sound complex, and, in many ways, it is. But it's also one of the most natural forms of learning. After all, most of us learn many things in life this way: trying something, seeing what happens (maybe not right away), and gradually figuring out what works best. The challenge is teaching computers to learn in the same way. If we can do that, computers will be able to solve a huge variety of problems.

2.3.2 Key Applications for Reinforcement Learning

Reinforcement learning can and has been applied in a huge variety of problems. Let's see some of them to show how the concepts of agents, actions, and rewards play out in different scenarios, from games to real-world systems.

Game playing has provided some of the most dramatic demonstrations of reinforcement learning's potential. In games such as chess or Go, the agent (the AI player) can take actions by moving pieces on the board. The reward is simple but powerful: winning the game. This seemingly straightforward setup led to remarkable breakthroughs, with AI systems learning strategies that surprised even world champions. What's surprising is that these systems learned not by studying human games but by playing millions of games against themselves, gradually discovering what works through trial and error.

The 2016 victory of AlphaGo over world champion Lee Sedol marked a turning point in AI, showing that machines could master even the most complex strategic games. More importantly, it demonstrated that AI could develop approaches that differed from centuries of human wisdom, finding novel strategies that experts had never considered.

In robotics, the applications become more physical and immediate. Consider a robot in a warehouse trying to pick up and sort packages. Its actions are physical movements—reaching, gripping, lifting, and placing. The reward might be successfully placing the right package in the correct location. Unlike in board games, robots must deal with the complexities of the physical world, where no two situations are exactly alike, and mistakes can have real consequences.

Resource management shows how reinforcement learning can tackle complex optimization problems in the real world. Consider a data center cooling system. The agent controls various cooling units and airflow systems, making constant adjustments to temperature settings and fan speeds across different zones. The reward could be reducing energy consumption while keeping all servers within safe temperature ranges. This balance between immediate needs and longer-term goals makes these applications particularly challenging. Traffic light control systems face similar challenges but on an even larger scale. They must learn patterns in traffic flow, predict changes, and coordinate multiple intersections—all while dealing with unpredictable human behavior.

These applications show how the same basic principles can apply to vastly different scenarios. Whether it's moving chess pieces, controlling robot arms or managing resources, the core idea remains: learning what works through trial and error and gradually improving based on the results. It's not all that simple though. There are many challenges we need to beat in order to make reinforcement learning work.

2.3.3 Challenges in Implementing Reinforcement Learning

Think of a child learning to play a new game. Should they keep using a strategy that seems to work (*exploitation*) or try something new that might work better (*exploration*)? This exploration versus exploitation dilemma is at the heart of reinforcement learning. Stick with what works, and you might miss better solutions. Try new things too often, and you might never develop reliable strategies. It's a delicate balance that every reinforcement learning system must manage.

Remember our driving example? The delayed feedback problem makes everything harder. Actions might have consequences that only become apparent much later. This connects directly to the credit assignment problem When something goes wrong (or right), which of our previous actions were actually responsible?

Some reinforcement learning problems are harder than others, often because of these timing and credit issues. Teaching an AI to play tic-tac-toe (where moves have immediate, clear consequences) is far easier than teaching it to play Go (where the impact of early moves might only become clear at the end of the game). In the real world, these challenges become even more pronounced. A robot learning to walk faces not just delayed feedback but also an almost infinite number of possible situations. The complexity increases exponentially when the environment is unpredictable or when actions can have multiple interacting consequences. Imagine teaching a self-driving car to navigate city traffic—every decision must account for dozens of moving objects, each with their own unpredictable behaviors.

What's remarkable is how closely these challenges mirror human learning. We all face the same dilemmas: when to stick with what we know versus when to try something new, how to learn from delayed feedback, and how to figure out which of our actions really led to success or failure. Understanding these parallels helps us develop better reinforcement learning systems while also giving us insights into human learning itself. These insights become particularly valuable as we work to create AI systems that can learn and adapt in ways that feel more natural and intuitive to the humans who interact with them. For decades, these challenges kept reinforcement learning largely confined to academic research. The theory was elegant and well-understood, but practical applications remained frustratingly out of reach. Then something unexpected happened. We discovered that neural networks—which were already revolutionizing other areas of machine learning—could be the missing piece of the puzzle. This combination proved surprisingly powerful and almost deceptively simple in solving many long-standing challenges. It was like finding the right key for a lock we'd been struggling with for years.

As you'll see in Chapter 10, this marriage of reinforcement learning and neural networks opened entirely new possibilities, finally bringing many theoretical ideas into practical reality. So, let's discuss this final part of our puzzle briefly in the next section.

2.4 Basics of Neural Networks

Neural networks are the powerhouses driving modern machine learning, and they appear in almost every cutting-edge AI system we've discussed. Actually, they're so important that we'll spend most of this book exploring them in detail starting from the next chapter. But for now, let's just get our feet wet with the core concept.

Remember our earlier example of recognizing handwritten digits? We mentioned that the system learned 128 different "things" about these digits. That might have seemed like a random number at the time, but there's something really interesting going on there. Each of these things was learned by a *neuron*—a simple processing unit in our neural network. The term neuron was originally inspired by neurons in the human brain, but while biological neurons receive signals and decide whether to fire based on those inputs, our artificial version is more like a simplified mathematical model. We started with inspiration from biology, but we've ended up somewhere quite different. Instead of trying to replicate the incredibly complex machinery of the brain, we've developed these artificial neurons to follow mathematical principles that turn out to work remarkably well for learning patterns.

> **Artificial Neural Networks**
>
> You'll notice that we use the term *neural networks* rather than *artificial neural networks* because the modern machine learning community dropped the "artificial" qualifier a while ago. In the machine learning community, "neural network" refers specifically to computational models, not biological ones. Using artificial neural network or its abbreviation ANN is considered outdated terminology in professional machine learning contexts.

Let's go back to our digit recognition example to see how this works in practice. Those 128 neurons were working together, each learning to recognize different aspects of the digits. One neuron might have learned to spot curves, another to detect straight lines, and others to recognize different angles or intersections. It's like having 128 tiny experts, each focused on one specific feature. But the real magic happens when we combine all of these pieces of information. These 128 individual observations work together to make the final decision about which digit was shown in the image.

That's the real power of neural networks—they can take many simple pieces of learned information and combine them into surprisingly complex and accurate decisions. And this is just the beginning of what they can do. In this section, we'll explore how they work in more detail.

2.4.1 Core Components of Neural Networks

Let's break down a neural network into its basic building blocks. Think of it like understanding how a car works—you don't need to know every technical detail, but understanding the main components helps you grasp how everything works together. We've already met our first component: the neuron. Each neuron is a tiny decision-maker. Just as a person might decide whether to bring an umbrella by looking at the clouds, wind, and weather forecast, a neuron makes simple decisions based on the information it receives. In our digit recognition example, one neuron might look at a small part of the image and decide how likely it is to contain a curved line.

But neurons don't work alone—they're connected to each other in interesting ways. Imagine a group of art experts trying to identify a painting. One expert might focus on the color palette, another on the brushstrokes, and a third on the composition. They share their observations with each other before making a final decision. Similarly, neurons are connected to each other, passing information along these connections. When one neuron makes a decision, it shares that information with other neurons that it's connected to. These neurons are organized into *layers*, like departments in a company. The *input layer* is like the reception desk—it receives the raw information. In our digit recognition system, this layer takes in the image of the handwritten digit. Next, come the *hidden layers*—think of them as different departments processing the information. Recall our 128 neurons. They formed a hidden layer, each learning to recognize different patterns in the digits. Finally, we have the *output layer*, like the executive team making the final decision. In our case, this layer decides which digit (0–9) it thinks it's looking at.

Let's see how information flows through this whole system. Here's the sequence when we show it a handwritten digit:

1. The input layer receives the image, breaking it down into individual pixel values.
2. These values flow through connections to our 128 neurons in the hidden layer.
3. Each of these neurons looks for its specific pattern and makes a decision.
4. These 128 decisions flow through more connections to the output layer.
5. The output layer combines all of this information to make the final digit *prediction*.

This flow of information, from input through hidden layers to output, is what allows neural networks to transform raw data (e.g., pixel values) into meaningful decisions (e.g., digit predictions). It's similar to how a large organization processes information— from initial data collection through various departments' analyses to final decision-making. See Figure 2.10 for how information flows from one layer to another in a neural network.

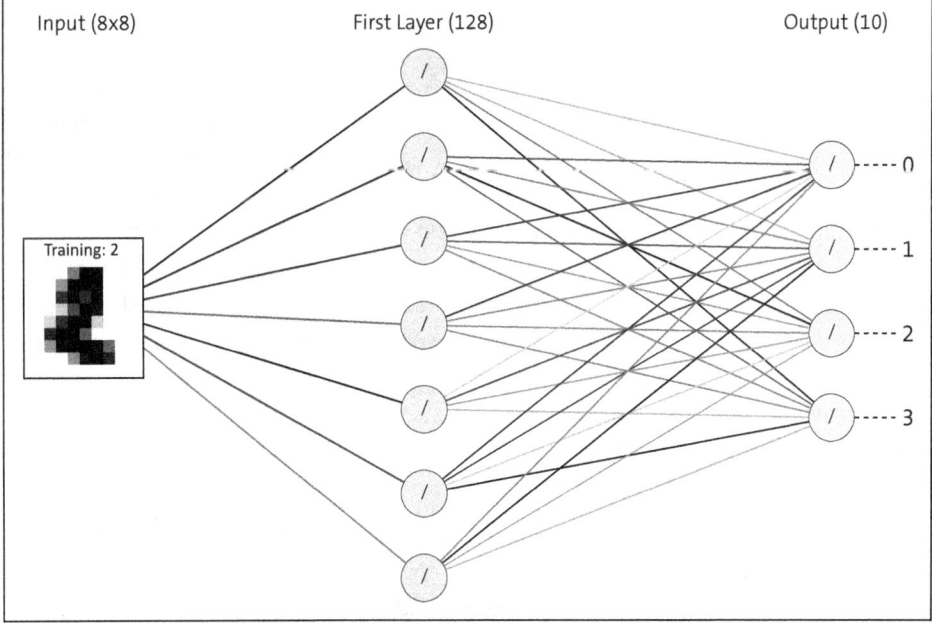

Figure 2.10 Layers in Neural Networks

What makes this system powerful is how each component plays its part while working together with others. Individual neurons make simple decisions, connections ensure information flows where it's needed, and layers organize this process into manageable steps. When we train the network, we're really teaching all these components to work together effectively, like training a team to operate smoothly.

2.4.2 The Unintuitive Process of Learning

How does a neural network learn? The process is really fascinating and, in some ways, mirrors how we might learn a new skill. When a neural network first starts out, it's like a blank slate with *random* guesses. Consider those 128 neurons we talked about, each one initially making arbitrary decisions about what patterns to look for in our handwritten digits. They're like novice art critics who haven't yet learned what to look for.

This is where the magic of supervised learning comes in. Remember how we talked about learning from examples where we know the correct answer? That's exactly what happens here.

We show the network thousands of handwritten digits, each one labeled with the correct number it represents. At first, the network makes lots of mistakes. A messy 6 might be mistaken for an 8, or a slightly tilted 1 might be seen as a 7. But here's the clever part: every time the network makes a mistake, it adjusts itself slightly. Those 128 neurons gradually refine what they're looking for. One neuron might realize it should pay more attention to curves at the top of the digit. Another might learn to focus more on straight lines at the bottom. The connections between neurons also adjust learning which patterns are more important for identifying each digit. Every mistake is an opportunity to learn.

A parallel to this might be a human learning to identify birds. At first, you might just make random guesses. But after seeing many examples of different birds, along with their correct names, you start noticing important details. Gradually, your observations of the shape of the beak, the color patterns, and the way they fly become more refined and reliable. Neural networks learn in a similar way, but they can process thousands of examples much faster than we can. Speed is their superpower. The key is that learning happens incrementally. The network doesn't suddenly jump from random guesses to perfect predictions. Instead, it gets a little bit better with each example it sees. Early on, the improvements might be dramatic—like finally realizing that all 1s tend to be tall and thin. Later improvements might be more subtle.

This is why having many good examples is so crucial. Just as we saw in supervised learning, the quality and quantity of training examples largely determine how well the system will learn. The more handwritten digits the network sees, and the more diverse these examples are, the better it becomes at handling new, unseen digits in the future. It's a simple principle with powerful results.

2.4.3 Clearing Up Some Misconceptions

Before we dive deeper, let's clear up some common misconceptions about machine learning that might trip us up later. These are ideas that have somehow taken root in the field but don't tell the whole story:

- First, there's a persistent belief that neural networks are exclusively tools for supervised learning. You'll hear this repeated in many introductory courses and even some textbooks. But it couldn't be further from the truth. Neural networks are more like a Swiss Army knife—incredibly versatile tools that can be applied to all types of machine learning. As we progress through this book, you'll see neural networks performing amazing feats in supervised, unsupervised, and even reinforcement learning scenarios.

- Another misconception is that unsupervised learning is just about grouping similar things together. While clustering is certainly a powerful application, it barely scratches the surface of what unsupervised learning can do. Neural networks have opened up entirely new possibilities in this space. They can learn to generate new images, understand the structure of language without any labels, and even discover patterns we never knew existed in our data.

These concepts might sound abstract right now, and that's okay. Machine learning is one of those fields where things really click when you start working with the code. That's why we're going to take a hands-on approach. Instead of drowning in theory, we'll set up a coding environment early on so you can start experimenting with Keras even before you fully understand how everything works under the hood. Don't worry if some concepts feel fuzzy at first—they'll become clearer as you build and experiment with real models.

Think of it like learning to cook. You might start by following recipes before you understand why certain ingredients work together or what chemical reactions are happening in your food. But as you keep cooking, these deeper insights naturally emerge from your hands-on experience. We'll follow a similar path, building your practical skills and theoretical understanding in parallel.

2.5 Setting Up Your Environment

Let's get you started with some hands-on machine learning! Now, you might be thinking about installing Keras on your computer, but setting up a local machine learning environment can be quite a process. You need to install the right version of Python, manage package *dependencies*, ensure your graphics drivers are compatible, and configure various environmental settings. Even when you get everything running, your computer might struggle with more complex models unless you have specialized hardware. But here's some good news: we can skip all that initial setup and jump straight

2 Introduction to the Core of Machine Learning

into the exciting part. Google *Colaboratory* (*Colab*) is a fantastic and free cloud-based environment that's all ready to go with everything we need for machine learning.

In fact, we can have you writing your first machine learning code in the next five minutes. Colab gives you access to powerful computing resources in the cloud, so you don't have to worry about whether your laptop is up to the task. Don't worry—later in the book, we'll walk through setting up everything on your local machine. But for now, let's focus on learning and experimenting with the code itself rather than getting bogged down in setup details.

To use Colab, you'll need to sign up for a free Google account (or use an existing one if you already have it.) Then, head over to *https://drive.google.com/drive/home*, and click on the **New** button on the left side of the screen (see Figure 2.11).

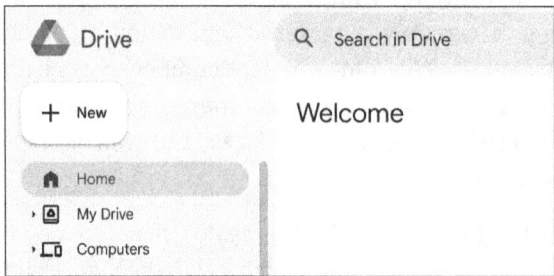

Figure 2.11 Opening Google Drive

In the dropdown (as shown in Figure 2.12), click the **More** option and then **Connect more apps**. A detailed view of what apps you can install appears.

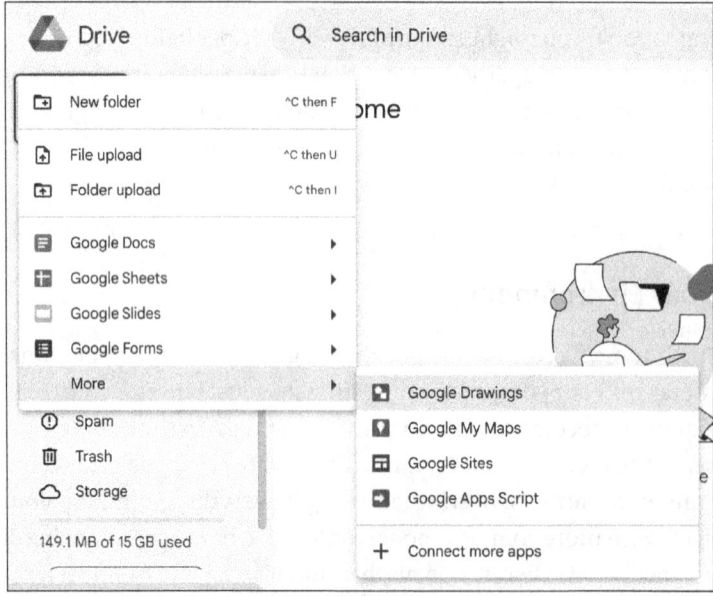

Figure 2.12 Connect More Apps to Google Drive

On the top of the newly open popup, you'll see a search option as a magnifying glass. Click on that and then search for "Colab" in the search box, as shown in Figure 2.13.

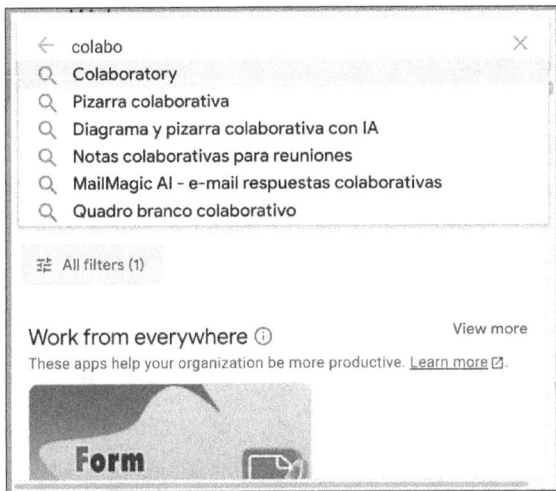

Figure 2.13 Searching for Google Colaboratory

Click on the **Colaboratory** icon. You can see the app logo in Figure 2.14.

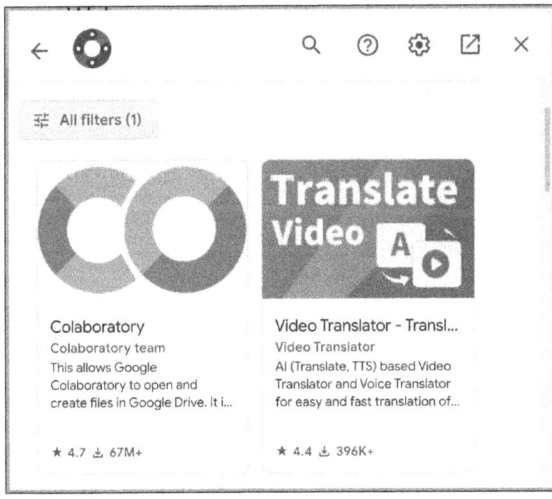

Figure 2.14 The Google Colaboratory Add-In

Click on **Install** (see Figure 2.15) and you'll be asked to grant Colaboratory permissions. This is an official app from Google itself so it's safe to use, but be sure to read through the permissions to know exactly what you're allowing.

Figure 2.15 Installing Add-Ins to Google Drive

Once all the permission granting is done and Colab is installed, you can head back over to *https://drive.google.com/drive/home*, click on **New • More** again, and now you'll now see **Google Colaboratory** (see Figure 2.16). Clicking on this will bring up a new Colab file. You're now ready to start writing your code. Installation and granting permissions to Colab was a one-time operation. The next time you want to use Colab, you can just create a new file using this last step.

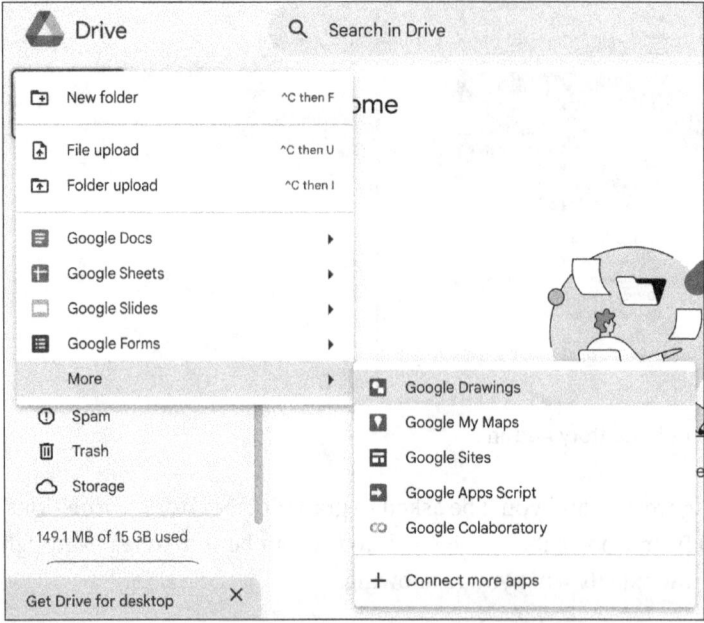

Figure 2.16 Google Colaboratory as an Additional File Type

When you open the file, the default theme will be applied to Colab. If you don't like theme, you can change it using the settings icon on the top right, as shown in Figure 2.17.

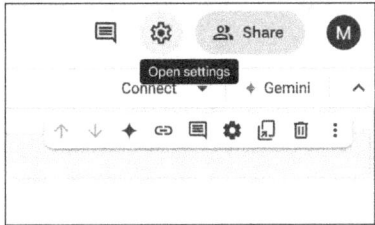

Figure 2.17 The Settings Icon in a Google Colab File

Take a look at Figure 2.18. In the **Site** tab, you can see the first option at the top that allows you to select the theme as dark or light. We'll use the light theme throughout this book whenever we use screenshots of Colab, but you're free to use whichever theme you prefer.

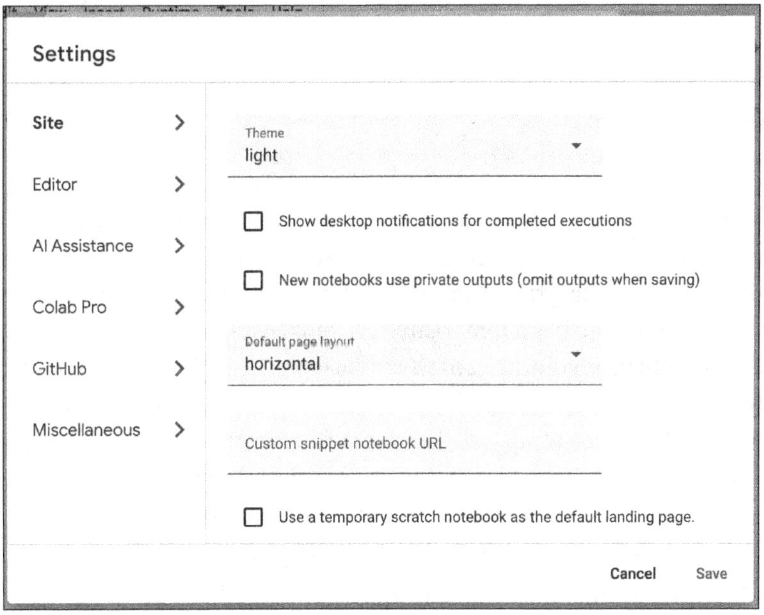

Figure 2.18 Selecting the Theme for Your Notebook

Finally, to actually use the code, you'll see a little gray box in the middle portion of the screen. Figure 2.19 shows two cells at the moment. To write code in a cell, just type or paste your code. Then, to run it, either click the **Play** button on the left of the cell or hold down the [Shift] key and press [Enter]. The code in the cell will be executed, and the output will be shown below the cell. For instance, the second cell called for a print statement. So, the value of the variable **x** was output below the cell. This is a great way of working with code and experimenting with parts of it without having to run the whole code again.

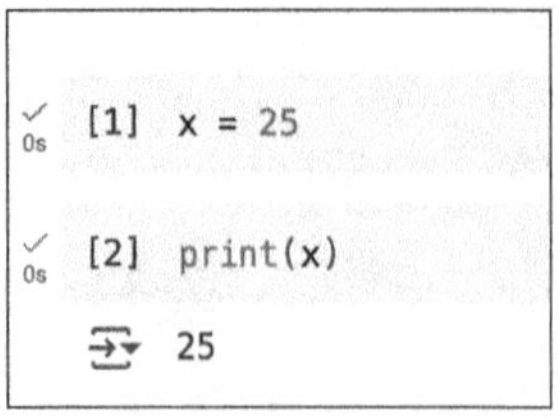

```
[1]  x = 25

[2]  print(x)
     25
```

Figure 2.19 Inserting and Executing Code in Cells

You can now take the code given in Listing 2.3 to Listing 2.9. Each listing's code can be pasted in a single cell and that cell run independently of others. By running this code, you've successfully helped the machine learn from the MNIST handwritten digits using a very powerful algorithm. In the next chapter, we'll start to peel off the layers of this black box and figure out exactly how the machine is finding the patterns in our 128 neurons.

2.6 Summary

We've taken our first steps into the fascinating world of machine learning. We've discovered how machines can learn in different ways—sometimes with clear guidance like a student with a teacher, sometimes by finding patterns on their own like a curious explorer, and sometimes through trial and error like a determined problem solver. We've met neural networks, which are remarkable systems inspired by but not bound by the human brain that have revolutionized what machines can achieve. From recognizing handwritten digits to making complex decisions, we've seen glimpses of their potential. And best of all, we've set up our own laboratory in the cloud with Google Colab, ready to start experimenting with these ideas ourselves. What we've covered so far provides a foundation for understanding how these systems work.

In the next chapter, we'll look more closely at the mechanics of learning, exploring the mathematics and techniques that make all of this possible. There's much more to discover as we continue our journey into machine learning.

Chapter 3
Fundamentals of Gradient Descent

Gradient descent is the cornerstone of modern machine learning, powering everything from simple predictions to complex AI systems. In this chapter, we'll build this fundamental algorithm from the ground up. This deep, foundational knowledge will serve as a guide to help you recognize how even the most advanced models are built on top of these guiding principles.

Think of gradient descent as the secret language of machine learning—a language that, once mastered, opens up a world of possibilities. Amazingly, this seemingly simple mathematical concept powers everything from image recognition to language translation, yet it's often treated as just another black box tool that "somehow makes things work." Understanding gradient descent at a deep, intuitive level transforms how you approach machine learning problems and provides insights that help you design, troubleshoot, and optimize models more effectively. There's more to this than just gaining the knowledge of gradient descent—it's also about feeling it in your bones, understanding its rhythm, and flowing with its nuances. From my personal time in the field, I can tell you the moment everything clicked—when gradient descent stopped being a mysterious algorithm and became a natural extension of my problem-solving toolkit—was when modern machine learning truly opened up to me.

Gradient descent is more than just a mathematical procedure; it's the heartbeat of modern deep learning itself. When you truly grasp it, something profound happens: the latest advances in deep learning start to feel less like impenetrable academic papers and more like logical extensions of principles you already understand. Complex frameworks such as Keras become transparent, and their design choices make perfect sense because you understand the fundamental harmony they have with the underlying core concepts. This deep understanding transforms how you interact with machine learning tools. You'll find yourself not just using them but truly wielding them with precision. When new tools emerge (and they will), you won't be intimidated because you'll see them for what they are: different interfaces to the same underlying principles you've mastered.

The best part is that you'll develop an almost supernatural ability to "hear" what the math and code are telling you. Debugging won't be about blindly trying solutions but about listening to what the gradients are saying, understanding where the learning process is struggling, and knowing intuitively how to guide it back on track.

But here's the thing—this level of understanding doesn't come from rushing through the concepts or memorizing formulas. It comes from building up your intuition step-by-step, layer by layer, like a master craftsperson honing their skills. That's why this chapter might be the most crucial one in this entire book. Each concept we'll cover, no matter how elementary it might seem at first glance, is a crucial piece of the puzzle that will eventually form your complete understanding of machine learning.

So, let's slow down, take a deep breath, and prepare to wade into the waters of gradient descent. We'll start with a simple example that might seem basic at first but—trust me—this gentle introduction is deliberately designed to build the rock-solid foundation you'll need for everything that follows.

Let's begin our journey into the heart of machine learning, one careful step at a time. For that, we're going to start with a basic definition of machine learning.

3.1 Understanding Gradient Descent

What exactly do we mean when we say *machine learning*? One of the most enduring and precise definitions comes from Tom Mitchell in his 1997 book *Machine Learning*:

> *A computer program is said to learn from experience E with respect to some class of tasks T and performance measure P, if its performance at tasks in T, as measured by P, improves with experience E.*

Let's unpack this definition using everyday language. Consider teaching a child to identify dogs. The task (T) is recognizing dogs, the experience (E) consists of showing the child many pictures of both dogs and other animals, and the performance measure (P) is how accurately they can identify dogs in new pictures they haven't seen before. Machine learning works similarly—we provide the computer with examples (experience), a clear task, and a way to measure how well it's doing. Let's begin explaining this process from scratch. We'll start with a simple example and gradually build on the premise by adding nuance to the discussion. Eventually, we'll get to the state-of-the-art model that almost all modern deep learning models are built on.

3.1.1 The Basic Setup

To make this even more concrete, let's consider a practical example: predicting house prices. Imagine you're a real estate agent who has been estimating house prices for years. You've developed an intuition about how a house's features relate to its market value. Machine learning can systematize this process, learning patterns from historical housing data to make accurate price predictions.

In our example, we'll start simple. Let's say we have data about houses (shown in Table 3.1) in a particular neighborhood, where each house is described by just one feature: its area in square meters. Our dataset might contain hundreds of pairs of information: the area of each house and its corresponding sale price. This becomes our computer program's experience (E). The task (T) is clear: predict the price of a new house given its area. Our performance measure (P) will be how close our predictions are to the actual sale prices.

Let's look at a dataset that shows what data we've collected. Each row of this dataset is called a single *data point*.

Area of the House (Square Meters)	Price of the House (Millions)
145	808
202	993
305	727
365	582
450	1,612

Table 3.1 Example Dataset for Housing Prices

The first column contains our *input data*, or *features*—the area of each house in square meters. This is the information we'll always have access to, even for new houses that come on the market. When a real estate agent encounters a new house, they can easily measure its area. Think of this as our *given information* or what we use to make our prediction. Each entry in this column tells a small part of the house's story: one house might be 145 square meters, another 450 square meters, and so on.

The second column contains our *output data*—the actual selling price of each house. These are the historical prices we know from past sales, and they're incredibly valuable because they help us learn the relationship between a house's area and its price. For houses in our dataset, we might see that a 305-square-meter house sold for 727 million while a 450-square-meter house sold for 1,612 million. This is what mathematicians and computer scientists call our *ground truth*—the real answers that we'll use to train our model. The collection of all rows, including input data and output data, forms our complete dataset. Because we're making predictions for outputs that are continuous in nature, you'll recall that this is called a *regression* problem.

> **Isn't This Just a Toy Example?**
> While this scenario might seem like a toy example, rest assured that we'll build it up to get to an understanding of state-of-the-art methods and tools. We want to take one baby step at a time to form a solid foundation.

What makes this setup fascinating is that we're trying to build a system that can learn from these known pairs of areas and prices to predict prices for houses we haven't seen before. Consider what happens when a new 250-square-meter house comes on the market—we have its area (our input), but we don't yet know its fair market price (the output we want to *predict*). Our machine learning model will use the patterns it learned from our existing data to make an educated guess about this new house's price. Of course, in reality, many other factors influence a house's price—the number of bedrooms, the age of the house, the neighborhood, nearby schools, and so on. We're starting with just the area to keep things simple, but later we'll see how to incorporate these additional features to make our predictions more accurate. This is just one of the assumptions we're making at the beginning of our understanding.

This predictive challenge mirrors how human experts work. An experienced real estate agent has seen many houses and their selling prices over the years. When they encounter a new house, they draw upon this experience to estimate its value. Our machine learning model will do something similar, but it will do so by finding mathematical patterns in the data rather than relying on intuition. Let's discuss how we can achieve this using a systematic approach.

3.1.2 Formalizing the Process: Training Phase

Understanding this relationship between inputs and outputs is crucial because it forms the foundation of our *prediction task*. In machine learning terms, we're trying to discover a reliable way to convert house areas into price predictions—essentially, we want to capture the underlying pattern that connects these two pieces of information. The beauty of this approach is that once we understand this pattern, we can make educated guesses about the price of any house, as long as we know its area.

Let's walk through how we turn machine learning from an abstract concept into a systematic process. Looking at Figure 3.1, you'll see a road map that breaks down our learning methodology into clear, manageable steps. Think of it as a recipe we'll follow to teach our computer to make predictions.

First, we do something that might seem counterintuitive—we split our dataset into two parts, like dealing a deck of cards into two piles. One pile becomes our *training set*, and the other becomes our *test set*. While Table 3.1 shows just a handful of examples for clarity, modern real-world applications deal with massive amounts of data—from thousands of examples in smaller projects to billions or even trillions in large-scale applications. But don't let these numbers intimidate you—the core concept stays exactly the same whether we're working with 10 houses or 10 million.

> **[»] Details of Training and Testing Split**
> For now, consider that the split is 50-50 between training and testing. There are more details involved with how we split the complete dataset and what issues might come up. We'll get to them as we proceed.

3.1 Understanding Gradient Descent

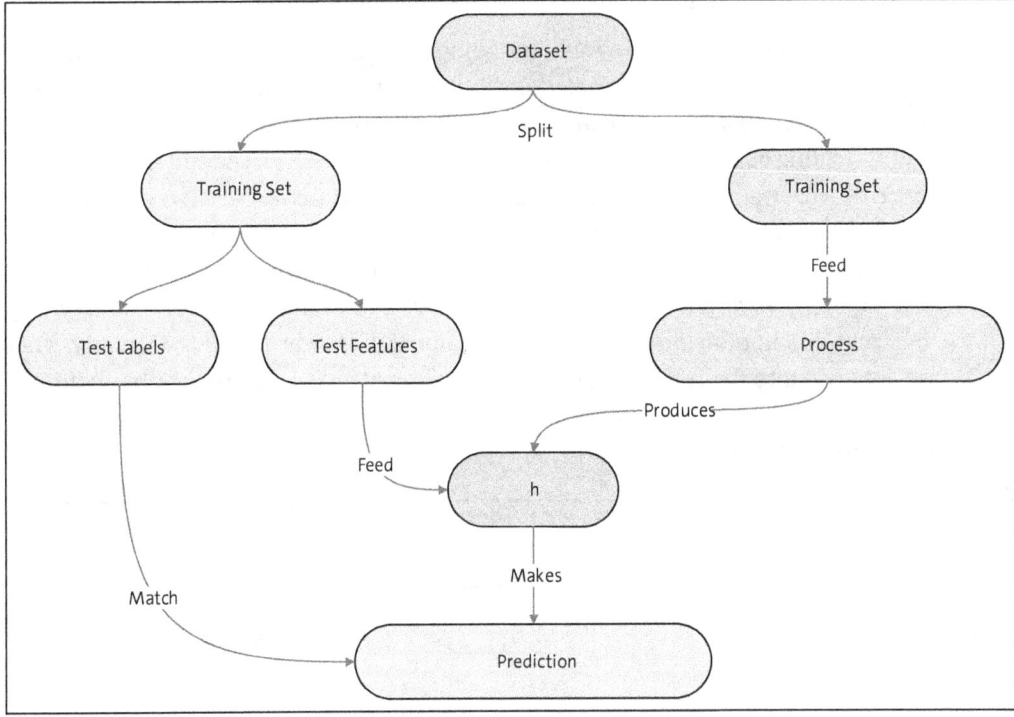

Figure 3.1 Typical Machine Learning Pipeline

Next, we focus entirely on our training set, using both the input features and output values to run them through our *process*. While we'll dive into the details of this process later, what's important to understand is that the process is like a factory that produces something special—a prediction system called the *hypothesis function* (h).

This hypothesis function h is like a well-trained expert who has learned to make educated guesses. Give it some input data about a house, and it will try to predict the house's price. It's similar to how an experienced real estate agent can estimate a house's value just by knowing its features.

Once h is ready, something interesting happens—we can actually set aside both the training data and the process that created h. It's like how once you've learned to ride a bicycle, you don't need the training wheels anymore.

Now comes the moment of truth. We take our *test set*—remember that pile of data we set aside earlier? We only show h the input features from this set, keeping the actual prices hidden. It's like giving our trained expert a pop quiz, showing house areas but not revealing the real prices. The function h makes its predictions for each house in the test set, and we compare these predictions with the actual prices we kept hidden. If h gets most of its predictions right, we can say it's doing a "good enough" job. Of course, what counts as most and what qualifies as good enough are subjective terms that we'll explore in detail later.

3.1.3 From Training to Deployment: Making It Work in the Real World

Once we're satisfied with how well our hypothesis function *h* performs—that is, once it's learned to make reliable predictions during our training phase—we can do something quite remarkable. We can discard the entire learning machinery, like taking down the scaffolding once a building is complete. The training set, test set, and all the complex processes that went into creating *h*? We don't need them anymore. All we keep is *h* itself, our perfectly tuned prediction machine. Think back to learning to ride a bicycle. During the learning phase, you need training wheels, a patient teacher holding the back of the bike, and perhaps an empty parking lot to practice in. But once you've mastered cycling, you can ride smoothly on any road without any of those learning aids. Similarly, when we deploy *h* into the real world, it stands alone, ready to make predictions about new houses that come on the market. To emphasize how alone *h* is in the wild, we show what's happening in Figure 3.2.

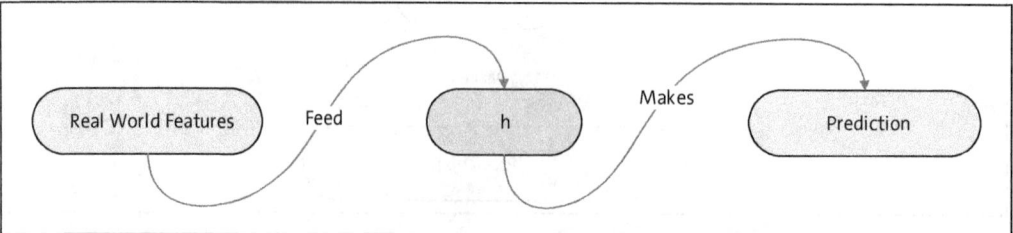

Figure 3.2 Making Predictions in the Wild

The contrast between the training and deployment phases is striking—it's like comparing the extensive preparation for a space mission to the elegant simplicity of the actual flight. When you hear about machine learning requiring massive computational resources, they're typically referring to the training phase, where all the heavy lifting happens. The deployment phase, by comparison, is computationally lightweight—often requiring thousands or even millions of times less computing power.

> **Lightweight Hypothesis Functions**
>
> There's a lot of work being done to make hypothesis functions as lightweight as possible. This is why we're now able to have very powerful models on small-factor devices, even as small as smartwatches!

But there's still a crucial piece of our puzzle missing. We haven't yet explored the mysterious "process" that produces *h*. This process is where the real artistry of machine learning lies. It's like knowing the beginning and end of a story, but now we need to discover the adventure in between. We'll do that next.

3.1.4 The Process of Learning

Let's begin by understanding the process from a human perspective. While computers don't actually "think" the way we do, understanding the human logic behind these steps will make the machine's approach much clearer.

Think of what happens when you try to solve this problem without a computer. You have a table of data (Table 3.1) showing house areas and their corresponding prices. The most natural way to visualize this relationship is to plot these points on a graph, with house areas on the x-axis and prices on the y-axis. Each house in our dataset becomes a point on this plot, creating a *scatterplot* that you can see in Figure 3.3.

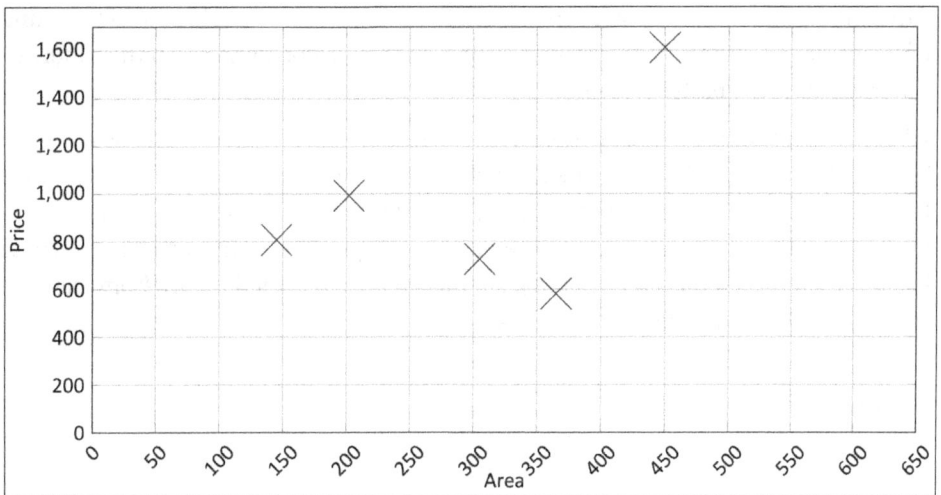

Figure 3.3 Scatterplot of the Dataset

Now comes the interesting part. We need to find a way to predict house prices based on their areas. The simplest approach is to assume that bigger houses cost more in a straightforward, linear way—like each additional square meter adds a fixed amount to the price. Mathematically, we're saying that the relationship follows a *straight-line equation*: $y = ax + b$, where a tells us how much each square meter adds to the price, and b represents a base price for having a house at all—called the *bias* term. As before, this is a simplifying assumption that we'll lift later.

To make our notation more consistent with what we'll see in machine learning literature, we'll make a small change. Instead of a and b, we'll use θ_1 and θ_0, respectively. In addition, because our predictions won't be perfect, we'll call our predicted price $\hat{y}$ to distinguish it from the actual price y. Our function thus becomes

$$\hat{y} = \theta_0 + \theta_1 x$$

This is our *model*. Think of it as a recipe with two special ingredients, θ_1 and θ_0, which are our *model parameters*. These parameters are like the knobs we can turn to adjust how our prediction line fits the data.

3 Fundamentals of Gradient Descent

> **[+] Learn How to Pronounce Latin and Greek Letters**
>
> It's very important to learn to pronounce Greek and Latin letters used in mathematics. When you can naturally say terms such as α (alpha), β (beta), or γ (gamma) instead of pausing to decode them, you can focus better on grasping the underlying mathematical concepts. This helps you parse the latest research much faster.

Here's where things might seem a bit counterintuitive—we're going to start by just making up some random values for these parameters. Yes, you read that right! We'll pick random values for the parameters, say $\theta_1 = 1.75$ and $\theta_0 = 25$. It's like throwing darts blindfolded, hoping to get somewhere close to the target. While this might sound unscientific, it's actually how many machine learning algorithms begin their journey toward finding the perfect values.

When we plot this line on our scatterplot, something interesting happens, as shown in Figure 3.4. Our prediction line cuts through our data points, but it's not a perfect fit—far from it, actually. You can see this clearly in the visualization, where we've drawn vertical bars connecting each actual house price to our predicted price on the line. These bars are like error meters, showing us just how far off our predictions are from reality—the longer the bar, the bigger our mistake.

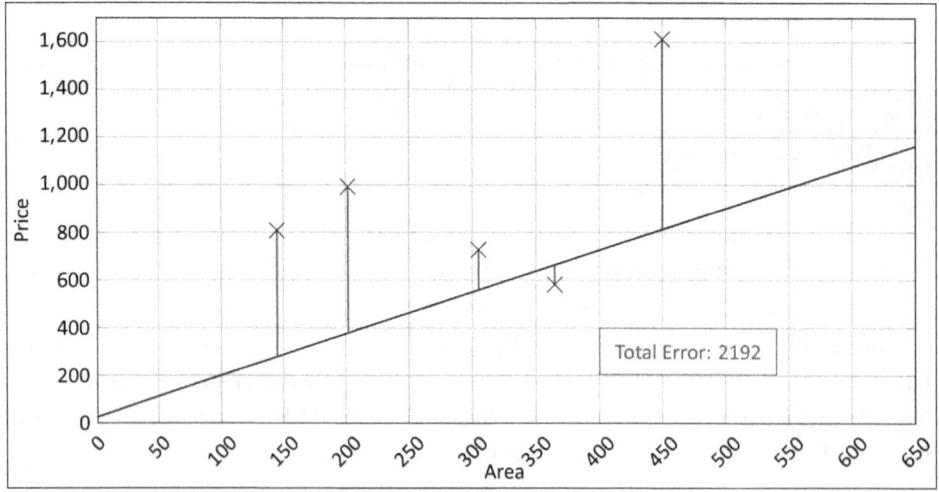

Figure 3.4 Error Bars on the Random Parameter Model

These vertical bars are probably the most important aspect of this figure. Some of our predictions are too high (when the line is above the actual price), and others are too low (when the line is below). It's like trying to guess someone's age—sometimes we overshoot, sometimes we undershoot, and occasionally we get lucky and guess just right.

So, the difference between our predictions for each data point and the ground truth is given by

$Error = \hat{y} - y$

Let's see how well (or poorly) our prediction line is doing by measuring the size of those error bars we saw earlier. If we add up all of these errors, we get a total of 2,192. But we need to be careful about negative errors. If a house price prediction is 50,000 too high or 50,000 too low, that's equally wrong. That's why we consider the absolute size of each error, regardless of whether we predicted too high or too low. Consider the case in which half our predictions were 50,000 too high and half were 50,000 too low. Adding these up would suggest we're doing perfectly, when we're actually off by quite a bit!

We could solve this by taking the absolute value of each error, but there's a more elegant solution that mathematicians and machine learning practitioners love: squaring the differences. When we square a negative number, it becomes positive, solving our sign problem beautifully. There are other benefits of using the square function that we'll revisit in the next section. So, for all data points (let's assume they are m in number) the error now becomes

$$Error = \sum_{i=1}^{m} (\hat{y} - y)^2$$

There's one more refinement we need. Assume we're comparing two models: one tested on 100 houses and another on 1,000 houses. Even if the second model makes the same average-sized mistakes, its total error would be much larger simply because it's dealing with more houses. The solution is to divide by the number of houses to get the average error per house and give this error function a name—*J*:

$$J(\theta) = \frac{1}{m} \sum_{i=1}^{m} (\hat{y} - y)^2$$

This elegant little formula goes by many names in the machine learning world—*loss* function, *error* function, *energy* function, *cost* function—but they all mean the same thing: it's our mathematical measuring stick for how badly our model is misbehaving. The lower this number, the better our predictions. There are other loss functions as well, but this is by far the most common one for problems like these. It's aptly called the *mean squared error (MSE)* function *parameterized* by the θ values.

Let's take a moment to clarify the difference between these two crucial functions in our machine learning system: the *hypothesis function* and the *loss function*. The hypothesis function *h* is like a price-predicting crystal ball. You feed it information about a house—in our case, its area—and it gazes into its mathematical depths to predict a price. The loss function *J*, on the other hand, is more like a report card or a quality control system.

It looks at all of our predictions together and grades how well we're doing. While the hypothesis function works with house features to make predictions, the loss function works with our model parameters (those θ values) to tell us how good our predictions are. Making predictions based on model parameters is called the *forward pass*.

Think of it this way: When we use the loss function, the actual house data (both features and prices) stays fixed—like answers in an answer key. What we change are the θ values, which are those mysterious numbers that control our prediction line. Each time we try different θ values, it's like submitting a different version of our answers, and the loss function tells us how many we got right. The goal is to find the parameter values that give us the best possible grade, that is, the smallest value for the loss function.

Let's look at what happens when we experiment with our model parameters. When we change θ_1 from 1.75 to 2.0, something interesting happens—our sum of absolute errors drops from 2,192 to 1,729, as shown in Figure 3.5. It's like adjusting the dials on a radio and getting a clearer signal.

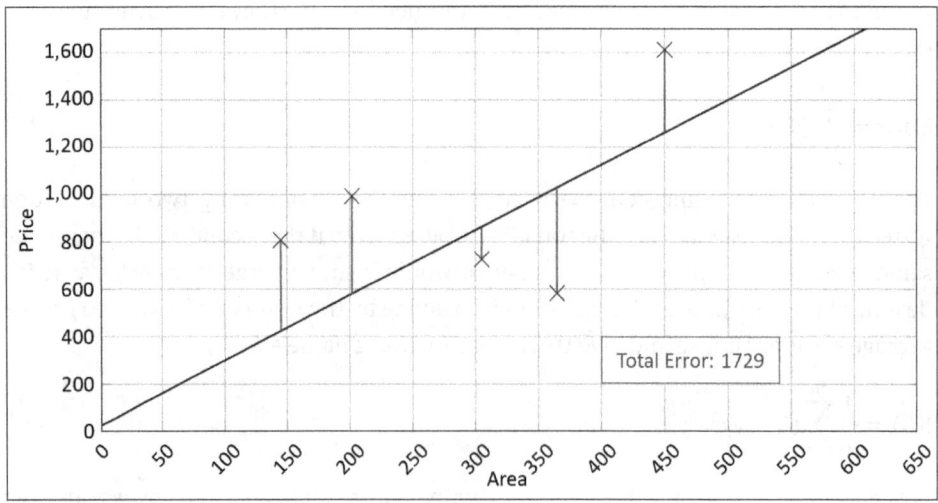

Figure 3.5 Model with Better Parameter Values

This observation gives us an important insight. Pause for a moment and consider what it means: If changing one parameter can reduce our error this much, there must be a "sweet spot"—some combination of parameter values—that gives us the smallest possible error.

> **[+] Interactive Tool for Line Fitting**
>
> We've made a tool for you to practice this line fitting and get the errors interactively. Try the demo on the book resources page at *https://recluze.net/keras-book*. Look under **Chapter 3** for the tool **03-03-mse-fit-demo**. As before, all the resources are available on the book's official web page at *http://rheinwerk-computing.com/6142* as well.

Now, we need a navigation system for our *parameter space*—a systematic way to navigate from our random starting point to these ideal values. Instead of blindly trying different numbers and hoping for the best, we need a clever strategy that can guide us toward better and better parameters.

In our next section, we'll uncover exactly how we can turn this insight into a practical algorithm.

3.1.5 Finding the Best Parameters: Minimizing the Loss

Before we dive deeper, let's recall that we're on a journey of understanding—we're thinking through this process the way a human might, not the way a computer actually does it. Keep this distinction in mind as we explore further, and we'll get to what the machine does at the end of this section.

Let's simplify our exploration by focusing on just one parameter for now: θ_1 (we'll bring θ_0 back into the picture later). Consider conducting a series of experiments. Each experiment involves picking a different value for θ_1, and then measuring how well (or poorly) our model performs with that value.

If we were to plot these experiments on a graph, with θ_1 values on the horizontal axis and the corresponding *loss* values on the vertical axis, we'd start to see a pattern emerge. For instance, when we picked the parameter value of 2, we got the error value of 10; when we picked the value of 9, we got the error value of 7. We keep doing this and getting some loss values, as shown in Figure 3.6. After a few experiments, we happened across the value of 6, which gave us the *minimum* value of the loss, that is, 1.75. This was a great result, but we arrived at it by chance. Instead, we need a systematic approach to find this point.

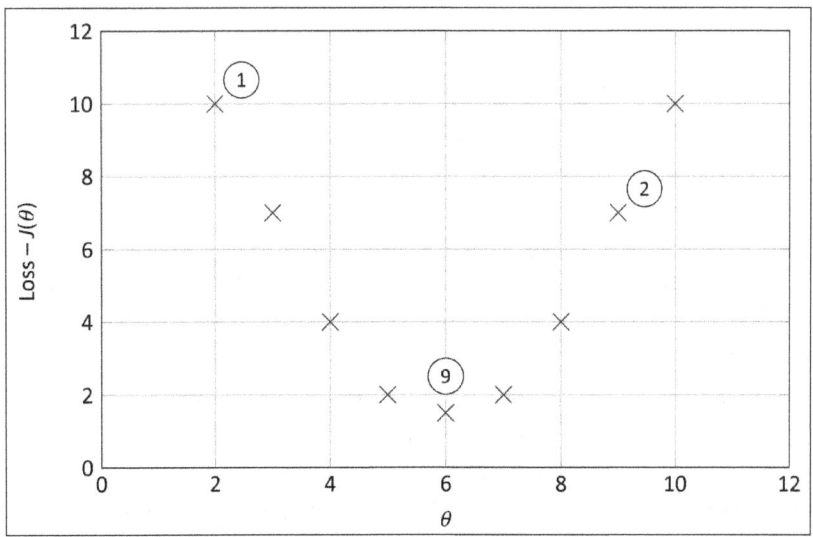

Figure 3.6 Landscape of the Error Function Based on Model Parameters

3 Fundamentals of Gradient Descent

This visualization reveals something fascinating about the relationship between our parameter choice and the resulting error. It's a map that shows us the terrain of possible solutions, complete with *hills* and *valleys*. The valleys (lower points) represent parameter values that give us lower errors, while the hills (higher points) show us parameter values we want to avoid.

When we look at this plot, we're seeing the landscape of possibilities for our model. Each point tells us a story about how a particular parameter value performs, and together, they paint a picture that will help guide us toward finding the best possible values for our model.

Examining our plot of parameter values versus loss, we discover something crucial about our loss function's shape—it forms what mathematicians call a *convex function*, appearing like a bowl that opens upward. This shape, as shown in Figure 3.7, isn't a coincidence. It's precisely why we chose to use squared errors rather than absolute errors earlier. Square functions always create this distinctive bowl-shaped curve.

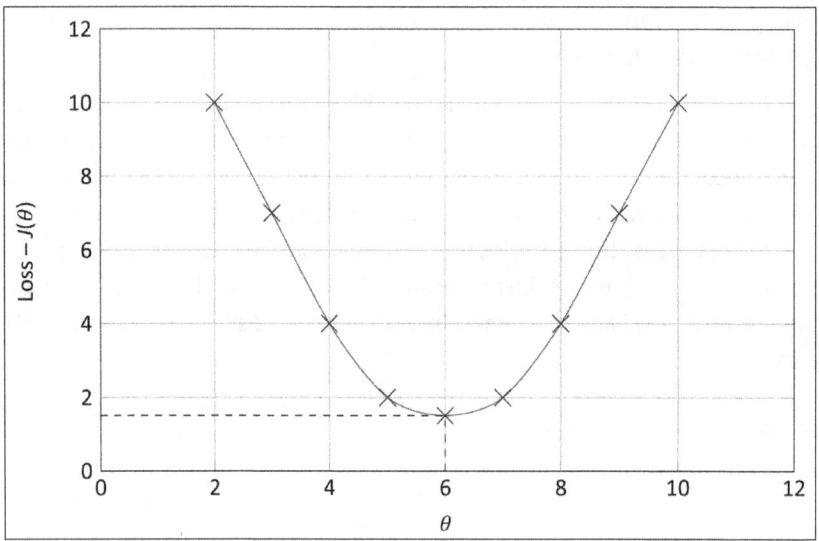

Figure 3.7 Underlying Shape of the Loss Plot

The key property of a convex function is that it has exactly one minimum point, and on both sides of this minimum, the function's values consistently increase. This shape exists independently of any points we plot—it's the underlying mathematical structure of our loss function $J(\theta)$.

Because we're working with a mathematical function, we can use calculus to understand how it behaves. Specifically, we can look at its *derivative*. Even if your calculus is a bit rusty, we only need to understand one key concept: The derivative of a function at any point tells us how the function's value will change when we adjust its input. In our context, the derivative of $J(\theta)$. reveals how our loss value will change if we increase θ by a small amount. Let's look at what happens when θ equals 2. At this point, shown in Figure 3.8, the derivative is negative.

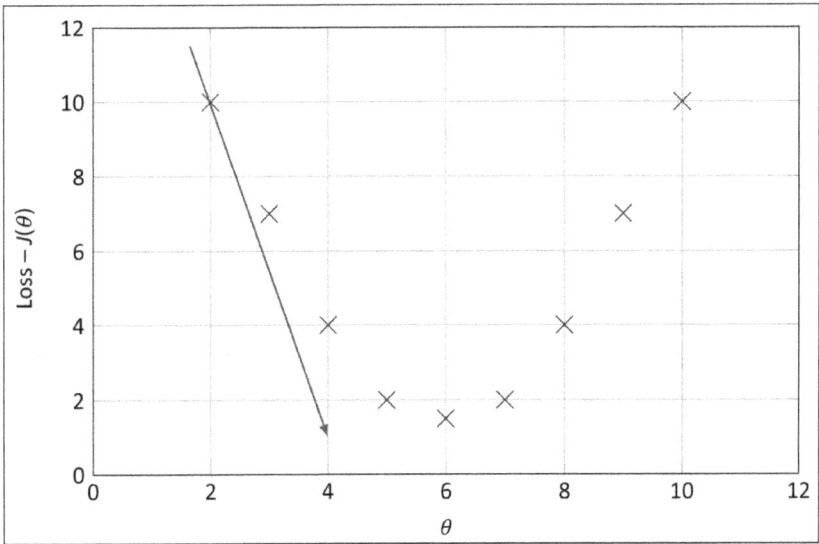

Figure 3.8 Derivative at Model Parameter Value at 2

This tells us something important: Increasing θ slightly will decrease our loss value. Therefore, if we randomly selected θ = 2 as our starting point, we can improve our model by increasing θ. We can formalize this using the following equation:

$\theta_{new} = \theta_{old} - \text{derivative}$

The derivative itself if written formally as

$$\frac{\partial}{\partial \theta_1} J(\theta_1)$$

Let's not worry about how we're going to calculate the derivative. It's not a big issue, and we usually get it for free. (We'll see how this automation is made possible in Chapter 5, Section 5.3.2.) The only important point to note is that it will be a scalar value, which will be negative if we are to the left of the target minimum. Let's continue with our discussion by specifying the update rule precisely:

$$\theta_{new} = \theta_{old} - \frac{\partial}{\partial \theta} J(\theta)$$

This observation gives us a systematic way to find better parameter values. When we calculate the derivative and find it's negative, we know we should increase θ to reduce our loss. The derivative acts as a signpost, pointing us toward parameter values that will improve our model's performance. That is why we have a negative sign in front of the derivative. Subtracting a negative value will effectively give us a new θ value that is going to be larger than the old one.

3 Fundamentals of Gradient Descent

This was just one case though. It's also possible that we initially picked the random value of θ = 10 (see Figure 3.9). If we calculate the derivative of the function Jθ at θ = 10, we get a positive value. Think about what this means for a moment. When we're standing at θ = 10 on our error landscape, we're like a hiker who has wandered too far up the right side of our bowl-shaped valley. The positive derivative is like a compass telling us we've gone too far in the positive direction.

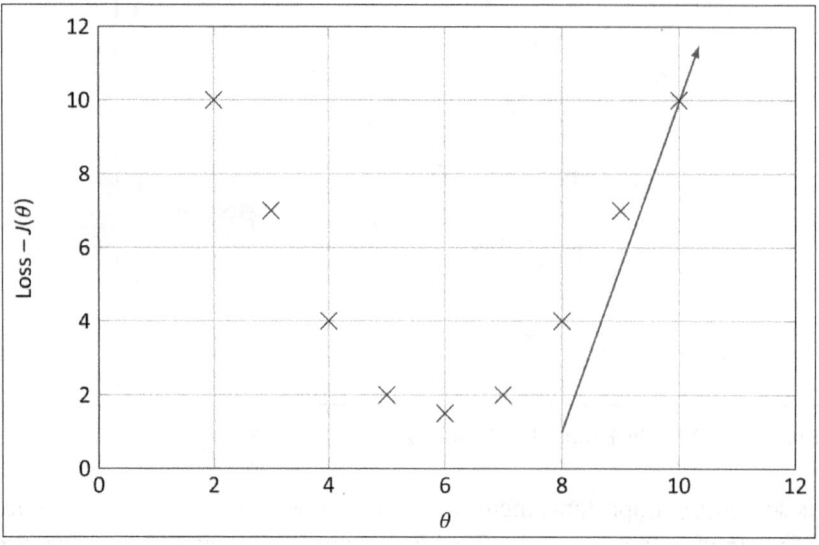

Figure 3.9 Derivative at Model Parameter Value 9

A negative derivative told us to increase θ, and a positive derivative works in the opposite way. It says that if we want to reduce the value of Jθ, we should reduce the value of θ. This means we should subtract this positive value of the derivative from the old value of θ to get the new value:

$$\theta_{new} = \theta_{old} - \frac{\partial}{\partial \theta} J(\theta)$$

The minus sign in front of the derivative becomes crucial. We're subtracting a positive number, which means our new θ value will be smaller than 10. Do you notice that this is exactly the same equation as we had for the first case? The beauty of this equation is that it works irrespective of what our initial random estimate was. If we follow the gradient to improve the value of θ, we're sure to reduce our loss—hence the name *gradient descent*. The process of calculating the derivative values—propagating the loss back to the parameter that caused it—is called the *backwards pass*.

So, this simple equation naturally guides us in the right direction regardless of whether we start too high or too low. If we're too high (e.g., at θ = 10), the algorithm tells us to decrease θ. If we're too low (e.g., at θ = 2), it tells us to increase θ.

This is why our choice to use squared errors was so clever—it gives us this bowl-shaped error *landscape* where we can always trust the derivative to point us in the right direction, like a reliable compass that never fails. Whether we're standing on the left slope or the right slope of the bowl, the derivative will always guide us toward that sweet spot at the bottom where our error is minimized.

To round off the discussion about our movement through the error landscape, let's take a look at one final scenario that's particularly special—what happens when the derivative becomes zero (or something very close to it). This is precisely the point in our space we've been searching for all along!

Think back to our hiking analogy. When you're walking in steep terrain, the steepness (our derivative) tells you which way to go. But what happens when you reach perfectly flat ground? That's exactly what zero derivative means—we've reached the valley floor of our bowl-shaped error landscape, and this is the value of θ that gives us the minimum amount of loss, even if the loss itself might not be zero. And here's the great part: In a convex function (the friendly bowl shape), this flat point is unique. It's like having a ball in a smooth bowl—no matter where you release it, it will always roll to the same spot at the bottom. When we find this point where the derivative is zero, we've discovered the perfect set of parameters that minimizes our prediction errors.

Now, you might wonder if need to find the exact point where the loss is *precisely zero*. In the real world, with all of its messy data and inherent noise, that's simply not possible. Instead, we're happy when our derivative gets very close to zero—it's like being "close enough" to the bottom of the valley that any further movement wouldn't meaningfully improve our predictions.

With just one equation and a random starting point, we've developed a reliable way to find the optimal parameters for our model. The elegance of this approach—that it converges to the best possible solution regardless of where we begin—is what makes gradient descent such a powerful tool in the machine learning toolkit. Gradient descent is a very powerful algorithm that comprises the core machinery behind all of modern machine learning. It's what guides all the latest and state-of-the-art models. If you fully understand gradient descent to its core, the other algorithms just make sense. At its core, the algorithm is simply this: start with some random value, and move around in the loss landscape guided by the derivative, so that you're always reducing the loss until you eventually get to the minimum. At this point, you have the model parameters that are best to make predictions about your outputs based on your inputs.

While doing all of this, we made a few assumptions. We've listed them all here so we can try to lift as many as we can:

- There is a single input feature.
- The relationship between the inputs and outputs is linear.
- The loss is a convex function. (We didn't really make this assumption, but we did come up with a loss function that was convex. We'll see later that we can have loss functions that aren't convex (i.e., bowl shaped), but the same algorithm still works out well.)

3 Fundamentals of Gradient Descent

Let's go ahead and lift the first assumption to see how it changes our model and algorithm.

3.1.6 Lifting the Assumptions of a Single Input Feature

Let's see what happens when our model grows beyond a single feature. Remember how we started with just the house area *(x)* in our equation? Our simple model looked like this:

$$\hat{y} = \theta_0 + \theta_1 x$$

But real-world data is rarely so simple. Think about buying a house—you don't just look at the size because you likely also care about the number of rooms, location of the house, its age, and so on. Let's add just the number of rooms to our dataset for now. We'll see that going from one feature to two will give us the tools to add as many features as we want. When we add this new feature to our dataset, we get something like Table 3.2.

Area of the House (Square Meters) – x_1	Number of Rooms (x_2)	Price of the House (Millions) – y
145	2	808
202	3	993
305	2	727
365	4	582
450	5	1,612

Table 3.2 Adding Features to the Dataset

Notice that features are now called x_1 and x_2 instead of just *x*. This will allow us to easily use this notation. In addition, our prediction equation—the model—naturally expands to accommodate it:

$$\hat{y} = \theta_0 + \theta_1 x_1 + \theta_2 x_2$$

Each θ value still plays its own special role: θ_1 tells us how much each square meter contributes to the price, while θ_2 reveals how much each additional room affects the value. Meanwhile, θ_0 continues to represent our base price.

With this expansion in our model to handle multiple features, we need to rethink how we visualize our *loss landscape*. Remember our earlier plot (see Figure 3.6), where we could track how the error changed as we adjusted a single θ value? Well, things are about to change a little bit but not too much.

With two parameters, θ_1 and θ_2, we can no longer capture our loss landscape in a simple *2D plot*. We need to step into the third dimension. Picture it this way: We're no longer walking along a curved line in search of the lowest point—we're now exploring a *surface*, like a landscape with hills and valleys.

This new visualization takes the form of what mathematicians call a *mesh plot*. Think of it as a topographical map, but one where we can see the entire surface at once. The x-axis represents possible values for θ_1, the y-axis shows values for θ_2, and the height (z-axis) tells us the loss value for each combination of these parameters. The z-axis is special as it predicts the loss that we want to minimize. Keep this in mind for later when we return to a discussion of higher dimensions in Section 3.1.7. Figure 3.10 shows what the mesh plot for a typical *square function* looks like.

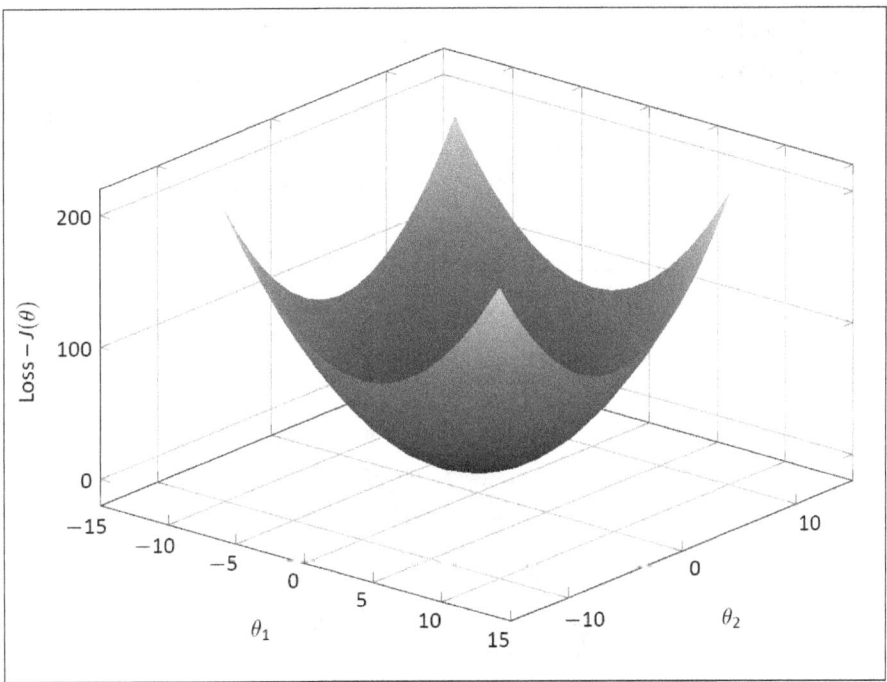

Figure 3.10 Mesh Plot with Two Model Parameters

For every point on this surface—that is, for every possible combination of θ_1 and θ_2—there's a specific loss value. There's a unique elevation for every location on a terrain map. Some combinations of parameters will lead to high loss values (peaks in our landscape), while others will give us lower loss values (valleys). Somewhere on this surface lies our *optimal* solution—the lowest point in the terrain where our model makes its best predictions.

We've reached a critical point in understanding the loss landscape. When we had just one parameter, the process was straightforward—we could follow a single derivative to find the minimum. But now, with multiple parameters shaping our terrain, we need a more sophisticated navigation tool.

Enter *partial derivatives*—a powerful concept that lets us tackle this complexity without completely reinventing our approach. A partial derivative answers a very specific question: If we change just one parameter while keeping all others fixed, how does our loss value respond?

Consider our house price prediction: We want to understand how changing just the area of the house affects the price, without touching any other knobs. That's exactly what a partial derivative tells us—how our loss changes when we adjust one parameter (e.g., θ_2) while leaving the others (e.g., θ_1) unchanged. In mathematical terms, when we calculate a partial derivative, we're taking a slice through our 3D landscape. Imagine standing at any point on our surface and looking straight along the θ_1 direction, and you'd see a plane that shows how the loss changes as you adjust θ_2, assuming θ_1 stays constant.

At the intersection of this plane and the loss landscape is a *curve*. This curve is like our old familiar 2D plot, just viewed from a particular angle in our new 3D world. Watch the little white line at the intersection of the plane and the mesh in Figure 3.11. Compare this white line with the curve in Figure 3.7. These look exactly the same. This gives us a good starting point.

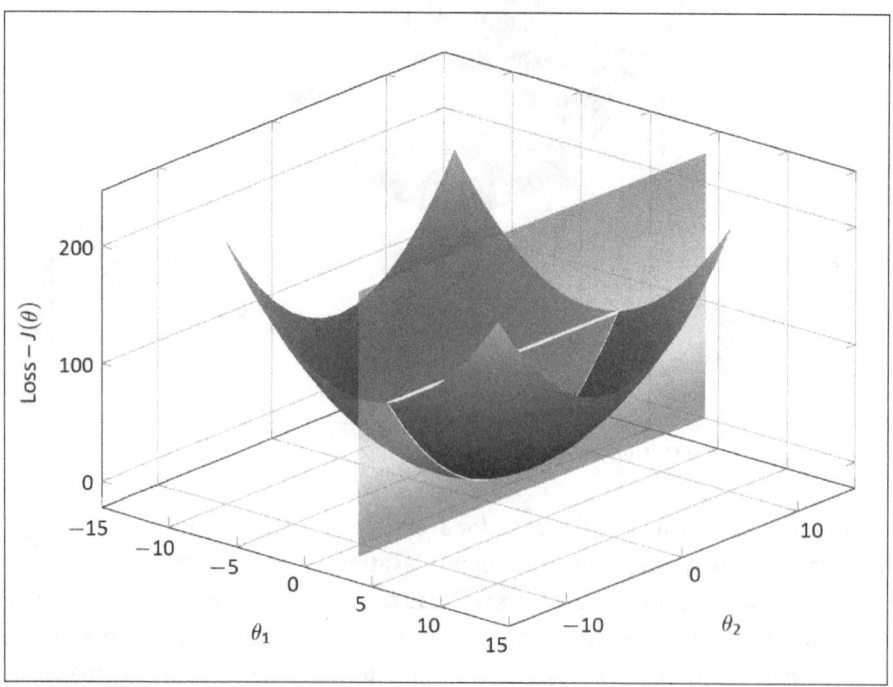

Figure 3.11 A Slice Through the 3D Landscape Demonstrating a Partial Derivative

Let's see how we can use these partial derivatives to navigate our more complex loss landscape. When we fix θ_1 and only adjust θ_2, we're essentially following a specific path down our surface—like tracking the slope of a single dimension while standing still in the other. The partial derivative with respect to θ_2 tells us exactly which way and how steeply this path descends.

This insight gives us a powerful strategy for updating both parameters simultaneously. For θ_2, we calculate the partial derivative $\partial/\partial\theta_2$, which measures how the loss changes with small adjustments to θ_2. We can then update θ_2 using our familiar update rule:

$$\theta_{2_{new}} = \theta_{2_{old}} - \frac{\partial}{\partial\theta_2}J(\theta)$$

Similarly, for θ_1, we calculate its partial derivative $\partial/\partial\theta_1$ and update it:

$$\theta_{1_{new}} = \theta_{1_{old}} - \frac{\partial}{\partial\theta_1}J(\theta)$$

These update equations work in perfect harmony. Each partial derivative guides its respective parameter in the direction that reduces the loss, and we can apply both updates in a single step. This means our overall update becomes a coordinated movement across our loss surface, with each parameter adjusting in its optimal direction.

Here's what makes this approach particularly elegant: When we make these independent adjustments to θ_1 and θ_2 based on their respective partial derivatives, something remarkable happens. If each change individually reduces the loss, the combined effect is guaranteed to reduce our overall loss as well. While the mathematical proof of this property is beyond our current scope, its implications are profound—we can confidently update multiple parameters *simultaneously*, knowing we're moving toward better predictions.

It's perfectly fine if the math seems too complicated. All we need to internalize is that each model parameter is updated independently of others. So, even if there are a thousand model parameters, there's nothing else to know or understand. The same formula that we used for a single model parameter works for any number of parameters! Let's try to write the formula in full now.

To do this, we're going to need the derivative of the loss function. This is something that might have been worrying you: the actual calculation of these derivatives. Here's the encouraging part—you don't need to become a calculus expert to use gradient descent effectively. The derivatives we need are readily available and well-understood. In fact, for the most common loss functions used in machine learning, mathematicians have already worked out the derivatives for us. You can check their work but it's really not needed. Modern machine learning code, tools, and frameworks handle these calculations automatically, letting us focus on the bigger picture of how to use them effectively.

Take our MSE loss function, for instance. While its name might sound intimidating, its partial derivatives have a surprisingly elegant form. For θ_1, the partial derivative is

$$\frac{\partial}{\partial\theta_1}J(\theta) = \frac{2}{m}\sum_{i=1}^{m}(h_\theta(x) - y)x_1$$

In this equation, the $h_\theta(x)$ is our model. This is the same as $\hat{y}$. We can further expand this to get rid of this h, but it's sufficient to keep in mind that it abstracts away the multiplication of model parameters with features. Now, the overall update rule becomes

$$\theta_{1_{new}} = \theta_{1_{old}} - \frac{2}{m}\sum_{i=1}^{m}(h_\theta(x) - y)x_1$$

This is called the *gradient descent step*, which is essentially updating model parameters based on the sum of differences between our predictions and the ground truth weighted by the corresponding input feature. You might notice that we're missing a term in this formula. Don't worry. We'll return to a discussion of this in Section 3.3. This formula encapsulates a powerful idea: It measures how much our predictions deviate from reality, weighted by the relevant feature. The beauty of this formula lies in its direct connection to our goal—minimizing prediction errors.

There's one crucial detail we need to address about updating our parameters: Timing matters. When we adjust multiple parameters, we need to update them all simultaneously, not one after another. This means keeping track of all of our old parameter values until we've calculated all the new ones.

The process works like this:

1. Store all current parameter values.
2. Calculate all new values using these stored old values.
3. Update all parameters at once with their new values.

This synchronization is very important. Assume we're updating θ_3, but we've already changed θ_2 to its new value. We'd be using the wrong version of θ_2 in our calculations, meaning the guarantees provided by the partial derivative rules no longer hold. We might be going in a direction that doesn't essentially reduce the loss—something we don't want. So, just make sure that each parameter update should be based on the same snapshot of all other parameters from the previous iteration. Let's include a little snippet of code to drive the point home.

In Listing 3.1, you can see how we would store temporary copies of all parameters. Only after we've calculated all the new values do we update the actual parameters. This careful orchestration ensures our gradient descent proceeds correctly toward the minimum of our loss function. At the end of this chapter, we'll use this concept along with all the missing parts to completely code gradient descent from scratch. For now, just make sure you understand this point and avoid the most common mistake made by people just getting started with gradient descent.

```
def gradient_descent_step(X, y):
    # Store current parameters
    old_parameters = self.parameters.copy()
```

```
# Compute all gradients using old parameters
gradients = self.compute_gradients(X, y)

# Update all parameters simultaneously
self.parameters = old_parameters - gradients
```
Listing 3.1 Simultaneously Updating All Model Parameters

There might be one thing in the back of your mind though: How do we take this from two dimensions (features) to even higher dimensions? Let's look at that briefly.

3.1.7 Higher Dimensions in Gradient Descent

The remarkable answer to the question, "how do we work with a larger number of dimensions", is that moving to higher dimensions requires almost no changes to our approach! Just as we moved from one parameter to two by adding another term to our equation and another partial derivative to our calculations, we can keep adding terms for each new feature. Our prediction equation simply becomes the following for n features:

$$\hat{y} = \theta_0 + \theta_1 x_1 + \theta_2 x_2 + \cdots + \theta_n x_n$$

We can elegantly simplify our equations by introducing a clever convention: setting $x_0 = 1$ for all data points. This allows us to express our entire prediction equation as a simple vector-matrix multiplication. Let's collect all of our input features in a matrix X. Each column is the collection of features of a single data point, and the number of columns in X is equal to the number of data points we have:

$$X = \begin{bmatrix} 1 & 1 & \cdots \\ x_{11} & x_{21} & \cdots \\ x_{12} & x_{22} & \cdots \\ x_{1n} & x_{2n} & x_{mn} \end{bmatrix}$$

So, in this case, we have n features and m total data points. All of our θ values can be combined in a single *row vector* Θ as well:

$$\Theta = [\theta_0, \theta_1, \ldots, \theta_n]$$

Notice the capital Θ to denote our vector. Our original long equation simply becomes

$$Y = \Theta X$$

Sometimes, the Θ is written as a *column vector* instead of a row vector as well. In that case, we simply transpose the vector, and the math works out fine.

This *vectorized* formulation offers two significant advantages. First, it provides mathematical elegance and power. By expressing our computations in terms of matrices and vectors, we can leverage the rich theoretical framework of linear algebra. This becomes particularly valuable as we progress to more sophisticated models where clean mathematical notation helps us understand and manipulate complex relationships more easily.

The second advantage is purely computational. Modern computing hardware, from CPUs to GPUs, is specifically designed to perform matrix operations with remarkable efficiency. When we express our computations in vectorized form, we can take advantage of highly optimized linear algebra libraries that tap directly into this hardware capability. Instead of writing loops to process each data point individually, vectorized operations can handle thousands or millions of computations simultaneously. This parallelization can speed up our computations by several orders of magnitude—what might take hours with manual loops can be completed in seconds with vectorized operations.

A challenging issue arises with the whole visualization that we did in, for example, Figure 3.10, if we try to do the same with higher dimensions: How do we visualize our loss landscape when we have more than two dimensions? Modern machine learning models often have millions or billions of parameters—far beyond what we can draw on paper or render on a screen.

Geoffrey Hinton, one of the pioneers of machine learning, offered a delightfully practical solution to this visualization problem. He suggested looking at a 3D loss landscape, for example, one that looks like Figure 3.12, and simply saying "14" very loudly. Suddenly, you're visualizing a 14-dimensional space!

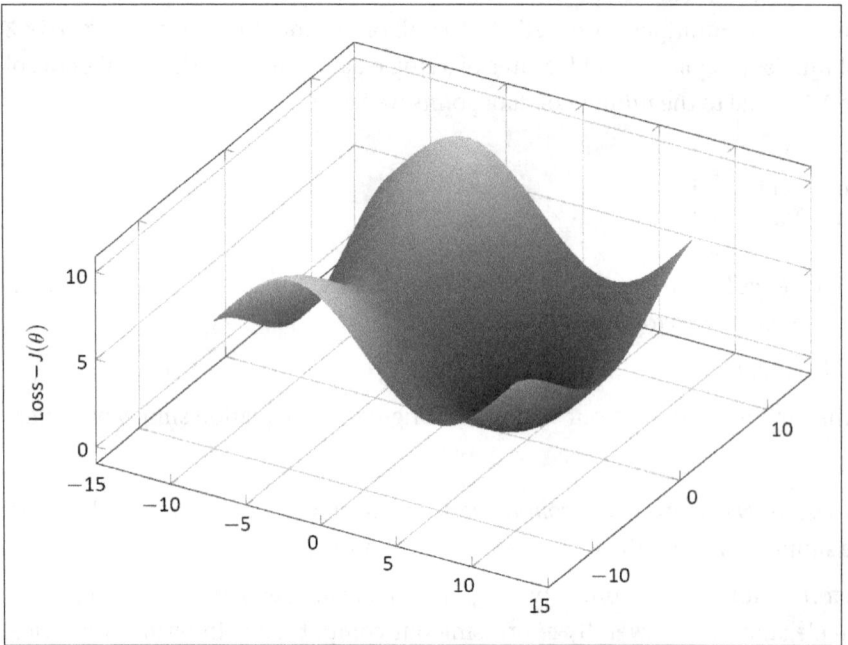

Figure 3.12 Visualizing Higher Dimensions

While humorous, this insight carries a deep truth: The exact dimensionality of the parameter space isn't what matters most. What's truly important is understanding that regardless of how many parameters we have—whether 14 or 14 million—we're still working with a single scalar value of interest: the loss. Think of it this way: the "ground" of our landscape might span an enormous number of dimensions, but we're only concerned with the "height"—our loss value—for our visualization. This height tells us how well our model is performing, and the gradient still points us in the direction of steepest descent.

The mathematics we've developed doesn't care about the number of dimensions. Our approach of calculating partial derivatives and performing simultaneous updates works exactly the same way whether we have two parameters or two billion. The only difference is in the size of our matrices, not in the fundamental principles of how we optimize them.

Coming back to the vectorization of code though, this point also ties up directly with our next section that deals with optimizing the whole process to take advantage of modern computational devices.

3.2 Types of Gradient Descent: Batch, Stochastic, Mini-Batch

We actually have a major issue in the formula for update that we came up with for gradient descent:

$$\theta_{1_{new}} = \theta_{1_{old}} - \frac{2}{m} \sum_{i=1}^{m} (h_\theta(x) - y) x_1$$

Recall that the summation is over all data points m. When we try to actually carry out this computation, we encounter a significant practical challenge with this gradient descent step formula. Let's examine the computational cost to understand why this is a serious problem.

Our update rule requires us to sum over all data points m in our dataset. In modern machine learning applications, m can easily reach a billion data points. For each of these billion points, we need to do the following:

1. Compute the prediction (involving all features).
2. Multiply this error by each feature value.

With a million features (actually below average in modern machine learning), each prediction step involves a million multiplications. Multiply this by our billion data points, and we're looking at 10^{15} operations—just for a single update step. Because we're carrying out this step operation using the whole batch of data, we'll call this *batch gradient descent*. Let's explain this in a little more detail, and in the process, also introduce an important mathematical/visualization tool—the contour plot.

3.2.1 Contour Plots for Visualization of Gradient Descent

While our 3D mesh plot in Figure 3.10 effectively showed how loss varies with different parameter combinations, it has several practical drawbacks. The surface itself becomes an obstacle to visualization—peaks can hide valleys behind them, making it impossible to see certain combinations of parameter values. If we want to track a specific point's trajectory during gradient descent, it might disappear behind a hill or beneath a valley. Even worse, when we want to compare multiple parameter combinations, some points might be completely hidden from view regardless of how we rotate the plot. This problem becomes particularly acute in regions of high curvature, precisely where we most need to understand the landscape's behavior. We can solve this visualization challenge by transforming our 3D plot into a more readable 2D representation.

The key insight is that we can flatten our 3D landscape while preserving all of its critical information. Instead of showing height directly, we can project our surface onto the ground plane and use *contour lines* to represent points of equal loss value. This is called a *contour plot*—an example of which can be seen in Figure 3.13. Each contour line connects all points that share the same loss value similar to elevation lines on a topographical map. (Ignore the lines with arrows for now. We'll come back to these in a minute.)

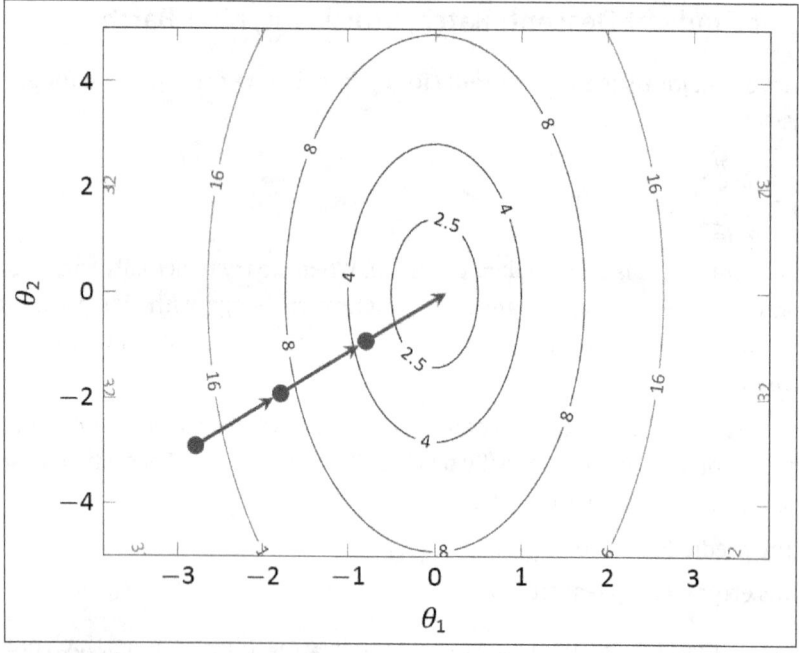

Figure 3.13 Contour Plot Mapping the 3D Landscape to the 2D Plane

This flattened representation offers several advantages:

1. We can clearly see all parameter combinations without any points being hidden behind the surface.

2. The spacing between contour lines tells us about the steepness of our loss landscape.
3. We can easily trace paths between different parameter combinations and understand how the loss changes along these paths.

Each contour line effectively marks a level set—all points where our loss function equals a specific value. By labeling these lines with their corresponding loss values, we maintain all the essential information from our 3D plot while making it much easier to analyze specific regions of our parameter space.

> **Working with Contour Plots**
>
> It's a great idea to find some contour plots and get used to what they look like. In addition, try creating them in the tool of your choice. Matplotlib has great examples of contour plots in its documentation. You can see some of them here: *http://s-prs.co/v614201*.

The contour plot shows us the steps of a batch gradient descent process in action (the lines with arrows). Starting from randomly initialized values of $\theta_1 = -3$ and $\theta_2 = -3.5$, we can trace how the algorithm navigates toward the minimum. At each step, the partial derivatives calculated from our formula guide us in the optimal direction. Over three iterations, we see the algorithm progressing directly toward the minimum point of the loss landscape.

This visualization illustrates both the strength and limitation of batch gradient descent. Its strength lies in taking the optimal path toward the minimum, as it uses information from all data points to determine each step. However, this thoroughness comes at a computational cost—each step requires processing the entire dataset, making the algorithm very slow when dealing with many features or data points.

To put batch gradient descent in perspective: even if we had a supercomputer capable of performing a trillion operations per second, each gradient descent step would still take more than 15 minutes. And remember, we typically need thousands or even millions of steps to reach convergence.

This computational burden makes our current approach impractical for real-world applications, even with the most powerful computing resources available. We need a more efficient strategy.

3.2.2 Improving the Efficiency of Batch Gradient Descent

Let's discuss a surprisingly effective solution to our computational challenge. Instead of calculating the loss over all data points at each step, we can choose a single, random data point and compute the loss for just that point. Although this approach might seem counterintuitive (after all, how can we make good decisions with such limited information?), it really well in practice.

3 Fundamentals of Gradient Descent

In this modified approach, each update step works like this:

1. Randomly select one data point from our dataset.
2. Calculate the loss using only this point.
3. Update our parameters based on this single-point calculation.
4. For the next step, pick another point (without replacement) and repeat.

This dramatically reduces our computational burden. Instead of performing billions of calculations for each step (summing over all data points), we now only need to calculate the loss for a single point. Our million multiplications no longer get repeated for a billion data points. That's a billion times speedup!

What about the issue of limited information used in each step? While each individual step might not point us in exactly the right direction (because we're only looking at one data point's gradient), we compensate for this by taking many more steps. Because we pick each data point at random and arrive at the next location based on chance, this is called *stochastic gradient descent (SGD)*. This meandering behavior can be seen on the contour plot in Figure 3.14.

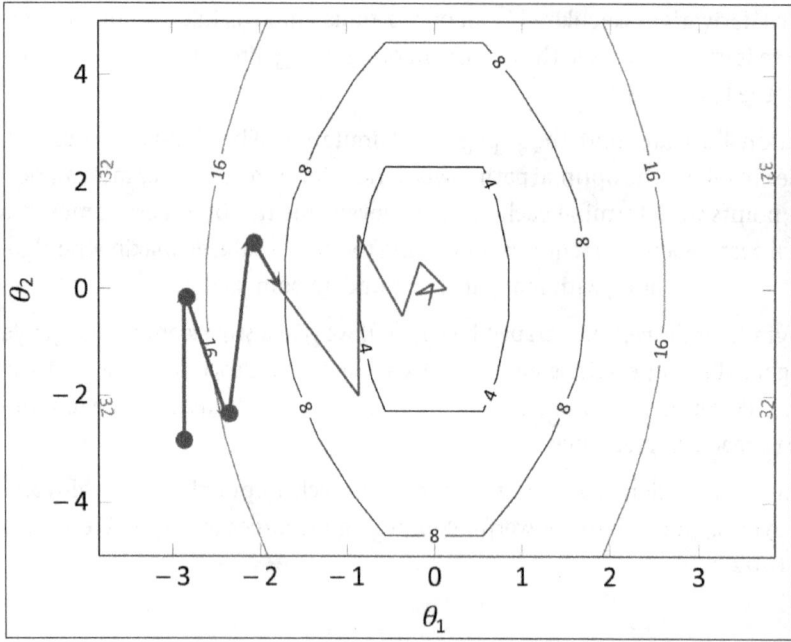

Figure 3.14 Gradient Descent Steps in Stochastic Gradient Descent

The randomness in our point selection helps ensure that we eventually consider all of our training data. This method essentially trades the precision of batch gradient descent for dramatically improved computational efficiency.

3.2.3 A Paradox in the Use of Stochastic Gradient Descent

Our discussion of SGD reveals an interesting paradox in practice. While it seems computationally efficient to process just one data point at a time, this approach actually underutilizes modern computing hardware. GPUs and specialized machine learning processors are designed for parallel computation—they excel at processing multiple data points simultaneously. When we process single data points sequentially, we're essentially leaving most of our computational power idle. Moreover, to process our entire dataset once (called an *epoch*), we need to perform m separate update steps—one for each data point. This sequential processing can end up being slower than we anticipated.

The solution lies in finding a middle ground, called *mini-batch gradient descent*. Here's how it works: We first shuffle our entire dataset randomly, and then divide it into small, equally-sized chunks called *mini-batches*. Each mini-batch typically contains data points that modern GPUs can process efficiently in parallel.

For example, if we have 10,000 training examples and choose a batch size of 100, we'll create 100 mini-batches of 100 examples each. We then process these mini-batches sequentially: compute the loss for all points in the current mini-batch, calculate the average gradient, and update our parameters. Once we've used a mini-batch, we don't reuse it until we've processed all other mini-batches to ensure that each data point contributes equally to training.

When we've processed all mini-batches once (completing the full epoch), we typically reshuffle the entire dataset and create new mini-batches for the next epoch. This reshuffling helps prevent our model from learning any accidental patterns in the order of our data.

This approach offers several advantages:

- Uses hardware parallelization capabilities
- Reduces the computational burden compared to full batch processing
- Provides more stable updates than SGD
- Allows for efficient memory usage and data throughput

The size of these mini-batches, known as the *batch size*, becomes a crucial hyperparameter in modern machine learning. A *hyperparameter* is a tunable variable that isn't a model parameter but affects the model's parameters, nonetheless. We'll see many more hyperparameters as we continue with our discussion in this book. The batch size hyperparameter helps us balance computational efficiency with the quality of our parameter updates.

When you start working with machine learning frameworks and libraries, you'll notice something potentially confusing: many tools label their gradient descent implementations as "SGD" (aka stochastic gradient descent), even though they're actually implementing mini-batch gradient descent. The key tell-tale sign is the presence of a `batch_size` parameter. This naming convention, while technically imprecise, has become standard in the field.

In any case, choosing the optimal batch size is more art than science, heavily dependent on the number of features, the number of data points, and specific hardware configuration, among many others. The decision involves balancing several factors.

If your batch size is too small (e.g., 8 or 16), you might experience the following:

- Updates becoming noisy and unstable
- Training path meandering more than necessary
- Underusing GPU parallel processing capabilities
- Significantly increased overall training time

If your batch size is too large (e.g., 1,024 or 2,048), you might experience the following:

- GPU memory becoming a bottleneck
- More time spent on memory transfers
- Significant overhead in data movement
- May slow down training despite seeming more efficient

The sweet spot depends on your specific GPU:

- Available memory
- Number of CUDA cores
- Memory bandwidth
- Cache size

This is why you'll often see recommendations to experiment with different batch sizes for your specific setup. Start with common values such as 32, 64, or 128, and then adjust based on both training performance and hardware utilization metrics.

One issue still remains though: Both mini-batch and SGD face a particular challenge near the minimum point of our loss landscape. Unlike batch gradient descent, which uses the entire dataset to compute precise updates, these methods will tend to oscillate around the minimum rather than settling precisely on it. This happens because each update is based on a subset of data, leading to slightly different gradient directions at each step.

We can address this meandering behavior by somehow slowing down as we approach the minimum. We can do this using optimization techniques that not only solve this particular issue but also lead to another critical problem we usually face when applying gradient descent. Let's discuss that issue in the following section.

3.3 Learning Rate and Optimization

Let's revisit the gradient descent step one more time. Irrespective of whether you're using batch, stochastic, or mini-batch gradient descent, we can face an issue with gradients not playing nice with our algorithm.

Consider a case where our minimum loss function can be achieved with a model parameter value of 6. Figure 3.15 shows such a loss plot. We don't know this minimum point yet, and we start with a parameter value randomly initialized at 5. When we calculate the derivative of the loss function at this point, we get a value of -2.5. The negative sign tells us we need to increase our parameter value, and the magnitude tells us how big of a step to take. Plugging this derivative into our update step, $\theta_{new} = \theta_{old} - (-2.5) = 5 + 2.5 = 7.5$, we get a new model parameter value of 7.5. That seems to be a problem. The initial random guess was only 1 unit away from the minimum, but this new model parameter value is now 1.5 units away. Instead of getting closer to our target, we've actually moved further away.

The situation becomes even more problematic on our next step. Because we're further away from our minimum (at 7.5 instead of 5), the loss function's convex nature dictates that we'll get an even higher derivative value here—albeit with the opposite sign. When we calculate the derivative at this point, we get a value of 3.5. The positive sign now tells us we need to decrease our parameter value.

Plugging this new derivative into the update formula, $\theta_{new} = \theta_{old} - 3.5 = 7.5 - 3.5 = 4$, we get a new model parameter of 4. This result reveals the full extent of our problem: We're now even further away from the minimum than what we started with in the first place. Our parameter value has oscillated from 5 to 7.5 to 4, with each step taking us further from our target of 6. Figure 3.15 shows what happens if we continue performing the update steps—we're led into a *divergence* away from the minimum.

This oscillation is more than a minor inconvenience; it's a fundamental challenge in gradient descent optimization. With each step, our parameter updates are getting larger instead of smaller, and we're moving further from our goal instead of closer to it. The very mechanism that should be guiding us toward the minimum is actually pushing us away from it.

What makes this particularly challenging is that the problem compounds itself. The further we move from the minimum, the larger our gradients become, leading to even larger steps in the wrong direction. Without some way to control these step sizes, our algorithm might never converge to the optimal solution.

Let's try to fix this problem by adding a small multiplier to the derivative term in our formula. Let's say we set the multiplier to *0.1*. This way, our initial step now becomes $\theta_{new} = \theta_{old} - (0.1 \times (-2.5)) = 5 + 0.25 = 5.25$, which is actually closer to our target parameter value, unlike our previous attempt that overshot to 7.5. That seems to be working out well. Let's update our step formula to accommodate this change:

$\theta_{new} = \theta_{old} - (0.1 \times derivative)$

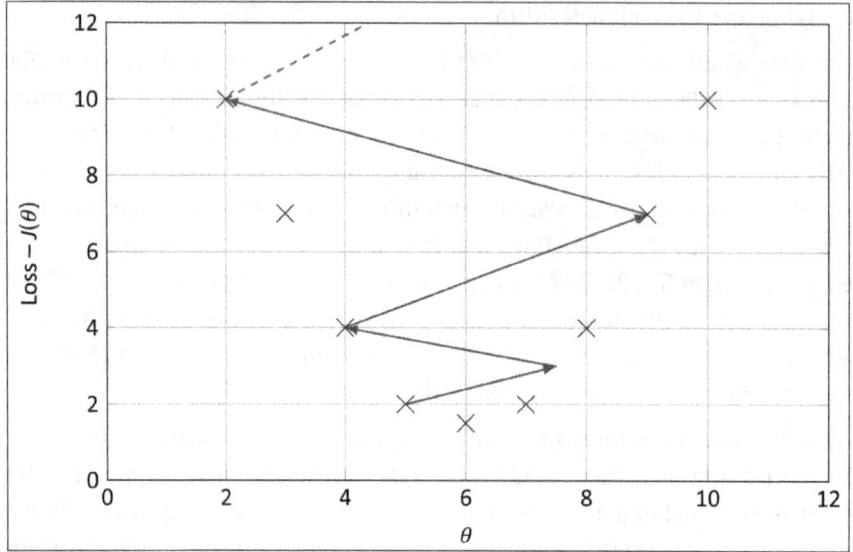

Figure 3.15 Divergence of Model Parameter Values in Gradient Descent

Before we leave this off and call it done, let's examine another scenario with a different dataset. We'll start with the same initial conditions—our parameter is randomly initialized to 3, and our target value (though unknown to us) is still 6. However, this time, our derivatives behave quite differently.

When we calculate our first derivative, we get a much smaller value: -0.2. This small derivative isn't due to any problems with our algorithm—it's simply a result of having a different dataset with different input and output values that produce naturally smaller gradients. Using our update formula of

$\theta_{new} = \theta_{old} - (0.1 \times (-0.2)) = 3 + 0.02 = 3.02$

we get a new value of 3.02. While this is a move in the right direction toward our target of 6, we've only taken a tiny step. Because we've barely moved from our starting position, our next derivative won't change much either. We're stuck in a situation where each step makes only minor incremental progress, as shown in Figure 3.16.

This slow progress becomes particularly problematic when we consider the computational cost. Remember that each step requires millions of multiplications, consuming both computational resources and time. When our algorithm needs hundreds or thousands of these small steps to reach the minimum, the cost adds up significantly. We can fix this problem by getting rid of the multiplier and getting the new value as 3.2, which is much better.

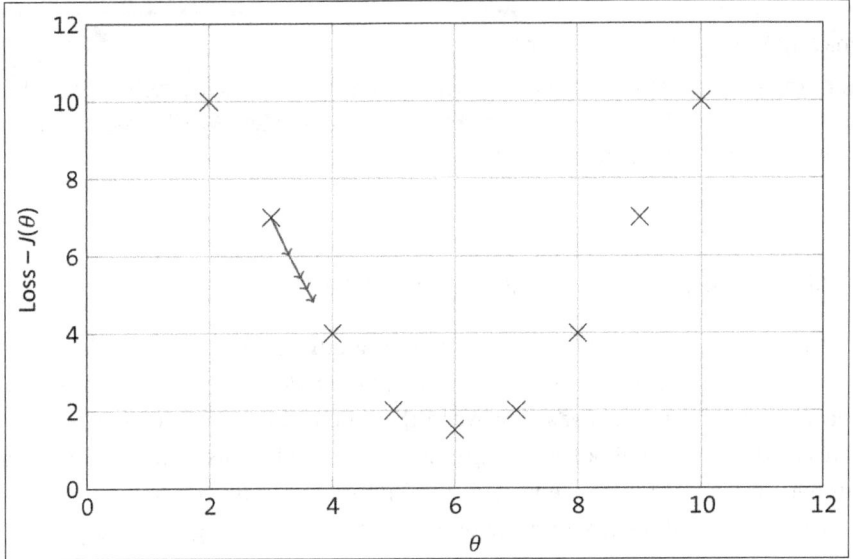

Figure 3.16 Slow Convergence During Gradient Descent Steps

We've discovered ourselves in a challenging situation: a fixed multiplier simply can't handle the diversity of real-world data. When the multiplier is present, it helps with some datasets but hinders others. Remove it, and we face the same problem in reverse. We need a more flexible solution.

This is where we must introduce another hyperparameter in the mix. Instead of fixing our multiplier or removing it entirely, we make it adjustable—something we can tune from outside our training process to accommodate different datasets. So, our final formula for gradient descent update step becomes

$$\theta_{i_{new}} = \theta_{i_{old}} - \frac{\partial}{\partial \theta_1} \alpha J(\theta)$$

This crucial hyperparameter is called the *learning rate*, denoted by α (alpha), and is arguably the most important hyperparameter in all of machine learning. Tuning the learning rate becomes a critical skill. When you start training and notice your model's parameter values or losses are diverging, it's a clear signal that your learning rate is too high—you'll need to dial it down. Conversely, if you observe the loss decreasing at a painfully slow rate while your derivatives remain far from zero, your learning rate is likely too low and needs a boost.

The art of selecting the right learning rate isn't something that can be fully automated—it remains a skill that machine learning experts develop through experience and practice. It requires careful observation of your model's behavior, and an understanding of how different datasets might respond to different learning rates. In the next section, we'll go ahead and put all this together to see how we can perform a simple machine learning process using Python code.

3 Fundamentals of Gradient Descent

> **Automating Hyperparameter Selection**
> There are tools available to help optimize hyperparameters. One such method is grid search, which helps you find optimum values of hyperparameters. We'll come back to these when we have more complex models.

3.4 Implementing Gradient Descent in Code

In this section, we'll write code to demonstrate how gradient descent works. This code will cover all the concepts we've discussed in this chapter along with some implementation-specific tips and tricks. Let's first write the whole code from scratch without using Keras so that you can link the underlying math to code. This is imperative for a deeper understanding of the concepts. Once we're through that, we'll solve the same problem using Keras and see what improvements are made by using the library.

3.4.1 Gradient Descent from Scratch

Let's begin by setting up our environment for implementing gradient descent. Fire up a Google Colaboratory notebook (refer to Chapter 2, Section 2.5, for how to do that). We'll break the code down so we can point out all the important aspects. We'll be using the numpy library to write vectorized code.

Notice the second line in Listing 3.2. You'll often see this in machine learning codes. This line sets the *random seed* that is used to generate all random numbers in the following code. If any piece of code starts with the same seed, the "random" numbers generated throughout the code are still going to be random but the setting of the seed ensures that every time you run the code, you'll get the same random numbers. This ensures the reproducibility of your results.

```
import numpy as np
np.random.seed(1337)    # Generate dummy data
```

Listing 3.2 Import and Initialize numpy

When we write np.random.seed(1337), we're not generating random numbers—we're creating a predetermined sequence that appears random but follows a specific pattern. This sequence influences everything from how we initialize our model weights to how we shuffle our training data, making it possible to recreate exact experimental conditions whenever needed. The number itself (whether it's 1337, 42, or any other integer) isn't special, but using the same seed guarantees that our random number generator will produce identical sequences every time, which is crucial for debugging, testing, and ensuring our results are reliable and reproducible.

3.4   Implementing Gradient Descent in Code

> **Get the Complete Code Listing**
> Access the complete code listing for this section on the book's web page at *http://rheinwerk-computing.com/6142* and on the resources page at *https://recluze.net/keras-book* in the **03-01-SGD-scratch** notebook.

The way we use random seeds changes depending on our goals: during development and research, we typically want consistent results to compare different approaches and debug issues, so we set specific seeds to create reproducible environments. This way, we can share our code with others who can run it and make sure they get the same results as we did. However, in production systems where we want our models to be robust and adaptable, we often skip setting seeds to embrace true randomness. This duality reflects a broader principle in machine learning: We need controlled environments to develop and validate our models, but we also need to ensure that they can handle the genuine randomness of real-world data. The power of random seeds lies in this flexibility—they give us the ability to switch between perfectly *reproducible* experiments and truly random operations.

With that out of the way, let's proceed to generate some dummy data using numpy's built-in functions. The first line in Listing 3.3 creates 100 random samples, and the second line defines the ground truth for the model parameters θ_0 and θ_1. We then go ahead and generate the output values y based on the inputs and also add a sprinkle of random noise to make our data more realistic. After all, in the real world, patterns are rarely perfect. The last line shows the shape of the inputs and outputs. This is one of the most important aspects of real-world engineering of machine learning models. At the moment, the shape of X and y is both (100, 1). This means 100 rows with 1 column each. It's a good idea to get in the habit of looking at shapes of different matrices in your code. As we venture into more sophisticated models with multiple features and complex architectures, keeping track of these matrix dimensions becomes increasingly important. Matrix *shape mismatches* are often the hidden culprits behind many debugging sessions, so developing an intuition for them early on is invaluable.

```
X = np.random.randn(100, 1)   # 100 samples, 1 feature
true_params = np.array([2.5, 3.7])   # True param values
y = true_params[0] + true_params[1] * X
        + np.random.normal(0, 0.1, (100, 1))
print("Shape of X:", X.shape, " Shape of y:", y.shape)
```

Listing 3.3 Generating Data and Considering Shapes

We then move on to complete our setup before we start with the actual training process. The first line in Listing 3.4 adds a column to our input matrix, converting it to the shape (100, 2). The first column is populated with all 1s using the np.ones function, and the np.hstack function does the heavy lifting of producing the final matrix in the correct

111

form. We then go ahead and initialize our model parameters with some random values. We use randn(2, 1) because we have two model parameters and need just one combination. Finally, we set the hyperparameters for learning_rate and batch_size along with the definition of the number of iterations we need for the gradient descent step. Now, we're ready to go ahead and perform the actual gradient descent steps and improve our initial random guess.

```
X_b = np.hstack((np.ones((X.shape[0], 1)), X))   # set x0=1
print("Shape of X_b:", X_b.shape)      # look at the shape again

# Initialize parameters
theta = np.random.randn(2, 1)
learning_rate = 0.01
n_iterations = 1000
batch_size = 32   # Mini-batch size
```

Listing 3.4 Adding Bias Term and Setting Up the Environment

We'll first *shuffle* the data as shown in the first three lines of Listing 3.5. This helps us do the randomization once and then sample batches in order instead of randomly sampling them during each step. Then, we start the loop that performs the gradient descent step iteratively. In the step, we calculate the start and end index of the batch so that the number of data points in our batch is equal to the batch_size hyperparameter. The line start_idx = (i * batch_size) % len(X) calculates where each batch should begin in our dataset through a clever use of the modulo operator. As we iterate through our data, we multiply the current iteration by our batch size to determine where we theoretically should start, but because we might exceed the length of our dataset, we use the modulo operator to wrap back around to the beginning when needed. For example, if we have 100 samples and a batch size of 32, our first batch would start at index 0, the second at 32, the third at 64, and the fourth at 96. When we try to calculate the start of the fifth batch, we'd get 128, but because that's larger than our dataset size of 100, the modulo operator gives us the remainder after division (28), effectively wrapping us back to continue processing from the 28th sample. This creates a continuous cycle through our data, allowing us to process multiple epochs while ensuring we use every sample in our dataset.

Once we have the batch ready in X_batch and y_batch, we simply perform what is called the *forward pass*, that is, carrying our matrix-vector multiplication of the features in X_batch with the model parameters theta. Numpy gives us a built-in syntax for performing this very common operation—the @ symbol. Applying this operation gives us the predicted values for features. We then calculate the loss using the derivative formula we discussed earlier in Section 3.1.6. It's highly recommended that you refer to the mathematical formula and the code side by side and convince yourself that these are the same operations—the code just used vectorized versions to ensure that we can calculate all gradients in a vector instead of calculating them separately. Print out the outputs of different parts of this equation to understand their values and shapes, as well as gain a deeper understanding of the math and code.

After computing the gradients, we can simply apply the gradient descent step. As you can see, the vector theta is being updated based on the learning rate and the gradients. An important point to note here is that numpy automatically takes care of simultaneous updates. Using vectorized code gives us this additional benefit along with the speed up.

Finally, after the specified iterations have been completed, we output the computed values as well as ground truth model parameters. Run the code and play around with different values so that you understand it well. Afterwards, we can go ahead and see how using Keras would make our life easy when solving the same problem using this extremely well-written library.

```
indices = np.random.permutation(len(X)) # Shuffle the data
X_b = X_b[indices]
y = y[indices]

for i in range(n_iterations):
    # Get current batch indices
    start_idx = (i * batch_size) % len(X)
    end_idx = min(start_idx + batch_size, len(X))

    # Get batch data
    X_batch = X_b[start_idx:end_idx]
    y_batch = y[start_idx:end_idx]

    y_pred = X_batch @ theta # Forward pass on batch

    # Compute gradients using batch
    gradients = (2/len(X_batch)) *
X_batch.T @ (y_pred - y_batch)

    theta -= learning_rate * gradients # actual step

print("Final parameters:", theta.flatten())
print("True parameters:", true_params)
```

Listing 3.5 Gradient Descent Step and Learning Parameter Values

3.4.2 Gradient Descent Using Keras

Once you've fully understood the from-scratch code, we can go ahead and convert the code to use Keras instead. We'll begin by importing the required packages and then

3 Fundamentals of Gradient Descent

setting up our data. (As before, you can access the complete code listing for this section on the book's web page at *http://rheinwerk-computing.com/6142* and in the resources page at *https://recluze.net/keras-book* in the **03-02-SGD-keras** notebook.)

Listing 3.6 is very similar to what we had before, except we have the Keras imports that we didn't have earlier. In addition, compare the data setup we did in Listing 3.4 with what we have here. When using Keras, we don't have to set up the bias term because Keras will take care of it for us automatically. Next, we'll set up the Keras "model."

```
import numpy as np
from tensorflow import keras
from tensorflow.keras import layers

# Generate dummy data
np.random.seed(42)
X = np.random.randn(100, 1)    # 100 samples, 1 feature
true_params = np.array([2.5, 3.7])    # True parameters [θ₀, θ₁]
y = true_params[0] + true_params[1] * X + np.random.normal(0, 0.1, (100, 1))

# Shuffle the data
indices = np.random.permutation(len(X))
X = X[indices]
y = y[indices]
```

Listing 3.6 Preamble of Gradient Descent Code with Keras

The specification of the Keras model that will carry out the learning for us is given in Listing 3.7. We'll get to Keras models in detail in a later chapter, but for now, a brief description of what a model is should suffice. A *Keras model* is a structure that takes numerical inputs and produces numerical outputs through a series of mathematical operations. At its most basic level, a model consists of *layers*, where each layer performs specific mathematical *transformations* on the data that passes through it. These transformations involve matrix operations such as multiplication and addition, along with optional nonlinear mathematical functions. This is exactly what we need for our gradient descent. The input layer is Shape(1,). That means it will take input where each data point will have just one feature. The trailing comma isn't a typing mistake—it turns the number into a tuple as required by Keras (we'll cover this further later). The Dense(1, use_bias=True) line means that we'll output a single value from this model (i.e., our prediction) and that Keras should automatically take care of the bias term for us. Again, we'll be discussing this in much more detail when we get to the use of Keras for much larger models. For now, this model is specified using these two lines and then compiled.

The model.compile() function sets up two essential components for training: the *optimizer* (controls how the model updates its internal parameters) and the *loss function* (measures how wrong the model's predictions are). It's like configuring the training rules before the actual training begins.

```python
# Create and compile the model
model = keras.Sequential([
    layers.Input(shape=(1,)),     # Input layer for 1 feature
    layers.Dense(1, use_bias=True) # Linear layer with bias
])

# Configure the training
model.compile(
    optimizer=keras.optimizers.SGD(learning_rate=0.01),
    loss='mse'  # Mean squared error
)
```

Listing 3.7 Setting Up and Compiling the Model

> **Finding More Keras Optimizers**
>
> There are many more optimizers available in Keras. All of them have gradient descent at their core but each one adds different optimizations. It's good to experiment with them and see what difference they make. You can find more about the available optimizers here: *https://keras.io/api/optimizers/*.

Finally, we can start the actual training process, as shown in Listing 3.8. This final bit goes ahead and actually performs the training based on the model hyperparameters specified earlier. You'll notice that we don't actually have a loop in here. The reason is that Keras will automatically take care of *parallelization* of our code and make any *optimizations* that are possible given our underlying environment. Once the whole training is done, which might take a bit of time, Keras will output the weights using the last two lines. The exact semantics of the shape of the weight variables will be discussed once we've looked at more complicated models. For now, just know that you'll get the θ_1 value out first and then the θ_0 value instead of the other way around.

With that, you're now equipped with a thorough understanding of one of the most important pillars of modern machine learning—gradient descent. I highly recommended that you run the code, experiment with it, give it a couple of days, and then re-read this chapter. It will help solidify the concepts and deepen your understanding of this very important concept.

```python
# Train the model
history = model.fit(
    X, y,
    batch_size=32,
```

```
    epochs=1000,
)

# Get final parameters
weights = model.get_weights()
print("Final parameters:", [weights[0][0][0], weights[1][0]])
print("True parameters:", true_params)
```

Listing 3.8 Train the Model and Output Weights

3.5 Summary

This chapter explored gradient descent, which is the foundational optimization algorithm at the heart of modern machine learning. We began with a simple house price prediction problem to understand how gradient descent iteratively adjusts model parameters to minimize prediction errors. We examined the computational challenges of processing large datasets and discovered how mini-batch gradient descent offers a practical balance between computational efficiency and convergence stability. The chapter also discussed critical implementation challenges, from oscillating parameter updates to the problem of selecting appropriate step sizes, helping us understand the crucial role of the learning rate hyperparameter. Through both a mathematical framework and practical code examples, we've seen how gradient descent, despite its apparent simplicity, requires careful tuning to effectively navigate the loss landscape and find optimal model parameters.

In the next chapter, we'll see how this technique we've developed for a very specific use case can also be used to solve problems of a completely different nature.

Chapter 4
Classification Through Gradient Descent

In this chapter, we'll build increasingly sophisticated models on top of our current understanding of gradient descent. We'll discover how to make yes/no decisions using logistic regression and then expand our capabilities to handle multiple categories simultaneously. Along the way, we'll encounter the essential concepts of softmax probabilities and cross-entropy loss. Through this journey, we'll see how the humble task of drawing lines through data points evolves into systems capable of making nuanced decisions across multiple categories, setting the stage for more advanced applications in neural networks.

In the previous chapter, we explored the foundations of gradient descent and watched our regression models evolve from randomly initialized parameters to finely tuned predictors. We examined how these models learned through iterative adjustment, minimizing loss functions to improve their predictions of continuous values such as house prices. Now, we're ready to apply this same powerful optimization strategy to a different and equally important class of machine learning problems: classification. Classification represents a distinct category of machine learning tasks focused on discrete outcomes rather than continuous predictions. Instead of answering quantitative questions that produce numeric values along a spectrum, classification models determine categorical assignments—which group does this input belong to? This fundamental shift in objective requires us to reconsider how we structure our models, define our outputs, and measure performance. Classification systems power critical applications across virtually every industry: email filtering, disease diagnosis, fraud detection, sentiment analysis, image recognition, and countless others. The ability to automatically categorize inputs with high accuracy enables automation of decision processes that would otherwise require human judgment at scale.

In this chapter, we'll start by exploring the basics of classification, seeing how we can repurpose our gradient descent knowledge to build models that make decisions. We'll begin with binary classification—problems with just two possible outcomes—and see how, with a simple twist, the regression techniques we've already mastered can be transformed into powerful classifiers. We'll also introduce neural networks as the logical next step in our journey. These layered computational structures fundamentally expand what's possible in classification, allowing us to model increasingly complex decisions that simple linear models can't capture. Neural networks provide the perfect bridge between the basic classification principles we'll explore and the sophisticated deep learning architectures that drive modern AI systems. Through this introduction, you'll gain both practical implementation skills and the conceptual foundation needed to understand how these powerful models learn to make decisions across countless domains.

Then, we'll expand our horizons to tackle more complex scenarios. What happens when we need to choose between more than two categories? We'll dive into multi-class classification, where our models must decide between several possible options—such as identifying handwritten digits (0–9) or categorizing news articles by topic.

Along the way, we'll encounter a critical concept that revolutionized classification: cross-entropy loss. Unlike the mean squared error (MSE) we used for regression, cross-entropy is specially designed for classification tasks. We'll explore why this matters and how it fundamentally changes the way our models learn. We'll also venture into multi-label classification, where items can belong to multiple categories simultaneously.

Finally, we'll put theory into practice by discussing classification models in Keras. We'll see how the theory we discuss in this chapter translates to very dense lines of Keras code, as well as how each decision we make in code must come from a thorough understanding of the founding principles. By the end of this chapter, you'll have a robust understanding of classification through gradient descent—a powerful combination that forms the foundation of many modern AI systems, from spam filters to medical diagnostics and beyond.

4.1 Classification Basics

Previously, we examined the regression approach to machine learning. To establish a foundation for understanding *classification*, it's helpful to summarize the key elements of the regression process.

In regression, we work with a model that assumed a linear relationship between inputs and outputs. For example, when predicting house prices, we used square footage as input features to predict a continuous numeric output (the price). The learning process follows a structured sequence. We begin by assigning weights (model parameters) to each input feature, which determine how strongly each feature influences the prediction. Because we don't know the optimal values in advance, we initialize these parameters randomly.

With our initialized model, we perform a forward pass by calculating predicted outputs using our current parameter values. We then measure the difference between our predictions and the actual values using a loss function, typically MSE. This quantifies how far off our predictions are from reality.

Next, we apply gradient descent by calculating derivatives of the loss function with respect to each parameter. These derivatives indicate how we should adjust each parameter to reduce the loss. We update all parameters simultaneously using the gradient information and a learning rate that controls step size.

We repeat this process iteratively—forward pass, loss calculation, gradient computation, and parameter updates—with each cycle producing incremental improvements in our parameters. This continues until the loss stabilizes at a minimum value (though not necessarily zero), indicating we've found the optimal parameter configuration given our data and model structure.

Once trained, these optimized parameters form a model that can generate predictions for new, previously unseen inputs. The model has effectively captured the underlying patterns in our training data. We can now evaluate its performance on a separate test dataset—data the model hasn't seen during training—to measure how well it generalizes. This critical step reveals whether our model has truly learned meaningful patterns or has simply memorized the training examples. A successful model will make accurate predictions on both training and test data, indicating that it has captured genuine underlying relationships rather than noise or coincidental correlations.

Let's apply the same methodology for classification in the following sections and see if it works for yes/no answers as well.

4.1.1 Classification Problem Setup

We'll start by examining a use case with profound real-world importance. Consider a doctor examining tumor data, faced with a critical question: Is this tumor malignant or benign? This decision carries immense weight, profoundly affecting the patient's treatment, prognosis, and peace of mind.

When doctors classify tumors, they're distinguishing between two fundamentally different conditions. *Benign* tumors are like unwelcome but relatively harmless houseguests. They grow slowly, don't invade neighboring tissues, and don't spread to other parts of the body. While they may cause discomfort or other symptoms depending on their location, they typically aren't life-threatening unless they press against vital structures. *Malignant* tumors, on the other hand, are like destructive intruders. They grow aggressively, invade surrounding tissues, and can break off to establish new colonies elsewhere in the body (metastasis). These tumors are what we commonly refer to as cancer and pose serious health risks.

This distinction—benign versus malignant—represents our *classification task*. We'll build a model that examines tumor data and predicts which category a tumor belongs to. While real medical diagnostics involve numerous factors—genetic markers, imaging results, patient history, and more—we'll begin with a simplified version that focuses on just one feature: tumor size. Size is often a significant indicator, with larger tumors generally (though not always) associated with higher risk of malignancy. This single-feature approach lets us visualize our classification problem in two dimensions and build an intuitive understanding before tackling more complex scenarios.

Our tumor classification problem is specifically a *binary classification* task because we're deciding between exactly two classes (malignant or benign), the classes are *mutually exclusive* (a tumor can't be both malignant and benign simultaneously), and the classes are collectively *exhaustive* (every tumor must be one or the other; there is no third option). We could frame this binary classification in several equivalent ways: malignant or benign, "Is this tumor malignant?" with answers of yes or no, Class 1 or Class 0, and true or false. This 1/0 representation is particularly useful in computing, which is why binary classification often uses these numerical labels. The binary nature of this problem makes it an excellent starting point for understanding classification concepts before moving to more complex multi-class problems.

Let's take some dummy data that seems reasonable for our tumor classification problem. We'll plot tumor size on the x-axis with some dummy values. On the y-axis, we'll have our *classification label*: 0 for benign tumors and 1 for malignant tumors. This data might look something like Figure 4.1.

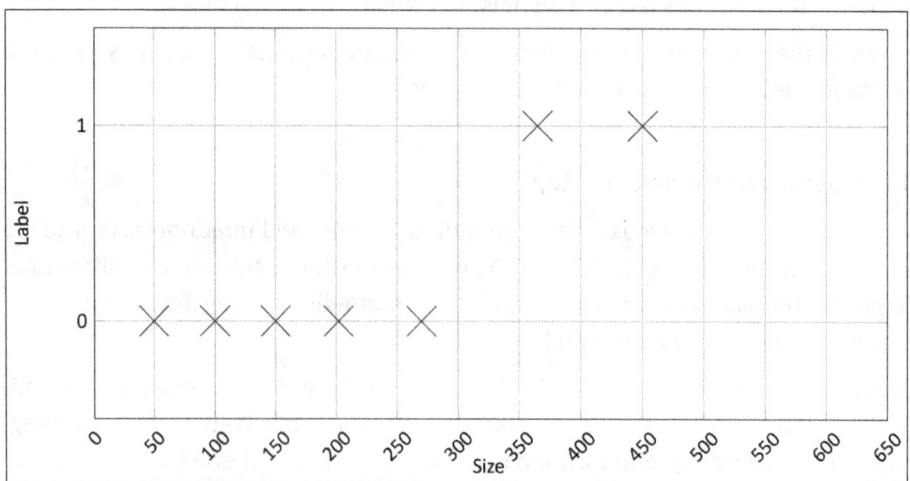

Figure 4.1 Plotting Tumor Data for Classification

When we visualize this data, the scatter plot looks quite different from what we saw in regression problems. The points only appear at y-values of exactly 0 or exactly 1, creating two horizontal lines of data points rather than the cloud of points we saw when predicting continuous values.

Despite the different appearance, there's a familiar pattern here. Generally, smaller tumors tend to be benign (clustered at y = 0), while larger tumors tend to be malignant (clustered at y = 1). This resemblance to our regression problems is promising—perhaps we could apply a similar approach here as well.

4.1.2 First Attempt Using Gradient Descent

What if we tried using our linear regression model from the previous chapter? We could fit a straight line through these points, just as we did before. Our hypothesis function might look something like

$$h_\theta(x) = \theta_0 + \theta_1 x$$

where x is the tumor size, and θ_0 and θ_1 are our model parameters that we'll optimize using gradient descent.

Let's try to follow this approach by picking random values for the model parameters and applying gradient descent. When we do that, we know that the final slope of the line will arrive at a point where the error is minimum and we'll get a line that goes through the points in a way to minimize the loss. The algorithm continues this process until the improvements become negligible—that's when we know we've reached (or at least approached) the bottom of the valley. This gives us the result shown in Figure 4.2. The solid slanting line shows our model.

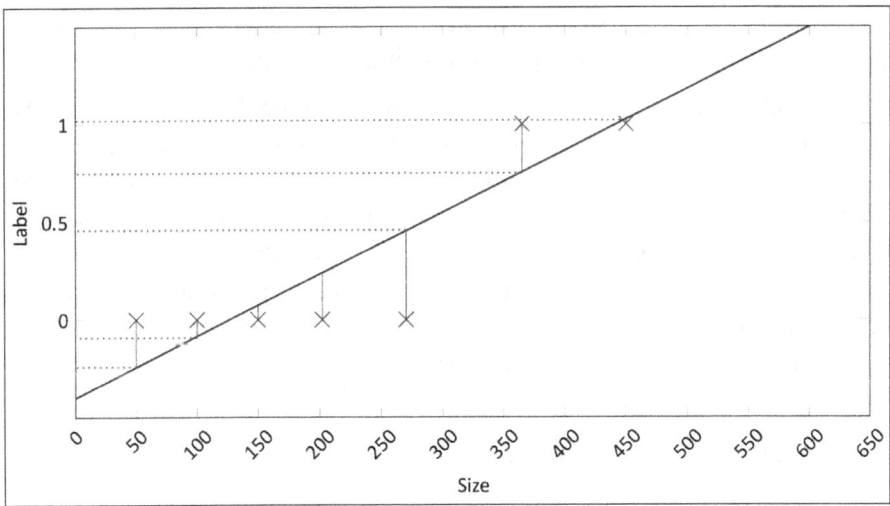

Figure 4.2 First Attempt at Classification Using the Regression Technique

We see that if we use our model now to make predictions, we immediately run into a problem. Looking at the graph, something doesn't quite fit. The solid line—our prediction model—extends far beyond the boundaries of what makes sense for our classification task.

For an input value of 50, our model makes a prediction of about -0.2. For our problem setup, this makes no sense at all. We're trying to classify tumors as either benign (0) or malignant (1), not predict mysterious negative values or values greater than 1! Remember that we're trying to predict the class—0 or 1. Nothing else makes sense in our context. A doctor wouldn't tell a patient their tumor is "-0.2 malignant" or "1.5 malignant." The diagnosis must be definitive: either malignant or benign.

Because the model is a continuous line and produces a real-value output, it's hardly ever exactly 0 or 1, so for almost all cases, the prediction it makes is wrong. Our predicted values might be 0.3, 0.7, or even outside the [0,1] range entirely. We've run into a mess applying the gradient descent approach directly. Our elegant mathematical tool seems to be speaking a different language than our classification problem requires. The linear model is answering, "How much?" when we're asking, "Which one?"

That doesn't mean we have to abandon gradient descent, start from scratch with some entirely different approach, or throw away all we've learned, however. As before, gradient descent is a very flexible algorithm. Let's try to make some changes so that we can keep using the extremely elegant framework that we've developed previously. We need to find a way to translate our continuous predictions into the binary language of classification while still leveraging the power of gradient descent.

If we formalize the problem, it's clear that we're facing two distinct challenges with our linear approach to classification:

- The first problem is that the output produced isn't constrained between 0 and 1. Our linear model happily ventures into negative territory or soars beyond 1, creating predictions that simply don't translate to probabilities or classifications.
- Even beyond that, the second problem is that the values aren't specifically 0 or 1. We don't want any values between 0 and 1; we want specifically 0 or 1. Even if we somehow constrained our model to stay within the [0,1] range, we'd still be left with wishy-washy predictions such as 0.37 or 0.82. But in the world of binary classification, there's no room for "maybe" or "somewhat"—a patient can't have a tumor that's sort of malignant. We need clear, definitive answers: yes or no, malignant or benign, 1 or 0. It's as if we've built a sophisticated analog dial when what we really need is a simple on/off switch. Our model needs to make a decision, not just estimate a value.

4.1.3 Second Attempt: Fixing the Issues in the First Attempt

Let's fix the first issue by introducing a special function that will force the predicted values to fall between 0 and 1. This transformation is exactly what we need—a way to tame our wild linear predictions and bring them into a reasonable range.

We have several options for this, but we'll start with a simple one known as the *sigmoid* function. The shape of the sigmoid function is shown in Figure 4.3. Look at its elegant S-curve shape—it's nature's way of transitioning smoothly between two states, like the way twilight gently bridges day and night.

The *domain* of the sigmoid function is negative infinity to positive infinity. We'll give its formula in just a minute, but the formula isn't that important. What's important is that for negative infinity values given as input, it returns the value 0, and for positive infinity, it returns 1. For input 0, it produces the value 0.5, and all the other values are based on the plot you can see.

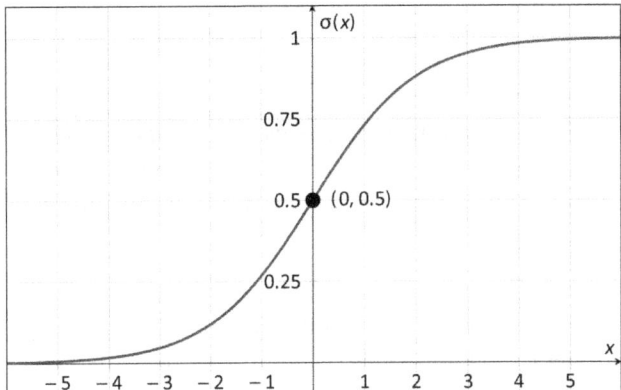

Figure 4.3 Sigmoid Function

The sigmoid function acts like a mathematical thermostat—no matter how extreme the input temperatures get (whether scorching-hot positive numbers or freezing-cold negative ones), it always regulates the output to stay within the comfortable range between 0 and 1. This property makes it perfect for our classification needs.

Think of the sigmoid as a translator between two different languages: the language of unbounded linear predictions and the language of probabilities. It takes our raw predictions, which might range from negative thousands to positive thousands, and converts them into values we can interpret as the probability of belonging to the positive class. The process of applying such a function so that the output lies between a specific range is called *squishing*, and the function itself is called a *squishing function* or an *activation function*.

To use the sigmoid function, we'll take our prediction and pass it to the sigmoid, and the output of this sigmoid will be our prediction instead. This redefines our hypothesis function as

$$h(x) = \sigma(\theta_0 + \theta_1 x)$$

where the sigmoid function is defined as

$$\sigma(z) = \frac{1}{1 + e^{-z}}$$

Think of this transformation as a mathematical passport control. Our linear function $\theta_0 + \theta_1 x$ can freely wander across the entire number line, traveling to extreme values in either direction. But before it's allowed to become our prediction, it must pass through the sigmoid checkpoint, which stamps it with a value between *0* and *1*.

Let's see this in action. Assume our linear model predicts a value of *-3* for a small tumor. Instead of accepting this nonsensical negative prediction, we pass it through the sigmoid:

$$\sigma(-3) = \frac{1}{1 + e^3} \approx 0.05$$

The same occurs if our linear model predicts 5 for a large tumor:

$$\sigma(5) = \frac{1}{1+e^5} \approx 0.99$$

The beauty of this approach is that extreme values get pushed toward the boundaries (0 or 1) without ever quite reaching them, while values near 0 get transformed to around 0.5—reflecting genuine uncertainty about which class the example belongs to. Keep this point in mind for later.

That solves our first issue. We've tamed our wild predictions, bringing them into the civilized realm between 0 and 1. However, the second issue still remains, and we aren't getting 0 or 1 as output. We want discrete values of exactly 0 or 1 instead of continuous values. The sigmoid didn't solve that issue; it just gave us better-behaved continuous values.

The solution to this is quite intuitive now, though. Consider standing at the midpoint of a seesaw—50% of the way from one end. If you're even slightly to the right of center, you'll slide all the way to the right end. If you're slightly to the left, you'll end up at the left end. We can apply this same principle to our predictions: if we get a value that is 0.5 or higher as a prediction, we'll just round it up to exactly 1, and if we get a value less than 0.5, we round down to 0.

Remember that we can't get a value above 1 or below 0 anyway, so we now have a very clean solution. We've transformed our continuous-output regression model into a binary classifier through a two-step process: first squeezing the predictions into [0,1] with the sigmoid, and then snapping them to either 0 or 1 with thresholding.

This value of 0.5 is called the *threshold*. The specific value of 0.5 makes sense if you look at it in terms of probability. We're saying that if there is more than a 50/50 chance of something, we consider it as truth, and if less than that, we consider it false. It's like making a decision based on a weather forecast—if there's a 60% chance of rain, you might take an umbrella, but if there's only a 30% chance, you might leave it at home.

Think about our tumor classification problem. If our model says there's a 75% chance a tumor is malignant, we classify it as malignant (class 1). If it says there's only a 23% chance, we classify it as benign (class 0). This reflects how we make decisions in uncertainty–we pick the outcome that seems more likely, even if we're not 100% certain. So, our final prediction becomes

$$\text{pred}(x) = \begin{cases} 1, & h(x) \geq 0.5 \\ 0, & \textit{otherwise} \end{cases}$$

The beauty of this approach is that it preserves valuable information during the learning process. While our final output is binary, the underlying sigmoid values give us a measure of confidence or probability that guides the learning. It's like having the benefit of nuanced thinking during deliberation but clarity and decisiveness when action is required.

Ok, great—that's both problems solved. Let's try to use this threshold approach on our first attempt and see what happens. We'll ignore the sigmoid function for a while for the sake of simplicity. If we try to apply our threshold to the original example we saw earlier, we get predictions as shown in Figure 4.4.

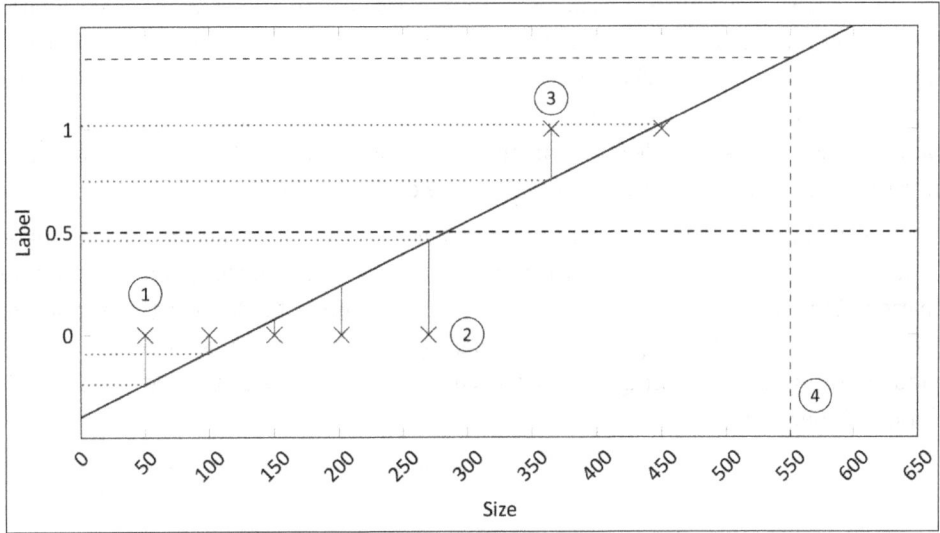

Figure 4.4 Making Predictions After Adding Thresholds

Let's examine how our model performs on specific examples in the figure, which reveals the elegant simplicity of our approach. Take a look at data point ❶. This small tumor sits firmly in the benign region. For this particular data point, our model made a raw prediction of -0.2 (ignoring the sigmoid transformation for now). When we feed this negative value through our sigmoid function, it produces a value well below 0.5. By applying our threshold rule, we get the final prediction of 0, which is correct! The model has correctly identified this small tumor as benign.

Similarly, for ❷, our linear function yields a value that gives us approximately 0.48. Again, this is less than our 0.5 threshold, so we predict 0 (benign) and again, we're correct! Notice how close this prediction is to the threshold despite being a relatively large benign tumor. This makes sense—larger tumors are generally more likely to be malignant, so even though this one is benign, the model expresses some uncertainty.

Finally, let's try with point ❸. This tumor is relatively large and, indeed, malignant. Our model's linear function produces a value that, after sigmoid transformation, becomes approximately 0.74. This is comfortably greater than our threshold, so our final prediction is 1 (malignant). Once more, we've correctly classified the tumor!

What about point ❹? This is an interesting case because it represents a prediction region rather than an actual data point. If we were to encounter a tumor of this size, our model would produce a prediction well above 0.5, leading to a malignant classification.

The further right we go on our graph, the more confident our model becomes in its malignancy prediction—which aligns with medical understanding that larger tumors generally pose higher risks.

Look at that—we got all the answers correct! Our model has successfully learned to distinguish between benign and malignant tumors based on their size. What's particularly beautiful about this approach is how we've layered transformations to turn a simple linear model into a powerful classifier. It's like we've given our mathematical model a way to make decisions just as a doctor might—combining evidence, assessing probability, and ultimately making a clear diagnostic call. And we've done all of this while preserving the elegant gradient descent framework we developed earlier.

Before we start celebrating though, let's just make sure by adding another data point to our dataset. Let's assume the doctors found a very large tumor in an unfortunate patient. This is shown in Figure 4.5 with the new data point ❶. Because the tumor size is so large, it's not surprising that the tumor is malignant. So, this new data point doesn't add anything to our existing knowledge. Our model should still be able to learn well from this new dataset.

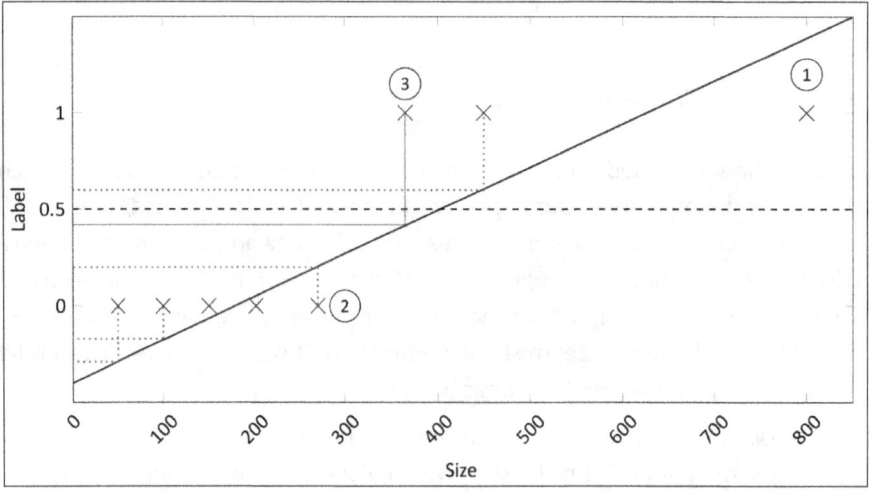

Figure 4.5 Adding an Extreme Outlier Data Point to the Dataset

As we can see, the model's parameters have been adjusted by gradient descent so that the slope is a lot gentler. It's like watching a seesaw adjust when a very heavy child sits on one end—the entire balance shifts. The reason is that if the slope kept the same as it was before, the error for this new data point at tumor size 800 would have been huge. Gradient descent (remember, it's not "smart" in the way humans are) only wants to minimize the loss, so it changed the slope value (model parameter) to minimize the overall error. The loss is now at its minimum for our dataset with this new, gentler slope. On paper, everything looks mathematically optimal. If you look at the point predicted by the model at x = 300, it's correct.

However, point ❸ (x = 375) is now predicted to have a value of 0.47, which is below our 0.5 threshold. So, our prediction is now benign, which is incorrect! Our model has forgotten how to classify a tumor that it previously identified correctly.

What went wrong? Adding a new data point to our dataset should have reinforced our learning. It's like giving a student more examples to study from—it should make them better, not worse. But instead, it has made our model worse. We're now making an incorrect prediction for a case we previously got right.

Spend some time thinking about why this might be. We followed the approach diligently; we minimized the loss but our model still made a mistake that it didn't make without this outlier point. It's as if the presence of this one extreme example has pulled our entire model out of alignment, like a magnet distorting a compass needle.

This phenomenon reveals something profound about machine learning models. Sometimes, being mathematically optimal doesn't translate to being practically useful. Our model is now technically "better" at minimizing overall error across all examples, but it's sacrificed accuracy on an important case to accommodate an extreme outlier.

The reason that the model doesn't work as intended is that our definition of "better" is incorrect. It's like training a dog to fetch sticks but then being disappointed when it doesn't guard your house. We've been optimizing for one goal while hoping to achieve another.

Think about it: The loss we're trying to minimize is the distance of the prediction (without the threshold) from the actual value—we're using MSE. Our model is dutifully doing exactly what we asked of it: minimizing the square of the difference between its continuous predictions and the true labels. That extreme data point at size 800 creates an enormous square error if not accommodated, so the model adjusted its parameters accordingly.

On the other hand, when we evaluate the model, we're measuring how many answers it got right after applying the threshold. We're counting correct classifications—a fundamentally different metric than the squared distance between continuous values! It's like training an archer to hit as close to the bullseye as possible but then scoring them only on whether they hit the target at all.

We told our model to minimize one loss, and we're evaluating it based on some other criteria. That's not fair to the model, and that's why it can't perform correctly. Consider being hired for a job where your performance review measures skills completely different from what you were told to focus on—it's a recipe for misalignment. This *misalignment* between optimization objective and evaluation metric is a fundamental challenge in machine learning. Our model is actually performing perfectly—it's minimizing the MSE across all data points, including our outlier. But because classification accuracy depends on the threshold, which isn't part of our optimization process, we get unexpected results. Let's try one more time, and this time, I promise we'll have the right answer.

> **[!] Better Isn't Always Better**
>
> Keep this lesson about misalignment in mind when doing practical machine learning. Just because a model performs well on the results you're measuring, it doesn't necessarily mean the model is good. You should always question whether you're making the right measurements as well as making the measurements right.

4.1.4 Third Attempt: Fixing the Loss Function

Because our loss function is an issue, we should try to fix that first and see how it goes. It's like realizing you've been using a ruler to measure weight—the tool simply doesn't match the task at hand.

Remember that our issue is that MSE measures something we're not really interested in. MSE is excellent at capturing the numerical distance between predictions and actual values, but in classification, we don't care about that distance in quite the same way. What we're interested in is getting the maximum number of classification answers correct. We want our model to say "benign" when a tumor is benign and "malignant" when it's malignant—regardless of how confident it is in that prediction (as long as it crosses our threshold). But we also need a function that has a clean derivative so that we can use gradient descent. Gradient descent is like a hiker following a path downhill—it needs a smooth slope to navigate effectively. If our function has sharp cliffs or flat plateaus, our hiker gets stuck or wanders aimlessly.

So, we need a function that is *differentiable* and gives a small value when the answer is correct and a large value when the answer is incorrect. Like a teacher's grading system that barely penalizes minor mistakes but severely punishes fundamental misunderstandings. The student (our model) would quickly learn to avoid those critical errors. If we sum over such a function, we'll get the loss or cost function that we want. It would give us a measure of how many classification errors our model makes, weighted by how confident it was in those wrong answers.

We can explore different options for such a function, but over time, a consensus has been developed on the best function. It's like how thousands of chefs over centuries eventually agreed that certain spice combinations just work better than others—wisdom born from collective experience.

Unlike before where we derived the loss function ourselves, this time, I'll just tell you what that function is, and I'm then going to try to convince you that this loss function does what we want. Sometimes, in machine learning, we stand on the shoulders of giants rather than rebuilding everything from first principles. So, here's a function that does exactly what we want:

$$f(x) = \begin{cases} -log(\,h(x)\,), & if\ y = 1 \\ -log(\,1 - h(x)\,), & if\ y = 0 \end{cases}$$

It's a *piecewise* function, which means it has two personalities. If the data point we're looking at is actually from a malignant tumor, we use the first case and calculate the loss as $-\log(h(x))$. If, on the other hand, the data point belongs to a benign tumor, the loss we use is $-\log(1 - h(x))$. Let's see a plot of both of these cases in Figure 4.6 to understand why this loss function does exactly what we need.

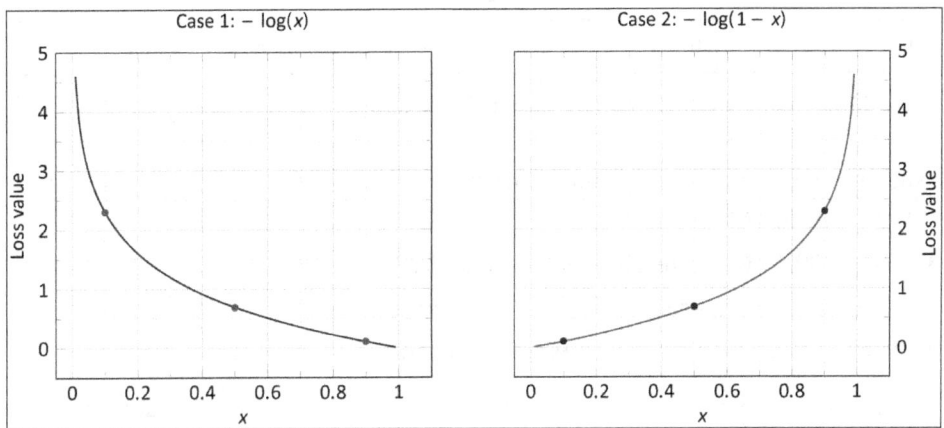

Figure 4.6 Values for the Negative Log Functions

Picture a doctor examining a scan of what is definitely a malignant tumor, meaning $y = 1$. If our model confidently says 99% chance of malignancy, that is, $h(x) = 0.99$, what happens to our loss? We get the loss of $-\log(0.99) \approx 0.004$. That's a tiny penalty that will help achieve a lower total loss value. The model gets a gentle pat on the back for being correct and confident. But what if our model looks at this same malignant tumor and declares only 10% chance of malignancy, that is, $h(x) = 0.1$? The loss increases to $-\log(0.1) \approx 1$. That's a much larger penalty! It's like the loss function is saying, "You were very wrong to be so confident about it being benign when facing a malignant tumor."

Another case is if our model is completely wrong and says 1% chance of malignancy or $h(x) = 0.01$. In this case, the loss skyrockets to $-\log(0.01) \approx 2$. It's the mathematical equivalent of a serious reprimand for missing a critical diagnosis.

Now, let's flip to the second case. When a tumor is truly benign (i.e., $y = 0$), our loss function becomes $-\log(1 - h(x))$. If our model correctly identifies a benign tumor with high confidence, saying "5% chance of malignancy" (i.e., $h(x) = 0.05$), we get the loss of $-\log(1 - 0.05) = -\log(0.95) \approx 0.02$. Again, a very small penalty for being correct and confident. But if our model incorrectly assigns a high probability to a benign tumor, saying "95% chance of malignancy" or $h(x) = 0.95$, our loss becomes $-\log(1 - 0.95) = -\log(0.05) \approx 1.3$. The model receives a substantial penalty for its misplaced confidence. These cases are summarized in Table 4.1.

Ground Truth	Prediction	Loss	Note
1	0.99	0.004	Correct prediction, low loss
1	0.01	2	Incorrect prediction, high loss
0	0.05	0.02	Correct prediction, low loss
0	0.95	1.3	Incorrect prediction, high loss

Table 4.1 Losses Incurred for the Two Ground Truth Cases

What makes this function brilliant is how the negative logarithm creates a natural, intuitive penalty system:

- Being slightly wrong results in small penalties.
- Being moderately wrong creates moderate penalties.
- Being confidently wrong triggers severe penalties.

Looking at the shape of $-\log(h(x))$, we can see why this works so beautifully. As x approaches 0, $-\log(h(x))$ shoots up toward infinity. This means that as our confidence in the wrong answer increases, the penalty or loss grows exponentially. It's like having a coach who barely notices small mistakes but becomes increasingly vocal as incorrect actions get more serious—exactly the kind of feedback our model needs to learn effectively.

There's one minor issue with this loss function, though. It's a piecewise function that is relatively difficult to differentiate and might lead to a more computationally intensive derivative. Piecewise functions require separate handling for different input ranges, which complicates both the mathematical analysis and the implementation in code.

In addition, it's much nicer to understand a *closed-form function*. A single, unified formula allows us to reason about the behavior of our loss across all possible inputs without having to consider different cases separately. This is especially valuable when we're trying to optimize the function through gradient descent. So, let's go ahead and use a trick to convert this piecewise function to closed form. This mathematical sleight of hand will give us a single expression that works everywhere.

We see that the first case is applicable when $y = 1$, so we'll multiply the formula for this case with y. The second case applies when $y = 0$, so we'll multiply that with $(1-y)$ and add the two resulting terms together:

$$\text{Cost} = y \cdot \left(-\log(h(x))\right) + (1 - y) \cdot \left(-\log(1 - h(x))\right)$$

This is elegant because of how the binary values of y interact with the formula. When $y = 1$ (malignant tumor), the first term remains as $-\log(h(x))$, whereas the second term becomes 0 because $(1-y) = 0$. When $y = 0$ (benign tumor), the first term becomes 0, and only the second term $-\log(1 - h(x))$ remains. This means the first term will cancel out when y = 0, and the second will cancel out when y = 1. These are switches that are perfectly coordinated—exactly one of them is always on, and the other is always off.

This elegant formula is called the *binary cross-entropy loss* in closed form—one of the most important loss functions in machine learning due to its significance for binary classification. Not only is it easier to differentiate mathematically, but it's also more straightforward to implement in code as a single, cohesive function rather than a series of conditional statements. That also means it can take advantage of all the vectorized code implementations and hardware.

This representation showcases how mathematical expressions can be manipulated to solve practical problems. We've taken a conceptually simple but computationally awkward piecewise function and transformed it into a smooth, differentiable expression that will work seamlessly with gradient descent—all while preserving the exact same behavior for our binary classification problem.

Finally, summing over all the data points and taking the common negative sign out of the two terms, we get the following final loss function for binary classification:

$$J(\theta) = -\frac{1}{m} \sum_{i=0}^{m} [y \cdot \log(h(x)) + (1 - y) \cdot \log(1 - h(x))]$$

As before, we need the derivative of this function; you can go ahead and compute it yourself or just take the word of mathematicians who have already calculated it. The derivative in this case turns out to be the following:

$$\frac{\partial}{\partial \theta_j} J(\theta) = -\frac{1}{m} \sum_{i=0}^{m} (h(x) - y) x_j$$

This looks suspiciously like the derivative for MSE, with a familiar structure that might remind you of our work in regression. But you must keep in mind that there's a squishing function hiding inside $h(x)$. This isn't a linear function anymore but a squished one, that is, $\sigma(\theta_0 + \theta_1 x)$. This sigmoid transformation fundamentally changes how the gradient behaves during optimization. While the mathematical form might appear similar at first glance, the presence of this nonlinear sigmoid function creates a completely different learning dynamic. Now, we can go back and use this as the loss function to reach a decision between the benign and malignant tumors. Because we use a loss function based on the logarithm (or log) in classification, we also refer to this type of machine learning as *logistic regression*.

4.1.5 Squishing Functions and Decision Boundaries

So now we have a loss function and a squishing function after our linear relationship. Let's discuss how this is going to help us make decisions in practice. Moving from theory to application helps us see how these mathematical tools solve real problems.

4 Classification Through Gradient Descent

This time, we're actually going to use two features to understand the concept. Assume we have a problem in which there are two input features, x_1 and x_2. The output is some binary class y. Perhaps in our tumor classification problem, x_1 might represent the tumor size, and x_2 might represent the tumor density measured in a scan. Together, these two features give us a more complete picture than either one alone.

In addition, assume that we went ahead and applied the gradient descent algorithm to find the optimal parameters. We'd start with some random values—perhaps $\theta_0 = -3$, $\theta_1 = -0.5$, and $\theta_2 = 0.1$. These initial values would likely produce poor predictions, with many tumors misclassified.

With each iteration of gradient descent, we'd take the following steps:

1. Calculate our current predictions using the sigmoid function.
2. Compute the binary cross-entropy loss to see how wrong we are.
3. Calculate the gradient (the direction of steepest increase in error).
4. Move in the opposite direction of that gradient.
5. Update our parameters accordingly: θ_0, θ_1, and θ_2.

Initially, these updates might be dramatic as our model flails about trying to find the right direction. As training progresses, the updates become more refined. After many iterations, the parameter updates become tiny, signaling that we've reached convergence—the valley floor of our error landscape. Let's say we got the following values for θ_0, θ_1, and θ_2 (recall that if we have two features, we'll have θ_0 as the bias and two other parameters, one for each feature):

$$\theta_0 = -3, \ \theta_1 = 1, \ \theta_2 = 1$$

This means that our model is making the following prediction:

$$h(x) = \sigma(-3 + 1 \cdot x_1 + 1 \cdot x_2)$$

And, if this h(x) is greater than 0.5 for some particular data point, we say that the data point is malignant. Let's dissect what this means geometrically. Remember that the sigmoid function equals 0.5 exactly when its input is 0. So, $h(x) > 0.5$ when

$$-3 + x_1 + x_2 > 0$$

Rearranging this by taking the 3 to the other side of the inequality, we get

$$x_1 + x_2 > 3$$

This is an elegant result! We get a line—our *decision boundary*—that separates malignant predictions from benign predictions. This is $x_1 + x_2 = 3$. Any point above this line (where $x_1 + x_2 > 3$) will be classified as malignant, and any point below it (where $x_1 + x_2 < 3$) will be classified as benign.

Let's plot this in two dimensions in Figure 4.7. The line $x_1 + x_2 = 3$ slices through our feature space at a 45-degree angle (because the coefficients for both x_1 and x_2 are equal at *1*). This line represents points where our model is precisely 50% confident—where it's most uncertain about the classification.

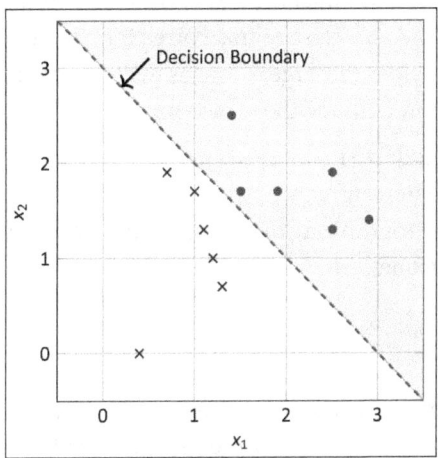

Figure 4.7 Decision Boundary Based on Hypothesis Function

As we move away from this line in the positive direction (increasing x_1 and/or x_2), our model becomes increasingly confident that tumors are malignant. Conversely, moving away from the line in the negative direction makes our model more confident that tumors are benign.

The bias term $\theta_0 = -3$ shifts this line away from the origin. If θ_0 were *0*, our decision boundary would pass through the origin *(0,0)*. The negative value pushes the line outward, requiring larger feature values to cross the threshold into malignancy. The model parameters combined control the rotation of the line. This line is called the decision boundary of our classifier.

What's remarkable is that we've created this perfect decision boundary by simply following the gradient of our loss function. The algorithm discovered on its own that tumors with combined feature values exceeding 3 should be classified as malignant. We never had to explicitly program this threshold—it emerged naturally from the data through our optimization process. This same process gave us a visual understanding of what our model is learning when it classifies data points.

This means that our model will now be optimized based on the correct answers because we're using binary cross-entropy. If a data point has a huge tumor size, as we saw earlier, we'll get a very small value for loss in binary cross-entropy, and it won't mess up our model at all.

What makes binary cross-entropy particularly effective for classification problems is how it handles *outliers* and extreme values. Unlike MSE, which would heavily penalize confident predictions that are numerically far from the target values, binary cross-entropy only cares about getting the classification right. Consider our previous problem in Figure 4.5 regarding a patient who had an extremely large tumor. With MSE, this single data point pulled our entire model out of alignment because MSE is obsessed with minimizing the numerical gap between predictions and targets. But binary cross-entropy approaches the problem differently. It asks: "Did you classify this correctly? And how confident were you?" If the model correctly classifies the large tumor as malignant with high confidence, the loss is tiny regardless of how extreme the tumor size is.

This *robustness* to outliers is what allows our model to maintain its performance on the entire dataset rather than being disproportionately influenced by extreme values. The decision boundary stays where it should be—optimized for classification accuracy across all examples, not just the most extreme ones.

4.1.6 Learning Process Summary

Stop here and make sure you understand all these concepts fully. These are our foundations, and the more comfortable you get with these, the easier the more detailed concepts will be for you to adapt. It's like mastering basic cooking techniques before attempting complex recipes—having a strong foundation makes everything else possible.

Here are the important steps of the workflow:

1. We get the dataset.
2. We decide on the loss function.
3. We initialize parameter values with some random ones.
4. We run gradient descent until convergence.

This workflow applies across countless machine learning models, from the simplest linear classifiers to the most sophisticated state-of-the-art ones.

The most important thing is to realize that our model is just some numbers in the form of model parameters, and we learn the best values for them so that the loss is minimized, that is, our answers are correct depending on our definition of the word. That's the elegance of machine learning—we're not explicitly programming rules; we're letting the algorithm discover the optimal parameters to minimize our carefully designed loss function. For classification problems, we've learned that binary cross-entropy is the appropriate loss function, paired with a sigmoid activation, to transform our linear model's output into a probability. We've seen how this allows us to create decision boundaries in our feature space, turning continuous inputs into binary decisions.

Keep in mind that the choice of loss function profoundly affects what our model learns. MSE optimizes for numerical closeness, while binary cross-entropy optimizes for correct classification decisions. Aligning your loss function with what you actually care about is crucial for building effective models.

In terms of implementation, we'll have the dataset in a matrix and the initial random model parameters in a vector. We'll feed these along with the loss function to a black box optimization algorithm. In this case, it's raw, core gradient descent. The black box will spit out the optimized parameter values, which we can then use as our hypothesis function to predict classes for future data points.

Think of it as assembling a machine from individual components. Our dataset is the raw material to be processed. Our model parameters are adjustable knobs that need to be fine-tuned. The loss function is our quality control metric. And gradient descent is the methodical worker that keeps adjusting those knobs until the output quality reaches its best possible level.

The beauty of this approach is its modularity. We can swap different components without changing the overall workflow:

1. Different datasets provide different challenges.
2. Different initial parameters might lead to different final solutions.
3. Different loss functions prioritize different aspects of correctness.
4. Different optimization algorithms might find solutions more efficiently.

However, the fundamental process remains the same: optimize parameters to minimize loss. This pattern repeats across virtually all machine learning algorithms, from the simplest linear models to the most complex neural networks.

Once training is complete, we're left with a set of optimized parameters that define our decision boundary. For new patients with unknown tumor classifications, we simply plug their feature values into our hypothesis function, apply the sigmoid transformation, and compare the result to our threshold. The resulting classification could help doctors make better treatment decisions, potentially saving lives.

> **Using the Learned Classification Models**
> Keep in mind that once everything is said and done, we have the model that we can plug our data into and get the prediction. In case of the logistic regression that we just derived, the formula is simply $h(x) = \sigma(\theta_0 + \theta_1 x)$.

While we've focused on raw gradient descent as our optimization method, it's worth noting that modern machine learning frameworks offer many sophisticated variations that can make the training process more efficient and effective. Algorithms such as Adaptive Moment Estimation (Adam), Nesterov-accelerated Adaptive Moment Estimation (Nadam), Root Mean Square Propagation (RMSprop), and Adaptive Gradient Algorithm (Adagrad) are all evolutionary improvements on the basic gradient descent approach. They're like specialized tools that have been refined for particular types of optimization challenges.

These advanced optimizers differ in how they adapt to the learning process. Some automatically adjust the learning rate for each parameter based on historical gradients. Others incorporate *momentum* to help overcome local minima. Some combine multiple techniques for even better performance. For example, Adam adapts learning rates for each parameter and incorporates momentum, making it particularly effective for problems with noisy or sparse gradients. Nadam enhances this further by incorporating the Nesterov-accelerated gradient, which looks ahead to where the parameters are going.

There's really no need to worry about the detailed workings of these algorithms. Despite their sophisticated internals, from a practical standpoint, these optimization algorithms serve the same role in our workflow. They're essentially black boxes that take our dataset, loss function, and initial parameters, and then output optimized parameters that minimize the loss.

What's great for practitioners like us is that switching between these optimizers often requires changing just a single line of code. We can experiment with different optimizers to see which one works best for our specific problem without altering our fundamental approach to model training. Regardless of which optimizer we choose, the core principle remains the same: We're searching for the model parameters that minimize our chosen loss function, allowing us to make accurate predictions on new data. The optimizers are simply different paths to reach the same destination, that is, a decision boundary based on some linear combination of our input features. The problem is that sometimes a decision boundary simply can't be created based on a linear relation between inputs and output. We'll explain that issue in the next section.

4.2 Nonlinear Relationships and Neural Networks

So far, we've successfully applied gradient descent to classification problems with linear decision boundaries. Our tumor classification example worked well because the data could be reasonably separated by a straight line. However, many real-world problems simply can't be solved with linear models, no matter how we adjust the parameters.

One of the most famous examples of this limitation is the *exclusive OR (XOR)* function. XOR is a fundamental logical operation that returns true only when inputs differ from each other. We can represent it with a simple truth table as in Table 4.2.

x_1	x_2	$x_1 \oplus x_2$
0	0	0
0	1	1
1	0	1
1	1	0

Table 4.2 Truth Table for the XOR Operation

In essence, XOR acts as an *inequality detector*—it outputs 1 when the inputs are different and 0 when they're the same. This operation is crucial in numerous computing applications, including the following:

- Error detection in data transmission
- Cryptography and data encryption
- Digital circuit design
- Parity checking in memory systems

Despite its simplicity, XOR presents a fundamental challenge for linear models. If we plot these four points on a coordinate plane (with input A and B as x and y coordinates), we face an interesting dilemma: it's impossible to draw a single straight line that separates the points where *XOR = 1* from points where *XOR = 0*.

To verify this, let's visualize the XOR function graphically through Figure 4.8. The points (0,0) and (1,1) should be classified as 0, while points (0,1) and (1,0) should be classified as 1. These points form the corners of a square, with opposite corners belonging to the same class.

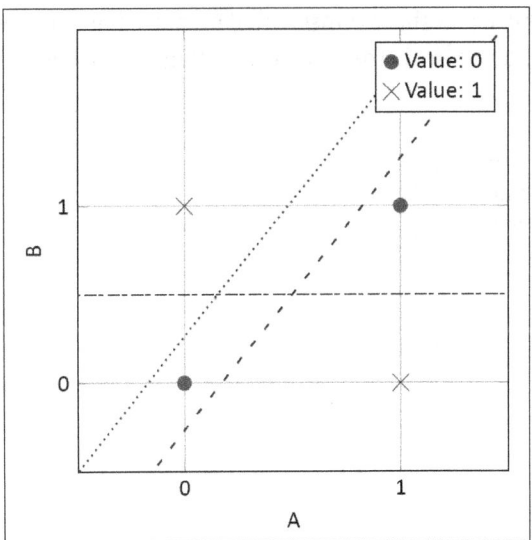

Figure 4.8 Impossible Decision Boundary for the Simple XOR Function

No matter how we position or rotate a straight line on this plane, we can't separate all four of these points correctly. This is a fundamental limitation of linear models—they can only create single straight-line decision boundaries.

4.2.1 Feature Transformations

Let's try something that might seem arbitrary at first, but the payoff will be worth it. Instead of trying to solve the XOR problem directly using the original features, let's transform our data a little. We're going to take our original input features x_1 and x_2 and create two new columns based on these. The first new feature is $x_1 \land \neg x_2$, which is true only when $x_1 = 1$ and $x_2 = 0$. The second new feature is $\neg x_1 \land x_2$, which is true only when $x_1 = 0$ and $x_2 = 1$. This will lead to the truth table shown in Table 4.3. We still have our original four data points, but we've changed how we represent them.

x_1	x_2	$\neg x_1$	$\neg x_2$	$x_1 \land \neg x_2$ (A)	$\neg x_1 \land x_2$ (B)	$x_1 \oplus \neg x_2$
0	0	1	1	0	0	0
0	1	1	0	0	1	1
1	0	0	1	1	0	1
1	1	0	0	0	0	0

Table 4.3 Adding New Features to the XOR Truth Table

Keep in mind that we aren't computing the exclusive OR of these new variables. The original inputs and original outputs are still the same—we've just created an intermediate representation of the inputs. We can plot these transformed points using our new features as the x and y coordinates, along with their corresponding XOR labels, as shown in Figure 4.9.

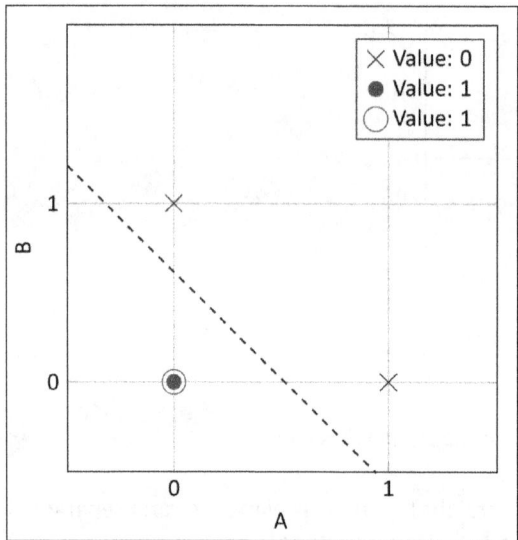

Figure 4.9 Modified Features Lead to a Clear Decision Boundary for XOR

When we plot our transformed data, something remarkable happens. Notice a couple of things: we aren't plotting x_1 and x_2 anymore. We're plotting A and B—the two new features—for each of the four data points. The values of XOR are coming from the original dataset.

In this new plot, the first three points are in the same places as before, but the fourth one, which was at (1, 1), has been shifted to (0, 0). It's as if the functions we chose have taken the piece of paper that the plot was drawn on and bent it in half diagonally, causing the fourth point to land right on top of the first. This is an important, foundational concept shown visually in Figure 4.10. Once you internalize this concept, the complexity of modern machine learning will click into place.

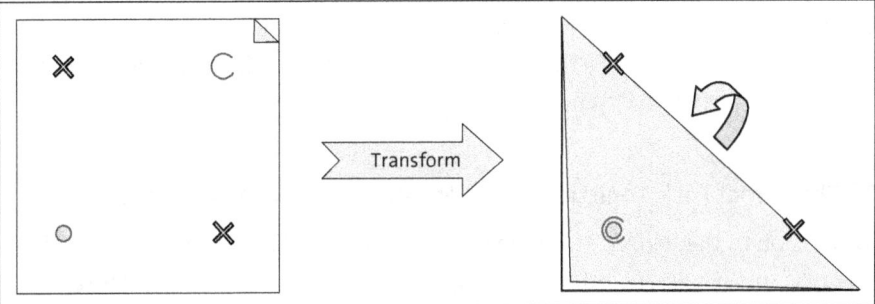

Figure 4.10 Transforming Feature Spaces

In this transformed space, we can easily draw a line, as shown in the figure, to differentiate between the two classes. This is the key insight—by transforming our features, we've made the previously inseparable problem linearly separable!

This warping of space illustrates the concept of feature transformation. In mathematical terms, we've applied a function that maps points from one space (our original x_1, x_2 space) to another space (our new A, B space). Mathematicians describe functions as mappings from a domain (the set of input values) to a range (the set of output values), where each input is associated with exactly one output.

What we've done here is create a *nonlinear transformation* that takes our original 2D input space and maps it to a new 2D space where the classification problem becomes *linearly separable*. Specifically, our transformation has "folded" the input space in such a way that points with the same XOR output value are now on the same side of a straight line.

This is a powerful concept in machine learning: When data isn't linearly separable in its original form, we can sometimes transform it into a new feature space where it becomes linearly separable. In our case, we've manually defined this transformation, but as we'll soon see, neural networks can learn these transformations automatically. Please spend some time on this to grasp the significance of this simple operation. We've essentially opened up a way to take any problem that isn't linearly separable at the moment and somehow make it so. This way, even if we have a problem in which the relationship between the inputs and outputs isn't linear, we can still apply our gradient descent algorithm after doing some housekeeping.

What we've done is actually quite profound. We've taken the original input features x_1 and x_2 and transformed them into feature A. We've done the same in another way to transform to feature B. Then, we input A and B to our logistic regression unit, which creates a linear decision boundary in this new feature space. This transformation to another space before performing classification is also known as the *kernel trick*—a fundamental concept in machine learning.

There's a gaping problem here though. Where did this particular formulation of A and B come from? These particular formulae for A and B were specific to this problem of XOR. How do we come up with these formulae automatically for new and novel problems? We'll answer that in detail in the next section.

4.2.2 The Kernel Trick, Logistic Regression, and All of Machine Learning

Our discussion of the logistic regression unit of Section 4.1.6 is summarized in Figure 4.11. Because this whole operation is very specific, we often represent it as just a circle with inputs to the circle's left and outputs to the right. This concise version is also shown in the figure.

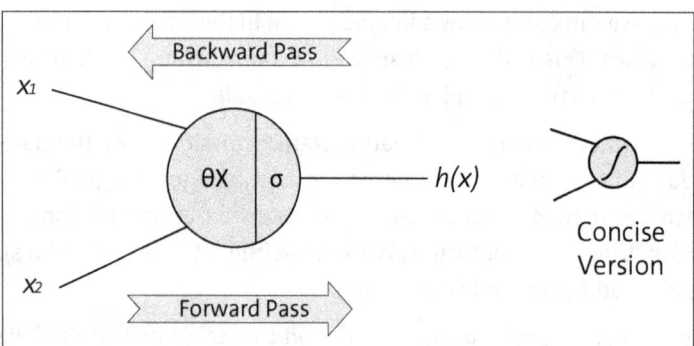

Figure 4.11 Logistic Regression Unit

We've stumbled upon something important in our XOR adventure. A single logistic unit, despite its elegance, has limitations—it can only draw straight lines through our feature space. For problems like XOR, we needed to manually craft clever transformations to make the problem linearly separable.

But here's the thing: What if we face hundreds or thousands of complex problems, each requiring its own unique transformation? Creating these kernel tricks by hand would quickly become overwhelming, like trying to craft a unique key for every lock in a giant castle. There has to be a better way. What if we could automate the process of feature transformation itself? Believe it or not, we already have something that can perform this transformation. So, we can actually get this transformation for free, as we'll discuss next.

Building a Transformation Factory

Rather than manually defining transformations, let's build a system that learns them automatically. Here's how we'll structure it. Consider we have two logistic units (let's call them *feature detectors*) that take our original features X_1 and X_2 as inputs. But—and this is the crucial part—these units aren't trying to solve our classification problem directly. Instead, their job is to transform our inputs into a new feature space.

Each of these units will learn to output a *transformed feature*. The first unit might output some value we'll call *A*, and the second unit will output a value *B*. These *A* and *B* values represent our data in a new space—a space we're not designing explicitly but rather letting the system discover on its own.

Once we have these transformed features, we feed them into a third logistic unit, as shown in Figure 4.12. This final unit takes A and B as inputs and performs the actual binary classification we care about. Notice how the derivatives for each layer are calculated based on the loss.

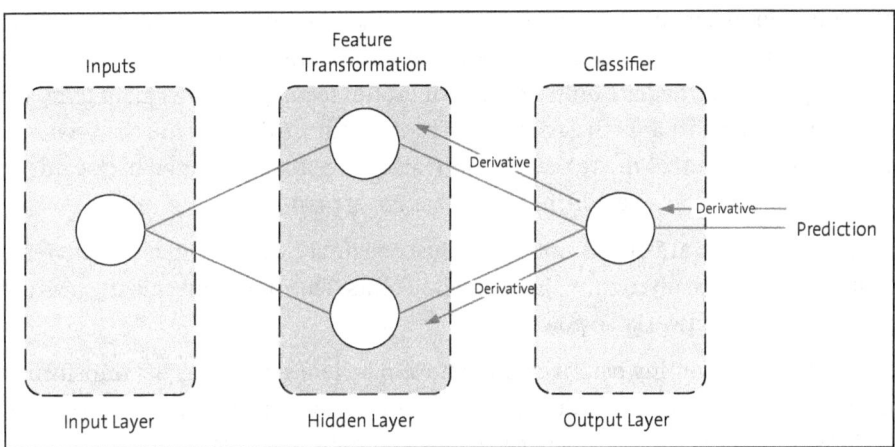

Figure 4.12 Learning Feature Transformations Automatically

What we've just described is the simplest possible neural network—a collection of logistic units, each one named a neuron, arranged in layers, where each layer feeds its outputs to the next layer:

1. The first layer contains two units that transform the original features.
2. The second layer contains one unit that makes the final classification.

The basic structure is that we have an *input layer*, a layer that performs transformations (typically referred to as the *hidden layer*), and an *output layer*. This structure has a remarkable property: It can learn the appropriate transformations needed to solve problems that a single logistic unit can't handle. It's like having assistants who prepare your ingredients before the head chef combines them into the final dish.

What's most exciting is that we don't need to specify what transformations the first layer should perform. Through the learning process, which we'll explore next, these units will discover useful transformations on their own—transformations that make the classification problem solvable for the final unit.

In essence, we're no longer handcrafting the process from input to output; we're building a system that figures out the best path by itself.

Learning Through the Ripple Effect

You might be wondering, "If we're not telling these intermediate units what transformations to learn, how do they figure it out on their own?" That's the million-dollar question, and the answer reveals the true elegance of neural networks.

These intermediate units learn through the same mechanism we've been using all along—gradient descent—but with a twist. Instead of operating in isolation, they learn as part of an interconnected system.

Think of our neural network as a river system. The final classification is the point where the river meets the ocean—that's where we ultimately care about the water flow (our predictions). The intermediate units are like tributaries feeding into this main river. We don't directly control what path each tributary takes, but we do measure the final outcome at the river mouth. We say that the derivative is being propagated backwards to the model parameters in an algorithm known as *backpropagation*.

When our predictions are off, we not only adjust the final classifier but also send feedback all the way upstream to those intermediate units. This ripple effect is the heart of what makes neural networks so powerful.

Here's the most captivating part: we don't know in advance what specific transformations those intermediate units will learn. We're not engineering them to perform AND operations, NOT operations, or any particular mathematical function. We're simply creating the conditions that allow them to discover what works best.

Before training, these units are like blank canvases. Their parameters (weights and biases) start as random values. However, as training progresses, they gradually fine-tune themselves to perform transformations that make the final classification task easier. We didn't know the exact parameter values our single logistic unit needed either. We just knew that whatever values minimized the final loss were the right ones. The same principle applies here, but at a larger scale. The intermediate units will land on transformations—whatever they may be—that help minimize that final loss.

This powerful approach is called *end-to-end learning*. We're optimizing the entire system as a unified whole, from raw inputs to final predictions. The beauty is that each component learns its role in service of the final objective.

When our network makes accurate predictions, it's not just because the final classifier is well-trained. It's because the whole system has harmonized—the intermediate units have learned to transform the input data in exactly the ways that make the classification problem solvable. This is the true power of deep learning: systems that discover their own path to solving problems, often finding solutions more elegant and effective than anything we might have engineered by hand.

The magic happens when we start training. Rather than trying to tune each neuron individually, we listen to the final loss and make adjustments throughout the whole chain. We run our inputs through the entire network, from the first layer of units to the final classifier. We compare the final output with what we wanted (the true labels) and calculate how far off we were (the loss). We determine how each parameter throughout the network contributed to that error. This last step involves what mathematicians call the *chain rule*—essentially, tracking how changes ripple backward through our network.

The best thing is that you don't need to worry about these complex derivative calculations! Modern machine learning frameworks such as TensorFlow and Keras handle all of this mathematical heavy lifting behind the scenes. These tools can handle the derivative of one function feeding into another, which feeds into another, and so on, keeping track of all the intricate mathematical relationships without you having to write a single derivative formula.

Based on these derivatives, we update all parameters simultaneously. We initialize all parameters randomly across the entire network, feed data forward through the network, compute the loss, calculate how each parameter affects the loss, adjust all parameters slightly to reduce the loss, and repeat until performance stops improving.

With each iteration, our intermediate units gradually discover useful transformations, and our final classifier learns to use these transformed features effectively. What emerges is amazing: a system that learns to transform data in exactly the ways needed to solve problems that were previously unsolvable with simple approaches. No human engineer specified these transformations—they emerged naturally from the learning process itself.

> **The Two Parts of Learning**
> Our end-to-end learning has two major parts: feature extraction and discrimination. The first part applies the transformations and the second performs the actual classification. This basic principle will come in handy in later, more complex models.

The Math Behind Layers

When we move from a single neuron to a network of neurons, our mathematical notation needs to level up as well. Remember how we used a vector θ to represent the weights of our single logistic unit? Now, we're dealing with multiple neurons working in concert, so we need a more powerful representation—enter the matrix.

Think of a matrix as an apartment building where each floor houses a different neuron. Each neuron has its own set of weights (connections to the previous layer), and we can organize these weights into a beautiful structure that makes computation not just possible but elegant.

Let's say we have a hidden layer with three neurons, each receiving two inputs (plus a bias). In our new notation, each row of our weight matrix represents all the weights for a single neuron:

$$\Theta = \begin{pmatrix} \theta_{11} & \theta_{12} \\ \theta_{21} & \theta_{22} \end{pmatrix}$$

Here, θ_{ij} represents the weight connecting the jth input to the ith neuron. Notice how each column captures everything about one neuron's connections. This arrangement is not only aesthetically pleasing but also computationally brilliant at the same time. When we want to calculate the outputs of all neurons in a layer, we can do a single matrix operation as

$$Z = \Theta X$$

where X is our input vector (including the bias term), and Z is a vector containing the pre-activation values for each neuron in the layer. We then apply our activation function (e.g., the sigmoid) to each element in Z to get our neuron outputs:

$$A = \sigma(Z)$$

This output is a vector, which is the combined output of all neurons in this layer. Instead of computing values for each neuron one at a time, which would take forever, you can direct them all simultaneously. This allows us to compute the outputs of all neurons in a layer in one swift mathematical motion.

This becomes especially important when we use hardware such as GPUs. Unlike CPUs, which are designed to do one thing at a time really well, GPUs excel at doing thousands of simple calculations simultaneously. They were originally built to render pixels on your screen in parallel, but as it turns out, they're perfect for neural network computations too.

When we organize our neural network using matrices, we're essentially preparing our calculations for this parallel processing power. Each row of calculations can happen independently and simultaneously on a GPU, dramatically speeding up training and inference times. What might take hours on a CPU can be reduced to minutes or even seconds. This matrix representation also makes the code cleaner and more intuitive. Instead of looping through each neuron and calculating its output separately, we can express the entire layer's computation in a single line of code. Modern deep learning frameworks such as TensorFlow and PyTorch, which form the backend for Keras, are built around these matrix operations, optimizing them even further behind the scenes.

The Next 10 or 100 Steps

So, we've successfully added the capability of handling nonlinear relationships between our inputs and output. Our little network with a hidden layer can now solve problems such as XOR that stumped our single-neuron approach. It's like we've taught our model to fold a piece of paper once to separate points that couldn't be divided by a straight line. But what if once isn't enough? What if we need to fold our paper at least four times before we can successfully separate the 0s from the 1s? The relationship between inputs and outputs in real-world problems can be incredibly complex, with twists and turns that a single transformation simply can't untangle.

Here's where the true elegance of neural networks materializes. If one hidden layer lets us perform one transformation, why not add another? And another after that? Each hidden layer in our network acts like a master origami artist, taking the paper (our data) and folding it in precisely the right way. The first layer might create a mountain fold, the second a valley fold, the third a squash fold, and so on. Each fold builds upon the previous ones, gradually reshaping our data space into a form where classification becomes possible.

What's remarkable is that each of these transformations follows the same simple template we've already discussed. Each neuron does the following:

1. Takes inputs from the previous layer
2. Weights the inputs according to their importance (i.e., assigns model parameter values)
3. Sums them up
4. Applies a nonlinear activation function
5. Passes the result forward

But when we stack these simple operations one after another, the combined effect is astonishingly powerful. It's like how a child can stack simple wooden blocks to create elaborate castles or how simple notes combine to form complex symphonies.

The first hidden layer might learn to detect basic patterns—edges, corners, or simple thresholds. The second layer combines these basic detections into more complex features, such as shapes or textures. The third layer might identify even more abstract concepts. With each layer, the network's understanding becomes more sophisticated and more nuanced. This is why we call it *deep learning*—depth refers to these multiple layers of transformation, each one bringing us closer to a space where our classification task becomes solvable. The complete architecture is referred to as a deep neural network. A deep neural network is a carefully choreographed sequence of folds that gradually reveals the inherent structure in our data.

Again, the best part is that we don't have to manually design any of these transformations. Just as with our single hidden layer, the network discovers them through the learning process, guided only by the final objective of minimizing our loss function. So, when faced with complex relationships in your data, remember that sometimes you need to go deeper. Adding more hidden layers isn't necessarily about making your network bigger; instead, it's about giving your network the power to discover and apply multiple successive transformations, each building upon the last, until the seemingly inseparable becomes beautifully, elegantly separable.

> **Which Is Better: More Neurons or More Layers?**
>
> Keep this important distinction in mind: More neurons in a layer can find more features at that level of complexity. More layers add levels of complexity. Sometimes, you'll need more width, sometimes more depth, and often a thoughtful combination of both. The best network architectures balance these two dimensions based on the specific problem at hand.

In the neural network world, we typically use the letter w instead of θ when referring to these connections. The word *weight* becomes our term of choice rather than model parameter, although they mean the same thing—they're the values our network adjusts during learning to improve its performance.

With multiple layers, we adopt a three-index notation: W_{ij}^h. This might look confusing, but don't let it deter you from understanding the underlying concept. Here's what each index means:

- h identifies which layer the weight belongs to.
- i points to the neuron in the current layer.
- j references the neuron in the previous layer that's sending its signal.

It's like a detailed address system: floor h, apartment i, receiving a message from neighbor j.

For each layer transition in our network, we have an entire matrix of weights. If we're going from layer 1 to layer 2, we have W^1; from layer 2 to layer 3, we have W^2; and so on.

Each of these matrices has a specific shape determined by how many neurons are in the connecting layers. If layer h has n neurons and layer $h+1$ has m neurons, then W^h will be an $m \times n$ matrix. Let's close this section by looking at the whole architecture in Figure 4.13. Notice that, in the output layer, we have a single neuron. This means we can only get one value out from our last logistic unit that we treat as a probability and thus get a single prediction of yes or no. What happens if we instead have more than one neuron in the output layer? We get something even more interesting that adds a ton of flexibility without making any major modifications to the semantics of what we've learned so far. We'll discuss this change in the next section.

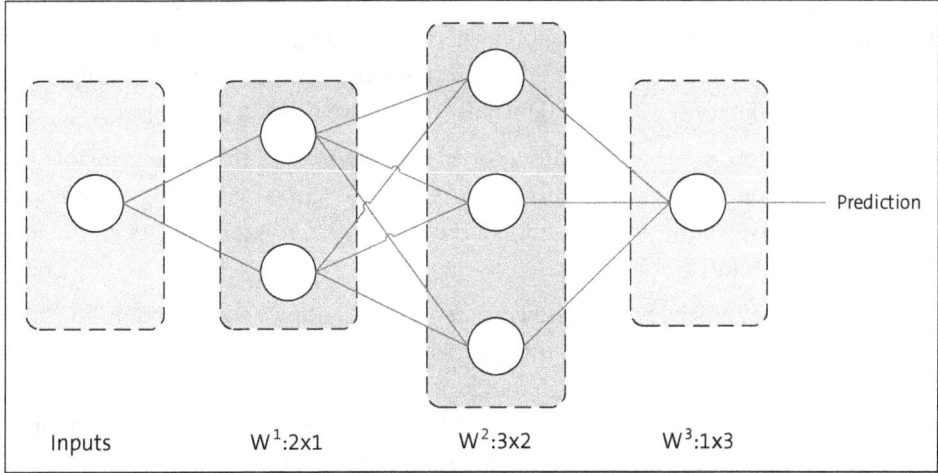

Figure 4.13 Deep Neural Network and Shapes of Parameter Matrices

4.3 Binary vs. Multi-Class Classification

Remember our first steps into neural networks? We started with a simple setup: a single neuron making binary classification. Like a light switch, it would output a number between 0 and 1—a probability that helped us make yes/no decisions. If the probability crossed 0.5, we'd say yes; otherwise, no. But real-world problems rarely fit into such neat binary boxes. Consider building a system to help a veterinary clinic identify different animals from photographs. You're not only asking, "Is this an animal?" You also need to know whether it's a cat, a dog, or a horse. It's like upgrading from that simple light switch to a sophisticated control panel, where each button serves a specific purpose. This is an example of the *multi-class classification* problem.

How do we tackle this more nuanced challenge? The solution lies in breaking down our complex question into a series of simpler ones, leveraging our understanding of binary decisions. This is like assembling a team of specialized experts, each trained to recognize one specific type of animal. We'll walk through this process in the following sections.

4.3.1 The One-vs-All Approach

We start with our cat specialist—a neuron focused entirely on identifying cats. When shown a cat picture, it outputs a high probability (close to 1) but stays quiet (close to 0) for any other animal or object. Through careful training, this neuron develops an expert eye for distinctly feline features, from whisker patterns to the unique way cats hold their bodies.

We only consult additional experts if needed. When presented with a new image, we first ask our cat specialist. If they respond with high confidence—say, "I'm 85% sure this is a cat!"—we've found our answer and stop there. But if they're uncertain, reporting something like "I'm only 30% confident this is a cat," we bring in our dog expert.

This second expert, another carefully trained neuron, follows the same principle but focuses on canine characteristics. It has learned to recognize everything from floppy ears to distinctive snout shapes, and like our cat expert, it expresses its confidence through a probability between 0 and 1.

Notice something crucial here: these confidence levels don't need to add up to 100%. Each expert is making their own independent assessment, just like real specialists might sometimes disagree or see different aspects of the same thing. If we show our system an ambiguous image—perhaps a stuffed toy with features of multiple animals— we might get high confidence scores from several experts. That's perfectly fine because each one is answering their own question, not trying to coordinate with the others.

This sequential decision-making process, called *one-vs-all* approach, mirrors how human experts might approach the problem: Start with the most likely category and only explore alternatives if needed. The approach offers the clear advantages of providing a transparent decision path and gracefully handling uncertainty. When none of our experts shows sufficient confidence, it might indicate something entirely new—perhaps an animal we haven't trained for, or maybe not an animal at all.

The Independent Nature of Expert Models

What makes this approach particularly interesting is the complete independence of our three specialists. Rather than sharing a single brain, each expert—our cat detector, dog detector, and horse detector—operates as a completely separate model, like different research teams working in different laboratories across the globe. This independence opens up unique possibilities for how we can develop and train these specialists.

Consider three different veterinary research centers, each specializing in a particular animal. The cat experts might be working with an extensive database of feline photographs collected from cat shows, veterinary clinics, and pet photographers. Meanwhile, across the country, dog specialists could be analyzing images from kennel clubs, working dog training programs, and even police K-9 units. And somewhere else entirely, equine researchers might be gathering data from horse races, breeding farms, and wilderness photography. Each team brings their unique perspective, expertise, and data sources to the table.

This independence extends beyond just the data. Each model can be developed using different computing resources, different teams, and even different technological approaches. The cat detection team might have access to powerful GPU clusters, allowing them to build an intricate neural network with dozens of layers. The dog specialists might prefer using cloud computing services, while the horse experts might have their own specialized hardware.

Speaking of intricate networks, let's peek inside what our cat expert might look like. Rather than being a simple neural network, it could be a sophisticated system with multiple layers of increasing complexity. The first few layers might act like digital photography experts, scanning for basic visual elements—edges, corners, and simple shapes. Think of them as artistic apprentices learning to sketch. Moving deeper into the network, we find layers that combine these basic observations into more complex features: one layer might specialize in identifying different types of cat ears, while another becomes an expert at recognizing various tail positions. Even deeper layers might learn to detect subtle patterns in fur textures or the unique way a cat's whiskers catch the light. At the highest levels, these layers work together to understand complex combinations of features—the specific curve of a sleeping cat, the characteristic pose of a cat ready to pounce, or the unique way a cat sits in a window.

At first glance, this independence seems like a stroke of brilliance. We can distribute the workload across different teams and resources. Different experts can focus on what they know best, using whatever tools and approaches work for their specific challenge. Training costs could be lower because we're running smaller, focused models instead of one massive system. It's like having specialized clinics instead of one enormous hospital.

The Hidden Cost of Independent Learning

There's a peculiar inefficiency lurking in this approach of using independent experts, and it becomes clear when we consider an art school analogy in which three talented students are each aspiring to become a specialist in drawing different animals. The first wants to master drawing cats, the second dogs, and the third horses. Now, picture these students being sent to three entirely separate art schools, each starting from scratch. Before they can even begin drawing their chosen animals, each student must independently learn the fundamentals: how to hold a pencil, how to sketch basic shapes, how to understand light and shadow. It's the same foundational knowledge being taught three times in three different places.

This redundancy mirrors exactly what's happening in our independent neural networks. Each model, in its early layers, is essentially learning the same basic visual vocabulary. Think about the features our animal detectors need to understand: eyes, noses, and fur textures. These aren't uniquely cat things, or dog things, or horse things—they're fundamental animal features that all our models need to grasp.

Let's break down what each model must learn step by step: First, every single model needs to understand edges and basic shapes. The cat detector spends computational power learning to identify curved lines and circles. Meanwhile, in another server somewhere, the dog detector is independently learning these exact same fundamental patterns. And yes, the horse detector is doing it too.

Then, each model moves on to combining these basic elements into features such as eyes. They're all separately discovering how edges can form circles, how circles can have darker centers, and how these patterns typically indicate an eye. It's like having three chefs in three different kitchens, all learning how to hold a knife correctly before they can even begin to specialize in their unique cuisines.

When our brain processes an image of an animal, it performs something interesting. First, it recognizes the universal features that tell us "this is an eye" or "this is a nose," and then it notices the subtle characteristics that tell us "this is specifically a cat's eye" or "this is distinctly a dog's nose." This natural hierarchy of visual recognition reveals an important truth about how our neural networks should approach animal identification.

The same applies to other features. Every nose has nostrils and a particular texture. Every set of fur has patterns of light and shadow. Every ear has curves and hollows that capture and channel sound. These universal features form the foundation of animal recognition, like the basic alphabet from which all words are built.

Once we understand these universal features, we start noticing the specialized characteristics that make each animal unique. A cat's eye has that distinctive vertical pupil that can dilate into an almost perfect circle in low light. The iris often has intricate patterns that you won't find in a dog's eye. The shape of the eye socket, the placement of the eyelids, even the pattern of fur around the eye—all of these elements combine to create what we recognize as uniquely feline.

Dog eyes tell a different story to a learning machine. They're typically rounder, with more visible whites, and they're set in the face at a different angle. Their pupils remain circular regardless of light conditions, and the surrounding facial muscles create those expressive eyebrows that help dogs communicate with humans. Some believe these are evolutionary adaptations that serve specific purposes. What makes this understanding crucial for our neural networks is that these features don't exist in isolation. The way a cat's eyes work with its nose, the relationship between a dog's ear placement and its snout, the proportions of a horse's facial features are the relationships creating the holistic patterns that make each animal instantly recognizable, even in silhouette.

This natural organization of features suggests something important about how our neural networks should be structured. Rather than having each network independently learn to recognize eyes, noses, and fur from scratch, wouldn't it make more sense to have a shared foundation that understands these universal features before branching off into specialized layers that learn the unique characteristics of each animal?

We've spent considerable time exploring this fundamental tension in modern machine learning because it's one of the most fascinating debates in our field: Should we build specialized models that excel at specific tasks or create vast, universal learners that attempt to understand everything at once? It's like choosing between training a team of expert specialists or developing a single Renaissance scholar who knows a bit of everything.

Let's take a look at the specialized approach first. Assume you're building a smart kitchen. You could have separate devices—one that's expert at identifying fresh vegetables, another that's perfect at measuring ingredients, and a third that monitors cooking temperatures. Each device would be relatively simple, quick to train, and excellent at its specific job. These specialized models offer several compelling advantages:

- **Faster training**
  With a focused dataset and a single task, these models can reach high-performance levels more quickly.
- **Lower resource requirements**
  Smaller models need less computational power and memory.
- **Easier debugging**
  When something goes wrong, it's simpler to identify and fix the issue.
- **Flexible deployment**
  You can update or replace individual components without touching the entire system.
- **Focused expertise**
  Like a master craftsperson, they can achieve extraordinary precision in their specific domain.

However, specialized models come with their own set of challenges:

- **Redundant learning**
  As we discussed earlier, they waste resources learning the same basic features repeatedly.
- **Limited context**
  They might miss important connections that come from understanding broader patterns.
- **Integration complexity**
  Getting multiple specialized models to work together smoothly can be challenging.
- **Maintenance overhead**
  Managing many different models can become a logistical headache.

Now, let's consider the alternative: the *universal learner* approach. Think of this as building a digital version of a human brain—one massive system that can process and understand virtually anything you throw at it. Modern large language models and vision transformers follow this philosophy. These comprehensive models offer intriguing benefits:

- **Shared understanding**
  The model learns fundamental patterns once and applies them across different tasks.

- **Cross-domain insights**
  It can discover unexpected connections between seemingly unrelated areas.
- **Efficient resource use**
  While the initial model is huge, it's often more efficient than running many smaller models.
- **Simplified deployment**
  There's just one model to maintain and update, rather than many.
- **Emergent capabilities**
  These models often develop surprising abilities that weren't explicitly trained for.

But this approach isn't without its drawbacks as well:

- **Enormous data requirements**
  These models need vast amounts of diverse data to learn effectively.
- **Computational intensity**
  Training requires significant computational resources and energy.
- **Complex training dynamics**
  Balancing performance across many tasks is challenging.
- **Black box nature**
  Understanding why the model makes specific decisions becomes more difficult.
- **Risk of interference**
  Learning new tasks might degrade performance on previously learned ones.

The reality is that both approaches have their place in modern machine learning. Small, specialized models might be perfect for *edge devices* or specific industrial applications where precision and efficiency are paramount. Meanwhile, large universal models might better serve general-purpose applications where flexibility and breadth of understanding are crucial.

As our field evolves, we're increasingly seeing hybrid approaches that try to capture the best of both worlds—models that start with a broad foundation of knowledge but can be efficiently specialized for specific tasks. It's like having a well-educated generalist who can quickly adapt to become an expert in particular areas when needed. However, what modifications do we need to make to our model so that it doesn't repeat learning but instead tries to learn the general features first in a shared manner? This question leads us to the second type of multi-class classification.

4.3.2 The Softmax Classifier

Before we dive into our improved approach to multi-class classification, we need to take a small but crucial detour to discuss something that might seem trivial at first: how we represent different categories with numbers. This might sound simple, but it reveals a challenge in how neural networks interpret the world.

Think about our animal classification problem. Because computers work with numbers rather than words, we can't simply tell our model to output cat, dog, or horse. Instead, we might be tempted to assign numbers to each category: let's say 1 for cats, 2 for dogs, and 3 for horses. Simple enough, right? Well, not really.

When we use numbers like this, we're accidentally introducing a hidden relationship that doesn't actually exist in our problem. Numbers, by their very nature, come with built-in *ordering*—2 is greater than 1, 3 is greater than 2, and so on. It's like accidentally arranging our animals on a hierarchy when we meant to put them in separate, equal categories.

Think back to Chapter 2, where we learned that neural networks discover patterns on their own, without us explicitly programming rules. This is where that characteristic becomes a potential pitfall. Our model might start treating these numerical labels as meaningful ordered values rather than arbitrary category markers. It might begin to think that dogs (category 2) are somehow greater than cats (category 1) or that horses (category 3) are superior to both—simply because we happened to assign those particular numbers. This isn't just a theoretical concern either. These numerical relationships could influence how our model learns and makes decisions. We could have just as easily labeled horses as 1, cats as 2, and dogs as 3—there's nothing inherently special about any of these assignments. They're as arbitrary as assigning house numbers on a street. So, let's fix them using a different kind of encoding.

One-Hot Vector Encoding

Instead of using simple numbers, we can employ a more sophisticated approach using vectors—think of them as digital name tags with multiple checkboxes. We'll create a vector with three positions (one for each animal), where each position can be either "off" (0) or "on" (1). It's like having a row of three light switches where only one switch can be on at a time.

For our cat category, we'll turn on just the first light: [1, 0, 0]. For dogs, we'll use the second position: [0, 1, 0]. And for horses, we'll illuminate the third: [0, 0, 1]. This elegant system is known as *one-hot encoding* because only one position is "hot" (set to 1) while all others remain "cold" (set to 0). This simple representation is therefore called *one-hot vector* encoding and forms a foundational concept in modern machine learning.

These vectors thus break free from the pitfalls of numerical ordering. Think about it: How do you compare [1, 0, 0] with [0, 1, 0]? Is one greater than the other? It's like trying to compare apples and oranges—they're simply different, not greater or lesser than each other. While these vectors do have mathematical properties we can measure, such as their magnitude (always 1 in this case because we're only using a single 1 and 0s), the vectors themselves resist any natural ordering. It's similar to how we might measure the height of three different trees, but understand that height alone doesn't make one tree "better" than another.

This representation carries another mathematically elegant property: Each label stands equally distant from the others in the mathematical space they occupy. Consider three points forming an equilateral triangle—each point is exactly the same distance from the other two. Our vectors work the same way. The distance between the cat vector [1, 0, 0] and the dog vector [0, 1, 0] is exactly the same as the distance between any other pair of animal vectors. This mathematical equality reflects the true nature of our categories—different but equal, with no inherent hierarchy or ordering. We can now go ahead and modify our original binary classification neural network to perform multiclass classification by modifying the final layer.

The Softmax Layer

Let's start with something we know works well: our cat-detecting neural network. Consider it a highly trained art critic who has spent years studying cats, learning every subtle detail from whisker patterns to tail positions. This network has developed an incredible eye for cat-like features through its many layers of processing—from basic shapes to complex cat-specific patterns. Now, instead of starting from scratch to detect dogs and horses, we're going to build upon this existing expertise. Think of it like expanding our art critic's repertoire. We keep all of those carefully trained layers that have become so good at understanding visual features, but we make one crucial modification at the very end of the network.

In the final layer, where we previously had a single neuron making the "cat or not" decision, we now place three neurons side by side. It's like having three specialists examining the same artwork simultaneously: our original cat expert plus new experts for dogs and horses. Each one will provide their confidence score about the image belonging to their respective category. When we show this enhanced network an image, something interesting happens. Each of our three output neurons makes its own prediction. Let's say we show it a picture that has some features of each animal. Our cat specialist examines the image and declares, "I'm 80% confident this is a cat" (outputting 0.8). Meanwhile, the dog expert pipes up with "I see some dog-like features here; I'd say there's a 60% chance it's a dog" (0.6). The horse specialist chimes in with less enthusiasm: "Only 30% chance this is a horse" (0.3).

At this point, you might notice something peculiar about these predictions. Our three experts seem a bit too enthusiastic. Their confidence levels add up to more than 100% (0.8 + 0.6 + 0.3 = 1.7). It's as if we asked three art critics about a painting's style, and one says it's mostly Impressionist, another says it's mainly Cubist, and the third sees significant elements of Abstract Expressionism. While each opinion might make sense in isolation, they can't all be right to these degrees simultaneously.

Now that we have our three predictions, we need to transform them into something that makes more logical sense. While our network has given us values between 0 and 1 (which is a good start), we want these values to behave like proper probabilities—in other words, they should add up to 1, just like the total probability of all possible outcomes should equal 100%.

Think of it like dividing up a pie chart. No matter how you slice it, all the pieces must add up to one complete pie. That's exactly what we want our predictions to do—each slice representing the probability of the image being a cat, dog, or horse, with all slices forming one complete picture.

This is where the *softmax function* comes to our rescue. It's like a mathematical recipe that takes our raw predictions and transforms them into proper probabilities. Here's how it works:

$$\text{softmax}(z_i) = \frac{e^{z_i}}{\sum_{j=1}^{K} e^{z_j}}$$

Don't let this formula intimidate you—it's actually doing something quite intuitive. For each prediction (z_i), we first make it positive and more extreme using the exponential function (e^{z_i}), and then divide it by the sum of all exponentially transformed predictions. It's like giving each prediction a fair share of the whole pie, while maintaining their relative strengths.

When we apply this to our earlier example, those somewhat confusing predictions (0.8, 0.6, and 0.3) get transformed into proper probabilities that sum to exactly 1. The probability of the input being a cat is now 0.41, being a dog is 0.34, and being a horse is 0.25. We can then simply apply the argmax on this to get the cat label. The beauty of this approach is that we've preserved the relative confidence of each prediction—the cat prediction is still the highest, followed by dog, then horse—but now they work together as a cohesive set of probabilities. The final layer of our network performing this softmax operation is called the *softmax layer*.

This elegantly sets us up for our next step: calculating the loss function, which we'll explore in the next section. But take a moment to appreciate how we've transformed our successful cat-detecting network into a multi-class classifier with just a few thoughtful modifications. We kept all the powerful feature-learning capabilities of our original network and simply added a sophisticated decision-making layer at the end.

4.4 Loss Functions: Cross-Entropy

Recall the construction we did for binary cross-entropy in Section 4.1.4. We used it to measure how well our network was doing at yes/no questions, like "Is this a cat?" It was like having a teacher who could only grade true/false questions, measuring the distance between our network's confident guesses and the actual truth. This worked beautifully for binary decisions, but now we need something more sophisticated for our multi-class predictions.

Think about how a teacher grades a multiple-choice test versus a true/false quiz. In a true/false quiz, being wrong is straightforward—you picked the wrong answer. But in a multiple-choice scenario, the nature of being wrong becomes more nuanced. You're right about one option and wrong about several others simultaneously. This is exactly the challenge we face when moving from binary to categorical cross-entropy, as we'll discuss in this section.

4.4.1 Categorical Cross-Entropy

Let's connect this to our recent discussion about softmax outputs. Remember how we transformed our network's raw predictions into proper probabilities that sum to 1? Now, we need a way to measure how good these predictions are. This is where categorical cross-entropy steps in, designed specifically to work with probability distributions across multiple classes.

Here's the mathematical recipe that makes this possible:

$$L = -\sum_{i=1}^{C} y_i \log(h_i(x))$$

Let's look at this formula piece by piece:

- y_i represents our true label in one-hot encoded form. Remember those vectors we talked about? If we're looking at a cat image, this would be [1, 0, 0].
- $h_i(x)$ represents our network's predictions about the ith data point after the softmax transformation, that is, those carefully balanced probabilities that sum to 1.
- The *log* function helps us capture the multiplicative nature of probabilities in an additive way. This is an important foundational point in machine learning. (We'll skip over this for the sake of brevity but if you're interested in the theoretical mathematics side of machine learning, it's highly recommended you explore this point in detail.)
- The minus sign at the front ensures our loss is positive (because the log of probabilities gives us negative numbers).
- We sum over all classes *(C)* to consider every possible category in our final score.

To see how this works in practice, let's walk through some examples. Let's examine how our loss function behaves in three different scenarios, starting with a nearly perfect prediction. When our network correctly identifies a cat image with high confidence, it might output something like [0.98, 0.01, 0.01], while our true label is [1, 0, 0]. Working through the loss calculation, we multiply each true label with the log of its corresponding prediction and sum them up:

$-1 \times log(0.98) - 0 \times log(0.01) - 0 \times log(0.01)$

Because our true label has 0s in the second and third positions, those terms vanish, leaving us with just $-1 \times log(0.98)$. This works out to approximately $-1 \times (-0.008) = 0.008$, a very small loss. The number is tiny because the $log(0.98)$ is very close to 0, reflecting how close our prediction was to perfection.

Now, consider a case where our network is on the right track but less confident. Given the same cat image, it might output [0.7, 0.2, 0.1]. Following the same calculation process with our true label [1, 0, 0], we compute:

$-1 \times log(0.7) - 0 \times log(0.2) - 0 \times log(0.1)$

Again, only the first term matters, giving us $-1 \times log(0.7) \approx -1 \times (-0.15) = 0.15$. This larger loss reflects our network's uncertainty: While it correctly identified the cat, its 70% confidence level leaves more room for improvement than our previous 98% confidence prediction.

Finally, let's look at what happens when our network gets it completely wrong, predicting [0.1, 0.8, 0.1] for our cat image. The calculation

$-1 \times log(0.1) - 0 \times log(0.8) - 0 \times log(0.1)$

gives us a much larger loss of approximately 1. This steep penalty makes perfect sense: Our network is 90% confident that the image isn't a cat (because it only assigns 0.1 probability to the cat category) when it's actually looking at one. This large loss sends a clear signal to our network that it needs to significantly adjust its parameters to avoid such confident mistakes in the future.

Just as a teacher can tell not only whether an answer is wrong but how wrong it is, categorical cross-entropy gives us a nuanced measure of our network's performance across all possible classes. It considers whether we got the right answer but also how confident we were in both our correct and incorrect predictions.

4.4.2 Pitfall to Avoid When Using Cross Entropies

One of the most common misunderstandings I encounter in applications of neural networks is about cross-entropy loss. Many newcomers to the field look at their network's architecture, see multiple neurons in the final layer, and immediately reach for categorical cross-entropy. But here's the truth: The number of output neurons doesn't determine which loss function you should use—it's the relationship between your predictions that matters.

Think about two different scenarios at a digital art gallery. In the first scenario, you're building a system to tag artwork with different attributes: "Contains Mountains," "Features Sunset," and "Includes Water." An image might show a beautiful mountain lake at sunset, earning all three tags. These attributes are independent of each other—having mountains doesn't prevent you from having a sunset. This is where binary cross-entropy shines, even with multiple output neurons.

Let's consider this in detail with a concrete example. Say you're analyzing a landscape photograph:

- True labels: [1, 1, 0] (yes mountains, yes sunset, no water)
- Network prediction: [0.9, 0.8, 0.1]

Each prediction stands on its own, independent of the others. The confidence about mountains (0.9) doesn't affect or compete with the confidence about sunset (0.8). We evaluate each prediction separately using binary cross-entropy because each one is essentially its own yes/no question. This is known as *multi-label classification*, and the loss used here is just binary cross-entropy.

Now consider a different scenario: classifying the primary art style of each piece—Impressionist, Cubist, or Abstract Expressionist. Categorical cross-entropy becomes essential because these categories are mutually exclusive. A painting can't be primarily Impressionist and primarily Cubist at the same time. When your network predicts [0.7, 0.2, 0.1] (70% confident it's Impressionist, 20% Cubist, 10% Abstract), these probabilities must sum to 1 because they represent different parts of the same whole—the artwork's primary style.

The key insight is that the choice between binary and categorical cross-entropy isn't about the architecture of your network—it's about the logical relationship between your predictions. Ask yourself: "Can multiple predictions be true simultaneously?" If yes, treat each prediction as an independent binary question. If your predictions are competing for a single truth—such as different art styles or animal categories—then categorical cross-entropy is your friend. In real-world applications, this distinction becomes crucial. A medical diagnosis system might use binary cross-entropy to detect multiple concurrent conditions (a patient can have both high blood pressure and diabetes), while an image recognition system using categorical cross-entropy would classify a car's make and model (a car can't be both a Toyota and a Honda simultaneously). When using categorical entropy though, you have an optional variant that is sometimes useful. Let's close our discussion by going through this last option.

4.4.3 Sparse Categorical Cross-Entropy

In categorical cross-entropy, we were able to apply softmax to get a cross-entropy loss that facilitates the option of giving a close-enough option. So, if you have probabilities of 0.5 and 0.4 for two options, the second option might also be useful to know.

Assume you're building a machine learning system to predict clothing sizes. If a customer is actually a size medium and your system predicts small, that's not entirely wrong—it's certainly better than predicting XXXL. Being "close" in this case carries valuable information. The same applies to many real-world predictions: age estimation, temperature forecasting, or color classification all benefit from this notion of approximate correctness.

But there are crucial situations where being close isn't good enough at all. Think about a bank's security system trying to verify your PIN—it doesn't matter if you're almost right. You either have the correct PIN or you don't. Sparse categorical cross-entropy comes into play here as a specialized tool for situations where we need exact matches and have many possible categories to choose from.

The term "sparse" might sound intimidating, but it's actually quite simple. Instead of using those one-hot encoded vectors we discussed earlier (e.g., [0, 1, 0, 0] for the second category), we just use a single number to indicate the correct category (e.g., "1" for the dogs). It's like the difference between filling out a multiple-choice quiz by marking every wrong answer with an X and the right answer with a checkmark, versus simply writing down the correct answer number. Both methods tell you the same thing, but one is much more efficient. This efficiency becomes critical when you're dealing with a large number of categories.

Assume you're building a system to recognize different species of birds—there might be thousands of possibilities. Using regular categorical cross-entropy, you'd need a vector with thousands of 0s and a single 1 for each training example. That's a lot of wasted memory! With sparse categorical cross-entropy, you just need one number per example. It's like the difference between carrying around a massive checklist of every bird species versus just writing down the name of the bird you saw.

But there's another crucial aspect to consider: When does being "close" matter? In some cases, it really doesn't. Take handwritten digit recognition—if someone writes an 8, predicting it's a 7 isn't partially correct just because 7 is numerically close to 8. They're entirely different symbols, and being off by one is just as wrong as being off by nine.

However, not all predictions are so black and white. Assume you're building a system to predict a person's age from their photo. If the true age is 24 and your system predicts 25, that's a pretty good guess! The closeness of the prediction carries meaningful information. The same applies to temperature forecasting—predicting 72°F when the actual temperature turns out to be 73°F is a far better prediction than guessing 90°F.

Color classification offers another example of when proximity matters. In the spectrum of colors, predicting a deep orange when the true color is red isn't as severe an error as predicting blue. These colors exist on a continuous spectrum, and their relationships to each other carry meaningful information.

This distinction between "exact match needed" and "close enough is fine" scenarios profoundly influences how we design and train our models. When using *sparse categorical cross-entropy*, we're telling our model that there's no middle ground—no partial credit for being close. Each prediction must stand or fall entirely on its own merit, without any consideration for how "close" it might have been to the right answer. When our network makes predictions using categorical cross-entropy, it's like having a panel of judges all voting simultaneously. Each judge (represented by values in our softmax vector) expresses their confidence level, and typically, we listen to the most confident judge—the highest value in our vector becomes our prediction.

Sometimes, we might want to consider the runner-up opinions, especially in creative tasks such as text generation. If you're writing a story and you always chose the most obvious next word, your writing would sound mechanical and predictable. Sometimes, you want to spice things up by picking a less obvious but still reasonable choice. It's like a chef who occasionally experiments with different spices rather than always reaching for the most common ones.

Let's say our network is predicting the next word in the sentence: "The cat sat on the _ __." Our softmax output might look like this: [mat: 0.5, rug: 0.3, chair: 0.15, table: 0.05]. While "mat" is the highest probability choice, we might occasionally choose "rug" or "chair" to add variety to our text. This flexibility doesn't mean we're ranking these words in any absolute sense—"rug" isn't inherently "better" than "chair" just because it has a higher probability. Each probability simply represents how well that word fits in this specific context. A summary of the three types of cross-entropy losses is given in Figure 4.14.

It's very important to understand that when we select from our softmax outputs, we're not introducing any ordering between our classes. When our network outputs [0.7, 0.2, 0.1] for our animal classifier, we're not saying dogs (0.2) are better than horses (0.1), we're just expressing the network's confidence for this particular image.

This approach is useful because it maintains the independence of our classes while giving us the flexibility to make nuanced decisions. Each prediction stands on its own merits in its specific context, free from any artificial hierarchy we might accidentally introduce through numerical labels. Let's go ahead and revisit our earlier example from Chapter 2 to see how we've already used these concepts in practice.

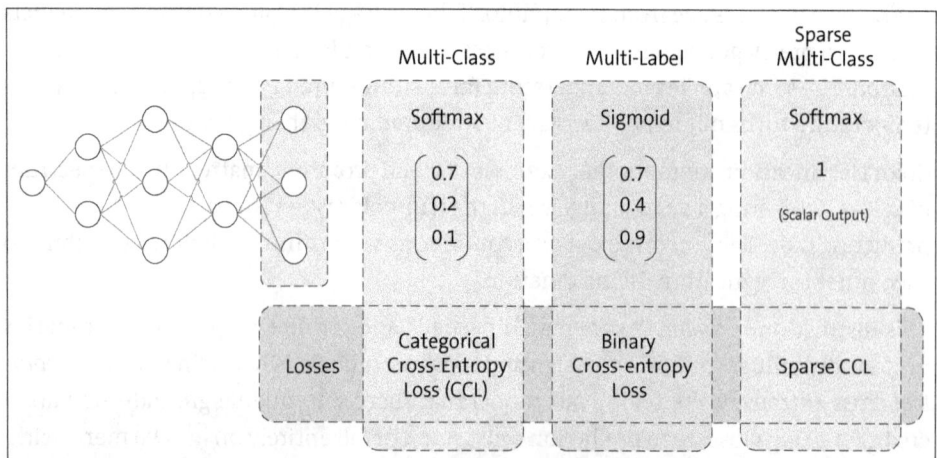

Figure 4.14 Binary, Categorical, and Sparse Categorical Cross-Entropy Losses

4.5 Building a Classifier with Gradient Descent

Recall the MNIST handwritten dataset example we introduced in the beginning of Chapter 2 and then coded using Keras without fully dissecting it. Here, we'll visit some of the more important parts of that code again and add a lot more explanation to it based on our much more in-depth understanding of neural networks and machine learning. We created the Keras model using the code given in Listing 4.1. You can see the complete code listing for this section on *https://recluze.net/keras-book* in the **04-01-classification-model** notebook as well as on the book's official web page at *http://rheinwerk-computing.com/6142*.

```
# Define the model
model = keras.Sequential([
    Dense(128, activation='relu', input_shape=(64,)),
    Dense(10, activation='softmax')  # 10 classes (digits 0-9)
])
```

Listing 4.1 Layers in the Keras Model

There are a few things that we can now clarify. We're using what's called a *Sequential* model, which is exactly what it sounds like—a sequence of layers where information flows in a straight line. Think of it as an assembly line in a factory, where each station (layer) performs its specific task before passing the product along to the next station.

This sequential nature is very powerful because it enforces a strict forward flow of information. When we feed input into our network, it travels through each layer in order, getting transformed along the way until it emerges as our final prediction. There's no skipping ahead or turning back—each piece of data must follow the same path through our neural assembly line.

Now, you might wonder about the term *feed-forward neural network* that's often used to describe these models. Don't let it confuse you with the concept of backpropagation that we use during training. Think of it this way: During normal operation, our network is like a waterfall—data can only flow downward through the layers. However, during training, we do something clever. After the data has flowed forward and made a prediction, we calculate how wrong that prediction was, and then we use this information to adjust our network's parameters. It's like having quality control inspectors at the end of our assembly line who send feedback to each station about how to improve their work. The model's prediction always remains sequential. We'll walk through this process in this section.

4.5.1 Layers in Code

Continuing on, both layers in our model are *Dense* layers, which means every neuron in one layer is connected to every neuron in the next layer—it's like having a fully connected social network where everyone knows everyone else. Our first Dense layer takes 64 inputs (specified by `input_shape=(64,)`) and transforms them into 128 outputs. Each of these 128 neurons is connected to all 64 input values, allowing the layer to learn complex patterns in our data. These dense connections are powerful because they allow our network to learn intricate relationships between all of our input features. Imagine a group of detectives working together, where each detective can see all the evidence (inputs) and contribute their insights to help solve the case. Every neuron gets to consider all the information available from the previous layer before making its contribution to the next layer's computation.

Our second dense layer has exactly 10 neurons, each one specializing in recognizing a specific digit from 0 to 9. Think of these neurons as a panel of expert judges, where each judge specializes in one particular digit. When an image comes through our network, each judge examines it and voices their confidence level: "How sure am I that this is the digit I specialize in?"

Finally, the softmax activation function steps in, acting like a diplomatic moderator for our panel of judges. The raw outputs from our 10 neurons might be all over the place—some high, some low, some negative. Softmax transforms these raw scores into a more useful format: probabilities that add up to 1.

For instance, when shown a clearly written "7", our network might produce outputs that softmax converts into something like the following:

- Digit 0: 0.01 (1% confidence)
- Digit 1: 0.02 (2% confidence)
- ...
- Digit 7: 0.85 (85% confidence)
- ...
- Digit 9: 0.03 (3% confidence)

Here's where our implementation gets clever with memory efficiency. Instead of storing our ground truth labels as full one-hot encoded vectors (i.e., storing 10 numbers for each image, with nine 0s and one 1), we're using a sparse representation. Our `digits.target` has a shape of `(1797,)`, meaning for each image, we're just storing a single number—the correct digit. It's like having an answer key that simply lists 7 instead of [0,0,0,0,0,0,0,1,0,0]. If you're following along in code, make sure you output the shape of `digits.target` (our output variable) and some of its values to ensure you understand what is being fed to the model. We've included the output in Listing 4.2. The shape of the variable is one-dimensional. That means it only holds 1,797 scalar values. To fully drive the point home, we output some values (from 100th value to the 109th value) and see that the values are indeed scalar and not one-hot vector encodings. This enables us to use sparse categorical cross-entropy because the output produced by the model must match with the output expected; that is, the shape and semantics of the output of the model and the ground truth must match.

```
print("Shape: ", digits.target.shape)
print(digits.target[100:110])
------
Output:
Shape: (1797,)
[4 0 5 3 6 9 6 1 7 5]
```

Listing 4.2 Studying the Shapes and Values of the Output Variables

4.5.2 Choice of Loss Function

In other words, this sparse representation is why we use sparse categorical crossentropy for our loss function. It's designed to work directly with these simple numerical labels rather than requiring the full one-hot encoded vectors. Think of it as our grading system being smart enough to understand "the answer is 7" without needing the longform version of "it's not 0, not 1, not 2 ... it's 7 ... not 8, not 9."

This efficiency not only saves memory but also makes our entire training process more streamlined. Our network can still learn just as effectively, but we're being economical with our resources, like a teacher who can grade tests just as accurately with a simple answer key as with a detailed rubric.

Notice the mysterious `relu` parameter in our first Dense layer? While we'll dive deep into the *Rectified Linear Unit (ReLU)* later in the book, for now, you can think of it as playing the same role as our familiar sigmoid function—it's the neuron's way of deciding whether and how strongly to fire based on its inputs. All activation functions are like each neuron's decision-making mechanism, determining how to respond to incoming signals. Just as we used sigmoid to transform our neuron's output into a nice, bounded range between 0 and 1, ReLU serves a similar purpose, albeit with its own unique characteristics.

For our current understanding, you can mentally substitute `sigmoid` wherever you see `relu` in the code, and your mental model will still hold true. Both functions serve as the "should I fire?" decision-maker for each neuron. They are like light switches—while they might work slightly differently (one dims smoothly, another clicks on and off), they both control how the light responds to input. We're using ReLU here because it has some nice properties that make our network learn better and faster, but those are details we'll explore in a future discussion. For now, what matters is understanding that this activation parameter is controlling how each neuron processes and passes along information, just like sigmoid did in our earlier examples. Next, we need to understand the number of parameters each layer has at least in this simple situation so that we can establish an intuitive sense behind the models.

4.5.3 Parameter Counts

As we discussed earlier, the number of model parameter (or *weights* in neural network parlance) depends on the number of neurons in each layer. Our input layer acts as the gateway to our network, accepting 64 individual values. Think of these as 64 separate entry points, each ready to receive a piece of information. This isn't a layer with trainable parameters—it's just our data's entry point into the network.

The real parameter counting begins with our first Dense layer. With 128 neurons, each one connected to all 64 inputs, we're creating a dense web of connections. Let's consider the math details:

- Each connection needs a weight parameter: 64×128 = 8,192 weights.
- Each neuron also needs a bias term: 128 additional parameters.
- The total for the first layer is 8,192 + 128 = 8,320 parameters.

Our second layer follows the same pattern, but now connecting 128 neurons to 10 output neurons:

- Connection weights: 128×10 = 1,280 parameters
- Bias terms for each output: 10 additional parameters
- Total for second layer: 1,280 + 10 = 1,290 parameters

Rather than doing these calculations manually each time, Keras provides a convenient summary() function that breaks down these numbers for us. Looking at Listing 4.3, we can see that our calculations match perfectly:

- First Dense layer: 8,320 parameters
- Second Dense layer: 1,290 parameters

```
model.summary()
---- Output (modified format for readability in print)
Model: "sequential"
Layer (type)        Output Shape         Param #
-----------         ------------         -----------
dense (Dense)       (None, 128)          8,320
dense_1 (Dense)     (None, 10)           1,290
-----------         ------------         -----------
Total params: 28,832 (112.63 KB)
Trainable params: 9,610 (37.54 KB)
Non-trainable params: 0 (0.00 B)
Optimizer params: 19,222 (75.09 KB)
```

Listing 4.3 Model's Summary Output by Keras

The summary also provides additional useful information. Notice how it shows the output shape (None, 128) for the first layer and (None, 10) for the second? The None here represents the batch size, which can be flexible, while the second number shows how many values each layer produces. What's particularly interesting in the summary is the breakdown of total parameters into different categories. While our hand calculations covered the trainable parameters (9,610), the summary shows us the full picture, including optimizer parameters. This comprehensive view helps us understand not just the architecture of our network but also the computational resources it requires.

We've just walked through a seemingly simple piece of Keras code, but like a masterfully crafted puzzle box, each line contains multiple compartments of complexity waiting to be understood. Keras code is remarkably dense—every parameter and choice of component carries significant weight. These few lines of code represent dozens of crucial decisions, each one requiring a deep understanding of the underlying mathematical and computational principles.

4.5.4 The Density of Keras Code

This density of decisions is precisely why you can't approach Keras like a cookbook, blindly following recipes and hoping they'll work. Making changes without understanding the foundations is like trying to modify a recipe without knowing how ingredients interact with each other—you might accidentally turn your soufflé into a pancake!

But now that we've built a solid foundation, let's turn this into an opportunity for exploration. Here are some experiments you can try, each one designed to deepen your understanding of how these components work together:

- **Cross-entropy type exploration**
  Think back to our discussion about binary versus categorical cross-entropy. Try switching between them and observing what happens. Does the network behave as you'd expect? If you see the accuracy plummet, can you explain why based on what we learned about these loss functions?

- **Experimenting with architecture changes**
  Add another Dense layer to your Sequential model. It's as simple as adding another line, but the implications are profound. Try different sizes—maybe add a layer of 64 neurons between our existing layers, as shown in Listing 4.4. Watch how the network's behavior changes. Is more always better?

```
model = keras.Sequential([
    Dense(128, activation='relu', input_shape=(64,)),
    Dense(64, activation='relu'),   # New layer!
    Dense(10, activation='softmax')
])
```

Listing 4.4 Adding a New Layer to the Keras Model

- **Experimenting with activation functions**
  Remember our discussion about ReLU? Try changing those activation functions. Swap `relu` for `sigmoid`, and observe the differences. Each activation function has its own personality. How does it affect your network's learning?
- **Effects of softmax**
  Here's a particularly interesting experiment: Replace the softmax activation in the final layer with sigmoid. Instead of getting nicely normalized probabilities that sum to 1, you'll get independent 0–1 predictions for each class. Think about why this might be problematic for classification tasks.

As you experiment, keep track of your network's *accuracy*, that is, how many predictions it gets right. We'll define accuracy more formally soon, but for now, think of it as your network's report card. Watch how each change affects this score, but more importantly, try to understand why each change has the effect it does.

Remember, the goal isn't just to try different combinations until something works. It's to build an intuition for how these components interact. This deep understanding is what will eventually allow you to architect neural networks that solve real-world problems effectively.

4.6 Summary

In this chapter, we've evolved from simple linear regression to the rich landscape of multi-class classification, discovering how neural networks can make sophisticated distinctions between multiple categories. We learned how to represent different classes using vectors rather than simple numbers, preventing accidental hierarchies in our data. The softmax function emerged as our mathematical lens, transforming raw neural outputs into meaningful probability distributions, while our exploration of cross-entropy loss—both binary and categorical—revealed the crucial distinction between independent events and mutually exclusive categories.

These theoretical concepts come to life in our Keras code, where each line represents crucial decisions that demand deep understanding. Your mastery of neural networks requires grasping these fundamental principles. Before moving forward, ensure you've internalized these core concepts: the independence of vector representations, the role of softmax in creating probability distributions, and the vital distinction between different types of cross-entropy loss. They're the foundation upon which all advanced neural network architectures are built.

In the next chapter, we'll dive deeper into Keras to fill in all the blanks that we've left until this point. The syntax and design decisions will be explained in full, giving you a better appreciation of how Keras can help you solve the problems discussed until now as well as lay the foundation for more complex models studied later.

Chapter 5
Deep Dive into Keras

In this chapter, we'll finally dive deep into Keras. After examining the core technologies that power Keras, we'll construct a practical image classifier that distinguishes between cats and dogs. Through this hands-on project, we'll implement each essential component of the machine learning workflow: data preparation, model design, training, evaluation with multiple metrics, and model preservation. By the chapter's end, you'll have mastered the fundamental patterns that form the foundation for all your future deep-learning projects.

So far, we've been working with neural networks from the ground up, painstakingly crafting gradient descent algorithms, designing loss functions, and manually calculating derivatives. It's been like learning to build a car by first understanding how each gear, piston, and spark plug works together. You get such a deep understanding of the underlying system that you can find bugs and improve your models at a much faster pace. But now, with that foundational knowledge firmly in place, it's time to accelerate your learning and productivity. Enter Keras—the framework that transforms how we build and train neural networks.

We've deliberately waited until this point to introduce Keras. To truly appreciate the elegance and power of this framework, you needed to first understand the complex mathematical machinery working beneath the surface. You now possess the insight to recognize what Keras is doing behind its clean, intuitive interface. Think of the previous chapters as learning to build a car by hand, understanding how each component functions, how the transmission transfers power, and how the electrical system operates. Now, you're ready to step into a modern automotive factory with specialized tools and assembly lines that make you vastly more efficient without sacrificing your understanding of automotive engineering.

In this chapter, we're going to explore Keras from multiple angles: First, we'll introduce the Keras framework itself—a little bit of its history, philosophy, and why it has become such a central tool in the machine learning ecosystem. You'll learn how Keras emerged during a critical transition period in deep learning and how its design principles have stood the test of time even as underlying technologies have evolved dramatically.

Next, we'll roll up our sleeves and set up Keras on your local machine. We won't just stop there—we'll return to the Google Colab environment we established earlier and fine-tune its configuration to create an optimal development environment. This dual approach ensures you can work effectively regardless of your preferred setup.

Then comes the exciting part—building your first model with Keras. But this won't be a superficial tour of application programming interface (API) calls any more. We'll peel back the layers to examine what's happening underneath, exploring the fundamental concepts that power modern machine learning frameworks. You'll discover the magic of symbolic computation that allows automatic differentiation to work so seamlessly. We'll reconnect with our old friend NumPy to understand how Keras interfaces with these essential numerical libraries. Rather than treating Keras as a mysterious black box, you'll gain crystal-clear insight into its inner workings.

As we progress, we'll tackle more sophisticated examples that implement the gradient descent and classification techniques you've already mastered. However, we'll now address the practical challenges that arise in real-world scenarios: How do you monitor a model during training? How do you save your work and resume it later? How do you extract meaningful information as your model learns? Through callbacks, model serialization, and other essential utilities, you'll develop a professional workflow that scales to complex projects.

What sets this chapter apart from many Keras tutorials is our commitment to understanding not just how to use Keras, but why it works the way it does. By the end of this chapter, you'll grasp the design philosophies that shaped Keras development and the careful trade-offs its creators made to balance simplicity with power. This deeper understanding will transform how you write your own code. You'll learn to work with Keras rather than around it, leveraging its strengths while avoiding common pitfalls. When you encounter problems—and all developers do—you'll have the knowledge to diagnose issues efficiently instead of resorting to trial-and-error fixes.

The journey ahead is both practical and illuminating. You'll gain powerful tools that accelerate your productivity while deepening your understanding of the principles that make deep learning work. Let's begin our exploration of Keras—the framework that will become your most trusted tool throughout the rest of this book and beyond.

5.1 Introduction to Keras Framework

Imagine having a universal translator that lets you express your ideas in English yet be perfectly understood by speakers of Swiss, Japanese, or Urdu. In the world of deep learning, Keras serves exactly this purpose—it's the translator that allows you to write your model once and have it run anywhere, on any major framework. When deep learning began gaining momentum, early practitioners faced a daunting challenge. Frameworks such as *Theano* offered tremendous power but demanded technical expertise that distracted from the actual goal: building intelligent systems. Working with these tools felt like trying to write poetry while simultaneously having to manufacture your own pen and paper.

Keras emerged as an elegant solution to this problem. Rather than replacing powerful frameworks such as Theano, it wrapped around them, providing a consistent, intuitive interface that shielded users from unnecessary technical complexities. You could focus on the architecture of your ideas—how many layers to use, what activation functions to apply—without getting lost in implementation details.

Next, we'll discuss the core design decisions behind Keras and how they led to the creation of this modern powerhouse of a framework.

5.1.1 The Philosophy Behind Keras: Making AI Human-Friendly

At its heart, Keras emerged from a profound belief that machine learning should be accessible to everyone, not just specialists. François Chollet, the creator of Keras, envisioned a framework that democratized deep learning by removing technical barriers. This philosophy manifests in what Chollet calls "progressive disclosure of complexity." Think about how you learn to use a new smartphone—you can make calls immediately, but then gradually discover advanced features as you need them. Keras works the same way. Simple tasks require just a few lines of code, while advanced capabilities remain available without overwhelming beginners.

The guiding principles of Keras—simplicity, modularity, and extensibility—aren't just technical design choices but reflections of a human-centered approach to technology. By prioritizing developer experience, Keras acknowledges that even the most sophisticated algorithms are useless if humans struggle to implement them effectively.

5.1.2 Evolution Through Adaptability

Keras's design prioritized adaptability from day one. Initially working primarily with Theano as its computational backend, Keras was architected to potentially work with any backend. This foresight proved invaluable as the landscape shifted. When Google's TensorFlow arrived on the scene, Keras demonstrated its greatest strength: users could switch their models to this new, powerful backend without rewriting their code—the same models, the same API, but now powered by different engines under the hood. I was teaching machine learning to a class of undergraduates at that time, and the students not only were able to understand TensorFlow in a one-hour lesson but were also able to port their Keras expertise to this new, powerful framework very easily.

The relationship between Keras and TensorFlow grew so synergistic that Keras was eventually integrated into TensorFlow as `tf.keras`, becoming its official high-level API. Yet Keras maintained its identity and philosophy of framework independence. The core

5 Deep Dive into Keras

strength of Keras is its modular nature. Its components work really well together but are loosely coupled. They also play very nicely with the ecosystem of supporting libraries and underlying engines. Look at Figure 5.1 for a brief overview of the components involved in a Keras setup.

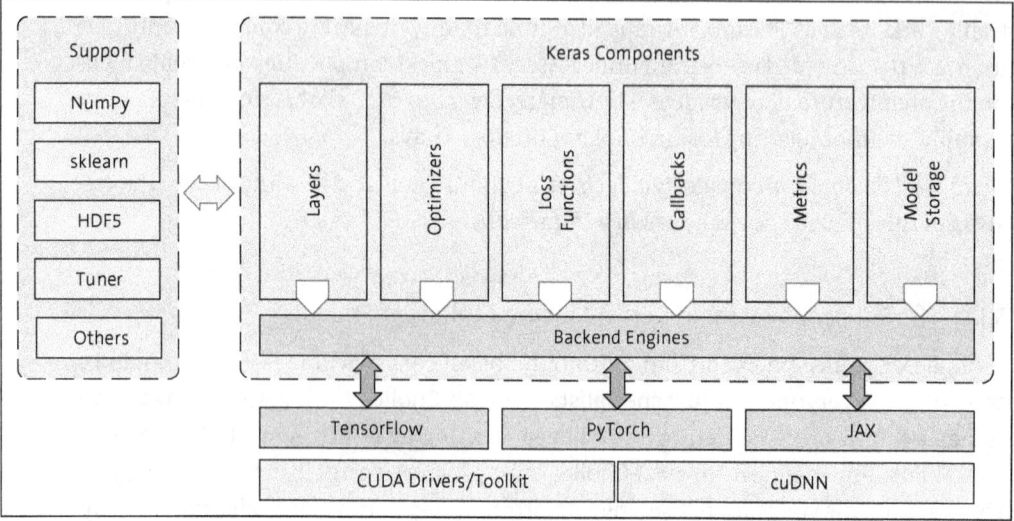

Figure 5.1 Keras Components and Ecosystem

This figure illustrates the Keras deep learning ecosystem, showcasing how various components work together to create a comprehensive framework for building and deploying neural networks.

At the center, we see the core **Keras Components**, which form the heart of any deep learning project and will all be discussed in this chapter:

- **Layers** provide the fundamental building blocks where data transformations occur.
- **Optimizers** control how the model learns from its mistakes during training.
- **Loss Functions** measure how far the model's predictions are from the truth.
- **Callbacks** execute important actions at specific points in the training process.
- **Metrics** help us evaluate model performance beyond just the loss value.
- **Model Storage** preserves our trained models for future use.

All of these components connect to **Backend Engines**, which serve as the computational foundation. Keras can wrap around **TensorFlow**, **PyTorch**, or **JAX** to perform the actual mathematical operations, giving users flexibility in choosing their preferred framework. Below these engines, we see hardware acceleration technologies such as **CUDA Drivers/Toolkit** and **cuDNN**, which enable neural networks to run efficiently on specialized hardware.

The **Support** column shows important auxiliary technologies that extend Keras capabilities. **NumPy** provides essential array operations, **scikit-learn** offers complementary machine learning tools, **HDF5** enables efficient storage and retrieval of model data, and **Tuner** from Keras automates the search for optimal model configurations. We'll get to some of these components in the following sections and leave a detailed discussion about the rest for future chapters in this book.

This ecosystem design allows Keras to serve as both a beginner-friendly entry point and a professional-grade framework for deep learning, with each component handling a specific responsibility while communicating seamlessly with the others. This communication becomes especially useful when we consider the multi-engine support provided by Keras.

5.1.3 Keras 3.0: The Multi-Engine Framework

With version 3.0, Keras took a giant leap toward its original vision of complete backend neutrality. Now, Keras can seamlessly work with three major deep learning frameworks:

- **TensorFlow**
  Google's comprehensive machine learning platform that excels in production deployment and distributed training. It offers a complete ecosystem that includes serving infrastructure (TensorFlow Serving), visualization tools (TensorBoard), and mobile deployment options (TensorFlow Lite). TensorFlow's static computational graph approach provides optimization benefits for fixed-architecture models, and its distributed training capabilities make it suitable for large-scale industrial applications. With features such as eager execution introduced in TensorFlow 2.0, it combines immediate execution for debugging with graph-based optimization for performance.

- **PyTorch**
  Facebook's framework known for its dynamic computational graph and intuitive, Pythonic design. PyTorch has gained enormous popularity in research settings for its flexibility, transparent debugging experience, and natural integration with Python control flow. Its *define-by-run* approach means computational graphs are built on the fly during execution, making complex model architectures with conditional operations more intuitive to implement. PyTorch's strong GPU acceleration, comprehensive libraries such as torchvision and torchaudio, and seamless integration with Python data science tools have made it the dominant framework in many research domains, particularly in natural language processing and computer vision.

- **JAX**
  A newer entrant from Google Research that combines *autograd* (automatic gradient) and Accelerated Linear Algebra (XLA) for high-performance numerical computing. JAX is particularly strong for research applications requiring function transformations and *just-in-time compilation*. Its functional design allows for powerful operations such as

automatic vectorization (vmap), automatic differentiation and parallelization (pmap) to be composed together. This makes JAX exceptionally well-suited for implementing advanced algorithms such as differentiable ray tracers, scientific simulations, and reinforcement learning systems. While less mature in its ecosystem than TensorFlow or PyTorch, JAX offers compelling performance advantages for certain mathematical workloads.

This multi-backend support means you can write your model once in Keras and run it on whichever framework best suits your current needs—without changing a single line of your model code.

Perhaps the most compelling aspect of Keras is how it future-proofs your skills and code. The deep learning landscape evolves rapidly, with new frameworks and optimizations appearing regularly. By writing code in Keras with best practices, you create models that ride these waves of innovation rather than being washed away by them. For example, if your organization initially standardized on TensorFlow but later needs PyTorch's research-friendly features for a specific project, your Keras models can transition without a complete rewrite. You can even switch frameworks through configuration changes, preserving your investment in model development.

5.1.4 Key Strengths of Keras

Keras has earned its reputation as the go-to high-level API for deep learning by offering several distinct advantages:

- **Intuitive API design**
  Keras follows the principle of progressive disclosure of complexity. Simple tasks remain simple, requiring minimal code, while complex architectures remain possible without artificial constraints. This human-centered design makes deep learning more accessible to beginners while still serving the needs of experts.

- **Consistency across backends**
  Whether running on TensorFlow, PyTorch, or JAX, Keras provides a consistent experience. This uniformity means you develop transferable skills rather than framework-specific knowledge that might become obsolete.

- **Rapid prototyping**
  The clean, modular design of Keras enables quick experimentation. You can assemble models like building blocks, swap components with minimal friction, and iterate rapidly on your ideas, significantly speeding up the research and development cycle. You've already seen examples of these simple pieces of code in previous chapters.

- **Production readiness**
  Despite its simplicity, Keras doesn't sacrifice performance or scalability. Models built with Keras can be deployed in production environments with the full performance benefits of the underlying backend.

- **Extensive ecosystem**
  Keras comes with pretrained models, datasets, and layers that accelerate development. The wide adoption of Keras has created a rich ecosystem of tutorials, examples, and community support that makes solving problems easier.
- **Modularity and extensibility**
  If you need functionality beyond what's included, Keras allows you to create custom layers, losses, and metrics. This extensibility means you're never constrained by the built-in components.
- **The community: Keras' hidden strength**
  When you adopt Keras, you're joining a global community of practitioners. This vibrant ecosystem of developers, researchers, and educators has become one of Keras's most valuable assets. The Keras community functions like a collaborative knowledge network, where solutions propagate rapidly. When someone solves a challenging implementation problem in Tokyo, their solution becomes available to a student in Santiago almost immediately through forums, repositories, and documentation. This collaborative spirit has led to community-driven extensions, specialized implementations for niche problems, and an abundance of learning resources that range from beginner tutorials to cutting-edge research implementations. The "wisdom of the crowd" effect means that common bugs get rapid fixes, documentation gaps get filled, and best practices emerge naturally through thousands of interactions. The Keras blog has some of the greatest documentation of latest models available for free.

  Perhaps most notably, the community's emphasis on accessibility aligns perfectly with Keras' own values. Experienced practitioners take time to mentor newcomers, complex concepts get translated into understandable examples, and the culture remains refreshingly welcoming compared to some more technical communities.

5.1.5 Keras in the Real World

The true test of any framework is what people build with it, and Keras powers an impressive array of real-world applications across industries:

- In healthcare, medical researchers at Stanford used Keras to develop deep learning models that can identify skin cancer with accuracy rivaling dermatologists. The framework's flexibility allowed them to experiment rapidly with different architectures while its clean API made collaboration between medical experts and AI specialists more seamless.
- Netflix uses Keras to power parts of its recommendation engine, processing massive datasets to understand viewer preferences. The ability to customize and extend Keras components has been crucial for adapting models to their specific needs while maintaining performance at scale.

- Conservation biologists deployed Keras models on edge devices in rainforests to monitor endangered species through audio recognition. Keras' ability to work with TensorFlow Lite made it possible to run sophisticated neural networks on low-power devices in remote locations.
- Automotive manufacturers employ Keras in developing advanced driver assistance systems, where its transparent design makes it easier to audit and validate models—a critical requirement in safety-critical applications.

These diverse applications share a common thread: Keras removed barriers between domain expertise and AI capabilities. Whether the practitioners were doctors, conservation scientists, or artists, Keras allowed them to translate their specialized knowledge into working AI systems without becoming deep learning experts first. You're already familiar with the deep learning framework, so you are in a much better position to harness the power of Karas—either by running it online or setting it up on your machine as we'll discuss next.

5.2 Setting Up Keras

A properly configured deep learning environment is your professional kitchen, letting you cook up amazing AI models without fighting your tools. Deep learning is computationally hungry. Without the right setup, your models might take days instead of hours to train, or worse, crash mysteriously after running for hours. The time you invest in proper configuration now will save you countless hours of frustration later.

For our Keras 3.0 installation, we'll need several key ingredients:

- **Python**
  This is our foundation language.
- **Backend framework**
  The frameworks include TensorFlow, PyTorch, or JAX (we'll focus on TensorFlow as it's the most beginner-friendly).
- **CUDA Deep Neural Network Library (cuDNN)**
  This is the special sauce that unlocks your GPU's power.
- **Keras**
  Our high-level interface makes building models intuitive.
- **Supporting libraries**
  Essential tools are available such as NumPy and matplotlib.

Before we dive in, let's gather some information about your system:

- What operating system are you running?
- What is your CPU model?
- Do you have a compatible GPU? (NVIDIA cards work best.)

- Do you have at least 10 GB of free disk space for different datasets?
- Do you have administrator access?

Having these answers ready will make your workflow much smoother. Let's begin building your AI laboratory! If you don't have a GPU, it's still okay to proceed. The models will run slower but you're learning right now, so it's fine. If you don't have the space, that might be a larger problem. In either of these two cases, you can continue to use Colab, which we discussed in Chapter 2. At the end of this section, we'll help you optimize that environment as well. So, let's begin the first step—Python.

> **Frustrations in Local Setup of Keras**
>
> If you find yourself tangled in installation wires and configuration knots, don't let frustration take over—Colab offers a ready-to-use environment where Keras is already configured and waiting for you. You can always circle back to setting up your local environment once you're comfortable with the ecosystem and have built a few models, allowing you to approach the installation process with more confidence and practical understanding. In this case, feel free to jump over to Section 5.2.5.

5.2.1 Setting Up Python

Keras 3.0 needs Python 3.9 or newer to work its magic, but Python 3.10 hits the sweet spot. While Python 3.11 and newer versions offer speed improvements, they sometimes clash with deep learning libraries. Think of Python 3.10 as the reliable foundation that everything else can safely build upon. Because this is a printed book, supported Python versions might change. It's recommended that you visit the official Keras page to find out which versions are recommended and just replace the version numbers in the commands discussed here. In any case, you have two major options for setting up Python, which we'll cover in the following sections. I prefer the second method, but both work fine.

Option 1: Direct Python Installation

Installing Python directly gives you precise control over your setup—perfect for those who like to understand every piece of their system. In the follow sections, we'll cover installation for Windows, MacOS, and Linux.

Windows Installation

For Windows, the process is straightforward:

1. Visit *www.python.org*, download the Python 3.10 installer, and execute it.
2. During installation, select the **Add Python to PATH** checkbox. This seemingly small checkbox saves enormous headaches later.

3. Select the **Disable path length limit** option, which is another small setting with big benefits.
4. After installation, open Command Prompt, and type the command to check the Python version, as shown in Listing 5.1.

```
python --version
pip --version
```

Listing 5.1 Checking the Python Version

If you see **Python 3.10.x** (or whatever the recommended version at the time is), you're on the right track! Also consider installing Visual C++ Build Tools, as some packages need them for compilation.

MacOS Installation

If you're using a Mac, you have two excellent options: Homebrew and the official installer from Python. Using Homebrew (the package manager that makes life easier on the Mac), you can just issue the following command:

```
brew install python@3.10
```

Alternatively, you can download the official installer from *www.python.org*. After installation, verify that Python is ready using the command given previously in Listing 5.1. Notice we use python here—macOS often has Python 2 installed as python, so be specific by using python3 if you get an incorrect version with the general command!

Linux Installation

Linux users, your system probably already has Python, but maybe not the version we need. Update your repository and install Python as suggested in Listing 5.2.

```
# For Ubuntu or Debian
sudo apt update
sudo apt install python3.10 python3.10-dev python3.10-venv python3-pip
# For Fedora, RHEL, or CentOS
sudo dnf install python3.10 python3.10-devel
```

Listing 5.2 Installing Python on Linux

Confirm your installation as given earlier in Listing 5.1.

Managing Multiple Python Versions

Sometimes, you need different Python versions for different projects. When you're working with different tools and codebases, you might need to work with different Python versions. While an in-depth discussion of this issue is beyond the scope of this book, here are some tools that discuss these issues and provide solutions (we recommend reading through their documentation to familiarize yourself):

- **pyenv**
  Lets you switch Python versions with simple commands.
- **Python Launcher (Windows)**
  Runs specific versions with `py -3.10`.
- **Virtual environments**
  Creates isolated spaces for each project.

Option 2: Using Anaconda for Python

My personal favorite, Anaconda gets you a fully equipped laboratory instead of building one tool at a time. It comes with Python and many scientific packages pre-installed. *Conda* environments give you superpowers:

- **Project isolation**
  Keep your projects from interfering with each other.
- **Easy sharing**
  One command creates the same environment on another computer.
- **Binary packages**
  It's often faster to install than compiling from source.
- **Simple management**
  Create, activate, and switch environments with clear commands.
- **Works everywhere**
  The same commands work on Windows, macOS, and Linux.

After installing Anaconda from *www.anaconda.com*, open your terminal (Linux or Mac) or Anaconda prompt (Windows), and create a specialized environment using the command:

```
conda create -n keras-env python=3.10
```

Then, activate your new environment using the following:

```
conda activate keras-env
```

Now you're working in a clean, controlled space perfect for deep learning experiments.

Verify that your Python installation is working correctly, as shown earlier in Listing 5.1. With Python working correctly, we're ready to add TensorFlow, the powerful engine that will drive our Keras models.

5.2.2 TensorFlow Installation and Points to Keep in Mind

Now that Python is properly set up, let's bring TensorFlow into our environment. TensorFlow will serve as the computational engine that powers our Keras models, handling all the complex matrix operations and gradient calculations behind the scenes.

Understanding TensorFlow Versions and Compatibility

TensorFlow releases follow a version numbering system that helps us understand compatibility. Here's what you need to know: TensorFlow 2.x represents a major shift from the original TensorFlow 1.x, with *eager execution* by default and tighter Keras integration. For our Keras 3.0 projects, we'll use TensorFlow 2.x (preferably 2.10 or newer). Version compatibility matters tremendously in the deep learning ecosystem. When selecting your TensorFlow version, consider these relationships:

- **Python compatibility**
  TensorFlow 2.10+ works with Python 3.7–3.10.
- **CUDA compatibility**
  Each TensorFlow version works with specific CUDA versions, which can be seen on the official TensorFlow installation pages.
- **cuDNN compatibility**
  Both TensorFlow and CUDA versions dictate which cuDNN version you need.

The safest approach? Install the latest stable TensorFlow version that's compatible with your Python version. This generally provides the best performance and access to newer features.

CPU vs. GPU Considerations

Before installing TensorFlow, you'll need to decide between CPU-only and GPU-accelerated versions. Let's talk about the differences. CPU-only TensorFlow works on any computer without special hardware requirements. It's significantly easier to install because you won't need to worry about CUDA or cuDNN configurations. This version is perfectly sufficient for learning TensorFlow fundamentals and experimenting with smaller models. The main drawback is performance: Training will be much slower, especially as your models grow in size and complexity.

GPU-accelerated TensorFlow requires an *NVIDIA GPU*, specifically from the GeForce GTX/RTX, Tesla, or Quadro series. You'll need to install additional CUDA and cuDNN libraries to enable GPU support. The extra setup complexity pays off tremendously in performance, offering 10–50 times faster training times for most models. This acceleration becomes essential when working with larger models and datasets, where training on CPU alone might take days instead of hours.

> **Practical Recommendation**
>
> If you have a compatible NVIDIA GPU with at least 4 GB of memory, go with the GPU version. The speed difference is substantial and will save you countless hours when training real models. If you're just getting started or working on a machine without a compatible GPU, the CPU version will still let you learn and experiment.

> If you don't have such a GPU, go through the setup and run a small model on your local machine, but then return to Colab to save time. Section 5.2.5 shows you how to do this. This way, you learn the details of installation but still don't waste countless hours waiting for your models to train.

Installation Methods

The recommended method is to install TensorFlow with pip. This is straightforward. Open your terminal or command prompt, and issue this command:

```
pip install tensorflow
```

Starting with TensorFlow 2.10, the same package supports both CPU and GPU. TensorFlow will automatically detect and use your NVIDIA GPU if you have the correct CUDA libraries installed (we'll cover that in the next section). If you want to specify the version, you can append it to the package name as

```
pip install tensorflow==2.18.0
```

If you're using Anaconda, Conda offers excellent package management. First, make sure you've activated your environment using

```
conda activate keras-env
```

and then issue the command

```
conda install tensorflow
```

A huge benefit of using Conda is that it can often handle CUDA and cuDNN installation automatically, eliminating a significant source of frustration.

Testing Your TensorFlow Installation

After installation, it's crucial to verify that everything works correctly. Let's run a quick test in Python. Create a new file test.py and paste the code given in Listing 5.3. Run the file using python test.py. (See the full listing for testing TensorFlow installations on the book resources page at *https://recluze.net/keras-book* in the **05-01-tensorflow-check** notebook and on the book's official web page at *http://rheinwerk-computing.com/6142*.)

```
import tensorflow as tf
# Check the version
print("TensorFlow version:", tf.__version__)

# Check for GPU support
print("GPU devices:", tf.config.list_physical_devices('GPU'))
```

```
# Run a simple operation to verify functionality
x = tf.constant([[1., 2.]])
y = tf.constant([[3.], [4.]])
z = tf.matmul(x, y)
print("Result of matrix multiplication:", z.numpy())
```

Listing 5.3 Checking TensorFlow Installation

If your installation is working correctly, you'll see the TensorFlow version, a list of available GPUs (empty list if you're on CPU-only), and the result [[11.]] from the matrix multiplication.

> **Common Installation Issues**
>
> You might encounter a few common issues during testing. An `ImportError` typically indicates TensorFlow wasn't installed correctly. If no GPU appears in the device list despite having compatible hardware, your CUDA installation likely needs attention. A version mismatch means you should verify you've installed the specific TensorFlow version you intended to use.

Now, your TensorFlow installation is ready! This powerful framework will handle the complex computations needed for your deep learning models, while Keras will provide a friendly interface for building and training them. Let's move on to enable CUDA support for our GPU.

5.2.3 Setting Up CUDA for GPU Acceleration

Deep learning models involve complex mathematical operations performed on large datasets. When training neural networks, your system needs to compute millions or billions of matrix multiplications and other operations. CPUs are designed for general-purpose computing and process tasks sequentially, which creates a significant bottleneck for these intensive calculations.

GPUs, on the other hand, excel at parallel processing. They contain thousands of smaller cores that can handle multiple computations simultaneously. This parallel architecture makes GPUs dramatically faster than CPUs for deep learning tasks, often providing speed improvements of 10-50x depending on the model and dataset. The acceleration provided by GPUs translates to numerous benefits. GPU acceleration enables faster model training times, reducing days of computation to hours or even minutes. It provides the ability to work with larger and more complex models that would be impractical on CPUs alone. Researchers and developers can perform more iterations for hyperparameter tuning and experimentation, greatly improving model quality. Additionally, GPU acceleration allows for quicker deployment and inference in production environments, enhancing real-world applications.

Let's go through the details of what you need to enable GPU acceleration for your Keras projects in the following sections.

Checking GPU Compatibility

Before proceeding with CUDA installation, verify that your GPU is compatible:

1. Identify your NVIDIA GPU model using the command `nvidia-smi`, or on Windows, check **Device Manager** under **Display adapters**.
2. Visit the NVIDIA CUDA-enabled GPU list (*https://developer.nvidia.com/cuda-gpus*) to confirm compatibility.
3. For deep learning, ensure your GPU has at least 4 GB VRAM (8 GB or more is recommended for larger models).
4. Check memory bandwidth and CUDA compute capability, as these factors affect performance.

> **AMD GPUs**
>
> *AMD GPUs* use a different framework called *ROCm*, not CUDA. In this chapter, we focus exclusively on NVIDIA GPUs as they are by far the most well-supported cards. Working with AMD GPUs isn't recommended, at least for beginners.

The CUDA Ecosystem Components

Understanding how each component fits within the GPU acceleration ecosystem helps troubleshoot potential issues:

- **NVIDIA drivers**

  The foundation of the stack, enabling communication between your operating system and the GPU hardware. These provide the basic functionality for all NVIDIA GPU operations.

- **CUDA Toolkit**

  A development environment that includes the *CUDA Runtime*, which is the interface between your applications and the GPU; *NVIDIA CUDA Compiler* (NVCC), which translates CUDA code to binary code executables on GPUs; CUDA Libraries, which are optimized functions for common operations (cuBLAS for linear algebra, cuDNN for deep learning, etc.); and development tools and debugging utilities.

- **cuDNN (CUDA Deep Neural Network Library)**
  A specialized library built on top of CUDA that provides highly optimized implementations of common deep learning operations such as convolutions, pooling, normalization, and activation functions. It's specifically designed to accelerate deep learning frameworks.
- **Deep learning frameworks (TensorFlow, PyTorch, etc.)**
  These high-level tools use cuDNN and CUDA under the hood to accelerate operations on NVIDIA GPUs.

Each layer depends on the layer below it, creating a stack where compatibility between versions is critical. So, let's go ahead and discuss all of them one by one.

Installing the NVIDIA Drivers

The NVIDIA drivers are the foundation for GPU acceleration. We'll discuss installation for both Linux and Windows next.

For Linux

If you're on Linux, determine the latest driver version compatible with your GPU using the command `ubuntu-drivers devices`, and then install the recommended driver using

```
sudo apt install nvidia-driver-XXX
```

Replace XXX with the recommended version number. If you're using a RedHat/Fedora flavor, you can use `dnf` instead. Reboot the system, and verify the installation using the command: `nvidia-smi`.

For Windows

Follow these steps for setting up the NVIDIA drivers for windows:

1. Visit the NVIDIA Driver Download page (*www.nvidia.com/Download/index.aspx*).
2. Select your GPU model and operating system.
3. Download and run the installer.
4. Follow the installation wizard, and restart your computer.
5. Verify installation by right-clicking on your desktop and selecting **NVIDIA Control Panel**.

Of course, modern Macs don't support NVIDIA GPUs, so you can't use this on the Mac. Your code will still run, it just won't be able to take advantage of the GPUs.

Installing the CUDA Toolkit

First of all, you must make sure to select the right version for your TensorFlow because it requires specific CUDA versions, so check compatibility (as listed in Table 5.1) before installation.

5.2 Setting Up Keras

TensorFlow Version	CUDA Version	cuDNN Version
TensorFlow 2.12.0	CUDA 11.8	cuDNN 8.6
TensorFlow 2.11.0	CUDA 11.2	cuDNN 8.1
TensorFlow 2.10.0	CUDA 11.2	cuDNN 8.1

Table 5.1 TensorFlow/CUDA Compatibility

This is the correct mapping as of the writing of this book. You should always verify the latest compatibility information in the TensorFlow documentation. We'll discuss installation for both Linux and Windows next.

For Linux

Download the CUDA Toolkit installer from NVIDIA's website by searching for "nvidia developer cuda toolkit installer linux". An example link is *https://developer.download.nvidia.com/compute/cuda/11.8.0/local_installers/cuda_11.8.0_520.61.05_linux.run*. Download the installer, make it executable, and run it using the following command:

```
chmod +x cuda_11.8.0_520.61.05_linux.run && sudo ./cuda_11.8.0_520.61.05_linux.run
```

Then, follow the installation prompts.

For Windows

Download the CUDA Toolkit installer from NVIDIA's website, run the installer, and follow the installation guide. Select **Custom installation**, and ensure that the driver option is unchecked if you've already installed drivers.

Installing cuDNN

cuDNN is NVIDIA's deep learning acceleration library that optimizes common operations in neural networks. You need to first register for an NVIDIA Developer account (required to download cuDNN). Then, visit the cuDNN download page (*https://developer.nvidia.com/cudnn*), and select the version compatible with your CUDA installation. Finally, based on your operating system, follow the instructions in the next sections for installation.

For Linux

Setting up cuDNN in Linux is also straightforward. Follow these steps to have everything working on a Linux machine:

1. Download the cuDNN tarball.
2. Extract it using `tar -xzvf cudnn-XX-linux-x64-v8.X.X.XX.tgz`.
3. Copy the files to your CUDA directory as given in Listing 5.4.

5 Deep Dive into Keras

```
sudo cp cuda/include/cudnn*.h /usr/local/cuda/include
sudo cp cuda/lib64/libcudnn* /usr/local/cuda/lib64
sudo chmod a+r /usr/local/cuda/include/cudnn*.h /usr/local/cuda/lib64/
libcudnn*
```

Listing 5.4 Copying the cuDNN Files to the Correct Locations

For Windows

In Windows, all we need to do is to copy the contents of the zip file to the correct location as follows:

1. Extract the cuDNN zip file.
2. Copy the contents of the *cuda* folder to your CUDA installation directory (typically, *C:\Program Files\NVIDIA GPU Computing Toolkit\CUDA\vxx.x*).

Verifying GPU Detection

After installation, verify that your deep learning framework can detect and use the GPU using the code given in Listing 5.5.

```
import tensorflow as tf
import time

# Create some random data
x = tf.random.normal([5000, 5000])

# Test on CPU
with tf.device('/CPU:0'):
    start = time.time()
    tf.matmul(x, x)
    end = time.time()
    print(f"CPU Time: {end-start:.4f} seconds")

# Test on GPU
if tf.config.list_physical_devices('GPU'):
    with tf.device('/GPU:0'):
        start = time.time()
        tf.matmul(x, x)
        end = time.time()
        print(f"GPU Time: {end-start:.4f} seconds")
```

Listing 5.5 Verifying GPU Detection

The GPU calculation should be significantly faster if acceleration is working correctly. This code creates a large 5,000×5,000 matrix multiplication task, which is computationally intensive but perfect for parallel processing. The time difference between CPU and GPU execution provides clear evidence of GPU acceleration.

5.2.4 Installing Keras

Finally, we can go ahead and install Keras itself. To install the standalone Keras package, you'll use pip, Python's package manager that we installed earlier and just issue this command:

```
pip install keras
```

This command fetches the latest stable version of Keras and installs it in your Python environment. It's clean, simple, and gives you the freedom to configure it as you wish. If you're working in a Jupyter Notebook, you can also install Keras with the magic command:

```
!pip install keras
```

The standard installation doesn't automatically install the backends, giving you the freedom to choose which ones to add to your toolkit.

> **Standalone Keras or TensorFlow Subpackage?**
> Keras was historically included as part of the TensorFlow package, providing an integrated experience for TensorFlow users. However, with Keras 3.0+, the official recommendation has shifted toward installing Keras as an independent package. This standalone installation enables Keras to function as a truly multi-backend framework, supporting not only TensorFlow but also PyTorch and JAX.

To truly unlock the cross-backend potential of Keras 3.0, you might want to install multiple backends. This is like being multilingual—your Keras code can now speak to different systems. While installing multiple backends isn't important when you're just getting started and it's recommended that you just stick to TensorFlow when you're just exploring Keras, we include the complete commands to install the other backends in Listing 5.6.

```
# Install PyTorch backend
pip install torch torchvision

# Install JAX backend
pip install jax jaxlib
```

Listing 5.6 Installing Multiple Backends for Keras

Having multiple backends installed gives you the freedom to experiment and switch between them based on your project's needs or to compare performance across different platforms. Once you have Keras and your chosen backends installed, you can specify which one to use as your default translator. This configuration is like setting your preferred language in a translation app. Keras 3.0 makes this straightforward with a simple API, as shown in Listing 5.7.

```
import keras
keras.config.backend = "tensorflow"   # Set TensorFlow as backend
# Alternatives: "pytorch", "jax"
```

Listing 5.7 Setting Up the Default Keras Backend Framework

This code tells Keras which computational engine to use behind the scenes. You can change this setting at any point in your script to switch backends, although it's best to set it early before creating any models.

Whew, that was a lot of technical setup to digest! Remember, you don't need to configure Keras on your local machine right away. If you find yourself getting stuck with driver versions or encountering cryptic error messages, take a step back. Colab offers a fantastic alternative where Keras, TensorFlow, and GPU support are already configured and ready to use. This gives you a chance to focus on learning the concepts and building models without wrestling with installation hurdles. Once you've gotten comfortable with how the ecosystem works and have built a few models, you can return to setting up your local environment with a better understanding of what you're trying to accomplish. Many experienced practitioners started their journey this way, using cloud notebooks before creating their own optimized setup.

5.2.5 Using a GPU in Google Collaboratory

As you know, installing Keras locally isn't the only option for working with deep learning. Colab provides a convenient alternative, especially if you don't have a high-performance GPU in your personal computer. Colab gives you access to powerful GPUs that significantly accelerate model training without any hardware investment on your part. You don't have to worry about complex installation procedures or keeping your software up-to-date, as Colab environments come preconfigured with the latest versions of all necessary libraries. The paid version of Colab does have time limitations; currently, it's six hours per session before needing to restart. However, this isn't a major obstacle because you'll learn how to save your model weights and resume training across multiple sessions.

To enable the use of a GPU in Colab:

1. Create a new Colab notebook or open up an existing one.
2. Click on **Runtime** in the menu bar, and then select **Change runtime type** from the dropdown menu, as shown in Figure 5.2.
3. In the popup dialog shown in Figure 5.3, select a GPU from the **Hardware accelerator** options.
4. Click **Save** to apply the changes.

You might get a warning that your progress will be lost. That is normal as Colab will set up a new environment for you from scratch when you change the runtime.

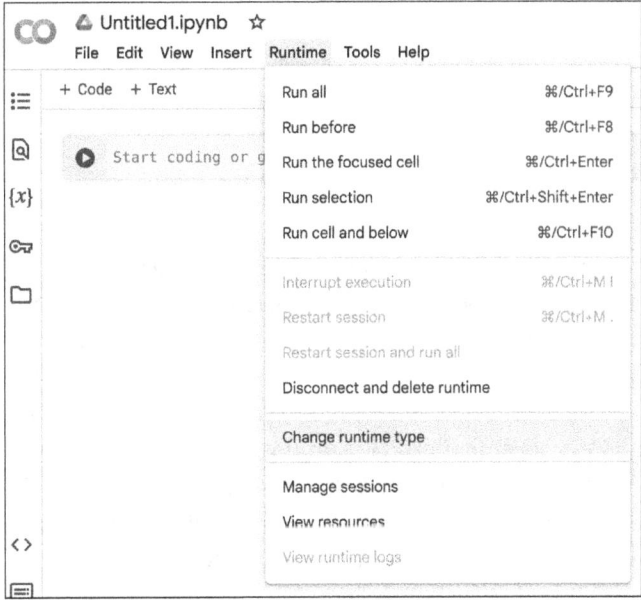

Figure 5.2 Changing the Runtime Type in Colab

Currently, Colab provides access to T4 GPUs, though the specific hardware may vary based on availability. You'll also see an option for *Tensor Processing Units* (*TPUs*) in the **Hardware Accelerator** section. While these specialized chips can offer exceptional performance, they're not recommended for beginners as they have compatibility issues with some code. Consider exploring TPUs after gaining more experience with deep learning.

With GPU acceleration enabled, you're free to choose either your local installation or Colab for the code discussed in this chapter—both will work equally well.

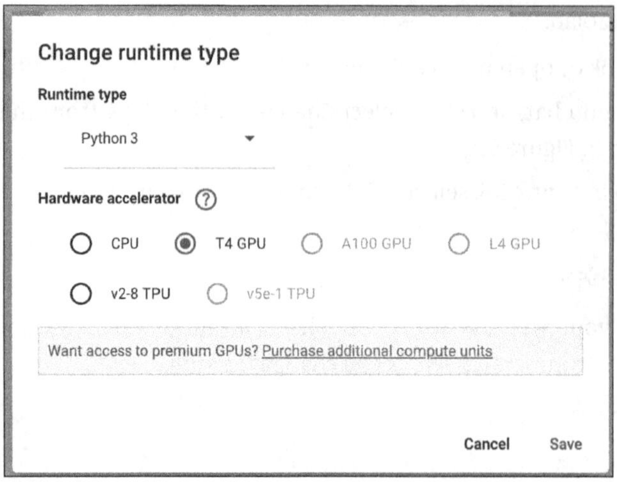

Figure 5.3 Setting Colab to Use a GPU

5.3 Building Your First Model

Let's now dive into the nitty-gritty details of Keras and machine learning code. For this, we need to understand the engine that powers modern machine learning: NumPy. Think of NumPy as the foundation upon which nearly all Python-based machine learning is built. Machine learning fundamentally revolves around *data manipulation*, mathematical operations, and optimization—lots of it. When you're training a neural network, you're handling millions of parameters, performing countless matrix multiplications, and computing gradients across massive datasets. Trying to do this with Python's built-in data structures would be like trying to transport a city's worth of people using only bicycles—theoretically possible, but impossibly slow.

5.3.1 Why NumPy Matters for Machine Learning

NumPy solves this problem by providing blazingly fast numerical operations that are computed in optimized C code rather than Python. It gives us the data structures and functions needed to work efficiently with the mathematical backbone of machine learning. As you build your first models, you'll see how these NumPy operations translate directly into the building blocks of neural networks. In the following sections, we'll discuss those aspects of NumPy that are most relevant to a proper understanding and to the code of machine learning and Keras.

Arrays and Matrices as Fundamental Data Structures

Let's start with the basics. In machine learning, data is king, and NumPy arrays are how we represent that data efficiently. These arrays store and operate on homogeneous data, making them perfect for representing the vectors and matrices that form the mathematical backbone of neural networks. See Listing 5.8 for examples of how to represent vectors and matrices through NumPy.

```
import numpy as np

# Creating a simple array (vector)
weights = np.array([0.5, 0.3, 0.2])
print(weights)   # [0.5 0.3 0.2]

# Creating a 2D array (matrix)
input_data = np.array([[1.0, 2.5, 3.1],
                       [0.8, 1.2, 2.7],
                       [3.2, 0.4, 1.8]])
print(input_data)
# Output
# [[1.  2.5 3.1]
#  [0.8 1.2 2.7]
#  [3.2 0.4 1.8]]
```

Listing 5.8 Data Representation in NumPy

Consider how a single neural network layer processes information: It takes a set of inputs (a vector), multiplies them by weights (another vector or matrix), and produces an output (yet another vector). Without efficient array operations, even a simple neural network would grind to a halt.

> **Practice with NumPy**
>
> You really should run these code listings even if they appear straightforward. For your ease, we've made these short listing available as cells in the **05-02-numpy** notebook on the book resources page at *https://recluze.net/keras-book* as well as on the book's web page at *http://rheinwerk-computing.com/6142*.

Broadcasting for Efficient Operations

One of the most powerful yet initially perplexing features of NumPy is *broadcasting*. In traditional programming, if you want to add a number to each element in an array, you'd write a loop. In NumPy, you simply add the number to the array, as shown in Listing 5.9, even if the shapes aren't correct in the strict mathematical sense.

```
# A matrix of input features for 3 examples with 4 features each
inputs = np.array([[2.0, 1.5, 3.1, 0.8],
                   [1.2, 3.2, 0.5, 2.8],
                   [0.9, 1.1, 4.2, 1.9]])
```

```python
# Adding a bias term to each input example
bias = np.array([0.5, -0.3, 0.1, 0.7])

# Broadcasting automatically aligns and adds bias to each row
adjusted_inputs = inputs + bias
print(adjusted_inputs)
# [[2.5 1.2 3.2 1.5]
#  [1.7 2.9 0.6 3.5]
#  [1.4 0.8 4.3 2.6]]
```

Listing 5.9 Broadcasting Basics

NumPy follows specific rules when broadcasting arrays together:

1. If the arrays don't have the same *rank* (number of dimensions), prepend the shape of the lower-rank array with 1s until both have the same length.
2. If the shape of the two arrays doesn't match in any dimension, expand the array with shape 1 in that dimension to match the other shape.
3. If the sizes disagree in any dimension, and none are 1, an error is raised.

For example, when adding our `bias` (shape (4,)) to our `inputs` (shape (3, 4)), NumPy does the following:

1. Expands `bias` to shape (1, 4) (rule 1).
2. Repeats the bias row three times to reach shape (3, 4) (rule 2).
3. Performs the addition with identical shapes.

This is very significant for neural networks. Broadcasting makes it possible to perform operations between arrays of different shapes, which happens constantly in neural network operations. For example, when adding a bias vector to the output of a matrix multiplication, broadcasting allows the bias to be correctly applied to each result without tedious reshaping or loop constructions. Without broadcasting, we would face significant issues both in code complexity and computational efficiency. See Listing 5.10 as an example of what we would need to do if broadcasting wasn't there to help us out.

```python
# Without broadcasting, here's what we'd need to do to add
# a bias vector
# Assume output_activations has shape
# (batch_size, num_neurons) = (32, 10) and bias has shape
# (num_neurons,) = (10,)

batch_size = 32
num_neurons = 10
output_activations = np.random.rand(batch_size, num_neurons)
bias = np.random.rand(num_neurons)
```

```python
# Approach 1: Explicit looping (very slow in Python)
result_loop = np.zeros_like(output_activations)
for i in range(batch_size):
    for j in range(num_neurons):
        result_loop[i, j] = output_activations[i, j] + bias[j]

# Approach 2: Manual reshaping (better but verbose)
result_reshape = output_activations + 
                        bias.reshape(1, num_neurons)

# Approach 3: NumPy broadcasting (concise and fast)
result_broadcast = output_activations + bias
```

Listing 5.10 Manual, Slow Replacement for Broadcasting and Vectorization

The explicit looping approach has a time complexity of *O(batch_size × num_neurons)* and becomes prohibitively slow for large networks. Even the reshaping approach requires extra memory allocations and cognitive overhead. In a neural network with millions of parameters processing thousands of examples, these inefficiencies would compound dramatically, making training impossibly slow. Broadcasting eliminates these issues by handling the shape alignment automatically and using optimized, vectorized operations under the hood.

Vectorization: The Secret to Performance

In machine learning, the difference between waiting minutes versus days for results often comes down to one concept: *vectorization*. If you want to compute the squared error across thousands of predictions, you could write a loop, but that would be painfully slow. Let's demonstrate this using code given in Listing 5.11.

```python
# The slow, looping way to compute error
def compute_error_slow(predictions, actual):
    error = 0
    for i in range(len(predictions)):
        error += (predictions[i] - actual[i]) ** 2
    return error

# The vectorized, lightning-fast way
def compute_error_fast(predictions, actual):
    return np.sum((predictions - actual) ** 2)

# With large datasets, the speed difference can be 100x or more
```

Listing 5.11 Vectorized vs Non-Vectorized Code

Why exactly is the looping approach so slow while vectorization is so fast? The reasons depend a lot on how modern computing hardware works. When the code is run on a CPU, the following occurs:

- Python loops have significant overhead for each iteration, as the interpreter must execute each step individually.
- Vectorized operations are implemented in highly optimized C or Fortran code that minimizes this overhead.
- CPU hardware includes Single Instruction, Multiple Data (SIMD) capabilities that can perform the same operation on multiple data points simultaneously.
- Vectorized NumPy functions automatically leverage these SIMD instructions
- Memory access is also optimized in vectorized code, following continuous memory patterns that maximize CPU cache efficiency

When code optimized for a GPU is run, the following happens:

- The performance gap becomes even more dramatic.
- GPUs are designed specifically for parallel processing, with thousands of cores meant to perform the same operation across different data points.
- A loop processes one element at a time, leaving most of the GPU's processing power idle.
- Vectorized operations can distribute work across all available cores, utilizing the GPU's full potential.
- GPUs can achieve 10–100 times speedups over CPUs for certain vectorized operations.
- Deep learning frameworks translate vectorized NumPy-like operations into optimized GPU instructions.

The reason vectorization is so crucial for machine learning is that neural networks process enormous amounts of data, and every training iteration requires multiple passes through the network. Vectorized operations harness the power of highly optimized, parallelized code under the hood. Without vectorization, deep learning as we know it would be impractical on standard hardware. This is one of the reasons why neural networks are able to achieve modern feats when a model as clever as Support Vector Machine (SVM) could not. The primary reason is that SVM simply couldn't be parallelized.

Linear Algebra: The Language of Neural Networks

If data is the fuel of machine learning, linear algebra is the engine. At the heart of any neural network lies the *dot product*—the operation that computes the weighted sum of inputs and weights, as shown in Listing 5.12.

```
# Inputs for a single example
x = np.array([2.0, 0.5, 3.0, 1.0])

# Weights for a neuron
w = np.array([0.1, 0.4, -0.2, 0.3])

# Computing the weighted sum (dot product)
weighted_sum = np.dot(x, w)

# Output might be around 0.25
print(f"Weighted sum: {weighted_sum}")

# For matrix operations, dot products become
#     matrix multiplication
# Inputs for a batch of 3 examples
X_batch = np.array([[2.0, 0.5, 3.0, 1.0],
                    [1.0, 1.0, 2.0, 2.0],
                    [0.5, 3.0, 1.0, 0.0]])

# Weights matrix for a layer with 2 neurons
W = np.array([[0.1, 0.4, -0.2, 0.3],
              [0.2, 0.1, 0.3, -0.1]])

# Computing activations for all examples at once
# Transpose W to match dimensions
activations = np.dot(X_batch, W.T)
print(activations)
# Weighted sum: 0.09999999999999998
# [[0.1   1.25]
#  [0.7   0.7 ]
#  [1.05  0.7 ]]
```

Listing 5.12 The Dot Operation in NumPy

This operation directly implements the core feed-forward computation we studied in Chapter 4. Recall that in a neural network, each neuron computes the following:

$$h(x) = \sigma\left(\sum (inputs \times weights) + bias\right)$$

The dot product efficiently implements the summation term $\sum(inputs \times weights)$, where x_1 through x_4 are the input features, and w_1 through w_4 are the corresponding weights for that neuron. This weighted sum z is then passed through an activation function to produce the neuron's output.

5 Deep Dive into Keras

When we scale to an entire layer of neurons, each with its own set of weights, matrix multiplication allows us to compute all neurons' outputs simultaneously. For a layer with n neurons receiving m-dimensional input, we have

$$Z = X \cdot W^T$$

where

- X is a *(batchsize × m)* matrix containing input features for multiple examples.
- W is an *(n × m)* matrix where each row contains the weights for one neuron.
- Z is the resulting matrix of weighted sums for each example and neuron.

This matrix operation simultaneously computes the weighted sum for every neuron in the layer across every example in the batch–a calculation that would require multiple nested loops if done sequentially. Let's see what this would look like with explicit loops, as shown in Listing 5.13, versus the matrix multiplication approach, as given in Listing 5.14.

```
# Input data: 3 examples, each with 4 features
X = np.array([
    [1.0, 0.5, 0.8, 0.2],   # Example 1
    [0.7, 0.3, 0.9, 0.5],   # Example 2
    [0.2, 0.8, 0.6, 0.1]    # Example 3
])

# Weight matrix: 2 neurons, each with 4 weights
#                (one per input feature)
W = np.array([
    [0.1, 0.2, -0.1, 0.3],  # Weights for neuron 1
    [0.2, 0.1, 0.3, -0.2]   # Weights for neuron 2
])

# Bias vector: 1 bias per neuron
b = np.array([0.1, -0.1])

# APPROACH 1: Using nested loops (slow and verbose)
Z_loops = np.zeros((X.shape[0], W.shape[0]))
# Initialize output matrix (3 examples, 2 neurons)

for i in range(X.shape[0]):  # For each example
    for j in range(W.shape[0]):  # For each neuron
        Z_loops[i, j] = 0
```

```
        for k in range(X.shape[1]):  # For each feature
            Z_loops[i, j] += X[i, k] * W[j, k]
        Z_loops[i, j] += b[j]  # Add bias
```

Listing 5.13 Approach 1: Forward Pass for a Dense Layer with Three Examples, Four Input Features, and Two Output Neurons

The matrix multiplication version accomplishes in just one line what it takes three nested loops to do in the explicit version. Under the hood, NumPy's matmul function (or the @ operator) implements highly optimized Basic Linear Algebra Subprograms (BLAS) operations that maximize CPU cache utilization and, when available, use parallel processing.

```
# APPROACH 2: Using matrix multiplication (fast and concise)
Z_matmul = np.matmul(X, W.T) + b

# Equivalently using the @ operator (Python 3.5+)
Z_matmul_alt = X @ W.T + b

print("Results match:", np.allclose(Z_loops, Z_matmul))
# Should print: True
```

Listing 5.14 Approach 2: Forward Pass for a Dense Layer with Three Examples, Four Input Features, and Two Output Neurons

After computing Z, we simply add the bias vector and apply the activation function to complete the layer's forward pass. Every forward pass through a neural network involves these matrix multiplications. The dot product efficiently computes how strongly each input affects each neuron's output, forming the basis of information flow in neural networks.

Random Number Generation: The Starting Point for Learning

Neural networks learn by gradually adjusting their weights from initial values. But what should those initial values be? This is where NumPy's random number generators come into play. Let's take a look at an example in Listing 5.15.

```
# Creating weights with small random values from
# a normal distribution
# For a layer with 5 inputs and 3 outputs
weights = np.random.normal(0, 0.1, (5, 3))
# mean=0, std=0.1, shape=(5,3)

print(weights)
# [[ 0.08  -0.03   0.12]
#  [-0.02   0.15  -0.07]
```

```
# [ 0.01  -0.09   0.05]
# [ 0.11   0.04  -0.06]
# [-0.14   0.02   0.10]]

# Creating bias terms initialized to zero
biases = np.zeros(3)
print(biases)  # [0. 0. 0.]
```

Listing 5.15 Creating Random Values

Starting with all 0s would cause all neurons to learn the same features, effectively reducing the capacity of your network. Random initialization *breaks the symmetry* and allows different neurons to learn different patterns. The right initialization can also help prevent problems such as vanishing or exploding gradients during training. There is an entire area of research in machine learning that deals with *optimized randomizers*. For now, we'll leave it as an exercise for you to go and explore the initializers. One good starting point might be to study the *Glorot* initialization.

Reshaping and Transformation: Aligning Data for Processing

Shapes are notorious for causing headaches in practitioners just starting with large-scale neural networks. The reason is that neural networks are extremely picky about data shapes. An image might start as a 2D array of pixels but needs to be flattened for a dense layer. Output predictions often need reshaping to match the format of target values. NumPy's reshaping operations make these transformations seamless. Let's take a look at an example in Listing 5.16 of how the same data can be represented using different shapes depending on which part of the network is consuming it.

```
# MNIST image data might be 28x28 pixels
digit_image = np.random.rand(28, 28)  # Simulated image data

# A dense neural network layer expects a flat vector
flattened_input = digit_image.reshape(1, 784)
# Shape becomes (1, 784)
print(f"Original shape: {digit_image.shape}, \
        Flattened shape: {flattened_input.shape}")

# After processing, predictions might need reshaping
predictions = np.random.rand(10, 1)  # Raw predictions
reshaped_preds = predictions.reshape(-1)  # Shape <- (10,)
print(f"Original preds shape: {predictions.shape}, \
            Reshaped: {reshaped_preds.shape}")
```

Listing 5.16 Different Shapes for the Same Data

Make sure to run these code snippets to get a thorough understanding of the concepts. These examples show vectors, matrices, and tensors. Before we go further, let's clarify what a tensor is. In machine learning, a *tensor* is simply a multidimensional array of values—a generalization of vectors and matrices to higher dimensions. NumPy arrays are the Python implementation of tensors. Let's take a look at Listing 5.17 to see how we can generalize the concepts of vectors and matrices to even higher dimensions. These (and even higher dimensions) are very common in neural networks. While the core concepts are simple, it takes a bit of time to fully internalize them.

```
# 0D tensor (scalar) - a single number
scalar_tensor = np.array(42)
print(f"Scalar tensor shape: {scalar_tensor.shape}, Rank: 0")
# Shape: (), Rank: 0

# 1D tensor (vector) - a list of numbers
vector_tensor = np.array([1, 2, 3, 4])
print(f"Vector tensor shape: {vector_tensor.shape}, Rank: 1")
# Shape: (4,), Rank: 1

# 2D tensor (matrix) - a table of numbers
matrix_tensor = np.array([[1, 2, 3], [4, 5, 6]])
print(f"Matrix tensor shape: {matrix_tensor.shape}, Rank: 2")
# Shape: (2, 3), Rank: 2

# 3D tensor - a cube or stack of matrices
# Often used for sequence data or a batch of images (grayscale)
cube_tensor = np.array([
    [[1, 2], [3, 4]],
    [[5, 6], [7, 8]]
])
print(f"3D tensor shape: {cube_tensor.shape}, Rank: 3")
# Shape: (2, 2, 2), Rank: 3

# 4D tensor - a batch of 3D data
# Often used for a batch of color images:
#   (batch_size, height, width, channels)
batch_images = np.zeros((32, 224, 224, 3))
# 32 RGB images of 224x224 pixels
print(f"4D tensor shape: {batch_images.shape}, Rank: 4")
# Shape: (32, 224, 224, 3), Rank: 4
```

Listing 5.17 Tensors and Higher Dimensions

Different neural network layers expect tensors with specific shapes, for example:

- Dense (fully connected) layers expect 2D tensors: (batch_size, features).
- Convolutional layers expect 4D tensors: (batch_size, height, width, channels).
- Recurrent layers expect 3D tensors: (batch_size, sequence_length, features).

Understanding shapes (and reshaping) is absolutely critical for machine learning. This seemingly simple operation solves a fundamental challenge in deep learning: different parts of a neural network architecture expect data in different shapes, yet the underlying data must flow seamlessly between them.

Consider a complete image classification pipeline:

1. Your input images are 2D arrays of pixels (or 3D with color channels).
2. Convolutional layers, which we'll study later, process this spatial data directly in its multidimensional form.
3. After extracting features, the network often needs fully connected layers that only accept flattened 1D vectors.
4. The output needs to be shaped correctly for your loss function (e.g., a specific shape for categorical cross-entropy).

Without reshaping, these components couldn't communicate with each other—each piece of your neural network would remain isolated. It's like having brilliant specialists who all speak different languages; reshaping acts as the translator that allows them to collaborate effectively.

Reshaping is also critical for batched operations, the backbone of efficient neural network training. When preparing mini-batches, we reshape individual examples into batched tensors, allowing the network to process multiple examples simultaneously rather than one at a time.

Without reshaping, we would need to rewrite entire algorithms to handle different tensor shapes or create custom layers for each shape transition—a maintenance nightmare and performance bottleneck. Reshaping allows us to adapt our data to our algorithms rather than the reverse, making neural networks both more flexible and more efficient.

Axis-Based Operations: Processing Along Dimensions

Neural networks often need to perform operations along specific dimensions of data, especially when dealing with batched inputs or multidimensional features. NumPy's axis parameter provides precise control over these operations. Let's discuss this with the example given in Listing 5.18.

```
# Batch of 4 examples, each with 5 features
batch_data = np.array([
    [1.2, 3.4, 0.5, 2.3, 1.1],
    [2.7, 1.8, 3.2, 0.8, 2.5],
```

5.3 Building Your First Model

```
    [0.3, 2.9, 1.7, 2.2, 3.1],
    [1.9, 0.7, 2.8, 3.3, 0.6]
])

# Normalizing each feature (column) across the batch
feature_means = np.mean(batch_data, axis=0)
# Mean of each column
feature_stds = np.std(batch_data, axis=0)
# Std dev of each column
normalized = (batch_data - feature_means) / feature_stds
print("Feature means:", feature_means)
# Feature means: [1.525 2.2   2.05  2.15  1.825]

# Computing sum of each example (row)
example_sums = np.sum(batch_data, axis=1)
print("Example sums:", example_sums)
# Example sums: [ 8.5 11.  10.2  9.3]
```

Listing 5.18 Axis-Based Operations

Let's break down what's happening with these `axis` operations line by line:

1. First, we create `batch_data`, a 2D array with shape (4, 5) representing 4 examples, each with 5 features.

2. When we call `np.mean(batch_data, axis=0)`, the `axis` value tells NumPy to compute the mean along the first axis (rows). For each column position, it averages all values in that position across all rows. The result is a 1D array of length 5, containing the mean of each feature across all examples. This is exactly what we need for feature normalization in a neural network.

3. Similarly, `np.std(batch_data, axis=0)`, computes the standard deviation of each feature across examples. It returns another 1D array of length 5. These standard deviations tell us how much each feature varies in our dataset, which is critical for proper normalization.

4. The broadcasting in `(batch_data - feature_means) / feature_stds`, subtracts the appropriate mean from each feature value and divides by the appropriate standard deviation. NumPy automatically applies these operations to every element in our batch through broadcasting. This results in a normalized dataset where each feature has mean 0 and standard deviation 1, which helps neural networks train more efficiently.

5. When we call `np.sum(batch_data, axis=1)`, the `axis` value tells NumPy to sum along the second axis (columns). For each row, it adds up all the values across columns. This returns a 1D array of length 4, containing the sum of features for each example. This

operation is similar to what happens when a neural network layer combines all of its inputs into a single activation value.

In batch processing, you frequently need to normalize data across examples (axis 0) or compute losses for individual examples (axis 1). Mastering axis-based operations allows you to efficiently manipulate multidimensional data without resorting to complex loops.

For neural networks specifically, axis operations are used in the following:

- **Batch normalization**
  Normalizing activations across the batch.
- **Loss computation**
  Averaging losses across examples.
- **Softmax calculations**
  Applying softmax to each example independently.
- **Pooling operations**
  Reducing dimensions along spatial axes.
- **Attention mechanisms**
  Computing weights along sequence dimensions.

With these NumPy fundamentals as our foundation, we're now ready to explore how symbolic computing builds upon them to create the powerful, flexible deep learning frameworks that make modern AI possible. Don't think of these NumPy operations as abstract concepts. They're the operations happening under the hood when you build and train neural networks with Keras. Understanding them gives you both the practical skills to manipulate data and the conceptual understanding to grasp how neural networks actually work.

5.3.2 Symbolic Computation: The Magic Behind Neural Networks

When most of us think about computing, we think about calculations with specific numbers (2 + 2 = 4) or perhaps solving more complex equations like determining the trajectory of a spacecraft. This is numerical computation—working directly with concrete values to get specific answers.

Symbolic computation takes a fundamentally different approach. Instead of immediately plugging in numbers, we keep the variables as symbols and manipulate the mathematical expressions themselves. It's like having a mathematical advisor who understands the deeper structure of equations rather than just calculating results.

In traditional programming, a variable such as x is a container that holds a specific value. Once you assign $x = 5$, the computer "forgets" about the variable name and just works with the number 5. In symbolic computation, however, x remains a symbol throughout the calculation—a placeholder representing any possible value. We maintain the entire analytical structure of the expression and can perform operations on it as a whole. Let's take a look at this with a concrete example: the expression $(x^2 - 1)/(x - 1)$.

In traditional programming, we might create a function that will compute the value of this expression and return that value. This works fine for most values of x. We know that `calculate(2)` outputs 3.0 and `calculate(5)` outputs 6.0, but what happens when we try $x = 1$? We get an error of division by zero! Our function breaks down completely! At $x = 1$, we end up trying to calculate 0/0, which is mathematically undefined. The computer can't give us a meaningful answer.

With symbolic computation, we take a different path. Instead of immediately calculating with specific values, we manipulate the expression $(x^2 - 1)/(x - 1)$ and recognize that $x^2 - 1$ can be factored as $(x - 1)(x + 1)$. Now, we can simplify by canceling the common factor $(x - 1)$ and get the final expression as $x + 1$. Through symbolic manipulation, we've discovered that our complex-looking expression is actually equivalent to the much simpler *x + 1* for all values of x where the original expression is defined. It's not really differentiable at that point, but it's infinitely close to the point, and the function behaves fine.

This symbolic simplification reveals something profound. While our original expression $(x^2 - 1)/(x - 1)$ has a hole at $x = 1$ (a removable discontinuity in mathematical terms), the simplified form *x + 1* shows us exactly what value the function should have at that point. Even though we technically can't evaluate the original function at precisely $x = 1$, symbolic computation reveals that the limit as x approaches 1 is exactly 2. We can write this formally as the following:

$$\lim_{x \to 1} \frac{(x^2 - 1)}{(x - 1)} = \lim_{x \to 1}(x + 1) = 2$$

This is X-ray vision that lets us see through the mathematical fog. When numerical computation hits a wall, symbolic computation can often find a way through by revealing the underlying structure of the formula. The formula itself is the primary object, not the numerical result. Only when we need a specific answer do we substitute values into our symbolic expression. Based on this basic rule, here are some key differences between numerical and symbolic computation:

- **Abstraction level**
  Symbolic computation works with abstract mathematical entities (variables, functions, operations), while numerical computation works with concrete values.
- **Derivatives**
  Symbolic computation can provide exact derivatives through mathematical rules, while numerical computation typically approximates derivatives using finite differences. This part is of high significance to us because, as you know, we need to compute gradients for our gradient descent algorithm. As our computations get more and more complex, the utility of symbolic computation increases exponentially.

- **Expression optimization**
  Symbolic systems can analyze and optimize expressions (e.g., simplifying $x \times 1$ to x) before computation, potentially improving efficiency.

- **Memory vs. computation**
  Symbolic computation often requires more up-front memory to store the computational structure but can be more efficient when repeatedly evaluated with different inputs.

- **Error propagation**
  Numerical computation accumulates rounding errors at each step, while symbolic computation postpones such errors until the final evaluation.

Let's take a look at a more code-oriented example of symbolic computing. Think of a straight line—the simplest relationship in mathematics. This line is described by the equation $y = ax + b$. Let's build this relationship using TensorFlow's symbolic approach, as shown in Listing 5.19. (Complete code listings for this section are available on the book's web page as well as on *https://recluze.net/keras-book* in the **05-03-symbolic-computation** notebook.)

```
import tensorflow as tf

# Define our coefficients
a = tf.constant(2.0)  # Our slope
b = tf.constant(3.0)  # Our y-intercept

# Create a variable that we'll use as our input
x = tf.Variable(0.0)

# Define our linear function
def compute_linear_function(input_x):
    return a * input_x + b
```

Listing 5.19 Symbolic Computation Example Using TensorFlow

What we've done here is create the mathematical framework, but we haven't actually computed anything yet. We'll explain the code to help you fully understand its strength. In TensorFlow, we work with two primary types of values: constants and variables. *Constants* are fixed values that remain unchanged throughout our computations, such as predefined coefficients or fixed parameters in our model. *Variables*, in contrast, are values that can be modified during computation, making them essential for scenarios where values need updating, such as model weights during training.

The true strength of TensorFlow emerges after we set up these basic components. We can write computations using regular Python syntax just as we've done in a function in the listing. It doesn't matter if we're performing basic arithmetic or complex matrix operations. Behind the scenes, TensorFlow constructs a *computation graph* that maps out how our operations connect. This graph enables TensorFlow to track and optimize our calculations efficiently. This combination of straightforward Python syntax and TensorFlow's computation graph system gives us the best of both worlds. We can write complex calculations naturally while leveraging TensorFlow's powerful optimization capabilities. After defining our variables and constants, we're free to implement any level of computational complexity our model requires, with TensorFlow handling the graph creation and evaluation automatically. Then, we can actually execute the graph by assigning values to the variables. Listing 5.20 shows an example of this in action.

```
# Let's set x to a specific value
x.assign(5.0)

# Compute y = ax + b
y = compute_linear_function(x)

print(f"When x = {x.numpy()}:")
print(f"y = {y.numpy()}")  # Will output 13.0 (2*5 + 3)
```

Listing 5.20 Executing the Computation Graph

When we run this code, TensorFlow follows our blueprint and calculates that when x is 5, y equals 13. It's like finally driving along that route we planned, and arriving exactly where we expected. Not only that, we can now go ahead and perform actions that would simply be impossible with numeric computation. For instance, we can calculate the derivatives of our operations automatically. Let's see how we can do this in Listing 5.21.

```
# To compute derivatives, we use TensorFlow's GradientTape
with tf.GradientTape() as tape:
    # We need to tell the tape to watch our variable
    tape.watch(x)
    # Compute y = ax + b
    y = compute_linear_function(x)

# Now we can compute the derivative dy/dx
dy_dx = tape.gradient(y, x)

print(f"dy/dx = {dy_dx.numpy()}")
# Will output 2.0 (derivative of 2x + 3 is 2)
```

Listing 5.21 Automatically Calculating Gradients from Computation Graph

The *GradientTape* implements what's known as automatic differentiation, which is fundamentally different from numerical approximations or symbolic differentiation. Here's how it works:

- **Computational graph**
  When you perform operations inside a GradientTape context, TensorFlow builds a computational graph that tracks every operation (addition, multiplication, etc.) and how variables flow through these operations.

- **Operation recording**
  Each mathematical operation has a corresponding derivative formula. For example, if $f(x) = x^2$, then $f'(x) = 2x$. TensorFlow records these operations and their known derivatives.

- **Reverse mode differentiation**
  When you call `tape.gradient(y, x)`, TensorFlow uses reverse mode differentiation, starting from the output (y) and working backward to the input (x). This is particularly efficient when you have many input variables and fewer output variables.

- **Chain rule application**
  TensorFlow automatically applies the chain rule of calculus, which states that if $y = f(g(x))$, then $dy/dx = (dy/dg) \times (dg/dx)$. It multiplies the local gradients along the computational path from y to x.

- **Gradient accumulation**
  If a variable appears multiple times in the computation (like in complex neural networks), TensorFlow automatically accumulates the gradients from different paths.

This capability is the secret sauce that makes neural networks possible, as it efficiently computes gradients for models with millions of parameters. This is because, as we've seen before in Chapter 2 and Chapter 3, neural networks are essentially complex mathematical functions with thousands or millions of parameters. If we were to calculate the derivatives ourselves, it would be extremely time-consuming and difficult.

> **Other Names for Automatic Differentiation**
>
> The concept of automatic gradient calculation through symbolic computation isn't unique to TensorFlow. Almost all modern frameworks have this capability under different names. Some call it *AutoDiff*, and some call it *AutoGrad*. The concept remains the same—keep track of the operations and construct derivative using the chain rule based on the sequence of operations.

When you use Keras to build neural networks, all of this symbolic computation happens behind the scenes. You don't need to manually set up the GradientTape or calculate derivatives—Keras and TensorFlow handle this for you. However, understanding this process helps demystify what's happening when your model learns from data. So the next time you call `model.fit()` to train your neural network, remember that under the hood, Keras is creating a complex mathematical blueprint and automatically navigating the terrain of derivatives to find the optimal path to a solution. So, with a deep understanding of the underlying mechanisms, we can revisit the Keras code once more with a more detailed example and fill in some more blanks.

5.4 Implementing Core Concepts in Keras: Gradient Descent and Classification

Now that we've explored the theoretical foundations of gradient descent and classification, let's bring these concepts to life with Keras. In this section, we'll build a complete image classification system that distinguishes between cats and dogs. This practical implementation will showcase how Keras elegantly handles the complex mechanisms we've discussed in previous chapters.

5.4.1 The Building Blocks of a Cat-Dog Classifier

Our task begins with a simple question: Can a neural network learn to tell cats and dogs apart? This seemingly straightforward task actually encompasses many core machine learning concepts:

- *Data preparation* and *normalization*
- Model *architecture design*, including layers and neuron numbers
- Training the machine through gradient descent
- Evaluating the learning with *classification metrics*
- Continuing our work through *model saving* and *loading*

Let's explore each of these elements through a hands-on implementation. The `import` section of our code is shown in Listing 5.22. Each library brings special capabilities that will help us build our image classifier. We start by importing TensorFlow and Keras. These form the foundation of our deep learning project. TensorFlow provides the computational engine that powers our neural network operations, while Keras offers a user-friendly interface that makes building complex models surprisingly straightforward. It should be noted that we're using TensorFlow here in two capacities—one as a backend engine that is called by Keras and the second as a collection of utilities. This is typical of a machine learning system, several components from distinct libraries are used to collectively achieve what we want instead of relying on a single library or rolling out our own function. The complete code listing for this section is available on the book's official web page as well as on *https://recluze.net/keras-book* in the **05-04-cats-dogs** notebook.

The imports from `keras` are the components we'll use to construct our neural network:

- `Sequential` gives us a straightforward container for stacking layers in order.
- `Flatten`, `Dense`, and `Dropout` are layer types that transform our data in different ways: `Flatten` converts our 2D image data into a format our `Dense` layers can process, `Dense`

layers create connections between neurons where learning happens, and `Dropout` randomly ignores some neurons during training, preventing our model from memorizing the training data. We'll get to a detailed discussion of `Dropout` in the next chapter.

Some of the other imports you see are important components that we'll discuss later in this chapter at appropriate times. Each of these imports plays a vital role in the machine learning workflow we're about to build—from preparing data, to constructing and training our model, to evaluating its performance and visualizing results.

```
import tensorflow as tf
import keras
from keras import Sequential
from keras.layers import Flatten, Dense, Dropout
from keras.utils import image_dataset_from_directory
from keras.callbacks import ModelCheckpoint
from keras.optimizers import Adam
import numpy as np
import matplotlib.pyplot as plt
from sklearn.metrics import accuracy_score, precision_score
from sklearn.metrics import recall_score, f1_score
import os
```

Listing 5.22 Imports for a Cats and Dogs Classifier

5.4.2 Getting and Fixing the Data

Of course, in order to perform machine learning, we're going to need some data. The `get_file` function call shown in Listing 5.23 reaches out to Google's storage repositories and retrieves a comprehensive collection of cat and dog images. This is a carefully curated dataset that researchers have already cleaned, organized, and split for machine learning tasks. We usually want to start with one of these instead of rolling out our own data.

What makes this especially valuable is that the dataset comes pre-divided into *training* and *validation sets*, saving us from the manual work of splitting the data ourselves. If you recall from Chapter 3, this separation is crucial—the training set teaches our model, while the validation set tests whether it has truly learned or merely memorized.

```
# Download and extract the dataset
url = 'https://storage.googleapis.com/mledu-datasets/
            cats_and_dogs_filtered.zip'
data_dir = keras.utils.get_file('cats_and_dogs.zip',
                            cache_dir=".",
                            origin=url,
                            extract=True)
```

5.4 Implementing Core Concepts in Keras: Gradient Descent and Classification

```
print(data_dir)
base_dir = os.path.join(data_dir, 'cats_and_dogs_filtered')

# Define directories
train_dir = os.path.join(base_dir, 'train')
validation_dir = os.path.join(base_dir, 'validation')
```

Listing 5.23 Getting the Dataset

When working with thousands of images, managing how they flow into our neural network becomes crucial. This is where Keras steps in as our data handler, elegantly handling the complex task of moving images from storage to our model during training. First, we define some basic parameters for our images in Listing 5.24. By setting img_width and img_height to 150 pixels, we're establishing a consistent canvas size for all of our cat and dog pictures. Think of this as standardizing the photo size so our neural network doesn't get confused by varying dimensions. Meanwhile, the batch_size of 32 determines how many images our model will process at once—like deciding how many flashcards to review in a single study session. The real heavy lifting of this code segment is the image_dataset_from_directory function. This powerful utility transforms a simple folder of images into a sophisticated, ready-to-train dataset with just a single function call. Behind the scenes, Keras is performing several critical tasks:

- **Automatic labeling**
  Keras reads the directory structure to determine which images are cats and which are dogs.
- **Resizing**
  Every image is automatically resized to our specified dimensions (150×150 pixels).
- **Batching**
  Images are grouped into batches of 32, optimizing memory usage and training speed.
- **Binary labels**
  The label_mode='binary' parameter tells Keras we're working with a two-class problem (cats vs. dogs), so it generates appropriate binary labels.
- **Reproducibility**
  By setting a seed value, we ensure that if we run our code multiple times, the data will be shuffled in exactly the same way.

What makes this particularly impressive is how much complexity Keras is hiding from us. Without this function, we would need dozens of lines of code to load images, resize them, convert them to arrays, create label vectors, batch them appropriately, and ensure they're in the right format for training.

Consider having to manually sort thousands of documents for a research project versus having an intelligent assistant who automatically organizes everything perfectly. That's what Keras is doing for us here—transforming a potentially tedious data preparation task into a single line of elegant code.

With our data now flowing smoothly through this digital conveyor belt, we're ready to focus on the more creative aspects of our model design.

```
# Image parameters
img_width, img_height = 150, 150
batch_size = 32

# Create datasets using Keras 3 API
train_dataset = image_dataset_from_directory(
    train_dir,
    image_size=(img_width, img_height),
    batch_size=batch_size,
    label_mode='binary',
    seed=123
)
```

Listing 5.24 Loading the Dataset Correctly with a Single Function Call

Throughout this book, we've emphasized one crucial concept: understanding the shape of our data as it flows through the neural network. This is the secret language that helps us visualize how information transforms as it moves through our model. Let's take a moment to inspect what our data actually looks like in this particular example. With just a few lines of code given in Listing 5.25, we can gain valuable insights into the structure of both our inputs and outputs. Keep in mind that we're only examining one batch before breaking out of the loop. This gives us a quick snapshot without processing the entire dataset.

When we run this code, we see `Inputs: (32, 150, 150, 3)`. The first dimension, 32, represents our batch size—we're processing 32 images at once. Think of this as examining 32 photos spread out on a table simultaneously. The next two dimensions, 150 and 150, reveal the spatial resolution of each image—150 pixels wide by 150 pixels tall. This consistent size ensures our model receives uniform inputs, regardless of the original photo dimensions. The final dimension, 3, is particularly interesting. It represents the three color *channels* in each image: red, green, and blue. Unlike our earlier work with MNIST where we had grayscale images (just one channel of intensity values), color images contain three separate layers of information. Each pixel has three values that together create the full spectrum of colors we see.

Meanwhile, our outputs have a shape of `(32, 1)`, indicating 32 scalar values, that is, one classification result for each image in the batch. These values will be between 0 and 1, representing the probability of an image being a dog.

```
# Iterate over the dataset
for inputs, outputs in train_dataset:
    print("Inputs:", inputs.numpy().shape)
    print("Outputs:", outputs.numpy().shape)
    break
# Output ------------------
# Inputs: (32, 150, 150, 3)
# Outputs: (32, 1)
```

Listing 5.25 Understanding the Shape of the Data

After setting up our training dataset, we need to create a separate *validation dataset* that will help us evaluate how well our model is learning. In Chapter 3, we called this the training set. There is a little ambiguity around these two terms—validation sets and test sets—that we'll get to in the next chapter. The split is shown in Listing 5.26. Just like its training counterpart, we use the `image_dataset_from_directory` function, keeping all parameters consistent to ensure fair comparison. This validation set serves as our model's pop quiz—data it's never seen during training but that will test its real understanding. When we feed raw pixel values into our neural network, we're essentially asking it to work with numbers ranging from 0 to 255. These large values can cause problems during training and might lead to unstable gradients or slow convergence.

The solution is elegantly simple: We scale all pixel values down to a range between 0 and 1. We accomplish this using Keras's Rescaling layer, dividing each pixel value by 255. This transformation doesn't change what the images look like to us, but it makes a world of difference to our neural network. Mathematically, this means a bright white pixel that was 255 becomes 1.0, while a pure black pixel at 0 remains 0. All the beautiful grays and colors in between now fall somewhere on this standardized 0–1 scale. Notice how we apply this normalization to both our training and validation sets using the map method. This ensures consistency across all the data our model encounters. The lambda function is a concise way to say "apply this normalization to the image data (x) but leave the labels (y) unchanged."

With our data now properly scaled and organized, we've created the ideal conditions for our neural network to learn efficiently. This seemingly small step of normalization can dramatically improve both the speed and quality of our model's learning process.

```
validation_dataset = image_dataset_from_directory(
    validation_dir,
    image_size=(img_width, img_height),
    batch_size=batch_size,
    label_mode='binary',
    seed=123
)
```

```
# Normalize images to [0,1] range
normalization_layer = keras.layers.Rescaling(1./255)
train_dataset = train_dataset.map(
    lambda x, y: (normalization_layer(x), y)
)
validation_dataset = validation_dataset.map(
    lambda x, y: (normalization_layer(x), y)
)
```

Listing 5.26 Creating the Validation Set and Normalizations

5.4.3 Performance Optimization and Model Specification

While we've focused on preparing our data correctly for learning, there's another critical aspect we shouldn't overlook: performance optimization. Deep learning is not only about accuracy but also about efficiency. The lines of code we're examining now might seem simple, but they represent powerful acceleration techniques that can dramatically speed up our training process. Let's see how we achieve this using Listing 5.27. First, we set AUTOTUNE = tf.data.AUTOTUNE, which is TensorFlow's way of saying, "I'll automatically figure out the optimal settings based on your available resources."

Then, we apply two powerful optimization techniques to both our training and validation datasets:

- **Caching**
  When we call .cache(), we're telling TensorFlow to keep our dataset in memory after it's loaded the first time. This saves the time and processing power that would otherwise be spent repeatedly loading and preprocessing the same images from disk.

- **Prefetching**
  The .prefetch(AUTOTUNE) method is where things get really interesting. While your GPU is busy training on the current batch of images, TensorFlow is already preparing the next batch in the background. It's like a restaurant kitchen where chefs are preparing your next course while you're still enjoying your appetizer—by the time you're ready, the next dish is already waiting. This technique prevents your GPU from sitting idle between batches, ensuring it's constantly fed with data.

Together, these optimizations create a smooth, efficient data pipeline that can dramatically reduce training time—sometimes turning hours of computation into minutes. What's particularly elegant about these methods is how they work invisibly in the background, automatically adapting to your specific hardware configuration without requiring complex manual tuning.

These performance boosters represent the difference between a merely functional deep learning system and one that's truly optimized for real-world applications. By implementing these few lines of code, we're ensuring our model can train as quickly as

5.4 Implementing Core Concepts in Keras: Gradient Descent and Classification

possible, allowing us to experiment, iterate, and refine our approach with minimal waiting time. This step is essential when working with large-scale models.

```
# Enable performance optimizations
AUTOTUNE = tf.data.AUTOTUNE
train_dataset = train_dataset.cache().prefetch(AUTOTUNE)
val_dataset = validation_dataset.cache().prefetch(AUTOTUNE)
```

Listing 5.27 Setting Optimization for Caching and Prefetching

Now comes the exciting part—designing our neural network's structure. Because we've already covered these concepts briefly in Chapter 4, let's focus here on how we're applying them specifically to our cat and dog classification problem in Listing 5.28. We're creating a *sequential* model—a straightforward stack of layers where data flows from bottom to top. Our network begins with a Flatten layer that transforms our 3D image data (150 × 150 × 3) into a 1D array of numbers that our dense layers can process. Remember, the Dense layer is just a collection of neurons arranged in a sequence, so we can only feed it a 1D vector.

After flattening, we build a hierarchy of increasingly refined understanding:

- A first dense layer with 256 neurons captures basic patterns and textures.
- A second dense layer with 128 neurons builds more complex combinations.
- A third dense layer with 64 neurons distills these into higher-level features.
- A final single-neuron layer with sigmoid activation makes the binary decision: dog (1) or cat (0).

The Dropout layers you see interspersed between Dense layers act as strategic noise-inducers. We'll cover this in detail in the next chapter, but for now, think of this layer as helping our model generalize better to new images rather than memorizing the training set.

Then, we get to the compile step that sets up the learning process itself. We've chosen the following:

- binary_crossentropy loss is perfect for our yes/no classification task.
- Adam optimizer is an adaptive gradient descent method that automatically fine-tunes learning rates.
- accuracy metric is a straightforward way to track how often our model is correct.

```
# Create model (dense layers only)
model = Sequential([
    Flatten(input_shape=(img_width, img_height, 3)),
    Dense(256, activation='relu'),
    Dropout(0.5),
    Dense(128, activation='relu'),
    Dropout(0.3),
```

```
    Dense(64, activation='relu'),
    Dense(1, activation='sigmoid')
])

# Compile model
model.compile(
    loss='binary_crossentropy',
    optimizer=Adam(learning_rate=0.0001),
    metrics=['accuracy']
)
```

Listing 5.28 Defining the Model Architecture

We've set a conservative learning rate of 0.0001 for Adam, helping our model take smaller, more careful steps toward the optimal solution. With this architecture in place, our neural network is ready to begin its learning—transforming from a blank slate into a sophisticated image classifier through the magic of gradient descent and backpropagation.

But before we do that, we need to take a little bit of a detour to explain the concept of accuracy metric just mentioned in a little more detail and discuss the pitfalls associated with this simple concept of, "are the answers correct?"

5.4.4 Evaluation Metrics

When our model has finished its training and starts making predictions, how do we know if it's any good? This is where *evaluation metrics* come into play—they're the report cards for our neural network, telling us not just whether it's right or wrong but how it's right or wrong. At the heart of classification evaluation is a powerful concept: comparing what our model thought was true against what was actually true.

The terminology here can be a bit confusing, so let's make sure we get it right. In binary classification, we say a prediction is *positive* when our model believes the target class is present. For example, with our cat classifier, a positive prediction means "yes, this is a cat," while a *negative* prediction means "no, this isn't a cat" (so it must be a dog in our binary scenario). Every prediction our model makes falls into one of four categories, which together form what's often called a *confusion matrix* (though there's nothing confusing about it once you understand it!), as shown in Table 5.2.

	Predicted Positive	**Predicted Negative**
Actual Positive	True positive	False negative
Actual Negative	False positive	True negative

Table 5.2 Confusion Matrix

We'll explain this in detail in the context of our cats and dogs classifier:

- **True positive (TP)**
  Our model said, "that's a cat," and it really was a cat. The model correctly identified the positive class. Like recognizing your beloved feline in a photo—both accurate and certain.

- **True negative (TN)**
  Our model said, "that's not a cat" (meaning it's a dog), and indeed it wasn't a cat. The model correctly identified the negative class. This is like correctly saying "that's not a cat" when shown a photo of a dog.

- **False positive (FP)**
  Our model said, "that's a cat," but it was actually a dog. This is called a *Type I error* or a *false alarm*. A very severe example of this would be a tumor identification system telling someone very healthy that they have a tumor—not life threatening but not something ideal either.

- **False negative (FN)**
  Our model said, "that's a dog," but it was actually a cat. This is a *Type II error* or a *miss*. This is like failing to recognize your own cat in a photo because they're partially hidden behind a curtain. In some cases, false positives are more severe and in other, we absolutely want to avoid false negatives at all costs.

Based on these, creating a confusion matrix is like sorting exam answers into different piles. We take our test dataset—images our model hasn't seen during training—and run each one through our classifier. For each prediction, we place it in one of four categories:

- When our model says "cat," and it really is a cat → True positive pile
- When our model says "not a cat," and it's not actually a cat → True negative pile
- When our model says "cat," but it's not a cat → False positive pile
- When our model says "not a cat," but it's actually a cat → False negative pile

After we've processed every image in our test set, we count how many predictions landed in each pile. These four numbers—our TP, TN, FP, and FN counts—become the building blocks for all our evaluation metrics.

The most intuitive question we can ask—"How often is our model correct?"—is answered by the metric called *accuracy*:

$$Accuracy = \frac{TP + TN}{TP + TN + FP + FN}$$

In plain language, accuracy is the proportion of all predictions (both positive and negative) that were correct. If our cat classifier correctly identifies 90 out of 100 images, it has 90% accuracy. While accuracy seems like the perfect measure, it can sometimes lead us astray. Assume we're building a model to detect a rare disease that affects only 1 in

100 people. A model that simply says "no disease" for every patient would achieve 99% accuracy—sounds impressive, right?

But wait! This model is essentially useless. It never identifies anyone who actually has the disease, which was the whole point of creating it in the first place. This model has high accuracy but completely fails at its primary task. This is the *accuracy paradox*, and it's particularly problematic when dealing with *imbalanced datasets*—situations where one outcome is much more common than the other. The accuracy paradox shows us why relying solely on accuracy can be misleading. In many real-world scenarios—from medical diagnostics to fraud detection—we're often more concerned with correctly identifying the rare cases than with overall accuracy. This is why data scientists have developed more specialized metrics that can give us a clearer picture of model performance, especially for imbalanced datasets or applications where certain types of errors are more costly than others.

Let's discuss two of these metrics that give us a more nuanced view of our model's performance. These metrics—*precision* and *recall*—are like different lenses through which we can examine our classifier's behavior, each revealing something unique about its strengths and weaknesses.

Imagine you have a friend who only points out cats when they're absolutely certain. They might miss some cats, but when they do say "that's a cat," you can almost always trust them. This friend has *high precision*. Precision measures how trustworthy our positive predictions are. In mathematical terms:

$$\text{Precision} = \frac{TP}{TP + FP}$$

This formula tells us the following: "Out of all the times our model predicted 'cat,' how often was it actually correct?" A model with perfect precision (1.0) never makes false positive mistakes—it might not catch every cat, but when it says something is a cat, it's always right. Precision is sometimes called *positive predictive value* (*PPV*), especially in medical contexts. It answers the critical question: "If my test comes back positive, how likely is it that I truly have the condition?"

Now consider another hypothetical friend who never wants to miss a single cat. They might occasionally mistake a small dog or a plush toy for a cat, but they'll spot nearly every actual cat they encounter. This friend has a high recall.

Recall measures how comprehensive our model is at finding all positive instances. Its formula is

$$\text{Recall} = \frac{TP}{TP + FN}$$

This formula asks the following: "Out of all the actual cats in our dataset, how many did our model successfully identify?" A model with perfect recall (1.0) never makes false negative mistakes—it finds every single cat, even if it occasionally misidentifies something else as a cat along the way. Recall goes by several other names: *sensitivity*, *true positive rate*, or *hit rate*. In medical diagnostics, high recall is crucial for screening tests where missing a condition could be life-threatening.

There is an issue though because precision and recall often exist in tension with each other. Consider adjusting the "confidence threshold" of our cat classifier: If we raise the threshold, demanding the model be very confident before declaring "cat," we'll likely increase precision (fewer false positives) but decrease recall (more missed cats). If we lower the threshold, making the model more willing to call something a cat, we'll likely increase recall (fewer missed cats) but decrease precision (more false positives).

This trade-off makes it difficult to compare models using precision and recall separately. If Model A has higher precision but lower recall than Model B, which one is better? The answer depends entirely on our specific needs, but if we need a single metric that balances precision and recall, we use what is called the *F1 score*—the harmonic mean of precision and recall:

$$F_1 = 2 \times \frac{\text{Precision} \times \text{Recall}}{\text{Precision} + \text{Recall}}$$

The F1 score reaches its best value at 1.0 (perfect precision and recall) and its worst at 0. Unlike a simple average, the harmonic mean penalizes extreme imbalances—a model with 1.0 precision but 0.1 recall would get a much lower F1 score than a model with 0.8 precision and 0.7 recall. This balanced perspective makes an F1 score particularly valuable when we need good performance on both metrics, neither sacrificing our ability to find all positive cases nor compromising on the trustworthiness of our positive predictions.

> **Accuracy vs. F1 Score**
> One of the best ways to identify low-quality research is to look at the metrics they report. For instance, if you see a research paper that mentions achieving high accuracy for their model but doesn't mention if their classes were imbalanced or not, it means you shouldn't trust their results. They may have a good model on a balanced dataset or a very poor model on a highly imbalanced dataset. The F1 score is a much better measure of overall balance in a model.

By adding precision, recall, and F1 score to our evaluation toolkit, we gain a much richer understanding of our model's performance than accuracy alone could ever provide. These metrics help us build models that aren't just accurate in a statistical sense but truly effective at solving the specific problems we care about. With that out of the way, let's return to our example where we had compiled our model and were about to begin training it. Listing 5.29 shows how we can add different metrics during training to see how the model is making progress.

```
# Compile model
model.compile(
    loss='binary_crossentropy',
    optimizer=Adam(learning_rate=0.0001),
    metrics=['accuracy', keras.metrics.Precision()]
)
```

Listing 5.29 Adding Precision to the Reported Metrics

5.4.5 Guiding the Training Through Callbacks and Checkpoints

Training a neural network isn't simply about setting it in motion and waiting for results. It's an evolving process that benefits from careful monitoring and timely interventions. This is where callbacks come into play. *Callbacks* are powerful tools that let us peek inside the training process and take action at critical moments.

They are automatically called by the Keras internal mechanism at specific points during training, such as after each batch of data or at the end of each epoch. They give us a remarkable ability to interact with a model while it's learning, rather than just waiting until training is complete. In the code given in Listing 5.30, we've created a list called callbacks that we pass to the fit method. This tells Keras to execute these functions at the appropriate times during training. While we're using just one callback here, we could include multiple callbacks in this list to perform various actions simultaneously. The one we're focusing on in the following sections is the ModelCheckpoint.

```
# Define callbacks
checkpoint_path = "model_weights.h5"
checkpoint = ModelCheckpoint(
    checkpoint_path,
    monitor='val_accuracy',
    save_best_only=True,
    mode='max',
    verbose=1
)

callbacks = [checkpoint]

# Train model
history = model.fit(
    train_dataset,
    epochs=25,
    validation_data=validation_dataset,
    callbacks=callbacks
)
```

Listing 5.30 Creating and Using a Callback

The ModelCheckpoint

The `ModelCheckpoint` callback is particularly valuable—it saves our model's weights at strategic points during training. In our implementation, we've configured it to do the following:

- Save weights to an *HDF5* file (with the *.h5* extension).
- Monitor the validation accuracy metric.
- Save only when the model achieves better validation accuracy than previously seen.
- Use `max` mode because we want higher accuracy values.
- Display progress updates with `verbose=1`.

This approach ensures we capture the model at its best performance point, not naively at the end of training. Sometimes, a model performs better in the middle of training before it starts making mistakes—our checkpoint callback catches this optimal state that we can return to later even if our model parameters have diverged to less than optimum values later. Storage of these large numbers of values is enabled by the HDF5 format. Keep this point in mind about the model going off track during training as it will be discussed in a lot of detail in the next chapter.

HDF5 Files: Preserving Your Model's Knowledge

The *HDF5* (Hierarchical Data Format version 5) file format is specially designed to store large numerical datasets efficiently. When we save our model weights to an *.h5* file, we're creating a compact representation of all the knowledge our network has acquired through training.

These files can be easily loaded later, allowing us to do the following:

- Resume training from a previous state.
- Deploy the model in production environments.
- Share the trained model with colleagues.
- Compare different versions of the same model.

Here, we've only scratched the surface of what callbacks can do. In later chapters, we'll explore additional callbacks that can do the following:

- Automatically stop training when performance plateaus (`EarlyStopping`).
- Dynamically adjust the learning rate (`ReduceLROnPlateau`).
- Log training metrics for visualization (`TensorBoard`).
- Create custom callbacks for specialized tasks.

Callbacks give us unprecedented control over the training process, transforming neural network training from a black box into a transparent, interactive experience where we can guide our model toward optimal performance. Once all of this is set, `model.fit` goes ahead and performs our gradient descent algorithm that you're all too familiar with by now.

5.4.6 Evaluating the Model

After training our model, we need to measure how well it actually performs on new, unseen data. This evaluation process is crucial—it tells us whether our model has truly learned to recognize patterns or has simply memorized the training examples.

The first step in our evaluation code given in Listing 5.31 is particularly important: The `model.load_weights(checkpoint_path)` command retrieves the weights that produced the best validation accuracy during training. Remember that our `ModelCheckpoint` callback saved these weights whenever the model improved. By loading these specific weights, we ensure we're evaluating the model at its peak performance, not just its final state. This approach acknowledges an important reality in neural network training: the best-performing model often emerges somewhere in the middle of the training process, not necessarily at the end.

```
# Restore best weights
model.load_weights(checkpoint_path)

# Evaluate and calculate metrics
y_true = []
y_pred = []
for images, labels in validation_dataset:
    # Get predictions
    predictions = model.predict(images)
    pred_labels = (predictions > 0.5).flatten()

    # Extend lists
    y_true.extend(labels.numpy().flatten())
    y_pred.extend(pred_labels)
```

Listing 5.31 Making Predictions and Collecting Results

Collecting True Labels and Predictions

Next, we set up empty lists (y_true and y_pred) to collect the ground truth labels and our model's predictions. We'll need these complete sets to calculate our evaluation metrics.

We then loop through our validation dataset batch by batch. For each batch of images, we do the following:

1. Use `model.predict()` to generate probability scores for each image. Because we used a sigmoid activation in our output layer, these probabilities range from 0 to 1, indicating how confident the model is that an image shows a cat.

5.4 Implementing Core Concepts in Keras: Gradient Descent and Classification

2. Convert these probabilities to binary predictions using the expression `predictions > 0.5`. This means any image with a score above 0.5 is classified as a cat (1), while anything below is classified as not a cat (0). The `.flatten()` method transforms the result into a simple list.
3. Add both the true labels and our predictions to their respective lists using the `extend()` method, convert the true labels to NumPy arrays with `.numpy()`, and flatten them to ensure they match the format of our predictions.

By the end of this process, we have two aligned lists: `y_true` containing the actual classifications of all our validation images and `y_pred` containing our model's predictions for those same images. These lists form the foundation for all the evaluation metrics we'll calculate next—accuracy, precision, recall, and F1 score.

Now, we transform our raw predictions into meaningful performance measures and preserve our trained model for future use. This simple code is shown in Listing 5.32. The code employs scikit-learn's functions to calculate four essential metrics:

- `accuracy_score` reveals the percentage of correct predictions across all samples.
- `precision_score` tells us how many of our cat predictions were actually cats.
- `recall_score` shows what fraction of actual cats our model found.
- `f1_score` balances precision and recall in a single metric.

```
accuracy = accuracy_score(y_true, y_pred)
precision = precision_score(y_true, y_pred)
recall = recall_score(y_true, y_pred)
f1 = f1_score(y_true, y_pred)

print(f"Accuracy: {accuracy:.4f}")
print(f"Precision: {precision:.4f}")
print(f"Recall: {recall:.4f}")
print(f"F1 Score: {f1:.4f}")

# Save model for later use
model.save('cat_dog_classifier.h5')
```

Listing 5.32 Reporting the Metrics and Saving the Model

Saving the Model and Weights for Later Use

The last line of this listing, that is, the `model.save()` function call, captures our entire model in an HDF5 file, including architecture, weights, compilation settings, and optimizer state. This complete snapshot allows us to reload the model later with a single line of code, ready for immediate use without rebuilding or retraining.

This saved model becomes a valuable asset we can deploy in applications, share with colleagues, use for transfer learning, or benchmark against future versions. With this final step, we've completed the machine learning lifecycle—transforming raw data into a quantifiably effective classification system.

Often, we want to see the progress our model has made during training, how it reduced the loss, and how our metrics changed during the process. We can do that using the code given in Listing 5.33. This is just basic matplotlib code that gets the data to plot from the `history.history` structure. We got this structure back from `model.fit()` back in Listing 5.30.

```
# Plot training history
plt.figure(figsize=(12, 4))
plt.subplot(1, 2, 1)
plt.plot(history.history['accuracy'])
plt.plot(history.history['val_accuracy'])
plt.title('Model Accuracy')
plt.ylabel('Accuracy')
plt.xlabel('Epoch')
plt.legend(['Train', 'Validation'], loc='upper left')

plt.subplot(1, 2, 2)
plt.plot(history.history['loss'])
plt.plot(history.history['val_loss'])
plt.title('Model Loss')
plt.ylabel('Loss')
plt.xlabel('Epoch')
plt.legend(['Train', 'Validation'], loc='upper left')
plt.tight_layout()
plt.show()
```

Listing 5.33 Plotting the Losses and Other Metrics

Don't worry if the actual values of our metrics look underwhelming for now. We're learning the intricacies of machine learning. Once we're through a couple of other models, you would be able to achieve state-of-the-art results by changing just a couple of lines of this code.

Finally, we can go ahead and load the model again. This will presumably be done after some time has passed since our training and we want to either resume our work or deploy the model to another machine. All we have to do is copy over the *.h5* file and run the code shown in Listing 5.34.

```
from keras.models import load_model

# Load the saved model
model = None # zeros out the whole model
model_new = load_model('cat_dog_classifier.h5')
```

5.4 Implementing Core Concepts in Keras: Gradient Descent and Classification

```python
model_new.load_weights('model_weights.h5')

# Evaluate and calculate metrics
y_true = []
y_pred = []

for images, labels in validation_dataset:
    predictions = model_new.predict(images)
    pred_labels = (predictions > 0.5).flatten()
    y_true.extend(labels.numpy().flatten())
    y_pred.extend(pred_labels)

accuracy = accuracy_score(y_true, y_pred)
precision = precision_score(y_true, y_pred)
recall = recall_score(y_true, y_pred)
f1 = f1_score(y_true, y_pred)

print(f"Accuracy: {accuracy:.4f}")
print(f"Precision: {precision:.4f}")
print(f"Recall: {recall:.4f}")
print(f"F1 Score: {f1:.4f}")
```

Listing 5.34 Reloading a Saved Model

We've now successfully completed an image classification system using Keras, going from raw cat and dog images to a trained neural network capable of distinguishing between these animals with impressive accuracy. We started with data preparation—downloading and organizing images, then normalizing and batching them for efficient processing. We constructed a neural network using Dense layers with strategic Dropout layers for regularization, compiled it with appropriate loss and optimization settings, and trained it while monitoring performance. Along the way, we implemented callbacks to save our best model weights, evaluated our model using sophisticated metrics beyond simple accuracy, and preserved our complete model for future use. This workflow represents the fundamental pattern you'll follow for countless deep learning projects.

Mastering these concepts is crucial because they form the foundation upon which all advanced Keras applications are built. The workflow you've learned—data preparation, model construction, training with callbacks, evaluation with multiple metrics, and model preservation—represents the core cycle of deep learning development. As you progress to more complex architectures such as convolutional neural networks (CNNs) and transformers, the basic patterns will remain consistent. Understanding how to properly normalize data, select appropriate loss functions, configure effective callbacks, and meaningfully evaluate results will serve you in every deep learning project you undertake. These are essential skills that enable everything from medical diagnostics to autonomous vehicles. By mastering these fundamentals in Keras, you're developing the conceptual framework and practical skills that will allow you to tackle increasingly sophisticated problems with confidence.

5.5 Summary

In this chapter, we began our exploration of the practical applications of Keras, establishing the complete ecosystem that drives modern deep learning. We learned how to use the computational power of GPUs via Colab, enabling us to create complex neural networks without the need for specialized hardware. We peeled back the curtain on Keras's foundation, examining how NumPy's efficient array operations and symbolic computation enable the magic of automatic differentiation that makes gradient descent possible. The heart of our exploration was a hands-on implementation of a complete machine learning pipeline—from data acquisition and preprocessing to model building, training with callbacks, evaluation with nuanced metrics, and deployment through model saving. Through our cat and dog classifier, we witnessed how Keras elegantly abstracts away complexity while still providing fine-grained control at every stage of development. This chapter didn't just teach techniques; it established the mental framework and workflow patterns that underpin all deep learning projects, providing you with the solid foundation needed to venture into more specialized architectures and applications in the chapters ahead, starting with techniques that help us build models that generalize better.

Chapter 6
Regularization Techniques

In this chapter, we'll explore the critical challenge that every machine learning practitioner faces: creating models that generalize well beyond their training data. You'll learn how to identify when our model is too complex or too simple. By the end of this chapter, you'll have acquired essential tools for building neural networks that capture meaningful patterns without being led astray by noise in the training data.

As we've explored in previous chapters, machine learning models can capture increasingly complex patterns in data. We've moved from simple linear models to powerful neural networks, each time expanding our ability to represent sophisticated relationships. But this growing power comes with a subtle danger. Just as a sharp knife requires careful handling, powerful models demand proper control to avoid cutting ourselves. In this chapter, we tackle one of the most fundamental challenges in machine learning: how to create models that are powerful enough to capture real patterns in our data without being so flexible that they memorize the noise and peculiarities of our specific training examples. This balancing act lies at the heart of successful machine learning.

When models are too simple, they miss important patterns—such as attempting to explain quantum physics using only words a five-year-old would understand. Imagine trying to describe the rich, multifaceted flavor profile of a gourmet meal using only the words "yummy" and "yucky"—you'd capture the basic sentiment but miss all the nuance that makes the experience special. This is what happens with underfit models; they capture the general direction but miss the subtleties that matter. This problem, called underfitting, limits the usefulness of our predictions, leaving us with a crude approximation where precision is needed.

Conversely, when models are too complex relative to the amount of training data available, they can achieve perfect performance on training examples while failing spectacularly on new data. It's the equivalent of a student who memorizes exam answers without understanding the underlying principles—they might ace the practice test but freeze when the real exam presents questions with even slight variations. Think of an overly complex model as a gossip who creates elaborate narratives based on coincidences: "John always wears blue on days when the stock market goes up!" Such a pattern might perfectly explain past observations while being utterly useless for predictions. This problem, called overfitting, often betrays itself through a growing gap

6 Regularization Techniques

between training and testing performance—a warning sign that our model is becoming too fixated on the particularities of our training examples rather than learning generalizable patterns.

> **How to Approach This Chapter**
>
> The content of this chapter might seem mathematically dense, but it's perfectly fine if you don't fully grasp every detail right away. I encourage you to read through the content even when concepts feel challenging. We'll revisit key takeaways at the end, and I promise these techniques will make more sense once you see them in action. In practice, they are really easy to use.

In this chapter, we'll see the tools that allow us to navigate between these extremes. These tools let us use sophisticated, high-capacity models while restraining their tendency to overfit. They act like guardrails, keeping our models on the road to good generalization rather than veering off into the ditch of memorization. The general term for this balancing act is regularization. Here, we'll explore several powerful regularization approaches. We'll begin with the L2 regularization method, which constrains our model weights to prevent them from growing too large. We'll then examine dropout, which makes regularization very easy in practice. Along the way, we'll develop an intuitive understanding of the bias-variance trade-off that underlies these techniques and learn practical skills for implementing regularization in Keras. By the end of this chapter, you'll possess the essential tools to build models that not only perform well on training data but generalize effectively to new, unseen examples—the true measure of machine learning success. Let's begin working on this plan by revisiting our old friend—the housing prices dataset.

6.1 An Overview of Overfitting and Underfitting: Do You Need More Data?

In Chapter 3, we built a simple housing price model with just one feature. We assumed a *linear* relationship and thus drew a straight line through our data points, capturing the basic relationship between house size and price. This linear approach served us well as an introduction, but it's a bit like trying to describe a complex landscape using only straight paths—sometimes, you need curves to capture the real terrain. We later introduced neural networks as one way to capture nonlinear relationships, giving our models the flexibility to bend and curve in response to the data. But there's another, often simpler approach we can take: *polynomial features*. In the following sections, we'll take a look at how we can add flexibility to our models. We'll also discuss the problems that arise as we try to do this and then introduce generalized solutions to these problems.

6.1.1 From Lines to Curves: Adding Polynomial Features

Let's consider what happens if the relationship between house size and price isn't a straight line but a curve. Think about it intuitively: Perhaps each additional square meter becomes more valuable as the house gets larger. A jump from 1,500 to 2,000 square meters might add more value than a jump from 1,000 to 1,500 square meters. If the ground truth is that the relationship between inputs and outputs is *quadratic*, but our model assumes linearity, we may get some data points right, but the rest might not fit that well. Figure 6.1 shows what this model will look like. While not that bad for these few data points, it will break down very quickly for larger x-axis values. (For the following discussion, try not to focus on the actual y-axis values. Try to look at the bigger picture by following the trend lines.)

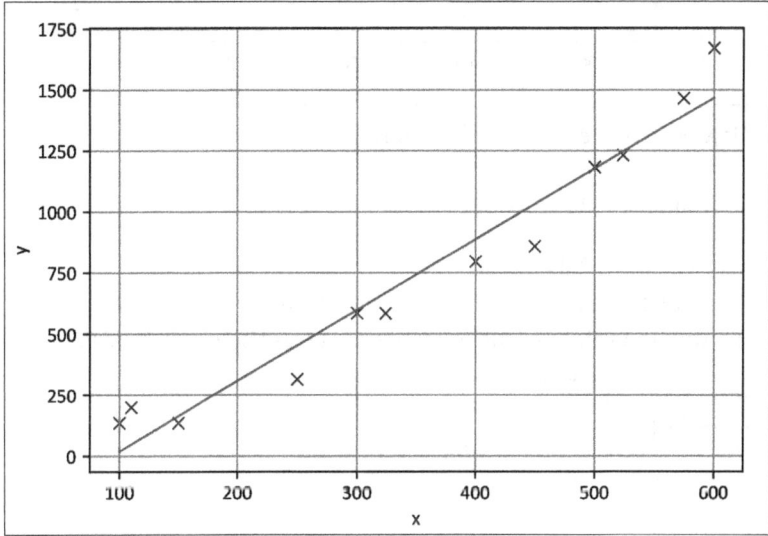

Figure 6.1 Linear Model for a Quadratic Relationship

We can model the underlying relationship between the area and the price as quadratic rather than linear. So, the ground truth is going to look like Figure 6.2. Compare these two models, and you'll notice that the quadratic line follows the data more closely, especially as the input values increase.

Notice how the squared values grow much faster than the original areas. As the house size increases linearly, the squared term increases quadratically. This gives our model the power to capture accelerating relationships—known as *polynomial regression*. This means that, instead of just using the original feature (house size), we create additional features by raising it to different powers. If our original model looked like this:

price = $\theta_1 \times area + \theta_0$

6 Regularization Techniques

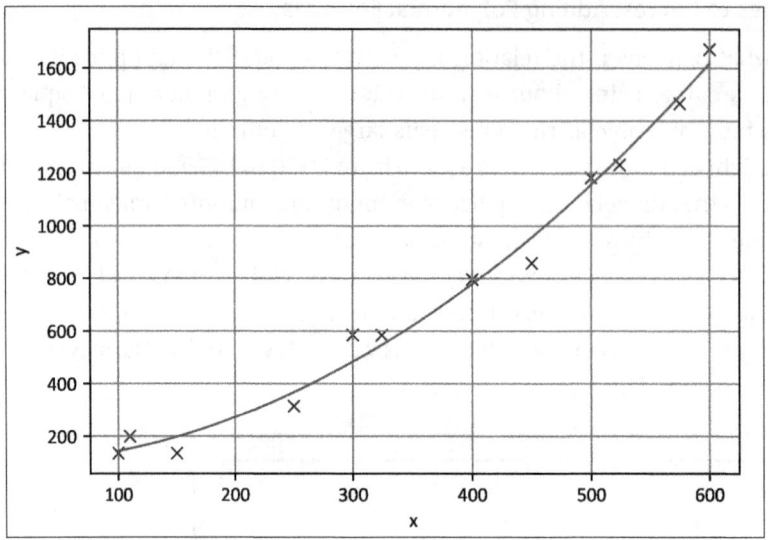

Figure 6.2 Quadratic Model for the Quadratic Relationship

A quadratic model would look like this:

price = $\theta_1 \times area + \theta_2 \times area^2 + \theta_0$

To incorporate this in our gradient descent algorithm, we simply have to add a new column containing the squared values of the original feature, as shown in Table 6.1.

Area	Area²	Price
145	21,025	808
202	40,804	993
305	93,025	727
365	133,225	582
450	202,500	1612

Table 6.1 Values of Features Squared

Adding this column is quite easy to implement in code, especially because this has to be done only once. We can augment our original dataset with these new pseudo-features and then save the modified dataset to the disk. After this step is done once, we only need to consider the modified dataset for polynomial regression. This can be easily achieved through Scikit-learn (sklearn) as well, as shown in Listing 6.1. Here, the Polyno-mialFeatures function acts like a feature multiplication workshop. When we set degree= 2, we're telling it to create all possible combinations of our features raised to powers up to 2. Because we only have one feature (house size), it simply gives us that feature and its square. The include_bias=False parameter tells it not to add a constant column of 1s

6.1 An Overview of Overfitting and Underfitting: Do You Need More Data?

because we'll do that in our model ourselves. (You can run the code through the **06-01-polynomial-features** notebook in the book resources on *https://recluze.net/keras-book* and on the book's official web page.)

When we call poly.fit_transform(X), the function analyzes our data to understand its structure and then creates the new polynomial features. The result, X_poly, now has two columns: the original house sizes and those same values squared.

```
import numpy as np
from sklearn.preprocessing import PolynomialFeatures

# Original feature - house sizes
X = np.array([[1000], [1500], [2000], [2500], [3000]])

# Transform into polynomial features
poly = PolynomialFeatures(degree=2, include_bias=False)
X_poly = poly.fit_transform(X)

print("Original features:")
print(X)
print("\nPolynomial features (degree 2):")
print(X_poly)
# Output --------------
# Original features:
# [[1000]
#  [1500]
#  [2000]
#  [2500]
#  [3000]]
#
# Polynomial features (degree 2):
# [[1.00e+03 1.00e+06]
#  [1.50e+03 2.25e+06]
#  [2.00e+03 4.00e+06]
#  [2.50e+03 6.25e+06]
#  [3.00e+03 9.00e+06]]
```

Listing 6.1 Adding Polynomial Features Using Sklearn

We've just transformed our single feature into two features—the original area and the area squared. This gives our model more flexibility to capture the curve in our data.

The beauty of this approach is that we can continue using our existing gradient descent algorithm from the previous chapters without any modifications. From the perspective of our gradient descent code, we're just adding another feature—it doesn't know or care

6 Regularization Techniques

that this new feature is derived from the original one. If we call our original feature x and our new squared feature v (where $v = x^2$), our model becomes

price $= \theta_1 \times x + \theta_2 \times v + \theta_0$

All our gradient descent equations from Chapter 3 still apply perfectly. We simply have one more weight (θ_2) to learn, but the fundamental process remains unchanged. The partial derivatives, weight updates, and convergence criteria all work exactly as before. In fact, we could add any number of transformed features (x^3, x^4, $log(x)$, etc.), and our gradient descent algorithm would still work without modification. This is one of the most powerful aspects of machine learning algorithms: once you understand the core principles, you can easily extend them to more complex scenarios by transforming your data rather than rewriting your algorithms.

6.1.2 Using Increasingly Complex Models

When we fit a linear model to our housing data, it can only draw a straight line. But when we add polynomial features, our model can fit curves of increasing complexity. Assume now that we don't know what the underlying relationship is. So, we go a step further and add 8 degrees of polynomial to our dataset instead of just the quadratic term. The relationship learned as a result will look like Figure 6.3. As you can see, the line follows even closer to the dataset but still misses out some points.

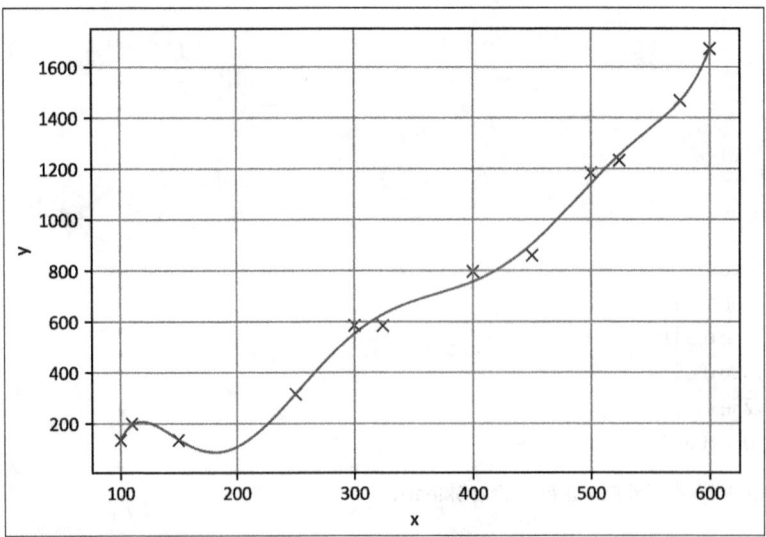

Figure 6.3 8th Order Polynomial Fitting the Same Data

Take this to an extreme and instead try to see what happens if we try to model a relationship that is a 12th degree polynomial. This leads to a curve fitting our data perfectly, as shown in Figure 6.4.

6.1 An Overview of Overfitting and Underfitting: Do You Need More Data?

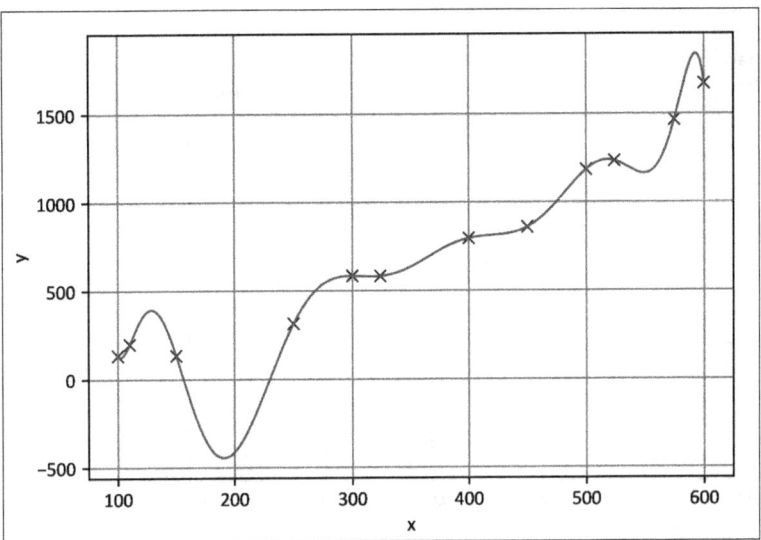

Figure 6.4 12th Order Polynomial Fitting the Same Data

The degree-1 model is just a straight line, while higher-degree models can capture increasingly complex patterns. As we increase the degree, our model becomes more flexible:

- Degree 1: A straight line that misses most points
- Degree 2: A gentle curve that follows the general trend
- Degree 8: A slightly more flexible curve that gets closer to all points
- Degree 12: A wildly oscillating curve that touches every point perfectly

Stop here and think about it! This flexibility is a double-edged sword. The degree-12 model fits our training data perfectly, but it's learned a complex pattern that's unlikely to generalize well to new houses. It's like memorizing the exact answers to the practice problems without understanding the underlying principles—you'll ace the practice test but struggle with new problems.

6.1.3 The Balance of Complexity

Adding more polynomial terms is like giving our model increasingly sophisticated vocabulary—but as with human language, more words don't always lead to clearer communication. Let's see what might go wrong if we try to use the 12th degree polynomial we modeled previously. To truly understand how our models perform, we need to look beyond the training data. That's where our test set comes in. These are data points our model hasn't seen during training—the equivalent of a pop quiz with new questions. Let's visualize how models of different complexities perform on both training and test data in Figure 6.5.

229

6 Regularization Techniques

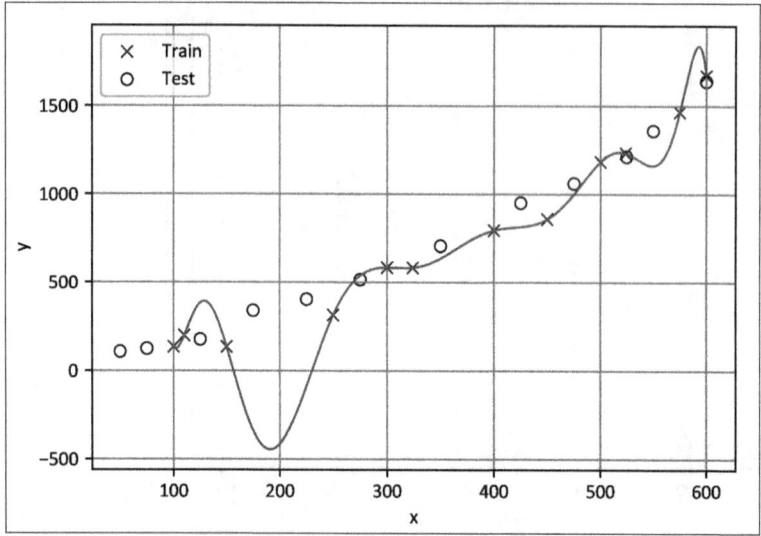

Figure 6.5 Introducing the Test Set into the Mix

Notice how the curve passes precisely through each *x* (training points), creating an elaborate, winding path. At first glance, this might seem impressive, like a student who can recite every example from the textbook perfectly. But the true test of learning isn't memorization; it's the ability to apply knowledge to new situations. When we look at how this model performs on the test data (the circles), we see concerning discrepancies. The most dramatic example occurs around *x = 200*, where our model takes a dramatic plunge to predict a value of -500. But the actual test point in that region shows a value closer to +400! That's a staggering 900-unit error.

> **[+] Interactive Demo for Polynomial Fits**
>
> You can play with an interactive demo we've created to show how the fit changes based on the order of the polynomial. You can find it on the book resources page at *https://recluze.net/keras-book* as **06-04-polynomial-fit-demo**.

Our model has become so fixated on perfectly fitting every training point that it's created an unnecessarily complex explanation. Between *x = 150* and *x = 250*, the model noticed there weren't many training points, so it took a wild dive downward before shooting back up. The algorithm didn't "decide" to do this out of malice or confusion; it simply found that this rollercoaster curve was mathematically the best way to hit all the training points perfectly. So, there are two extremes that we can face while deciding on model complexity. We'll discuss these in the following sections and provide guidance for finding the sweet spot in between.

When Models Are Too Simple: Underfitting

Looking at our degree-1 polynomial (the straight line), we notice it performs poorly on both the training and test data. The model fails to capture the curvature that clearly exists in our housing price relationship. The training error is high, and the test error is similarly high.

As mentioned earlier, this problem is called *underfitting* or *high bias*. It's like trying to explain quantum mechanics using only elementary school vocabulary—the tools are simply inadequate for the task at hand. Our model has made an overly simplistic assumption about the world (that housing prices are a linear function of area), and this bias prevents it from learning the true pattern. This assumption is built into the very structure of our linear model. By restricting ourselves to a straight line, we're essentially declaring that each additional square meter adds exactly the same value to a house, regardless of the home's size. This ignores economic realities such as premium pricing for larger homes, the diminishing utility of extra space beyond certain thresholds, or the way different size brackets might appeal to different market segments.

This is what high bias means in machine learning: The model has strong preconceptions about what the underlying function should look like, and these preconceptions are wrong. No amount of additional training data will help a linear model fit a quadratic relationship—it simply doesn't have the capacity to represent the curve. In practice, it's usually easy to recognize a situation in which a model is underfitting. This is usually when we have very high training loss. This means that our model can't even memorize the data points given to it, let alone predict unseen data points.

> **The Term High Bias**
> Underfitting is called high bias because the model is biased toward its own simplistic understanding rather than adapting to the true complexity of the data. It doesn't want to learn from its mistakes.

When Models Are Too Complex: Overfitting

On the other end, look at the degree-12 polynomial. It creates a wildly oscillating curve that passes almost perfectly through each training point. The training error is nearly zero—our model has essentially memorized the training data! But look what happens with the test points: The model's predictions are way off. The test error is enormous compared to the training error. Our model has learned the peculiarities of our specific training examples rather than the general relationship between house size and price.

To understand this distinction, we need to recognize that any dataset contains two components: the *underlying pattern* we want to learn and the *random noise* or peculiarities specific to our sample. The underlying pattern is the true relationship that generalizes across all houses—perhaps house prices increase with area following a gentle quadratic curve due to fundamental market dynamics. This pattern applies to houses we haven't seen yet and represents the predictive knowledge we're actually seeking. The peculiarities, on the other hand, are random fluctuations specific to our training examples—perhaps one house in our training set sold for slightly more because it had a particularly nice view, or another sold for less because the owner needed to move quickly. These factors aren't captured in our features and appear as random noise in our data. When we use a degree-12 polynomial to fit just a handful of training points, we're giving our model enough flexibility to memorize not just the general trend but also these random fluctuations. It's like creating an elaborate theory to explain why one student got a 93% and another got a 91% on a test, when the difference might just be that one student guessed correctly on one more question.

As mentioned at the beginning of the chapter, this problem is called *overfitting* or *high variance* and is recognizable through a key indicator: very low error on training data along with significantly higher error on test data that leads to a large performance gap between training and test results. Overfitting occurs when our model is so flexible that it captures not only the underlying pattern but also the random noise in our training data.

> **The Term High Variance**
> Overfitting is called high variance because the parameters the model learns are going to vary a lot depending on which data points are used to train it. If we pick odd-numbered data points, the model will learn their peculiarities, which will be quite different from the specific noise in even-numbered data points.

Finding the Sweet Spot

The degree-2 polynomial seems to strike a good balance. It's flexible enough to capture the curvature in our data without going overboard with unnecessary complexity. Both the training and test errors are relatively low, and they're similar to each other—a sign that our model is generalizing well.

This illustrates a fundamental principle in machine learning: the *bias-variance trade-off*. As we increase model complexity, the following happens:

- Bias tends to decrease (the model can represent more complex patterns).
- Variance tends to increase (the model becomes more sensitive to the specific training examples).

Our goal is to find the sweet spot that balances these two sources of error. It's like stretching a rubber band to wrap around a package—not stretching enough leaves it too loose to hold anything together (underfitting), while stretching it too far causes it to snap (overfitting). We need just the right amount of tension that secures the package without breaking the band.

6.1 An Overview of Overfitting and Underfitting: Do You Need More Data?

Important Issues to Keep in Mind

Remember that our algorithms don't "see" the data the way we do. When we perform gradient descent, the machine isn't visualizing curves or making aesthetic judgments about whether or not the line fitting the points is visually appealing. It's simply calculating a loss value and following the gradient downward, regardless of whether the resulting function is sensibly smooth or wildly oscillating between data points. This presents a fundamental challenge that becomes even more pronounced with high-dimensional data. In our housing example, we could visualize the relationship between house size and price in a simple 2D plot. But what if our model includes dozens or hundreds of features? We lose our ability to eyeball the complexity of the relationship, and even experts can't intuitively judge when a model with 50 features is using a relationship that's too complex.

We have two broad approaches to address this challenge. The first, which we touched on in Chapter 3, is to simply make an assumption about the appropriate complexity for our problem. We might decide based on domain knowledge that a quadratic relationship makes sense for housing prices or that certain features should interact while others shouldn't. This approach works well when we have strong *prior knowledge*, but it's not always available.

The second approach—which we'll focus on throughout the rest of this chapter—is to develop automated methods that help us determine the right level of complexity. We need mechanisms that allow our models to find that balance themselves without requiring us to manually specify the perfect polynomial degree or feature interactions.

Another critical point to remember is that we must make these decisions without peeking at the test set. Using test data to choose our model's complexity is like getting access to exam questions before a test—it defeats the purpose of having an independent evaluation. This form of *data leakage* can give us an overly optimistic view of how well our model will perform on truly new data when deployed in the real world.

> **Always Keep the Test Set Separate**
> When writing code, always make sure that you aren't using your test data points during training. While this might lead to good laboratory results, the model's real success—that in the real world—will become unpredictable!

These techniques are guardrails that keep our models on the right path. Rather than manually deciding when to stop adding complexity, we'll develop systems that naturally prefer simpler explanations unless there's strong evidence for complexity. As we explore these techniques, remember that finding the right complexity isn't simply a theoretical concern. It's fundamental to creating models that work in the real world. A model that perfectly memorizes training data but fails on new examples is like a student who can recite textbook pages verbatim but can't apply the concepts to new problems.

6 Regularization Techniques

Our goal is to build models that truly learn the underlying patterns. They shouldn't be limited to the specific examples we've shown them.

But how do we come up with complex models that capture intricate patterns in our data without overfitting? That's where regularization comes in—a technique that allows us to use complex models while preventing them from overfitting. We'll explore this powerful approach in the next section.

6.1.4 Regularization Term

Now that we've identified the problem of overfitting, let's discuss a powerful solution that doesn't require us to manually restrict our model's complexity. Rather than deciding in advance which polynomial degree to use, what if we could use a high-degree polynomial but somehow encourage it to behave more reasonably? This is where *regularization* enters the picture. Regularization works by modifying our loss function—the very compass that guides our model during training. Let's recall the loss function we used in Chapter 3 for our housing price prediction:

$$J(\theta) = \frac{1}{m}\sum_{i=1}^{m}(\hat{y}_i - y_i)^2$$

This function only cares about one thing: how well our predictions ($\hat{y}_i$) match the actual values (y_i) in our training data. It's like a teacher who grades solely on getting the correct final answers without considering how students arrive at those answers or whether their methods would work on different problems.

We're going to add a term to this loss function. It might seem strange at first but we'll explain the "why" of this in a minute. This is called the *L2 regularization term* and is given as

$$\lambda \sum_{j=1}^{n} \theta_j^2$$

where λ (lambda) is a new hyperparameter that controls complexity, and θ_j represents each parameter in our model. Our complete regularized loss function becomes

$$J(\theta) = \frac{1}{m}\sum_{i=1}^{m}(\hat{y}i - y_i)^2 + \lambda \sum_{j=1}^{n} \theta_j^2$$

We can modify the value of this parameter at will. First, let's consider what happens when we set $\lambda = 0$. In this case, our loss function simplifies to

$$J(\theta) = \frac{1}{m}\sum_{i=1}^{m}(\hat{y}i - y_i)^2 \quad + \quad 0 \cdot \sum_{j=1}^{n} \theta_j^2 \quad = \quad \frac{1}{m}\sum_{i=1}^{m}(\hat{y}_i - y_i)^2$$

We're right back to our original mean squared error (MSE)! The regularization term completely disappears, and our model is free to use any parameter values it wants, however large, to fit the training data perfectly. This is the world we were living in before, where our polynomial models could create those wild oscillations to pass through every training point. See the solid lined contour plot in Figure 6.6 for what the loss function landscape will look like in this case. In this example, we get $\theta_1 = 10, \theta_2 = 5$ for our model parameter values that minimize the loss. If you want a refresher on contour plots, please refer to Chapter 3, Section 3.2.1.

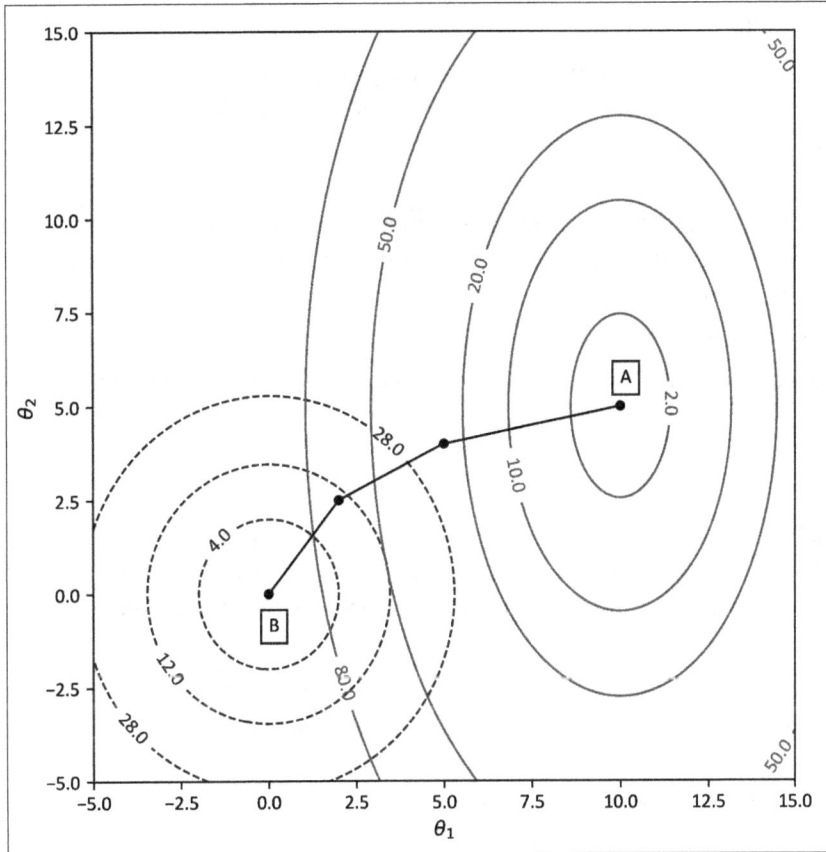

Figure 6.6 Effect of Lambda on Contour Plots and Minimizing Values

But what happens at the other extreme? Let's say we set λ to a very large value, such as *1000*. Now our loss function becomes overwhelmingly dominated by the regularization term. The whole MSE loss function doesn't matter much at all because the regularization term has all the say in the final value of the overall loss.

> **Effect of Weights: A Simple Example**
> To fully internalize what λ is doing, consider
> $p = q + W \cdot r$

> where *W* (our hyperparameter) is a very large number such as *1000*. If *q* = *10* and *r* = *0.01*, then
>
> $p = 10 + 1000 \cdot 0.01 = 10 + 10 = 20$
>
> Now, if we change *q* by adding *2* to it, *p* becomes
>
> $p = 12 + 1000 \cdot 0.01 = 12 + 10 = 22$
>
> On the other side, if we merely change *r* a tiny bit to *0.015*, then
>
> $p = 10 + 1000 \cdot 0.015 = 10 + 15 = 25$
>
> Even though we made a much smaller relative change to *r*, its effect on the final result is more significant than a large change in *q* because it's being multiplied by that large weight *W*.

When λ is very large, the MSE term becomes almost irrelevant compared to even tiny changes in the regularization term. Our gradient descent algorithm will focus almost entirely on minimizing the regularization term, with little regard for how well the model actually fits the training data.

So, what does minimizing the regularization term mean? Let's look at it again:

$$\lambda \sum_{j=1}^{n} \theta_j^2$$

When we're trying to minimize this expression and λ is fixed (even if it's large), the only way to make it smaller is to reduce the sum of squared parameters. And the absolute minimum value this sum can take is *0*, which happens when all parameters θ_j are exactly *0*.

This is a critical insight: When λ is extremely large, our model will push all parameters toward *0*, effectively ignoring the actual relationships in the data. Even increasing one parameter to a small value like *0.1* would add $0.1^2 = 0.01$ to our sum, which gets multiplied by our large $\lambda = 1000$ to add *10* units to our loss—a huge penalty compared to the minor improvement it might make to the MSE term. Refer back to Figure 6.6. In this case, when λ is very large, the only thing that affects the loss is the regularization term, which is forcing the loss to have a minimum when both parameter values are *0* (or extremely close to it). Remember, the MSE term is what's looking at our data points and measuring how well our predictions match the actual house prices. It's the part of our loss function that connects our model to reality. If we ignore it by setting λ too high, we're essentially telling our model, "Don't worry about predicting house prices accurately—just make sure all of your parameters are as close to 0 as possible!"

With all parameters at or near *0*, our model would predict virtually the same value for every house, regardless of its size or any other features. This is the opposite extreme of overfitting—we've now created a model that's too simple to capture any meaningful patterns in our data, which is a classic case of underfitting.

6.1 An Overview of Overfitting and Underfitting: Do You Need More Data?

To summarize, consider what this means: With a large value of λ, we're saying that all values of thetas should be 0. This feels like a mathematical quirk, but it's a fundamental reshaping of our model's underlying behavior.

In turn, we're saying that none of the inputs have any relation to the output, so we completely disconnected the inputs from the predicted output. That is obviously not what we want. Imagine trying to predict house prices without considering size, location, or any features at all! It would be like a real estate agent who gives the same price estimate for every house, regardless of whether it's a mansion or a studio apartment. But now, we have two extremes: when λ is 0, the regularization term has no effect, but when it's very large, the regularization term dominates and says that *none* of the inputs have any effect on the output. We've mapped out two boundaries of a spectrum—on one end, our model is free to create wild, complex relationships that overfit our training data, and on the other, it's constrained to the point of uselessness, ignoring all input features entirely.

What we want is to somehow control this by slowly turning the knob of λ. Think of λ as a sensitivity dial on a sophisticated instrument. Turn it too low, and the readings become chaotic and unreliable—our model overfits. Turn it too high, and the instrument becomes unresponsive—our model underfits. Somewhere in the middle lies the sweet spot where our model is responsive to genuine patterns in the data without being overly sensitive to noise.

6.1.5 Adjusting the Complexity Knob

Refer back to Figure 6.6 one last time. When we change the λ from 0 to a very high value, this takes us slowly from point A in the figure to point B. During this movement, some of the inputs are ignored less, and some are ignored more. Point A sits far from the origin, with large values for both θ_1 and θ_2. This is where our model lands when λ equals 0—we're only concerned with minimizing the MSE, and we've found parameters that fit our training data exceptionally well. But these large parameter values suggest a complex model that might be overfitting.

Point B, in stark contrast, sits much closer to the origin. As λ increases, our model is pulled toward this simpler configuration with smaller parameter values. It's like watching a ball rolling downhill, but the landscape itself is changing as we adjust λ, creating a stronger gravitational pull toward the origin.

The black solid line connecting A and B traces the path our model takes as we gradually increase λ. Notice how the path isn't simply a straight line toward the origin—it follows a nuanced trajectory influenced by both the shape of our error surface (solid contours) and the regularization penalty (dashed circles).

This visualization reveals something profound about regularization: It doesn't simply shrink all parameters equally. As we move from A toward B, some parameters decrease more rapidly than others. The model is making strategic sacrifices, reducing the influence of some inputs while preserving others that provide more explanatory power per unit of regularization cost.

It's similar to how a company might respond to budget cuts—not by reducing every department equally, but by strategically preserving critical functions while scaling back areas that deliver less value relative to their cost. Our model likewise preserves the most efficient parameters while penalizing those that contribute less to predictive accuracy relative to their size. When this happens, the model might decide to make the weights associated with 4+ degree polynomials 0, thus effectively reducing the model complexity. We have a way of adjusting complexity by just changing one value—the λ hyperparameter! This movement between A and B represents the fundamental trade-off at the heart of regularization—finding the sweet spot where our model is complex enough to capture true patterns in the data but simple enough to avoid fitting noise.

But how do we figure out where to stop? How do we know which value of λ gives us that perfect balance between simplicity and accuracy? Here's the plan: We already know that we can't trust our training loss. So, we're going to set λ to a particular value, run the whole training/testing, and come up with a final test loss value for this particular λ value. Then, we'll change the λ value and rerun the whole thing again—this time getting a different overall test loss. Remember, training loss isn't important. That is always going to be minimum for an overfitting model.

So, we end up with this experimental process:

1. Choose a specific λ value.
2. Train a model using this λ value on your training data.
3. Evaluate the model's performance on test data to get a test loss.
4. Record the λ value and corresponding training and test losses.
5. Select a different λ value.
6. Repeat steps 2–5 for multiple λ values.
7. Plot the relationship between λ values and training and test losses.
8. Identify the λ value that produces the lowest test loss.

Each experiment gives us a point on a curve showing the relationship between regularization strength and generalization performance. It's like a scientist methodically testing different conditions to find the optimal formula. A typical example of this plot is shown in Figure 6.7.

The figure illustrates the dance between underfitting and overfitting as we adjust our regularization strength. Take a careful look at the x-axis—it's actually plotted in reverse, with high λ values on the left (*1.0*) decreasing to 0 on the right. This reverse scaling helps us visualize the journey from simple to complex models as we move from left to right.

6.1 An Overview of Overfitting and Underfitting: Do You Need More Data?

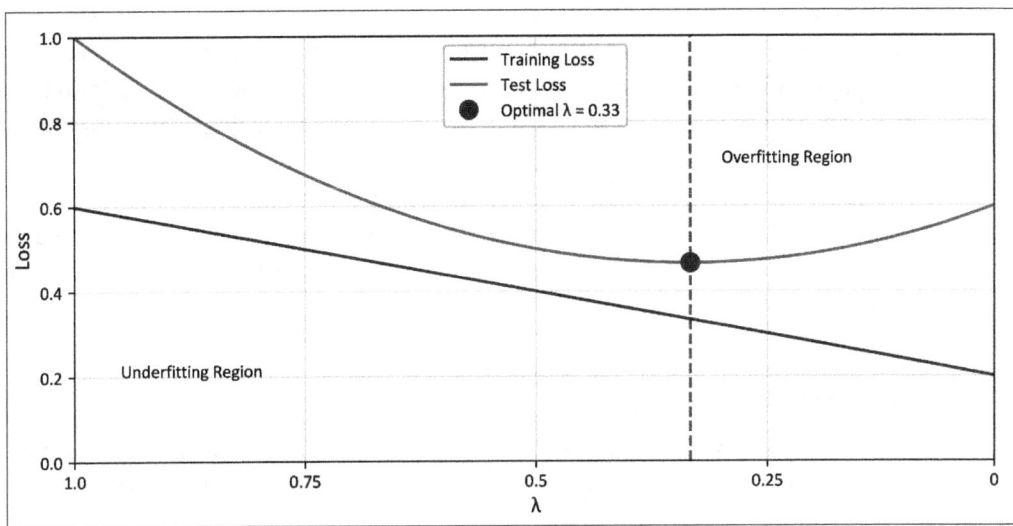

Figure 6.7 Effect of Regularization on Training and Testing Losses

When we start on the left side with a high λ value, we're firmly in the underfitting region. Here, our model is playing it too safe—most parameters are essentially 0, and the model is largely ignoring the patterns in our data. The training loss is high because our model isn't capturing much of anything, and the test loss is equally disappointing.

As we begin to decrease λ, moving rightward on the graph, both training and test loss start to decrease. This is the sweet spot of learning—our model is gradually embracing more complexity, discovering genuine patterns in the data rather than just playing it safe. The model is learning something meaningful from the data, and this learning generalizes well to new examples. But notice what happens as we continue moving right, further decreasing λ below 0.33. The training loss continues its downward trajectory—after all, with less regularization, our model can fit the training data more and more precisely. It's like a student who memorizes the textbook examples perfectly.

However, the test loss reaches its minimum at $\lambda = 0.33$ (marked by the dot and vertical dashed line) and then begins to climb back up. This inflection point tells us something crucial: our model has started to overfit. It's no longer just learning meaningful patterns; it's starting to memorize the quirks and noise in our training data. Like an actor who rehearses so rigidly that they can't adapt to unexpected circumstances, our model is losing its ability to generalize.

The Goldilocks Zone

The visualization in Figure 6.7 perfectly captures the Goldilocks principle in machine learning: With high λ, our model is too simple (underfitting); with low λ, it's too complex (overfitting); but at $\lambda = 0.33$, it's just right—complex enough to learn meaningful patterns but not so complex that it gets distracted by noise.

This vertical line marks this optimal value—the point where we should stop reducing λ and declare this as our best model. It has just the right amount of complexity, balancing the need to fit the training data against the risk of memorizing noise. It's found the sweet spot between ignoring the data (underfitting) and believing everything it sees (overfitting). What's particularly elegant about this approach is that we didn't need to manually specify how complex our model should be. Instead, by systematically exploring different λ values and measuring their effect on test performance, we let the data itself tell us the right level of complexity. This data-driven approach to model selection is at the heart of modern machine learning.

6.1.6 Do I Need More Data

Now we're in the position to answer that difficult question in machine learning practice: Should I get more data to make my machine learn better. The answer is "not always." It's a bit like wondering if you need more ingredients to improve your cooking—sometimes the secret lies not in more stuff but in better technique—or in our case, complexity. Data collection is expensive, especially *labeled data*. Getting it in good quality is more difficult. Think of labeled data like handcrafted furniture—each piece requires careful human attention, making it valuable and time-consuming to produce. So, we need to make sure our model actually needs more data before spending the time and effort to collect it.

Before rushing to gather more examples, you first need to use whatever data is available and plot your complexity curves, as we saw in the previous subsection. This diagnostic step is like a doctor running tests before prescribing medication—we need to understand the nature of our model's struggles before we can properly address them. Then, figure out if you're in the overfitting or underfitting zone. If you're underfitting, first increase the complexity of the model by adding higher order polynomials or add complexity in other ways. This is like upgrading from a bicycle to a car when you're consistently arriving late—sometimes, the tool itself lacks the necessary capacity.

In neural networks, adding more complexity means adding more layers or more neurons. Each layer in a neural network acts like a team of specialists that processes information in increasingly sophisticated ways. The first layer might identify simple patterns such as edges or colors, while deeper layers combine these basic observations into complex concepts such as "this is a face" or "this sentence expresses disappointment." Adding more neurons is like hiring more specialists within each team, allowing for more nuanced processing.

For example, if you're building an image recognition system that struggles to distinguish between cats and dogs, adding more layers might help it recognize hierarchical features—moving from identifying basic shapes to understanding fur patterns to recognizing distinctive facial structures. Similarly, adding more neurons might help it become sensitive to subtle distinctions within these categories. When you do this, you'll get a better understanding of which zone you're in. It's like turning up the volume on your model's capabilities to see if that solves the problem.

If you're underfitting, first make your model complex enough. Don't rush to collect more data if your current model isn't even capable of capturing the patterns in your existing dataset.

If you're overfitting, then more data can usually help because as you get more data, there are more patterns and complexity, and your model can actually find these patterns and decrease the loss. Think of it like teaching someone to recognize bird species—if you only show them three examples of eagles, they might fixate on irrelevant details such as the background color. But show them hundreds of eagles in different positions, lighting conditions, and environments, and they'll start to focus on the truly distinctive features. Additional data in the overfitting zone works because it forces your model to find generalizable patterns rather than memorizing specific examples. It's the difference between a student who has memorized a few specific math problems versus one who has solved enough variations to understand the underlying principles that apply to all problems of that type.

This strategic approach to model improvement—diagnosing the problem first, then addressing it appropriately—saves you from the costly mistake of gathering unnecessary data or the frustration of trying to force a too-simple model to perform complex tasks. It's about working smarter, not just collecting more. Once all of this is done and we have the best value for our hyperparameter λ, one final test for our model remains.

6.1.7 Reporting the Final Results: Validation Set

Hold on a minute—we've just uncovered a subtle but critical issue in our approach. Remember how we used our test set to pick our hyperparameter value λ? There's a hidden danger here that we need to address.

Think about what we've been doing so far. We used our training data to learn the model parameters, which makes perfect sense. Then, we used what we've been calling our "test set" to find the best λ value. But here's the catch: By using this test data to make decisions about our model structure, we've actually leaked information from the test data into our design process. Like peeking at part of the final exam while studying, it compromises the integrity of our evaluation.

Just as we used the training set to pick the model parameters and then used the test set to ensure that these parameters were correct, we need to somehow ensure that the λ picked using the test set is also correct. We need a truly independent evaluation that hasn't influenced any of our choices. For this, we're going to need some more data, but we used up all our data during training and testing (*hyperparameter optimization*). What now? Don't worry. We can still keep the same process but start slightly differently. In step 1 of our machine learning process, we won't split the data into just training and testing. We'll split it in three parts: *training*, *validation*, and *test* sets. The training part is self-evident—it's the data our model learns from directly, adjusting its parameters to minimize the loss function. The second set that we use to select the best hyperparameters—what we've been calling the test set until now—will be called the validation set. It's like a practice exam that helps us fine-tune our study strategy before the real thing.

Once the hyperparameter best value is decided, we'll use this singular value to run the whole model on the third set, which we'll now call the test set as it's doing the final testing of our model. This test set is our genuine, untouched evaluation data—it's like the sealed final exam that only gets opened when everything else is ready and once you take that, you can no longer retake it.

Whatever result is achieved by the model on the test set, we report that. If it's not good, our model didn't learn well. This lets us ensure that when the model is deployed in the real world, it will likely perform just the same as it did on the test set. By maintaining this discipline, we build models that don't just perform well on data they've seen before but are truly prepared for the challenges of the real world.

> **[+] The Three-Way Split of Data**
>
> This three-way split creates a clean separation of concerns:
> - Training data teaches our model parameters.
> - Validation data helps us choose the best hyperparameters.
> - Test data gives us a true evaluation of our final model's performance.
>
> This is the scientific method for machine learning: We form hypotheses with the training data, refine them with the validation data, and then conduct the final experiment with the test data.

To summarize this section, regularization gives us a powerful way to manage complexity, acting like a judicious editor that simplifies our model by gradually silencing unnecessary parameters—encouraging our model to generalize from patterns rather than memorize specific examples. However, this approach does come with its own challenges, particularly in computational efficiency. To calculate the best value of λ, we have to run the entire training multiple times in a loop, which might not be feasible for large-scale modern machine models.

In the next section, we'll explore an alternative technique that maintains this core philosophy of controlled complexity while addressing this limitation. The concepts we've learned until now will still be useful, but we'll find an easier way to reduce unnecessary complexity.

6.2 Dropout: Concept and Implementation

Take a look at Figure 6.8. Even though a large part of the object in front of the keyboard is obstructed, you can tell that it's unmistakably a banana. Despite this obstruction, your brain can still identify it easily. You've learned the concept of "banana-ness" so well that you don't need to see the entire fruit to recognize it even if it's surrounded by other unrelated objects.

Figure 6.8 An Obstructed Image

Neural networks, however, often struggle with this kind of robust recognition because they tend to fall into a trap known as co-adaptation.

6.2.1 The Problem: Co-Adaptation and Overfitting

Co-adaptation occurs when neurons in a network become overly reliant on one another during training. Recall that in neural networks, neurons are all connected to each other in adjacent layers, with each neuron feeding its learned features forward to neurons in the next layer. This interconnected structure is powerful, but it creates an environment where *co-dependencies* can easily form. Instead of learning independent, robust features, neurons develop intricate dependencies. This is like a study group where, instead of each student developing their own understanding of the material, they become completely dependent on each other's notes and can't function independently. In our neural network, this means that specific neurons start to compensate for the mistakes or biases of others. Rather than learning complementary features that contribute to overall understanding, they form a fragile ecosystem where each neuron's output critically depends on precisely what other neurons are doing.

Think about our banana image again. A robust network should recognize bananas based on multiple independent features: color, shape, texture, and size. But a *co-adapted network* might develop strange dependencies where one neuron only activates if another specific neuron has activated in a particular way. Remove or alter any piece of this intricate puzzle, and the whole recognition system falls apart.

6.2.2 The Road to Memorization

This co-adaptation leads directly to overfitting, which is perhaps the most persistent challenge in machine learning. Instead of learning generalizable patterns from the training data, the network essentially memorizes the training examples. When this happens, our network performs brilliantly on training data but fails miserably when faced with new examples. The network has learned the peculiarities and even the noise in the training set rather than the true underlying patterns that would allow it to generalize to new situations.

The consequences of this co-adaptation and overfitting are far from academic. In real-world applications, they manifest as follows:

- Medical diagnosis systems that work perfectly in the lab but fail with patients from different demographics
- Facial recognition systems that can't handle varying lighting conditions
- Recommendation engines that can't adapt to shifting user preferences
- Autonomous vehicles that struggle with unfamiliar road conditions

In each case, the network has learned to rely too heavily on specific patterns in the training data rather than developing robust, generalizable features.

The elegant concept of *dropout* was designed to solve this very issue. The strategy is pretty simple: Randomly ignore some neurons during training. That's it—that's the whole strategy. By randomly "dropping out" neurons during training, we force the network to build *redundancy* and *resilience*. Each neuron can no longer rely on any other specific neuron being present, so it must learn features that are robust even when parts of the network are missing—just like how you can still recognize a banana even when most of it is covered by a white rectangle.

Let's consider how this simple concept cleverly simulates an *ensemble* of different networks to combat this problem of co-adaptation and overfitting.

6.2.3 The Ensemble Intuition Behind Dropout

Picture yourself assembling a team for an important project. Would you rather have one brilliant but temperamental expert who might not show up on the day of your presentation, or would you prefer a diverse group of solid performers who can cover for each other if someone falls ill? Most of us would choose the reliable team. This is precisely the wisdom behind dropout, as we'll discuss in the following sections.

Training an Ensemble in Disguise

At its heart, dropout is a clever illusion. While it appears we're training a single neural network, we're actually training thousands of different networks simultaneously. It's like having a theater company where, for each rehearsal, some actors randomly call in sick, forcing the remaining cast to adapt and cover their roles. When we apply dropout to a neural network with 100 neurons and set our dropout rate to 0.5 (meaning each neuron has a 50% chance of being temporarily disabled), we're effectively creating 2^{100} different possible network configurations. That's more possible networks than there are atoms in the observable universe! Each training iteration randomly samples one of these possible networks, trains it for that batch, and then moves on to another random configuration.

This is the magic of dropout: instead of training one massive, co-dependent network, we're training an implicit ensemble of thinner, more robust networks that are forced to work independently.

Connection to Other Ensemble Methods

The concept behind dropout is part of a broader class of machine learning techniques known as *ensemble methods*. In traditional machine learning, algorithms such as *random forests* improve on basic *decision trees* by training on different subsets of features and examples to create a more robust predictor.

If you're interested in exploring these connections further, looking into traditional ensemble methods such as *bagging* and *boosting* can be illuminating. These approaches explicitly train multiple complete models and then combine their predictions. What makes dropout special is that it achieves a similar ensemble effect implicitly within a single model, making it computationally efficient while still capturing the wisdom of the crowd.

The key insight that connects dropout to classical ensemble methods is this: *Diversity* in learning leads to *robustness* in prediction. By forcing different parts of the network to function independently, we create a system where errors tend to cancel out rather than compound. This ensemble intuition helps explain why, counterintuitively, deliberately handicapping our network during training by randomly shutting off neurons actually leads to better generalization. It's the neural network equivalent of the old saying, "What doesn't kill you makes you stronger."

In the next section, we'll dive into the nuts and bolts of how dropout is actually implemented, including the subtle but important scaling adjustments needed to make the technique work properly.

6.2.4 Dropout Mechanics

Now that we understand the "why" behind dropout, let's roll up our sleeves and explore the "how." Like many brilliant ideas, dropout's implementation is surprisingly straightforward, though there are some subtle but crucial details that make all the difference, as we'll discuss in the following sections.

Two Phases: Training vs. Inference

Think of dropout as wearing two different hats—one during training and another during testing (or inference). This dual personality is key to understanding how dropout works in practice. During training, dropout behaves like that one friend who's always canceling plans at the last minute. As your network processes each batch of data, dropout randomly selects neurons and says, "Sorry, can't make it today!" These neurons are temporarily removed from the network, their outputs are set to zero, and they don't contribute to the forward pass or receive updates during backpropagation.

When it's time for the real show—the inference phase where your model makes predictions on new data—dropout suddenly becomes completely reliable. All neurons show up for work with no exceptions. This Jekyll and Hyde behavior serves a purpose. During training, the random dropout forces the network to build redundancy and resilience. During inference, we want our model to use all of its resources to make the best possible predictions.

The Mathematics of Scaling: Balancing the Books

There's a mathematical wrinkle in this that we need to iron out though. If we're turning off, say, 50% of our neurons during training but keeping them all on during inference, won't this create a mismatch in the scale of activations? This is a crucial detail that the original dropout paper addresses. Let's walk through it with a simple example.

Consider the case where you have a layer with 4 neurons, each outputting the value 2. The total output sum is 8. Now, if we apply dropout with a rate of 0.5, we randomly disable 2 of these neurons, leaving us with a sum of 4 instead of 8. That's a significant reduction!

To compensate for this reduction during training, we scale up the remaining activations by dividing by *(1 – dropoutRate)*. In our example, we'd multiply the output of each remaining neuron by *1/(1 − 0.5) = 2*. So, our 2 remaining neurons now output 4 each, giving us a total of 8 again—matching the expected magnitude of the full network.

Inverted Dropout: The Modern Approach

In early implementations, the scaling correction was applied during inference—all activations were multiplied by *(1 − dropoutRate)*. However, this approach has a disadvantage: You need to apply different operations depending on whether you're in training or inference mode.

Modern implementations use what's called *inverted dropout*, which flips this approach. With inverted dropout, the following occurs:

- During training, we scale up the remaining activations immediately.
- During inference, no scaling is needed at all.

This shift simplifies deployment by ensuring the inference-time computation is exactly what you'd expect without dropout. In code, inverted dropout looks like that given in Listing 6.2.

```python
def inverted_dropout(x, dropout_rate):
    if training:
        # Generate binary dropout mask
        mask = np.random.binomial(1, 1-dropout_rate, size=x.shape)
        # Apply mask and scale
        return (x * mask) / (1 - dropout_rate)
    else:
        # During inference, no changes
        return x
```

Listing 6.2 Inverted Dropout in Python

6.2.5 Finding the Sweet Spot: Dropout Rates in Practice

Not all layers are created equal when it comes to dropout. Through years of experimentation, the deep learning community has developed some rules of thumb for dropout rates:

- **Input layers**
  Light dropout (0.1–0.2) or none at all.
- **Hidden layers**
  Moderate dropout (0.3–0.5).
- **Very deep networks**
  Increasing dropout for deeper layers.
- **Convolutional layers**
  Lower dropout rates than fully connected layers.

- **Recurrent layers**
  Apply dropout carefully, often with specialized techniques.
- **Transformers**
  Typically use dropout rates around 0.1.

We'll get to the *convolution* and *transformer* layers in Chapter 7 and Chapter 9, respectively. The intuition here is that earlier layers learn more general features that are less prone to overfitting, while deeper layers learn more specialized features that benefit more from regularization.

It's worth noting that dropout isn't a one-size-fits-all solution. Some architectures (particularly very deep ones or those with skip connections) may benefit from different dropout strategies or alternative regularization techniques. Like any good recipe, the key is experimentation and adaptation to your specific ingredients.

In the next section, we'll put theory into practice by visualizing exactly what happens to your network's internal representations when dropout is applied. Seeing is believing, and these visualizations will give you a concrete understanding of how dropout reshapes your model's learning process.

6.2.6 Implementing Dropout in Pure Python

Let's roll up our sleeves and build dropout from the ground up. There's something magical about implementing an algorithm yourself—it transforms an abstract concept into something tangible that you can poke, prod, and truly understand. Like learning to bake bread from scratch instead of buying it at the store, coding dropout yourself gives you insights you'd never get from just importing a library. So, let's create a simple Dropout layer in pure Python using Listing 6.3. The complete code for this section can be seen in the book resources page on *https://recluze.net/keras-book* in the notebook **06-02-dropout-scratch** and on the book's official web page.

```
import numpy as np

class Dropout:
    def __init__(self, dropout_rate=0.5):
        self.dropout_rate = dropout_rate
        self.mask = None
        self.training = True

    def forward(self, inputs):
        if not self.training:
            return inputs
```

```python
        # Create a binary mask: 1 for keep, 0 for drop
        self.mask = np.random.binomial(1, 1 - \
                    self.dropout_rate, size=inputs.shape)

        # Apply the mask and scale by the keep probability
        # This is inverted dropout
        outputs = (inputs * self.mask) / (1 - self.dropout_rate)

        return outputs

    def backward(self, grad_outputs):
        # Apply the same mask to the gradient
        return (grad_outputs * self.mask) / (1 - self.dropout_rate)

    def set_mode(self, training):
        self.training = training
```
Listing 6.3 Dropout Layer in Pure Python

Let's take a moment to digest what's happening here. Our Dropout class is like a gatekeeper that neurons must pass through. During training, this gatekeeper randomly stops some neurons from passing (setting them to 0) and gives a boost to the ones that make it through (scaling them up). During inference, the gatekeeper steps aside and lets everyone through without any interference.

To really understand what's happening, let's create a toy example and see dropout in action through Listing 6.4. Consider having a small layer with just 10 neurons, and we're applying a dropout rate of 0.5, meaning roughly half the neurons will be deactivated.

```python
# Let's create a toy example
np.random.seed(42)  # For reproducibility

# Create an input with 10 neurons, each with activation 1.0
inputs = np.ones(10)
print("Original inputs:", inputs)

# Apply dropout during training
dropout = Dropout(dropout_rate=0.5)
dropout.set_mode(training=True)
outputs_training = dropout.forward(inputs)
print("Mask:", dropout.mask)
print("Training outputs:", outputs_training)
```

```
# Now, let's see what happens during inference
dropout.set_mode(training=False)
outputs_inference = dropout.forward(inputs)
print("Inference outputs:", outputs_inference)
# Outputs ---------
Original inputs: [1. 1. 1. 1. 1. 1. 1. 1. 1. 1.]
Mask: [1 0 1 0 1 1 0 1 1 0]
Training outputs: [2. 0. 2. 0. 2. 2. 0. 2. 2. 0.]
Inference outputs: [1. 1. 1. 1. 1. 1. 1. 1. 1. 1.]
```

Listing 6.4 Using the Custom Dropout Layer

Notice how during training, some outputs become 0 (where the mask is 0), while others are scaled up to 2.0 (because our `dropout_rate` is 0.5, we divide by 0.5, which doubles the value). But during inference, all outputs remain at their original values of 1.0. This is inverted dropout in action—scale during training, and do nothing during inference. Keras uses something very similar to this layer in its implementation. So, you can use Keras's implementation but be aware of some common pitfalls that you might encounter when using it, as we'll discuss next.

6.2.7 Common Pitfalls and Debugging Tips

When implementing dropout, there are a few common pitfalls to watch out for:

- **Forgetting to switch modes**
  The most common mistake is forgetting to set `training=False` during inference. This would apply random dropout to your production predictions! Always double-check your mode transitions.
- **Incorrect scaling**
  If you scale during inference instead of training (noninverted dropout), you might accidentally apply the scaling factor twice. This manifests as predictions that are consistently too small by a factor of *(1 – dropoutRate)*.
- **Too much dropout**
  Applying too high a dropout rate, especially in smaller networks, can prevent the model from learning anything useful. If your model is underfitting, try reducing the dropout rate.
- **Memory leaks**
  In our simple implementation, we store the dropout mask for backpropagation. In a production system with many layers, these masks can consume significant memory. Consider clearing them after backpropagation if you're rolling out your own `Dropout` code.

- **Uneven distribution**
  Random dropout should follow a *binomial distribution*. If you implement it with a simple threshold on uniform random values, check that you're not introducing subtle biases.

To debug dropout issues, try these approaches:

- Temporarily set the dropout rate to 0 and see if your model converges. If it does, gradually reintroduce dropout.
- Print the proportion of 0s in your dropout masks to verify it matches your intended dropout rate.
- Look at activations before and after dropout to ensure the scaling is working correctly.
- Check gradients flowing through Dropout layers during backpropagation to confirm they're properly masked and scaled.

Now that you've built dropout from scratch, you'll have a much deeper appreciation for what's happening under the hood when you use high-level libraries such as Keras or TensorFlow. Speaking of which, in the next section, we'll see how to implement dropout in Keras with just a few lines of code, using all the insights we've gained so far.

6.3 Other Regularization Methods: L1 and L2 Regularization

So far, we've explored dropout as a powerful technique to combat overfitting. Think of dropout as randomly silencing neurons during training, forcing the network to build redundancy and more robust feature detection. But dropout isn't the only tool in our regularization toolkit.

In this section, we'll turn our attention to two other important regularization techniques: L1 and L2 regularization. These approaches tackle overfitting from a different angle—instead of randomly shutting down neurons, they put constraints on the magnitude of the weights themselves. At their core, both L1 and L2 regularization add an extra term to our loss function that penalizes large weights. The mathematical approach is straightforward. If our original loss function is denoted as $L_{original}$, our regularized loss becomes

$$L_{regularized} = L_{original} + \lambda \times \text{penalty}$$

where λ controls how much we care about the penalty compared to our original objective. A larger \lambda means we're more concerned about keeping weights small than minimizing the original loss.

6.3.1 L1 Regularization (Lasso)

L1 regularization, also known as *Least Absolute Shrinkage and Selection Operator* (*Lasso*), adds a penalty equal to the absolute value of the weights. Mathematically, we add this term to our loss function:

$$L_{L1} = L_{original} + \lambda \sum_{i=1}^{n} |w_i|$$

This simple modification creates some interesting and highly useful effects on how our model learns. The most notable characteristic of L1 regularization is its tendency to push weights exactly to 0, effectively removing certain features from consideration. While L2 regularization reduced the weights of less important features, L1 regularization is like a harsh judge who says, "If that feature isn't absolutely necessary, don't use it at all."

Consider a situation where you're trying to predict house prices, and you have dozens of potential factors—number of bedrooms, location, square footage, age of the house, and even more specific details like the color of the front door or the type of doorknobs. L1 regularization helps you identify which features actually matter by setting the weights for irrelevant features (perhaps doorknob type) to exactly 0, while maintaining nonzero weights for important features such as location and square footage. This *feature selection* property makes L1 particularly valuable when you suspect many of your inputs might be irrelevant or redundant. It leads to *sparse models*—models that only use a subset of available features—which are often easier to interpret and can be more efficient to deploy.

On the other hand, L2 regularization also has a nice mathematical property: It works particularly well when you have correlated features. For example, if both "square footage of living space" and "number of rooms" help predict house prices (and they're obviously related), L2 will distribute the importance between them rather than arbitrarily picking one over the other.

6.3.2 Elastic Net: Combining L1 and L2

What if we want the best of both worlds—the feature selection properties of L1 and the stability benefits of L2? Enter *Elastic Net*, which simply combines both penalties:

$$L_{Elastic} = L_{original} + \lambda_1 \sum_{i=1}^{n} |w_i| + \lambda_2 \sum_{i=1}^{n} w_i^2$$

Elastic Net gives us fine-grained control over regularization by letting us adjust the relative importance of L1 versus L2 penalties. It's like having two different teachers in the classroom—one focused on removing distractions entirely and another on making sure everyone participates without anyone being too dominant.

This hybrid approach shines in scenarios where you have many correlated features but suspect only some are truly relevant. For instance, in genomic studies where thousands of genes might be measured but only a small subset actually influence the trait being studied, Elastic Net can identify groups of related important genes rather than arbitrarily selecting one representative from each correlated group (as L1 might do) or spreading importance too thinly across all of them (as L2 might do).

6.3.3 When Not to Use Regularization

Now that we've explored the powerful regularization techniques in our arsenal, let's talk about when to put these tools back on the shelf. Regularization isn't always the answer, and knowing when to avoid it is just as important as knowing how to apply it. The first and most obvious scenario is when you're already facing underfitting. Adding regularization to a model that's struggling to capture the underlying patterns in your data is like giving low calorie diet advice to someone who's already underweight. If your training loss is high and your model can't even fit the training data well, adding regularization will only make the problem worse.

Very small datasets present another case where regularization might do more harm than good. When data is scarce, your model needs to squeeze every bit of information from the limited examples available. Regularization, by design, restricts the model's capacity to fit the data perfectly. With very small datasets, the risk of overfitting is naturally reduced because there's simply not enough data to memorize. In these cases, you might find better performance by letting your model use its full capacity without regularization constraints.

There are also situations where *interpretability* is your primary concern. L1 regularization can be helpful here because it creates sparse models, but sometimes you need to retain all features to understand their relationships properly. For instance, in medical research, removing certain variables through aggressive regularization might eliminate factors that doctors need to see, even if they only have a small effect on the prediction. It's like a detective needing to consider all the evidence, not just the most compelling pieces.

Interpretability is a very active area in modern deep learning research. Modern models work so well that they are being used in almost all domains you can think of. The issue though is that researchers and engineers aren't really sure why and how the results are so accurate. *Interpretable machine learning* is an active area of research that aims to create tools and methods for figuring out why the models are working and what different learned weights mean, as well as to improve the models in the process.

> **Interpretable Machine Learning**
> I find the area of interpretability to be highly interesting. While it's beyond the scope of this book, I highly recommend you check out the work by Chris Olah (*https://colah.github.io*) as an introduction to this area.

Finally, consider your *domain knowledge*. If you have strong prior information that all of your features are relevant, regularization might be unnecessarily discarding valuable information. For example, if you've carefully crafted a set of financial indicators for stock prediction based on years of economic theory, you probably don't want regularization to arbitrarily zero out some of these hard-won features.

6.3.4 Practical Considerations: Dropout vs. L1/L2 in Neural Networks

While understanding L1 and L2 regularization is essential for your machine learning toolkit, it's worth noting that dropout has become the dominant regularization technique for deep neural networks in practice. The reason lies in computational efficiency and convenience. Both L1 and L2 require careful tuning of the regularization strength parameter λ. Finding the optimal λ often involves running multiple training sessions with different values, which becomes prohibitively expensive for large neural networks that might take days or weeks to train once, let alone multiple times.

Dropout, by contrast, typically works well with default settings (dropping 20%–50% of neurons), making it more of a plug-and-play solution. It also integrates naturally into the forward and backward passes of neural network training without requiring additional hyperparameter searches.

That said, in smaller models or when you have specific goals such as feature selection, L1 and L2 regularization remain valuable tools. Many practitioners even combine techniques—using dropout between layers while also applying a small L2 penalty on the weights. It's important to understand the strengths and limitations of each approach, allowing you to make informed choices based on your specific needs rather than blindly applying the same technique to every problem. Regularization is both art and science—it requires not just mathematical understanding but also intuition developed through practice. This intuition can be developed by repeatedly using it in actual code. So, let's do such an implementation in the next section.

6.4 Applying Regularization in Keras

In this section, we'll take a look at how both L2 regularization and dropout can be applied in Keras. Adding these powerful regularization techniques to your models is surprisingly straightforward.

6.4.1 Implementing L2 Regularization in Keras

In Keras, adding L2 regularization is refreshingly easy, as shown in Listing 6.5.

```
from tensorflow.keras.layers import Dense
from tensorflow.keras.regularizers import l2

# Adding L2 regularization to a Dense layer
x = Dense(units=128,
          activation='relu',
          kernel_regularizer=l2(0.001))(inputs)
```

Listing 6.5 L2 Regularization in Keras

That's it! One parameter, and you've transformed your model. The `kernel_regularizer=l2(0.001)` argument tells Keras to apply L2 regularization to the layer's weights (but not biases) with a regularization strength of 0.001. You can visit the official documentation of Keras to take a look at other options that are available for regularization here: *https://keras.io/api/layers/regularizers*.

6.4.2 Dropout in Keras

Finally, let's see how we can implement dropout in Keras with just a few lines of code. The beauty of Keras is that it handles all the complex machinery we explored earlier, wrapping it in a clean, intuitive interface that lets us focus on designing our architecture rather than managing implementation details.

In Keras, adding dropout to your model is remarkably straightforward. It's akin to installing a quality control checkpoint between assembly lines in a factory. Each checkpoint randomly inspects some products and lets others pass through without scrutiny. Take a look at how we can create a simple neural network with Dropout layers in Listing 6.6.

```
from tensorflow.keras.models import Sequential
from tensorflow.keras.layers import Dense, Dropout
import numpy as np

# Create a simple model with Dropout
model = Sequential([
    # Input layer
    Dense(128, activation='relu', input_shape=(784,)),
    # Dropout layer with 30% rate
    Dropout(0.3),
    Dense(64, activation='relu'),
    Dropout(0.2),
    Dense(10, activation='softmax')
```

```
])

# Compile the model
model.compile(
    optimizer='adam',
    loss='sparse_categorical_crossentropy',
    metrics=['accuracy']
)
```

Listing 6.6 Dropout Usage in Keras

That's it! With these few lines, we've created a network that automatically applies dropout during training. Notice how we place the Dropout layers directly after the Dense layers we want to regularize. The parameter we pass to Dropout(0.3) indicates the fraction of neurons that will be randomly deactivated during each training pass—in this case, 30% of the neurons in the first hidden layer. When we train this model, Keras handles all the complexity behind the scenes. During each training batch, it randomly selects different neurons to deactivate, forcing the network to learn redundant representations. But here's the really clever part—when it comes time to make predictions, Keras automatically switches to using all neurons with appropriately scaled weights. You don't need to write any additional code to handle this transition between training and inference behaviors.

There are a few key options for the Dropout layer that are worth knowing:

- rate

  This is the fraction of units to drop, typically between 0.2 and 0.5. Higher values mean more aggressive regularization but might slow down learning.

- seed

  If you want reproducible results, you can set a random seed to ensure the same neurons are dropped in each run.

- noise_shape

  This allows you to specify the shape of the binary dropout mask, giving you finer control over which dimensions get dropped.

For most applications, you'll only need to adjust the dropout rate. Finding the right rate often requires experimentation. Too little dropout, and your model might still overfit. Too much, and it might struggle to learn patterns at all. One common pattern is to use higher dropout rates for larger layers (those with more neurons) and lower rates for smaller layers. You can also experiment with gradually increasing dropout rates as you go deeper into the network.

Let's see a complete example of training our model with dropout in Listing 6.7.

```
# Generate some dummy data
# 1000 examples, 784 features
x_train = np.random.random((1000, 784))
# 10 classes
y_train = np.random.randint(10, size=1000)

# Train the model
history = model.fit(
    x_train, y_train,
    epochs=10,
    batch_size=32,
    validation_split=0.2
)
```

Listing 6.7 Using Dropout with Dummy Data

When you run this code, you'll notice something interesting in the training metrics. Initially, the training accuracy might be lower than you'd expect for such a simple dataset. That's dropout at work! During training, the network is deliberately handicapped by having some neurons turned off. Sort of like if you ask a football team to play with only 7 players instead of 11, performance will suffer.

But when evaluation time comes on the validation set, Keras automatically switches to using all neurons with scaled weights. You'll often see a significant gap between training and validation performance, with validation surprisingly outperforming training. This isn't a bug—it's a sign that dropout is working as intended. Your model is learning to be robust.

Remember, the whole point of regularization isn't to achieve the highest possible training accuracy. It's to build a model that generalizes well to new, unseen data. Dropout helps us achieve that by preventing our network from becoming too reliant on any specific neuron.

In the next section, we'll explore other forms of regularization available in Keras and discuss strategies for combining them effectively.

6.4.3 Beyond Basic Dropout: Specialized Variants

While standard dropout has proven remarkably effective across many architectures, it sometimes introduces unexpected complications in specific network designs. As we continue to refine our understanding of neural networks, researchers have developed specialized dropout variants to address these unique challenges, as we'll discuss in the following sections.

6 Regularization Techniques

The Challenge with Self-Normalizing Networks

Try to picture a beautiful fountain with multiple levels, where each basin perfectly controls the water flow to the next. The water pressure at each level stays consistent, creating a harmonious system. Now, what happens if you randomly block some of the water spouts? The carefully balanced pressure throughout your fountain system becomes disrupted.

This is similar to what happens in *self-normalizing neural networks* that use *Scaled Exponential Linear Unit (SELU)* activation functions. These networks are designed to maintain a specific statistical distribution of activations across layers—essentially keeping the "water pressure" steady throughout the network. Standard dropout disrupts this carefully balanced system by randomly zeroing activations.

AlphaDropout: Preserving the Statistical Character

AlphaDropout offers an elegant solution to this problem. Instead of simply setting values to 0, which shifts the mean and variance of the layer, AlphaDropout replaces dropped values with noise that maintains the layer's statistical properties. Let's take a look at a model using AlphaDropout in Listing 6.8.

```
from tensorflow.keras.models import Sequential
from tensorflow.keras.layers import Dense, AlphaDropout

# Create a self-normalizing neural network with AlphaDropout
model = Sequential([
    Dense(128, activation='selu', input_shape=(784,)),
    AlphaDropout(0.2),  # Maintains mean and variance of SELU activations
    Dense(64, activation='selu'),
    AlphaDropout(0.3),
    Dense(10, activation='softmax')
])
```

Listing 6.8 AlphaDropout in Keras

What makes AlphaDropout special is its mathematical design. When using SELU activations, our networks develop a particular statistical property—activations tend to have a mean of 0 and variance of 1 (what statisticians call the *standard normal distribution*). It's specifically designed for networks using SELU activation functions. If you're building a self-normalizing neural network with SELU activations, AlphaDropout should be your regularization method of choice. Its parameter options are virtually identical to standard dropout as shown in Listing 6.9.

```
AlphaDropout(
    rate=0.2,   # Fraction of units to drop
    seed=None   # Optional random seed for reproducibility
)
```

Listing 6.9 AlphaDropout Settings

Like standard dropout, AlphaDropout operates only during training and is automatically disabled during evaluation and prediction phases.

Other Specialized Dropout Variant

The Keras ecosystem offers several other specialized dropout variants worth knowing about:

- SpatialDropout 1D/2D/3D
  These variants drop entire feature maps rather than individual neurons, particularly useful in convolutional networks where adjacent pixels have high correlation. We'll return to these in the next chapter after we've covered another type of layer.
- GaussianDropout
  Instead of binary dropout, this applies multiplicative Gaussian noise to the inputs, achieving a similar regularizing effect with different statistical properties.
- GaussianNoise
  This adds zero-centered Gaussian noise to the inputs, serving as a form of data augmentation rather than dropout.

Each variant addresses specific challenges in neural network architectures, giving you a powerful toolkit for regularization. As we explore more complex architectures in later chapters, we'll see how these specialized techniques can be vital for optimizing advanced models.

The power of Keras is that it makes these sophisticated techniques accessible through simple, consistent interfaces. Just like how modern cameras abstract away the complex physics of photography, Keras allows you to focus on the architecture of your network while it handles the mathematical intricacies under the hood. One issue remains though: How do we figure out the dropout rate when we're actually applying dropout?

6.4.4 Finding the Perfect Dropout Rate: A Systematic Approach

So far, we've talked about dropout rates as if they were values you should just intuitively know. "Use 0.2 here, maybe 0.5 there"—but where do these numbers come from? The truth is, finding the ideal dropout rate is a bit like finding the perfect amount of spice in a recipe. Too little, and your model might still overfit. Too much, and your network might struggle to learn anything useful. In practice, the optimal dropout rate depends on many factors:

- The complexity of your dataset
- The depth and width of your network
- The amount of training data available
- The specific patterns your network needs to learn

Rather than relying on rules of thumb or intuition, we can let systematic experimentation find the answer for us. This is where *Keras Tuner* comes to our rescue, allowing us to try multiple dropout rates and see which one yields the best performance.

Let's see how we can systematically search for the best dropout rate using Keras Tuner. You might have to install the Tuner using the following command:

```
pip install keras-tuner
```

If you're in Google Colab or in a Jupyter Notebook, you can issue the magic command:

```
!pip install keras-tuner
```

The code begins by loading the MNIST dataset, as shown in Listing 6.10, which contains images of handwritten digits. We've done this a few times already, so we'll go through it quickly. The complete code for this section can be seen in the book resources page on *https://recluze.net/keras-book* in the notebook **06-03-keras-tuner**.

```python
import keras
import keras_tuner as kt
from keras import layers
from keras.datasets import mnist
from sklearn.model_selection import train_test_split

(X, y), (_, _) = mnist.load_data()
X = X.reshape(X.shape[0], -1) / 255.0
X_train, X_test, y_train, y_test = \
    train_test_split(X, y, test_size=0.2)
```

Listing 6.10 Loading and Shaping the Data

This section loads the MNIST dataset, normalizes pixel values to [0,1] by dividing by 255, and splits data into 80% training and 20% testing sets with a fixed random seed for reproducibility. Then, we get to the heart of *hyperparameter tuning*, which is defining which aspects of the model should be adjustable. The `build_model` function shown in Listing 6.11 creates a neural network with *tunable* hyperparameters.

```python
# A model-building function that takes a hyperparameter object
def build_model(hp):
    model = keras.Sequential()
```

6.4 Applying Regularization in Keras

```python
    # Add our first layer with tunable units
    model.add(layers.Dense(
        units=hp.Int('units_1', min_value=32, \
                max_value=128, step=32),
        activation='relu',
        input_shape=(784,)
    ))

    # Add our first Dropout layer with a tunable rate
    model.add(layers.Dropout(
        rate=hp.Float('dropout_1', min_value=0.2, \
                            max_value=0.4, step=0.1)
    ))

    # Output layer
    model.add(layers.Dense(10, activation='softmax'))

    # Compile the model
    model.compile(
        optimizer=keras.optimizers.Adam(),
        loss='sparse_categorical_crossentropy',
        metrics=['accuracy']
    )

    return model
```

Listing 6.11 Building the Model for Parameter Tuning

This function defines a neural network with the following:

- A tunable first Dense layer where the number of neurons (units) can vary from 32 to 128 in steps of 32
- A tunable Dropout layer where the dropout rate can range from 0.2 to 0.4 in steps of 0.1
- A fixed output layer with 10 neurons (one for each digit) and softmax activation

The hp parameter is a hyperparameter object provided by Keras Tuner that allows us to define searchable parameters. For each parameter, we specify the following:

- A unique name (e.g., units_1 or dropout_1)
- The range of values to search (min_value to max_value)
- The *step size* between values

The model uses Rectified Linear Unit (ReLU) activation for the hidden layer and softmax for the output layer, which is standard for classification tasks. It's compiled with the

Adam optimizer and sparse categorical cross-entropy loss function that we've already seen previously.

Once we've defined our model structure and tunable parameters, we set up and run the hyperparameter tuning process, as shown in Listing 6.12.

```
# Create a tuner with our model builder
tuner = kt.Hyperband(
    build_model,
    objective='val_accuracy',
    max_epochs=10,
    factor=3,
    directory='dropout_tuning',
    project_name='mnist_dropout'
)

# Start the search
tuner.search(X_train, y_train,
            epochs=30,
            validation_split=0.2)
```

Listing 6.12 Keras Tuner Execution

This sets up a *hyperband tuner*, which will find the optimum values for us, with these specifications:

- Uses our `build_model` function as a template
- Aims to maximize validation accuracy (`val_accuracy`)
- Trains models for a maximum of 10 epochs in its final round
- Prunes underperforming models at a level of aggression determined by the factor of 3

The results of optimum hyperparameter search are stored in a directory called `dropout_tuning` under the project name `mnist_dropout`. The search method initiates the hyperparameter search by following these steps:

- It trains on our training data (`X_train, y_train`).
- Each model can train for up to 30 epochs.
- Of the training data, 20% is set aside for validation during training.

After the search completes, we retrieve and use the best hyperparameters through the code shown in Listing 6.13.

```
# Get the best model and hyperparameters
best_hps = tuner.get_best_hyperparameters(num_trials=1)[0]

print(f"Best dropout rate for first layer: {best_hps.get('dropout_1')}")
print(f"Best units for first layer: {best_hps.get('units_1')}")
```
Listing 6.13 Getting Keras Tuner Results

The code extracts the best hyperparameters based on validation accuracy. The get_best_hyperparameters method returns an ordered list of hyperparameter sets, with the best-performing configuration first. By specifying num_trials=1, we retrieve only the single best result. The code then prints both the optimal dropout rate and the optimal number of units (neurons) found for the first layer. The output of the tuner when I ran it is shown in Listing 6.14. For this run, the best dropout rate it found was 0.2 and the number of units was 128. These values represent the hyperparameter combination that produced the highest validation accuracy during the tuning process.

```
Trial 3 Complete [00h 00m 10s]
val_accuracy: 0.9292708039283752

Best val_accuracy So Far: 0.9512500166893005
Total elapsed time: 00h 00m 35s

Search: Running Trial #4

Value          |Best Value So Far |Hyperparameter
128            |128               |units_1
0.3            |0.2               |dropout_1

Best dropout rate for first layer: 0.2
Best units for first layer: 128
```
Listing 6.14 Results of the Keras Tuner

We can use these optimal values to construct our final model, knowing it has been specifically tuned for this dataset with the ideal dropout rate and neuron count in the first layer. As you can see, we can optimize any parameter value, including the number of neurons in the layer. This systematic approach to *hyperparameter optimization* gives us confidence that we're using dropout effectively, neither under-regularizing nor over-regularizing our model.

> **[+] Hyperparameter Tuning in Production**
>
> Once the best values of hyperparameters are found, the model is updated to just use the parameter values instead of the hp parameter. For instance, the preceding model will be replaced with the following:
>
> ```
> model.add(layers.Dense(128, activation='relu', \
>                        input_shape=(784, )))
> model.add(layers.Dropout(0.2))
> ```

6.5 Summary

This chapter examined regularization techniques to solve overfitting in machine learning models. It began by explaining overfitting and underfitting through examples of polynomial features and model complexity, introducing regularization terms as a solution. The chapter then explored dropout, presenting its concept, mechanics, and implementation details to prevent neurons from co-adapting. It covered L1 and L2 regularization methods, explaining their mathematical foundations and appropriate use cases. The final section demonstrated practical implementations of these techniques in Keras, including code examples for L2 regularization, Dropout layers, and specialized dropout variants. The chapter concluded with systematic approaches for finding optimal dropout rates for different model architectures. We're now in the position to move on to more advanced models and techniques, starting with a powerful type of model in the next chapter that's especially good at working with images. All the fundamentals we've built until now will be reused time and again to support us.

Chapter 7
Convolutional Neural Networks

In this chapter, we'll explore convolutional neural networks (CNNs) and their revolutionary impact on computer vision. We'll investigate how convolutional layers detect patterns while minimizing parameters and explore some other types of commonly used layers. Through our case study, you'll see how these components unite into a powerful image classification system, gaining the intuition needed to design effective CNNs that process visual information similar to our own visual systems.

Look at your smartphone recognizing faces in photos, or think about how self-driving cars detect road signs. Behind these remarkable capabilities lies a special kind of neural network we'll be exploring in this chapter—one that revolutionized how computers understand images.

So far, we've built a foundation with fully connected neural networks—those intricate webs where every neuron connects to every other neuron in adjacent layers. We've seen how they learn by adjusting weights through gradient descent and how techniques such as dropout can prevent them from memorizing instead of learning. However, when confronted with images, these networks face significant challenges. Consider trying to process a typical smartphone photo with a fully connected network. Each photo contains millions of pixels, and connecting every pixel to every neuron creates an enormous number of connections. Not only does this require massive computational resources, but as we'll see in this chapter, the dense connectivity pattern makes these networks particularly vulnerable to a severe problem we'll explore shortly.

This chapter introduces a specialized architecture that overcomes these limitations. We'll start from familiar ground—our understanding of fully connected networks—and gradually build up to this new approach that has transformed computer vision. First, we'll examine why traditional fully connected networks struggle with image data, particularly focusing on how their structure leads to not only a computation issue but also that of hampering learning itself.

Next, we'll explore the building blocks of this new architecture. Rather than connecting every input to every neuron, we'll discover how using localized connections and sharing weights across the image creates a more efficient and effective system for processing visual information. We'll build up an understanding of the key components that make this possible without assuming any prior knowledge of how they work.

7 Convolutional Neural Networks

Perhaps the most practical section comes next, where we tackle what I call the "shapes problem." This is the perplexing issue that confuses nearly every beginner: keeping track of how data dimensions transform as information flows through these networks. I've seen students spend hours troubleshooting their models simply because a tensor's dimensions were incorrect, so we'll develop a powerful but concise strategy for understanding and managing these transformations.

Finally, we'll bring everything together in a comprehensive case study, building a complete image classification system. You'll create a model that serves as the foundation for real-world applications.

By the end of this chapter, you'll have both a conceptual understanding of this powerful neural network architecture and the practical skills to implement it. The concepts you learn here will serve as building blocks for even more advanced techniques in later chapters.

7.1 Introduction to Convolutional Neural Networks

Let's begin our exploration of the neural network architecture that gave computers the ability to see and understand the visual world by looking at what's wrong with our existing architectures.

7.1.1 The Limitation of Fully Connected Networks

We've spent the previous chapters exploring *fully connected networks* (*FCNs*), those interconnected layers of neurons where every neuron connects to every other neuron in adjacent layers. These networks have shown remarkable capabilities, but they come with inherent challenges that limit their effectiveness, especially as they grow deeper. Let's now step back and consider what happens in these fully connected networks during training. At the heart of our neural network's learning process lies the sigmoid activation function (or similar "squishing" functions) that we've been using. This function takes any input and transforms it into a value between 0 and 1, creating a smooth, S-shaped curve that helps our neurons make decisions about whether to "fire" based on their inputs.

While this squishing property is useful for stabilizing our network's outputs, it introduces a subtle but significant problem. Not only does the sigmoid function restrict outputs to between 0 and 1, but its derivative, which tells us how quickly the function changes at any point, is also bounded between 0 and 1. In fact, the derivative of the sigmoid function reaches its maximum value of 0.25 at the center (when the input is 0) and trails off toward 0 as we move away in either direction. This becomes clear when we look at the mathematical expression for the sigmoid function and its derivative, both of which are shown in Figure 7.1.

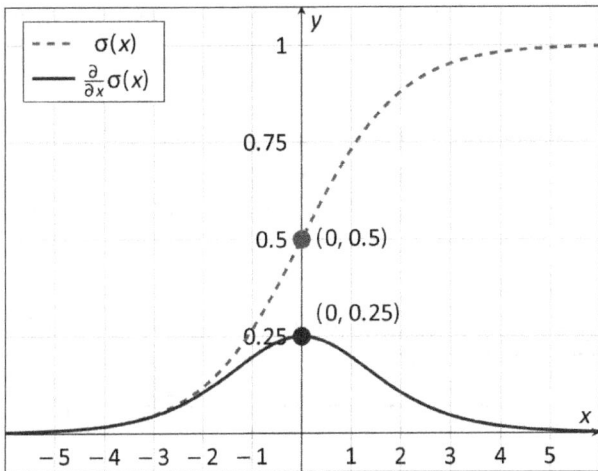

Figure 7.1 Sigmoid Activation Function and Its Derivative

The sigmoid function is defined as $\sigma(x) = 1/(1 + e^{(-x)})$ and its derivative as $\sigma'(x) = \sigma(x) \cdot (1 - \sigma(x))$. When we plot this derivative, we see that it peaks at *0.25* when $x = 0$ (i.e., $\sigma(0) = 0.5$) and approaches *0* as x moves toward either positive or negative infinity. This means that the sigmoid function is most sensitive to changes when its input is near *0* and becomes increasingly insensitive as the input moves away from *0*.

This limitation might seem minor at first glance, but it becomes critically important when we consider how neural networks learn through backpropagation. When our network makes a prediction and calculates the resulting error, that error needs to be communicated backward through the network to adjust the weights. This process relies on the *chain rule* from calculus, which tells us how to calculate derivatives through composed functions. To understand this clearly, consider that each layer in a neural network actually takes inputs, performs some operations (multiplying, adding, and applying an activation function), and produces outputs. Mathematically, we can view each layer as a function that transforms data.

If we have a network with three layers, we can denote the network as a composition of functions:

$f_{network}(x) = f_3(f_2(f_1(x)))$

Here, f_1, f_2, and f_3 are the functions computed by each layer. When we want to find how sensitive our network's output is to a change in a particular weight in the first layer, we need to apply the chain rule.

Let me illustrate this with a simple calculus example. Consider two simple functions: *g(x) = x²* and *h(u) = 3u + 1*. If we compose them to get $f(x) = h(g(x)) = 3(x^2) + 1$, and we want to find $f'(x)$, the chain rule tells us the following:

$f'(x) = h'(g(x)) \cdot g'(x)$

Let's compute this as follows: $h'(u) = 3$ as the derivative of $3u + 1 + 3$ and $g'(x) = 2x$ as the derivative of $x^2 = 2x$, so

$f'(x) = 3 \cdot 2x = 6x$

In neural networks, this same principle applies, but with many more functions composed together. The gradient of the loss with respect to a weight in the first layer requires us to multiply derivatives across all subsequent layers. This last sentence is all that matters here. There are several multiplications of derivatives with each other. In a multilayer network, we're essentially dealing with functions of functions, as shown in Figure 7.2. To calculate how a weight in an early layer affects the final output, we need to multiply the derivatives of each layer in between.

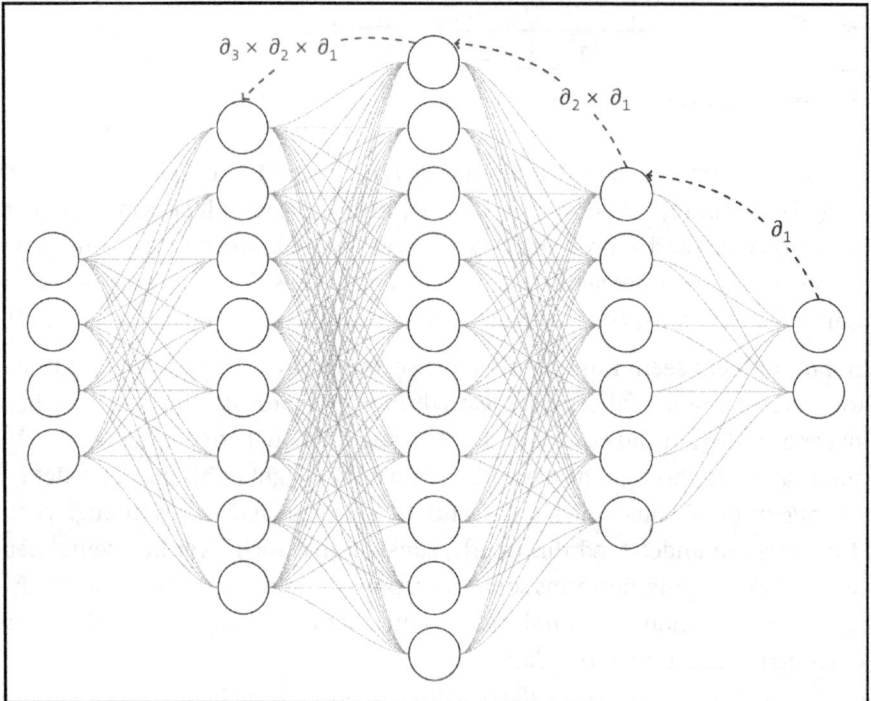

Figure 7.2 Fully Connected Network (FCN) with Back Propagation

If each derivative is at most 0.25 (as with the sigmoid function), what happens when we multiply several of these small numbers together? The result gets exponentially smaller. If we have, say, 10 layers, and we multiply 0.25 ten times, we end up with a minuscule value of approximately 0.0000095—effectively 0 for computational purposes. This phenomenon is what we call the *vanishing gradient problem*. As we backpropagate through many layers, the gradient, which guides our weight updates, becomes so small that the earliest layers in the network receive practically no *update signal*. It's as if we're trying to whisper a message through several walls, and by the time it reaches beyond the last wall, the sound is so faint that it's completely inaudible.

Fully connected networks are particularly susceptible to this vanishing gradient problem because of their dense connectivity pattern. Each neuron connects to every neuron in the previous layer, which means during backpropagation, gradients flow through all these connections, getting multiplied by small derivatives at each step. The math becomes even more troublesome when we consider that these multiplications occur not just layer by layer but connection by connection. In a fully connected layer with 100 neurons connecting to another layer with 100 neurons, we have 10,000 connections, each potentially contributing to gradient shrinkage.

7.1.2 The Learning Standstill

The practical implication of vanishing gradients is straightforward but severe: Our network stops learning effectively. Remember that in gradient descent, we update our parameters (weights and biases) using

$$\theta_{new} = \theta_{old} - \alpha \cdot \frac{\partial}{\partial \theta} J(\theta)$$

where α is our learning rate and is multiplied with the gradient of our loss function with respect to our parameters. If this gradient approaches 0 due to the vanishing gradient problem, the second term in our update rule also approaches 0, leaving us with

$$\theta_{new} = \theta_{old} - 0$$

In other words, no meaningful update occurs. The weights in the earlier layers of our network remain virtually unchanged, regardless of how much error the network produces. It's as if these layers are frozen in time, unable to learn from their mistakes. What makes this situation particularly frustrating is that it creates a learning dead end that's remarkably resistant to our usual solutions. At this point, it doesn't matter how large our loss value is because even enormous prediction errors won't generate sufficient gradients to travel back through all of those layers. It doesn't matter how much data we have because even with millions of perfectly labeled examples, the network can't extract meaningful patterns if the learning signal can't reach the early layers. It doesn't matter if we simplify or complicate our model architecture as the fundamental mathematics of the vanishing gradient remains the same.

Perhaps most insidiously, the weights in these early layers remain stuck at or near their random initialization values. Remember that when we first create a neural network, we typically initialize weights with small random values. Without meaningful updates, these random patterns never evolve into useful feature detectors. It's trying to build a sophisticated image recognition system but being unable to progress beyond random pixel shuffling in the critical early processing stages.

This a practical barrier that prevented researchers from successfully training deep neural networks for many years. No matter how much data we feed into such a network, if the gradients vanish, the network can't learn from that data.

7.1.3 Solving the Vanishing Gradient Problem

Now that we understand the vanishing gradient problem that plagues deep networks with sigmoid activations, let's discuss the first major innovation that helped overcome this limitation. The solution came from a surprisingly simple idea: What if we used a different activation function?

Enter the *Rectified Linear Unit (ReLU)*, which we've mentioned earlier in the book. Unlike the sigmoid function with its elegant S-curve, ReLU is refreshingly straightforward:

$$f(x) = max(0, x)$$

That's it! ReLU simply takes an input and returns either that same value (if it's positive) or 0 (if it's negative). Visually, it looks like a bent line that hugs the horizontal axis for negative inputs and then shoots upward at a 45-degree angle for positive inputs. ReLU and its derivative are shown in Figure 7.3.

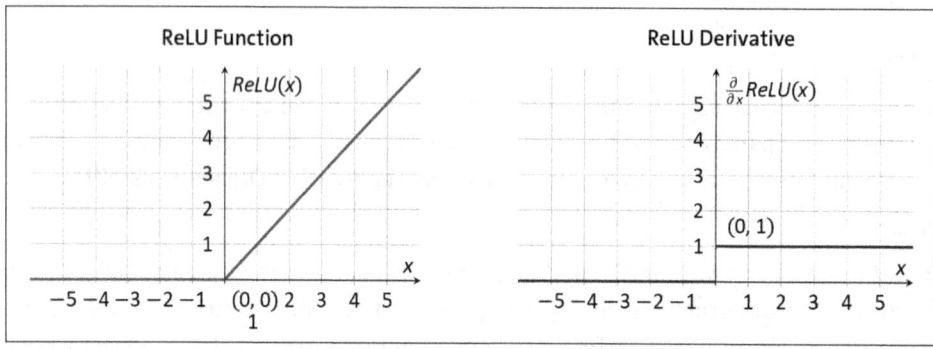

Figure 7.3 ReLU and Its Derivative

This simple function sparked a revolution in neural networks. But why? Let's discuss its advantages. First, ReLU tackles the vanishing gradient problem head-on. Its derivative is either 1 (for positive inputs) or 0 (for negative inputs). When backpropagating through a ReLU network, those 1s don't shrink the gradient signal like the tiny values from sigmoid derivatives. You might wonder that ReLU isn't squishing values between 0 and 1. Is it still a squishing function? Absolutely! While the term *squishing function* made intuitive sense for sigmoid, the broader concept is about introducing nonlinearity. ReLU still creates a threshold (at 0) that determines whether a neuron fires or not. Without this nonlinearity, our neural networks would just be glorified linear regression models, unable to learn complex patterns. This is one of the reasons that the term *activation function* has gained more traction over the squishing function term recently.

ReLU brings additional benefits too. It's computationally efficient—just a simple max operation between 0 and the input. This matters when you're running networks with millions of neurons. It also tends to create sparse activations (many neurons output 0), which helps with feature specialization and efficient representation. However, ReLU isn't perfect. Its most notorious issue is the *dying ReLU problem*. If a neuron's weights are updated such that its inputs always fall in the negative range, it will always output 0, and its gradient will also be 0, meaning it stops learning entirely.

To address these (and some other) limitations, researchers developed several variants. One of the most successful is the *Gaussian Error Linear Unit* (*GeLU*), which has become a favorite in modern language models and other cutting-edge networks. The GeLU function can be approximated as

$$\text{GELU}(x) = x \cdot \Phi(x) = x \cdot \frac{1}{2}\left(1 + \text{erf}\left(\frac{x}{\sqrt{2}}\right)\right)$$

Intimidating formula aside, GeLU essentially creates a smooth curve that resembles ReLU but with some key differences. Instead of a hard cutoff at 0, GeLU gradually attenuates negative values based on their magnitude, informed by ideas from probability theory. Take a look at Figure 7.4 for what GeLU appears like.

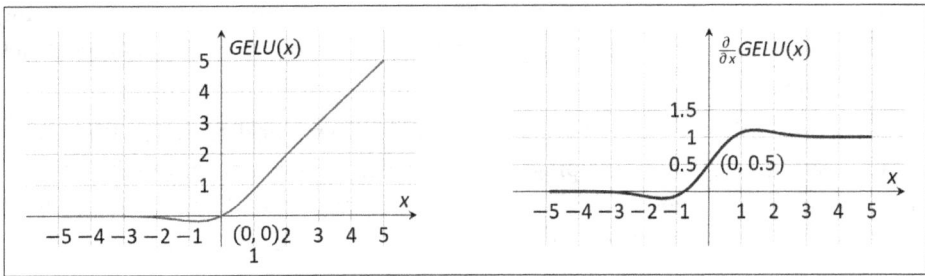

Figure 7.4 GeLU and Its Derivative

Consider a neuron using GeLU as being more thoughtful about negative inputs. Rather than immediately dismissing them as ReLU does, it considers: "How negative is this input? Should I let some of this information through?" This nuanced approach allows GeLU networks to learn more complex representations.

GeLU offers several advantages:

- Avoids the dying neuron problem by allowing gradients to flow even for negative inputs
- Makes optimization more stable via its smooth nature
- Often leads to better performance in very deep networks
- Maintains computational efficiency similar to ReLU

Effect of Minor Changes in Deep Learning Models

The success of GeLU in models such as BERT and GPT demonstrates how seemingly small changes to activation functions can have outsized impacts on model performance. It's a reminder that in neural networks, the details matter tremendously. A single character change in your Keras code can make a world of difference.

Both ReLU and GeLU represent crucial steps in solving one part of the deep learning puzzle. By choosing activation functions with better gradient properties, we've cleared one obstacle on our path to building effective deep networks. However, the problem of densely connected layers still remains, and that's where the architecture changes we explore next come into play.

7.1.4 Dense vs. Sparse Connections

We've tackled one part of our vanishing gradient problem by introducing activation functions that don't shrink our gradients to microscopic values. But we still have another challenge to address: the everything-connected-to-everything architecture of fully connected networks. Think about what happens in these densely connected networks: Each neuron is like a social butterfly with hundreds or thousands of connections. While this might seem powerful—giving each neuron access to all possible information—it actually creates a tangled web that's difficult to train effectively. It's like being at a party where everyone is shouting at once; extracting meaningful signals becomes nearly impossible.

To solve this problem, we need to look back at a mathematical operation that's been around for centuries, quietly helping scientists in fields from signal processing to physics: the *convolution operator*. The word *convolution* might sound intimidating, but it's actually a beautifully simple idea with a profound impact on how we process information. Unlike matrix multiplication, where every element interacts with every other element, convolution creates local, focused interactions—more like conversations between small groups of neighbors rather than a chaotic town hall meeting.

The Convolutional Operation Explained

Let me walk you through how convolution works with a concrete example. Assume we have a 5×5 image represented as a matrix:

$$\begin{pmatrix} 10 & 20 & 30 & 40 & 50 \\ 15 & 25 & 35 & 45 & 55 \\ 20 & 30 & 40 & 50 & 60 \\ 25 & 35 & 45 & 55 & 65 \\ 30 & 40 & 50 & 60 & 70 \end{pmatrix}$$

Now, let's say we have a small 3×3 *filter* (sometimes called a *kernel*):

$$\begin{pmatrix} 1 & 0 & -1 \\ 1 & 0 & -1 \\ 1 & 0 & -1 \end{pmatrix}$$

To apply convolution, we slide this filter across our image, one step at a time. So, at step 1, *[1 0 -1]* will be placed over the matrix values of *[10 20 30]*, which is the second row of the filter on the first three elements of the second row of the matrix, and so on. At each position, we multiply each filter value with the image pixel it overlaps and then sum all these products. It's similar to placing a stencil over different parts of our image and calculating a weighted average of the covered pixels. A visual representation of this operation is shown in Figure 7.5.

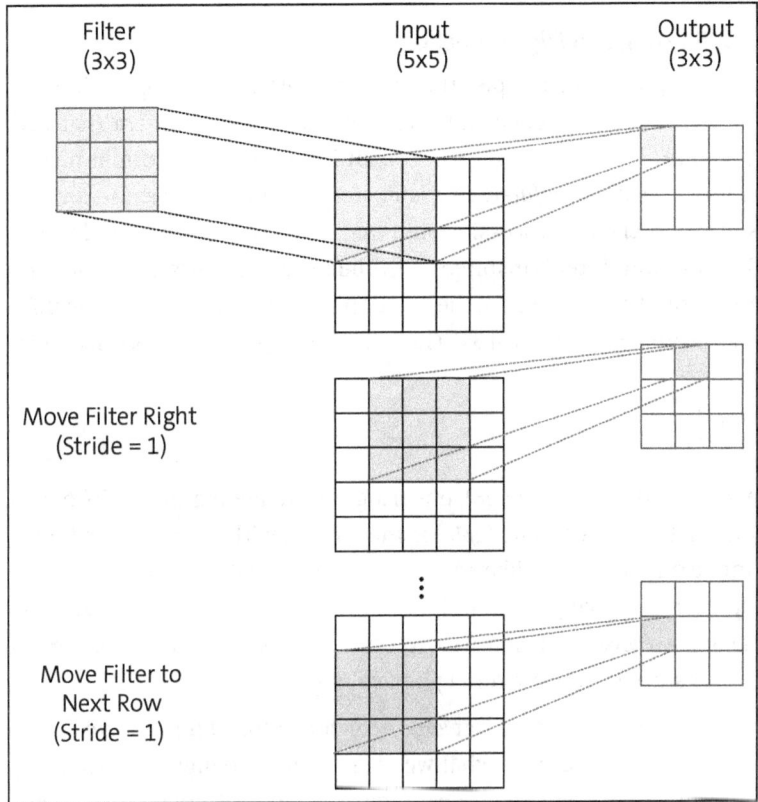

Figure 7.5 The Convolution Operator with a Single Kernel

Let's make this concrete by computing the first output value. We position our filter over the top-left 3×3 portion of our image:

$[10\ 20\ 30] \times [1\ 0\ -1] = (10 \times 1) + (20 \times 0) + (30 \times -1) +$

$[15\ 25\ 35] \times [1\ 0\ -1] = (15 \times 1) + (25 \times 0) + (35 \times -1) +$

$[20\ 30\ 40] \times [1\ 0\ -1] = (20 \times 1) + (30 \times 0) + (40 \times -1)$

$= 10 + 0 - 30 + 15 + 0 - 35 + 20 + 0 - 40$

$= -60$

That's our first *convolution output* value that goes in (0, 0) of our output matrix! We then slide our filter one step to the right and repeat the process. The full output of this convolution operation would be a 3×3 matrix (smaller than our original 5×5 because the filter needs to fit completely within the image). There will be no "hanging" filter operations. That's all there is to the operation. This simple operation leads to some powerful results. You can see an interactive demo we've created for you to play around with different values on the book's web page and on the resources page at *https://recluze.net/keras-book* in the **07-05-conv-demo** link.

The Magic of Filters: Blurring and Edge Detection

Let's take a look at some practical examples that demonstrate why this operation is so powerful. Think of convolution filters as different lenses for your camera that each reveal something unique about the image you're examining. First, let's look at how we can blur an image using a *Gaussian filter*. This isn't simply for creating those dreamy, soft-focus photos; it's a crucial preprocessing step that helps reduce noise and unnecessary details. A Gaussian blur filter is inspired by the bell-shaped *Gaussian distribution*. It creates a gentle, weighted average where the center pixels matter most and the influence fades as we move outward. A typical 3×3 Gaussian blur filter might look like this:

$$\begin{pmatrix} 1/16 & 2/16 & 1/16 \\ 2/16 & 4/16 & 2/16 \\ 1/16 & 2/16 & 1/16 \end{pmatrix}$$

Notice how these values add up to 1 (*16/16*), ensuring our image maintains its overall brightness. In addition, the center value (*4/16*) is the largest, with the values gradually decreasing as we move outward—just like a tiny bell curve. It's asking a group of neighbors about their opinion but giving more weight to those who live closest to you. The result is a more balanced, consensus view that smooths out extreme opinions—much like how Gaussian blur smooths out extreme pixel values.

Now, for another example, let's take a look at *edge detection*. While blurring can help us remove unnecessary details, sometimes what we really want is to highlight where significant changes occur in an image—the edges. Edge detection is fundamental to how computers understand images, helping them distinguish between objects and recognize shapes.

One of the most common edge detection filters is the *Sobel operator*. Unlike our previous examples, Sobel filters are directional, so they detect edges oriented in specific directions. Here's a Sobel filter for detecting horizontal edges:

$$\begin{pmatrix} 1 & 2 & 1 \\ 0 & 0 & 0 \\ -1 & -2 & -1 \end{pmatrix}$$

This filter highlights places where pixel values change significantly as we move from top to bottom. When there's a significant change from top to bottom, we get a negative value in output. This negative value indicates a strong edge because there's a significant change from lighter values at the top to darker values at the bottom. If we were to apply this filter across an entire image, we'd end up with a new image that highlights horizontal edges, with stronger edges appearing as brighter or darker pixels depending on the direction of the change.

Where Are These Numbers Coming From?

Don't worry about where these specific numbers in the filters are coming from. It took experts from computer vision and image processing field decades to come up with these precise values. We'll see in a while that we don't have to come up with all of these values. In fact, we don't even have to know them. We're just looking at them as an example to fully understand the underlying operation.

These edge-detecting filters serve as the foundation for how convolutional networks "see" images. In the early layers of these networks, many of the filters naturally evolve to become edge detectors during training, automatically discovering what we manually designed in these examples.

Implementing Raw Convolution Through Code

Let's take a look at some code that helps us understand how this works exactly. Listing 7.1 shows the code that retrieves and processes an image for analysis. It fetches an image from a URL using the `requests` library and then opens the image data with the *Python Imaging Library* (*PIL*). The code converts the color image to grayscale using the `'L'` mode, which simplifies the image by representing each pixel as a single intensity value rather than RGB channels. Finally, it transforms the grayscale image into a numerical array that can be manipulated mathematically. This numerical representation in `img_array` enables various computational operations such as edge detection, feature extraction, or input preparation for machine learning models. The complete code for this section can be seen in the book resources page on *https://recluze.net/keras-book* in the notebook **07-01-conv-basics**, as well as on the book's official web page.

```
from io import BytesIO
import numpy as np
import matplotlib.pyplot as plt
from PIL import Image
import requests
# Load image from URL
url = "https://recluze.net/kb/building.jpeg"
response = requests.get(url)
img = Image.open(BytesIO(response.content))

# Convert to grayscale
gray_img = img.convert('L')
img_array = np.array(gray_img)
```

Listing 7.1 Loading an Example Image for Applying Convolutions

7 Convolutional Neural Networks

Next, we create a function shown in Listing 7.2 that takes a filter (or kernel) and applies convolution to it. We'll code the actual convolution filter later. This listing simply takes the input and the convolved output and displays it in a plot.

```python
# If you want to try different kernels, you can use this function
def apply_kernel(kernel_name):
    if kernel_name in kernels:
        result = apply_convolution(img_array, kernels[kernel_name])
        result = np.clip(result, 0, 255).astype(np.uint8)

        plt.figure(figsize=(10, 5))

        plt.subplot(1, 2, 1)
        plt.imshow(img_array, cmap='gray')
        plt.title('Grayscale Image')
        plt.axis('off')

        plt.subplot(1, 2, 2)
        plt.imshow(result, cmap='gray')
        plt.title(f'After {kernel_name.replace("_", " ").title()} Filter')
        plt.axis('off')

        plt.tight_layout()
        plt.show()
        return result
    else:
        print(f"Kernel '{kernel_name}' not found.")
```

Listing 7.2 Taking a Kernel, Applying Convolution, and Displaying the Result

To actually perform the convolution, we first prepare the inputs as shown in Listing 7.3. The code creates an empty output array with the same shape as the original image to store the results. It then applies *padding* to the input image using NumPy's pad function with constant mode, which adds 0s around the borders. This padding is necessary because when the kernel is positioned at the image edges, it would otherwise extend beyond the image boundaries. The added 0 buffer allows the convolution operation to process every pixel in the original image while keeping the spatial relationships intact.

```python
def apply_convolution(image, kernel):
    # Get dimensions
    image_height, image_width = image.shape
    kernel_height, kernel_width = kernel.shape

    # Calculate padding
    pad_height = kernel_height // 2
```

```
pad_width = kernel_width // 2

# Create output array
output = np.zeros_like(image)

# Apply padding to the input image
padded_image = np.pad(image, ((pad_height, pad_height), \
                (pad_width, pad_width)), mode='constant')
# ... function continues in next listing
```
Listing 7.3 Preparing Inputs for Convolution

The convolution operation itself is shown as continued code within this function in Listing 7.4. This code implements the core of the convolution operation through nested loops. For each pixel position in the original image, it performs these steps:

1. Extracts a region from the padded image that's the same size as the kernel, centered on the current pixel position. This region of interest moves pixel by pixel across the entire image as the loops progress.
2. Applies the kernel to this region through element-wise multiplication (region * kernel), which means each corresponding element in the region and kernel are multiplied together.
3. Calculates the sum of all of these multiplied values and assigns this single value to the corresponding pixel in the output image. This sum represents how well the pattern defined by the kernel matches the local image region.

The operation continues until every pixel in the original image has been processed, resulting in a transformed image where certain features (determined by the kernel) have been enhanced or detected. Notice that we're using multiple loops here which will be inefficient.

```
    # Apply convolution
    for i in range(image_height):
        for j in range(image_width):
            # Extract the region of interest
            region = padded_image[i:i+kernel_height, \
                                    j:j+kernel_width]
            # Apply the kernel
            output[i, j] = np.sum(region * kernel)

    return output

kernels = {
    'identity': np.array([[0, 0, 0], [0, 1, 0], [0, 0, 0]]),
    'edge_detection': np.array([[-1, -1, -1], [-1, 8, -1], [-1, -1, -1]]),
```

```
    'sharpen': np.array([[0, -1, 0], [-1, 2, -1], [0, -1, 0]]),
    'gaussian_blur': np.array([[1, 2, 1], [2, 4, 2], [1, 2, 1]]) / 16,
}
result = apply_kernel('gaussian_blur')
```

Listing 7.4 Applying the Convolution

When we apply this `gaussian_blur` kernel to an input, the output will be slightly blurred, as shown in Figure 7.6.

Figure 7.6 Applying a Convolution Filter to an Image

> [!] **Inefficiency in the Implementation**
>
> In practice, we'll never roll out our own convolution code but use pre-optimized libraries such as TensorFlow to calculate it for us. These libraries will use the underlying GPU libraries such as the CUDA Deep Neural Network library (cuDNN) to apply highly efficient parallel processes to do the calculation.

An efficient implementation of this operation becomes necessary as input size increases. With fully connected networks, each additional input pixel creates new connections to every neuron in the next layer, causing an exponential growth in parameters. In contrast, convolution maintains a fixed number of parameters regardless of input size.

Number of Parameters in Convolution

The parameter efficiency of a convolution operation comes from two key properties. First, each filter has a fixed size (typically small, e.g., 3×3 or 5×5) that doesn't change when the input grows. Second, we reuse exactly the same filter weights across the entire input by sliding the filter across it. To make this concrete, when we apply a 3×3 filter to a small 28×28 image (like those in the MNIST digit dataset), we use just nine parameters per filter. If we then move to high-resolution 1000×1000-pixel images, a fully connected layer would require millions more parameters, but our convolutional filter still uses just those same nine parameters. This remarkable property means that our filter learns to detect patterns regardless of the image size. The filter that detects edges in a thumbnail will detect the same edges in a billboard-sized image. This scaling property has profound implications for how neural networks process visual information. It's why we can train a network on smaller images but apply it to larger ones without retraining every parameter.

There's also an interesting connection between convolution and other mathematical operations. If you look at the code implementation of convolution, you might notice striking similarities to the dot product operation. Both involve element-wise multiplication followed by summation. In fact, convolution can be implemented efficiently using highly optimized matrix multiplication routines, which is one reason modern GPUs excel at these operations.

This connection to fundamental linear algebra operations hints at deeper mathematical patterns that recur throughout machine learning. The same principles that make convolution so effective will reappear later when we explore *transformers*—the architecture behind today's most powerful language models. Both approaches leverage the power of focused, weighted combinations of inputs, just applied in different contexts and with different structural constraints.

7.1.5 From Convolution to Neural Networks: The Conv2D Layer

Now that we've explored the power of convolutions, this section will showcase how this useful operation transforms our neural networks. So far, we've become familiar with dense layers, where every neuron connects to every neuron in the next layer to create our fully connected architecture. But now, we're going to introduce a new member to our neural network family: the *two-dimensional convolutional layer* (Conv2D layer). The key difference lies in how these layers process information. Instead of the familiar matrix multiplication operation ($\theta \cdot X$) we used in dense layers, Conv2D uses our newly discovered convolution operation. This shift completely transforms how information flows through our network. The layer doesn't treat each pixel as an isolated input. Instead, it applies filters across the entire image, scanning for patterns just like we did in our manual convolution examples.

Each filter slides across the input image, performing its convolution operation at each position. The result isn't a single number but another "image"—a feature map that highlights where certain patterns appear in the original image. If our filter is designed to detect vertical edges, the output feature map will light up wherever those edges appear. The shape of these feature maps depends on several factors: the input size, filter size, and something new called *stride*, which is how many pixels we move each time we slide our filter.

Strides in Convolution

Let's work through some examples to make this concrete: Assume we have a simple 5×5 input image and a 3×3 filter, as shown in Figure 7.5. If we use a stride of 1—moving our filter 1 pixel at a time—how large will our output feature map be? Starting at the top-left corner, we can place our 3×3 filter on positions (0,0), (0,1), (0,2), then to the next row as (1,0), and so on. The last valid position would be (2,2) because if we go any further, our filter would extend beyond the input image. This gives us a 3×3 output feature map (3 positions horizontally × 3 positions vertically).

Now, let's scale up to a real example. If we have a 28×28 MNIST digit image and apply a 3×3 filter with a stride of 1, our output shape becomes 26×26. We can position our filter at coordinates (0,0) through (25,25) for a total of 26 positions in each dimension. The output shrinks because our filter needs to fit completely within the input image.

What if we use a stride of 2, jumping 2 pixels each time we move our filter? This cuts our output dimensions roughly in half. With our 28×28 input, the filter positions would be (0,0), (0,2), (0,4), and so on. This gives us a 13×13 output feature map. We can verify this: starting from 0, we can jump by two 13 times (0, 2, 4, 6, 8, 10, 12, 14, 16, 18, 20, 22, 24) while keeping our 3×3 filter within the 28×28 image. Keras calculates this output size automatically for us, but you need to spend some time understanding it to avoid any debugging headaches when you get to larger networks.

Multiple Filters: Introducing Channels

Let's do something a little more interesting and closer to the real world. Instead of using just one filter, we can apply multiple filters to the same input image. Each filter searches for different patterns—perhaps one detects horizontal edges, another vertical edges, and a third diagonal lines. If we apply two different filters to our 28×28 MNIST image with a stride of 1, we get two 26×26 feature maps. We can stack these maps together to form a 26×26×2 output. That third dimension—2, in this case—is what we call the number of *channels* or *feature maps*. This is conceptually similar to how color images work. A color image typically has three channels—red, green, and blue—stacked together. But in convolutional layers, each channel represents a different feature detected by a unique filter.

So far, we've been applying convolution with filters that have predefined values—like our edge detectors and blurring filters. What if we let the network itself discover what filters would be most useful?

From Handcrafted to Learned Filters

When we build a CNN, we don't actually program the specific values in each filter. Instead, we initialize these values randomly—essentially starting with filters that do nothing meaningful. One filter might randomly brighten some pixels and darken others in a chaotic pattern. Another might create strange, haphazard distortions.

This might seem counterintuitive. How can random noise possibly help us classify images? The key lies in what happens next. Once we've initialized our random filters, we push an image through our network on the *forward pass*. The image gets transformed by each convolutional layer, with each filter extracting meaningless patterns (at first). Eventually, after passing through several layers, the network makes a prediction about what's in the image.

In the beginning, this prediction is almost certainly wrong. A network trying to recognize handwritten digits might confidently declare that an 8 is actually a 1—it's basically guessing randomly because its filters haven't learned anything yet.

But here's the crucial part: We calculate how wrong the network is (the loss), and then we propagate this error backward through the network. This *backward pass* calculates how each filter value contributed to the mistake, and then—drawing on the gradient descent algorithm, we explored in Chapter 3—it adjusts these values in the direction that reduces the error. Remember the core idea of gradient descent? We find the gradient (the direction of steepest increase) of our loss function with respect to each parameter and then take a small step in the opposite direction to reduce the loss. This same principle applies here, but now our parameters are the values in each convolutional filter.

After this adjustment, each filter changes slightly, moving a tiny bit closer to extracting a pattern that's actually useful for the classification task. We repeat this process thousands of times with many examples, and gradually, these initially random filters transform into purposeful feature detectors.

Some filters might indeed evolve to detect edges, just like the ones we designed manually. Others might become specialized for detecting curves, corners, textures, or more abstract patterns that humans might not even recognize as meaningful. The network discovers what features are most useful for distinguishing between the classes it's trying to predict. It's like letting a group of artists discover their own styles through trial and error. We don't tell them exactly what techniques to use; they figure out what works best through practice and feedback.

The Beauty of Sparse Connectivity

This approach of learning filters through gradient descent combines the best of both worlds. We get the benefit of *sparse connectivity*—each output value depends only on a small neighborhood of input pixels, not the entire image. This dramatically reduces the number of parameters compared to fully connected networks. At the same time, we're using better activation functions such as ReLU that don't squash gradients to microscopic values. This helps the learning signal propagate more effectively through the network, especially as we add more layers. The result is a network that can learn much faster and more effectively than its fully connected counterparts. It uses the inherent structure of images—the fact that nearby pixels are related and distant ones less so—while still discovering the features that matter most for the specific task at hand.

Take a moment to appreciate how this connects to our earlier discussions of gradient descent. The intuition we developed in Chapter 3—about parameters adjusting to minimize loss—is exactly what powers these convolutional networks, just with a different set of parameters and a different network structure. The fundamental principles remain the same, showcasing the beautiful coherence of deep learning's mathematical foundations. We were able to take our grayscale images and extract several feature maps out of them.

The Third Dimension: Understanding Image Channels

We've been working with simple matrices as our inputs, which is perfect for grayscale images where each pixel has just one intensity value. But the colorful world around us demands more nuance. This is where channels come in as the secret ingredient that adds richness and depth to our image representation.

Think about how your digital camera captures a sunset. It doesn't simply record a single value for each point in the scene. It captures separate intensity values for red, green, and blue light. These three primary colors, when combined in different proportions, can recreate virtually any color our eyes can perceive.

In an RGB image, each pixel location doesn't hold just one value but three:

- The red channel records how much red light is present at each position.
- The green channel captures the intensity of green.
- The blue channel measures the blue component.

When we stack these three matrices together, we get a more complex structure—a 3D array that computer scientists call a *tensor*. It's having three separate but aligned grayscale images, each one capturing a different aspect of the scene.

An Example of Channels

In a picture of a vibrant red apple, the pixels in the red channel have high values, while the green and blue channels show lower values. The sky portion of an image might have high values in the blue channel but lower values in the red and green channels. You can actually look at these channels in any photo editor. I highly recommend you try doing that to get a deeper understanding of how they are represented.

This shift from a simple matrix to a *multi-channel tensor* requires us to think more carefully about how we organize our data. If our image is 28×28 pixels with three color channels, we're actually working with a tensor of 28×28×3 values.

But things can get a bit tricky here. Different deep learning frameworks organize these dimensions in different ways. TensorFlow and Keras (which we'll be using) typically organize image tensors in the shape of (Height, Width, Channels), also known as the *HWC format*, so our example is a tensor of shape (28, 28, 3). Some libraries use the (Channels, Height, Width) convention, also known as *CHW*. In that case, the shape will become (3, 28, 28). This ordering might seem arbitrary, but it's crucial to keep track of it. When you're debugging shape errors (and trust me, you'll encounter these), the first thing to check is whether you're aligning your dimensions correctly. Always pay attention to the shape information in error messages and when printing tensor shapes as this information is a compass that helps you navigate through the multidimensional landscape of your model.

Filtering Across Channels

Now that our inputs have become three-dimensional, our filters need to match. When working with a multi-channel input, each filter must have the exact same number of channels as the input, as shown in Figure 7.7. For an RGB image, this means each filter becomes a 3D tensor itself. If we're using a 3×3 filter, it actually becomes a 3×3×3 filter—the extra dimension corresponding to the three color channels. You can visualize this as three separate 3×3 filters, one for each channel, working together as a single unit. How does convolution work with these 3D tensors? The process, shown concisely in Figure 7.7, is a natural extension of what we've already learned. When applying a three-channel filter to a three-channel input, we do the following:

1. Position the filter at a location in the image.
2. For each channel, multiply the filter values with the corresponding input values.
3. Sum up all of these products—not just across each 2D slice but across all three channels.
4. Recognize that this single scalar value becomes one element in our output feature map.

Contrast this operation with the single channel one shown previously in Figure 7.5. This is a crucial point: No matter how many channels our input has, a single filter always produces a single channel in the output. The multi-channel filter collapses the input channels into a single value at each position. In addition, if the input is a three-channel one, the filter applied to it must also be a made of three channels.

If we want our output to have multiple channels, we need to apply multiple filters. Each filter creates its own feature map, and these maps stack together to form a multi-channel output.

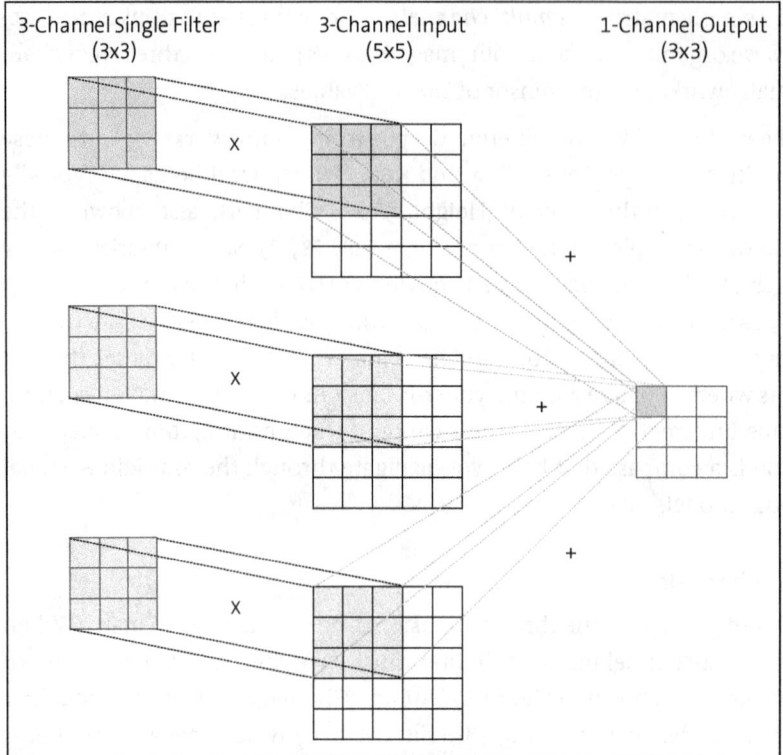

Figure 7.7 Applying Convolution on a Three-Channel Input

Let's consider the concrete example shown in Figure 7.7. We have an RGB input image of shape (5, 5, 3)—5 pixels tall, 5 pixels wide, with three color channels. We apply a 3×3×3 filter with a stride of 1. The output will be a single-channel feature map of shape (3, 3, 1). If we add a second 3×3×3 filter, we get two single-channel feature maps, which stack to create an output of shape (3, 3, 2). An interactive demo of three-channel convolution is available on the book resources page at *https://recluze.net/keras-book* in the link **07-06-conv-3-channel-demo** and on the book's official web page.

Refer back to the code given in Listing 7.1 to Listing 7.4. We can modify it to use a three-channel input instead. When we load the image, we make a minor change so that the image isn't converted to grayscale. The code will look like Listing 7.5.

```
# Load image from URL
url = "https://recluze.net/kb/building.jpeg"
response = requests.get(url)
img = Image.open(BytesIO(response.content))

# Convert to numpy array (now with 3 channels)
img_array = np.array(img)
```

Listing 7.5 Loading a Three-Channel Image to NumPy

The `apply_kernel` function remains the same as in Listing 7.2 but the `apply_convolution` function is updated to that shown in Listing 7.6. This code creates a convolution operation that performs channel mixing across a color image. It starts by preparing a single-channel output array, recognizing that we're transforming a multi-channel image into a grayscale result. Each color channel gets padded individually and stored in a list for processing. Then comes the heart of the operation—nested loops that visit every pixel position in the image. At each position, the code extracts regions from all color channels, applies the same kernel to each one, and accumulates their contributions into a single sum.

This *channel mixing* is like blending ingredients in a recipe—information from red, green, and blue channels combines to create a single, rich output value. When a filter finds a pattern in any channel, it contributes to the final result. This approach mirrors how convolutional layers in neural networks often work, where a single filter produces one feature map by integrating information across all input channels.

The final output represents the collective response of all channels to the filter pattern, capturing the essence of the image's structure in a single grayscale representation.

```python
def apply_convolution_rgb(image, kernel):
    # Get dimensions and calculate padding as before.

    # Create single-channel output array
    output = np.zeros((image_height, image_width))

    # Pad each channel
    padded_channels = []
    for c in range(channels):
        padded_channel = np.pad(image[:,:,c],
                            ((pad_height, pad_height),
                             (pad_width, pad_width)),
                            mode='constant')
        padded_channels.append(padded_channel)

    # Apply convolution with channel mixing
    for i in range(image_height):
        for j in range(image_width):
            pixel_sum = 0
            # Sum contributions from all channels
            for c in range(channels):
                # Extract the region of interest
                region = padded_channels[c][i:i+kernel_height,
                                            j:j+kernel_width]
                # Apply the kernel and add to running sum
                pixel_sum += np.sum(region * kernel)
```

```
            # Store the combined result
            output[i, j] = pixel_sum

    return output
```

Listing 7.6 Applying Three-Channel Convolution

Running this code, we see that the input image isn't really blurred but somehow a mix of all three input channels, as shown in Figure 7.8.

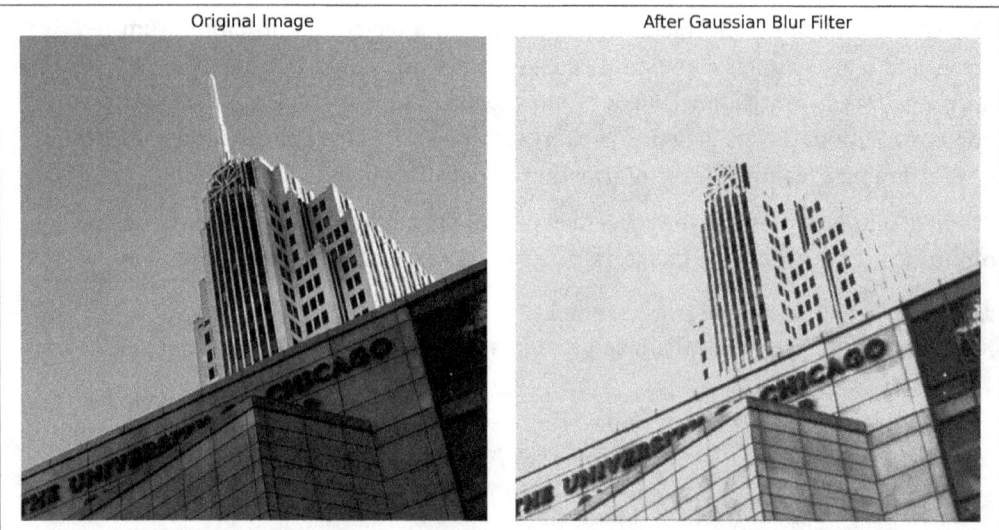

Figure 7.8 Applying a Three-Channel Convolution Filter

However, the more important task here is to understand the shapes of inputs and outputs. Listing 7.7 shows what we get for the shapes of inputs, kernels, and output.

```
print(img_array.shape)
print(kernels['identity'].shape)
print(result.shape)
# Output
# (473, 473, 3)
# (3, 3)
# (473, 473)
```

Listing 7.7 Shapes of Different Arrays When Performing Convolution

You'll notice that the resulting output is 473×473 because the input was 473×473×3, that is, with three channels. We also had a filter of size 3×3 with a stride of 1. With *padding*, our input became 474×475, and striding over it with a value of 1 would produce 473 steps in a row. If we were to go ahead and apply another filter here, the result would become

473×473×2. If we have a hundred filters, the output would be 473×473×100. Please stop here and make sure this calculation makes sense. It will be of paramount importance as we move to more complex models.

> **Number of Channels vs. Number of Filters**
>
> This distinction between the number of channels in a filter and the number of filters is subtle but important. A single filter must match the input channel depth, but it always creates just one output channel. The number of filters determines how many channels your output will have.

In practice, CNNs typically use dozens or even hundreds of filters in each layer, creating rich, multi-channel representations that capture diverse aspects of the input. Each filter specializes in detecting different patterns, giving the network a comprehensive vocabulary for understanding images. Let's explore this in further detail in the context of Keras next.

> **Practice with Shapes**
>
> It's a really good idea to make sure you know exactly how these numbers work. These shapes are an essential part of any modern network, and you should internalize these calculations. As an example, try to calculate the output shape if we change the stride here to 2. Later, in Section 7.4, you'll see how to make Keras show exact shapes to us.

7.2 Convolutional Layers, Pooling Layers and Fully Connected Layers

The convolution layer is used in a particular way inside neural networks. Figure 7.9 shows a typical *convolutional neural network* (*CNN* or *ConvNet*).

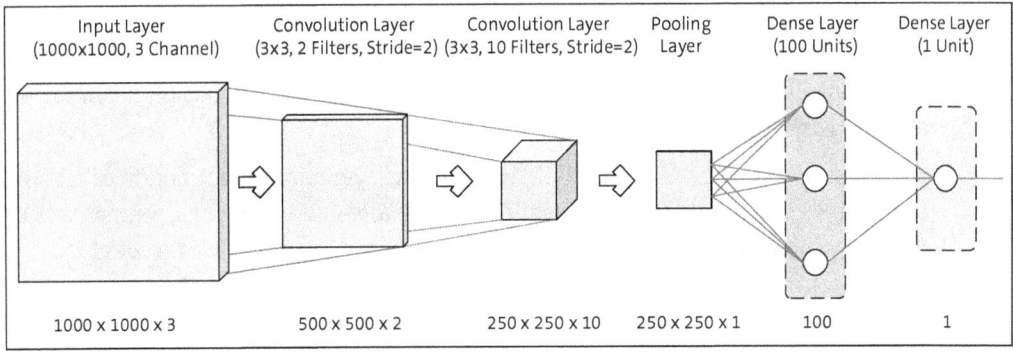

Figure 7.9 A Convolutional Neural Network

We'll cover the different types of layers shown in this figure one by one in the following sections. There are even some points we need to consider for the fully connected or dense layers that we missed earlier. So, let's begin from the top where we have the input image being fed to the first convolution layer.

7.2.1 Core Implementation of a Convolutional Layer

The code for creating this is shown in Listing 7.8. Let's walk through what's happening under the hood. First, we create our digital canvas—an image array with a specific shape using NumPy's `random()` function. This results in a 4D tensor of shape (1, 1000, 1000, 3).

This 4D tensor might look complex, but let's break it down:

1. The first dimension 1 is our batch size. Neural networks typically process multiple images simultaneously for efficiency, but here, we're feeding just one image through our layer. Think of batch size as the number of paintings an art critic examines at once—in this case, just a single masterpiece we just created.
2. The next two dimensions 1000, 1000 represent the height and width of our image in pixels—a high-resolution square image with a thousand pixels in each direction.
3. The final dimension 3 represents the color channels, similar to red, green, and blue in a standard digital photo, combining to create the full color spectrum.

In real-world applications, we'd typically use larger batch sizes (e.g., 32 or 64) to process multiple images in parallel, which dramatically speeds up training. The batch size of 1 here keeps our example simple and focused on understanding the convolutional operation itself.

Now comes the interesting part—our convolutional layer—through `layers.Conv2D`. Here, we're creating two pattern detectors (our filters), each looking through a small 3×3 window. But notice that special `strides=(2, 2)` parameter. With a stride of 2, our filter doesn't examine every possible position in the image. Instead, after looking at one 3×3 patch, it skips 1 pixel and moves to the next position.

The `padding='same'` parameter tells the system to add a border of 0s around our image so that the filter can be centered on edge pixels too. This ensures our filter can see the entire original image, including the edges.

We process our image through this layer using the layer name as a function call. This is a powerful feature of Keras where every layer is *callable* allowing for easy experimentation. So, when we compute the output simply through `conv_layer(dummy_data)` our 1,000×1,000 image transforms into a 500×500 representation. The output shape becomes (1, 500, 500, 2)—half the original dimensions in both height and width.

What if we changed the padding to `'valid'` instead of `'same'`? With *valid padding*, we don't add any 0s around the border of our image. This means our filter can only be applied where it fits completely within the original image. The 3×3 filter needs to stay entirely inside the boundaries, so it can't be centered on edge pixels. The result would be even more dimension reduction. The output shape would become slightly smaller than half the original—approximately (1, 499, 499, 2) instead of (1, 500, 500, 2).

This subtle but important difference shows how padding choice affects not only edge preservation but also your network's architecture dimensions. It's not necessary to memorize how to do these calculations, but it does help to internalize what difference they make to the shapes as these are critical for large-scale networks. In any case, irrespective of the padding type, what's fascinating is that this random *downsampling* allows our filters to simultaneously learn to detect meaningful patterns while reducing the image size. Please take a look at the **07-02-layer-types** notebook on *https://recluze.net/keras-book* for the full code listing.

```
import numpy as np
import tensorflow as tf
from keras import layers

# Create dummy data, shape:(batch_size, height, width, channels)
# Here we have 1 sample of 1000x1000 image with 3 channels
dummy_data = np.random.random((1, 1000, 1000, 3))

# Create a Conv2D layer with 2 filters of size 3x3
conv_layer = layers.Conv2D(
    filters=2,              # Number of output filters
    kernel_size=(3, 3),     # Size of the convolution window
    strides=(2, 2),         # Stride of the convolution
    padding='same',         # Try 'valid' too
    activation=None,        # No activation function
    use_bias=True           # Include bias term
)

# Feed the data through the conv layer
output = conv_layer(dummy_data)

# Print input and output shapes
print(f"Input shape: {dummy_data.shape}")

print(f"Output shape: {output.shape}")
# Output --------
# Input shape: (1, 1000, 1000, 3)
# Output shape: (1, 500, 500, 2)
```

Listing 7.8 Applying the Conv2D Layer from Keras

Finally, let's consider the number of parameters our network has to learn here as well as the shapes of our tensors. First, our code in Listing 7.9 and its output tell us there are exactly 56 parameters in our convolutional layer, that is, $3 \times 3 \times 3 = 27$ weights per filter. Plus, each filter gets its own bias term—think of this as the filter's baseline sensitivity. With 27 weights, one bias per filter, and two filters total, we get $(27 + 1) \times 2 = 56$ total parameters.

As before, we must examine the shapes more closely so that we can see exactly how these parameters are organized. In a Conv2D layer, weights are stored in a *4D tensor* with shape (`height`, `width`, `input_channels`, `filters`). The indexing weights [:, :, 0, 0] are detailed here:

- [:, :]: Take all rows and columns (the full 3×3 kernel)
- 0: From the first input channel only
- 0: Of the first filter only

So, this code line shows the 3×3 kernel matrix for the first filter's first channel, essentially displaying how that filter responds to the first channel of the input.

> [!] **Inside-Out Arrangement**
>
> Notice how this indexing of weights might seem inside out. You need some practice to get used to it. For instance, try to output all channels for this filter. What needs to change in the code? This arrangement of parameters is one of the most confusing aspects of modern machine learning but is necessary for efficient computations involving these high dimensional values.

```
# Print the number of parameters in the conv layer
print(f"Number of parameters: {conv_layer.count_params()}")

# Access and print the weights and biases
weights, biases = conv_layer.get_weights()
print(f"Weights shape: {weights.shape}")  # (3, 3, 3, 2)
print(f"Biases shape: {biases.shape}")    # (2,)

# Print a sample of the weights (first filter, first channel)
print("Sample weights (first filter, first channel):")
print(weights[:, :, 0, 0])

# Output ----------
# Number of parameters: 56
# Weights shape: (3, 3, 3, 2)
# Biases shape: (2,)
# Sample weights (first filter, first channel):
```

```
#  [[-0.26943967  0.28036994  0.04134339]
#   [-0.2828865  -0.29205972  0.3480931 ]
#   [ 0.15567559  0.29582542 -0.2867809 ]]
```

Listing 7.9 Number and Sample of Parameters

What's truly remarkable about this number is how efficiently convolutional layers handle high-resolution images. If we were to replace our convolutional layer with a single dense (fully connected) layer containing just two neurons, the parameter count would explode astronomically. Each dense neuron would need one weight for every single pixel value for one channel in our input image. This means *1000 × 1000 × 3 = 3,000,000* weights per neuron. Add a bias term for each neuron and with two neurons, this totals *(3,000,000 + 1) × 2 = 6,000,002* parameters that need to be learned. All derivatives are going to be divided into these million boxes leaving each box with essentially nothing—compounding the vanishing gradient problem. With ConvNets though, we have just 56 parameters, ensuring each parameter gets a good share of the loss's derivative and allowing it to update in a meaningful way. What's more, that's more than 100,000 times more parameters with just two neurons! This number will also increase as the input size increases. The number of parameters in ConvNet layer doesn't depend on the input size and will thus scale up efficiently. This parameter efficiency is one of the most beautiful aspects of convolutional layers—they can process massive images while maintaining a compact, focused set of parameters that look for the same patterns everywhere instead of learning separate patterns for each pixel position.

7.2.2 The Hidden Superpowers of Convolutional Layers

When we talk about convolutional layers, we often focus on their parameter efficiency, that is, how they can handle massive images with just a handful of weights. But that's just the beginning of their magic. These layers possess several remarkable properties that make them uniquely suited for understanding visual information. In the following sections, we'll go over some of these in brief as they will be very important for our discussions in later chapters as we move to even more powerful models.

Translation Invariance: Why Your Cat Is Still a Cat, No Matter Where It Sits

Consider what happens when you show cat photos to a young child. Whether the cat is lounging in the center of the frame, hiding in the corner, or perched on a windowsill, the child easily recognizes it. This ability to recognize objects regardless of their position seems obvious to us, but it's actually a sophisticated cognitive skill—one that convolutional layers naturally replicate. This property, called *translation invariance*, is perhaps the most powerful feature of convolutional layers. Because the same filter weights slide across the entire image, a pattern detector that recognizes whiskers in the top-left corner will also recognize whiskers in the bottom-right. The filter doesn't care where the pattern appears; it responds the same way everywhere.

In a traditional dense neural network, moving an object just a few pixels would completely change which neurons process it, requiring the network to essentially relearn the same pattern in different locations. It's like having thousands of specialized art critics, each only able to evaluate one tiny spot on a canvas. With convolutional layers, we instead have a few expert critics who examine the entire artwork, bringing their expertise to every inch of the painting.

Hierarchical Feature Learning: From Simple Lines to Complex Concepts

Another superpower of convolutional networks emerges when we stack multiple layers together. Early layers typically detect simple features such as edges, corners, and color gradients—the basic building blocks of visual information. As we move deeper into the network, something remarkable happens: These simple features combine to form increasingly complex representations.

It's similar to how you might learn a language. First, you recognize individual letters, then common words, then phrases, and eventually entire narratives. A deep convolutional network undergoes a similar progression:

- Layer 1 might detect edges and simple textures.
- Layer 2 might combine these to recognize shapes and contours.
- Layer 3 might identify more complex patterns such as eyes or wheels.
- Deeper layers might recognize entire objects or scenes.

This *hierarchical feature learning* happens naturally from the network architecture, creating a sophisticated visual understanding system that mirrors how our own visual cortex processes information—from simple to complex, concrete to abstract.

Preserving Spatial Relationships: The Power of Locality

When you look at a face, you don't often recognize individual features. You understand the relationships between all these features and their nuances: eyes above the nose, nose above the mouth, ears on the sides. These *spatial relationships* are crucial for visual understanding. Convolutional layers excel at preserving these spatial relationships. Unlike dense layers that flatten the input and lose all spatial information, convolutional layers maintain the 2D structure of images. This allows them to capture not only what features are present but also where they are in relation to each other. This property, known as *spatial locality*, means that nearby pixels are processed together, preserving their relationships. It's like reading a book and understanding not only the individual words but also how they're arranged into sentences and paragraphs to convey meaning.

Built-In Regularization: Less Overfitting, More Generalization

Perhaps most surprisingly, convolutional layers come with a built-in resistance to overfitting. By forcing weights to be shared across all positions, we're imposing a strong constraint on what the network can learn. This weight sharing acts as a form of regularization, preventing the network from becoming too specialized to training examples and helping it generalize better to new images. It's like teaching someone to identify birds by focusing on general patterns such as "wings, beak, feathers" rather than memorizing specific examples. The constraint actually improves learning by focusing on what truly matters rather than individual pixel values.

When we combine these properties—parameter efficiency, translation invariance, hierarchical feature learning, spatial awareness, and natural regularization—we get a system that's remarkably well-suited for visual understanding tasks. It's no wonder convolutional networks have revolutionized computer vision and become a cornerstone of modern AI. In a while, we'll put these layers to use in a larger network, but first, let's finish off out discussion about the rest of the layer types shown earlier in Figure 7.9.

7.2.3 Pooling Layers: The Image Simplifiers

Consider a beautiful, high-resolution photo of a mountain landscape. While those millions of pixels create a stunning image, your brain doesn't need every single pixel to recognize that that's a mountain. You naturally focus on key features—the triangular peak, the snow cap, the tree line—while ignoring much of the fine detail. Pooling layers in neural networks perform a similar kind of visual simplification. They're like the art of summarizing—keeping the essence while reducing the word count. Their job is elegantly simple: Take a patch of an image, pick out the most important information, and discard the rest.

In the most common type—*max pooling*—we slide a small window (typically 2×2 pixels) across our image and keep only the brightest (maximum) pixel value from each window. It's like saying, "If there's something interesting in this neighborhood, let's remember it, but we don't need to remember every house on the block." This process dramatically reduces the image size while preserving the most prominent features. If a strong edge or bright spot exists somewhere in that 2×2 window, it will survive the pooling operation, ensuring important visual information isn't lost.

Other pooling variants exist too. *Average pooling* takes the average value of each window (like capturing the general mood of a scene rather than just the standout moments), while *min pooling* keeps the minimum values (useful for dark features against bright backgrounds). We'll discuss code for max pooling in the next section, but it's straightforward to modify it to use another variant.

Max Pooling in Action

In Listing 7.10, we're creating a max pooling layer with a 2×2 window. The strides=None means our window will move 2 pixels at a time (defaulting to the *pool size*), and padding= 'valid' means we're not adding any extra padding around the edges as before. Our massive 1,000×1,000 image shrinks down to just 500×500—a 75% reduction in data! This is because each 2×2 patch becomes a single pixel in the output, effectively halving both the width and height.

```
import numpy as np
from keras import layers

# Create dummy data - shape: (batch_size, height, width, channels)
# Here we have 1 sample of 1000x1000 image with 3 channels (like RGB)
dummy_data = np.random.random((1, 1000, 1000, 3))

# Create a MaxPooling2D layer
pool_layer = layers.MaxPooling2D(
    pool_size=(2, 2),      # Size of the pooling window
    strides=None,          # Default is same as pool_size
    padding='valid'        # No padding
)

# Feed the data through the pooling layer
output = pool_layer(dummy_data)

# Print input and output shapes
print(f"Input shape: {dummy_data.shape}")
print(f"Output shape: {output.shape}")

# Print the number of parameters in the pooling layer
print(f"Number of parameters: {pool_layer.count_params()}")

<output>
Input shape: (1, 1000, 1000, 3)
Output shape: (1, 500, 500, 3)
Number of parameters: 0
</output>
```

Listing 7.10 Max Pooling Layer in Keras

Notice that the number of channels (three) stays the same. Pooling operates independently on each channel, preserving the color information while reducing the spatial dimensions. And perhaps most remarkably, this entire operation requires zero learnable parameters. Unlike convolutional or dense layers with their adjustable weights, pooling layers perform a fixed mathematical operation—no learning required. They're like simple machines with a single function: reduce size while preserving important features.

The Fading Role of Pooling Layers

While pooling layers were once a staple in neural network architecture, they've gradually fallen out of fashion in cutting-edge designs. They're still perfectly functional, but researchers discovered more sophisticated alternatives. Modern networks often opt for *strided convolutions* instead. By using a convolutional layer with a stride of 2, we can achieve the same downsampling effect while giving the network the freedom to learn exactly how that downsampling should happen. Rather than blindly taking the maximum value, a strided convolution can learn which patterns are most important to preserve. Think of it as the difference between having a fixed rule for summarizing text (e.g., "always keep the first sentence of each paragraph") versus training someone to identify and preserve the most important ideas. The latter is more flexible and potentially more effective.

Another issue with traditional pooling is that it can be too aggressive in discarding information. Sometimes, those "unimportant" pixels actually contain subtle patterns that would be useful for classification. Strided convolutions and other modern approaches can be more nuanced in what they keep and what they discard.

The philosophy in modern neural network design has shifted toward letting the network learn every step of the process rather than hard-coding operations such as pooling. This gives these networks more flexibility to adapt to the specific challenges of each dataset.

That said, pooling layers haven't disappeared entirely. They remain valuable tools in many applications, especially when computational efficiency is a priority or when a simple, reliable downsampling method is all that's needed. These remain in the form of global pooling.

7.2.4 Global Pooling

If standard pooling is like creating a smaller, simplified version of an image, *global pooling* is like distilling that image down to its absolute essence—a handful of numbers that capture its most distinctive features. Imagine you're an art critic trying to describe a famous painting. You could describe it region by region, noting the colors, shapes, and textures in each part (similar to what we get after regular pooling). Or, you could step back and capture its signature characteristics—"vibrant use of blue" or "swirling brushstrokes"—a few key descriptors that summarize the entire work.

Global pooling does exactly this. It takes an entire feature map—no matter how large—and reduces it to a single value per channel. It's like asking, "What's the most important feature in this entire channel?" or "What's the average characteristic across this whole feature map?" While this might sound counterintuitive, it does actually work really well in practice on many problems. So, it's great to keep these types of pooling in your machine learning toolbelt. And, just as with regular pooling, we have several flavors of global pooling:

- **Global max pooling**
  Takes the maximum value from each feature map. It's essentially asking, "What's the strongest activation anywhere in this feature map?" This is particularly useful for detecting the presence of specific features, regardless of where they appear.

- **Global average pooling**
  Calculates the average of all values in each feature map. This captures the overall *intensity* or prevalence of each feature across the entire image. Think of it as measuring the general mood or atmosphere rather than picking out specific highlights.

- **Global sum pooling**
  Adds up all values across each feature map. This is less common but can be useful in certain specialized networks where the cumulative effect of a feature matters more than its average or maximum strength.

The most remarkable aspect of global pooling is how dramatically it transforms our data. Let's look at the code example shown in Listing 7.11 to understand this point. When we apply global pooling to our 1,000×1,000 pixel image with three channels, our million-pixel image (1,000×1,000) has been distilled down to just three numbers—one maximum value for each of our three channels! This is a truly radical transformation, compressing all spatial information into a handful of values. In addition, like regular pooling, this operation requires zero learnable parameters. It's a fixed mathematical operation—find the maximum (or average) value in each channel—with no adjustable weights or biases.

```
# Create a GlobalMaxPooling2D layer
global_pool_layer = layers.GlobalMaxPooling2D()

# Feed the data through the pooling layer
output = global_pool_layer(dummy_data)

# Print input and output shapes
print(f"Input shape: {dummy_data.shape}")
print(f"Output shape: {output.shape}")

# Print the number of parameters in the pooling layer
print(f"Number of parameters: {global_pool_layer.count_params()}")

# Output
```

```
# Input shape: (1, 1000, 1000, 3)
# Output shape: (1, 3)
# Number of parameters: 0
```

Listing 7.11 Global Pooling in Code

Why Global Pooling Matters

This extreme dimensionality reduction serves several crucial purposes in modern neural networks:

- **Creating fixed-size representations**
  No matter the input size—whether your image is 100×100 or 1,000×1,000—global pooling will always produce the same output shape (just the number of channels). This makes it perfect for handling variable-sized inputs.

- **Bridging convolutional and dense layers**
  After a series of convolutional and pooling operations, we often want to connect to fully connected (dense) layers for final classification. Global pooling creates the perfect bridge, transforming our spatial data into a simple vector that dense layers can work with.

- **Enforcing translation invariance**
  By keeping only the strongest activation regardless of location, global max pooling ensures our network doesn't care where a feature appears—only whether it appears at all. This helps our network recognize objects even when they move around in the frame.

- **Reducing overfitting**
  By dramatically reducing the parameter count of subsequent layers, global pooling serves as a powerful *regularizer*, helping prevent our network from memorizing training examples rather than learning generalizable patterns.

> **Importance of Global Pooling**
> Unlike regular pooling layers, which have somewhat fallen out of favor, global pooling remains a cornerstone of modern network architectures. Networks such as ResNet, Inception, and EfficientNet all use global average pooling in their final layers.

Whether we're building a network to classify images, detect objects, or generate new visual content, global pooling provides that crucial step of simplification—helping our networks see the forest, not just the trees. What's more, we have some fine-tuned control over how we perform this pooling. Let's see what other variations we have for this.

7 Convolutional Neural Networks

Selective Pooling: Controlling What Gets Simplified

Consider analyzing data from a *time series*, where preserving the time dimension while condensing other features is crucial. This is where *axis-specific pooling* comes into play as it lets us decide exactly which dimensions to simplify and which to preserve. Before we jump into custom pooling, let's recall what those axis numbers actually mean in our image data. A typical image tensor has four dimensions that are arranged like this:

- Axis 0: Batch dimension (number of images)
- Axis 1: Height (rows of pixels)
- Axis 2: Width (columns of pixels)
- Axis 3: Channels (color information)

Standard pooling operations reduce both height and width together. Global pooling collapses both completely. But what if we want to be more selective? To do this, instead of using prepackaged pooling layers, we can craft our own using the *lambda layer* in Keras, combined with TensorFlow's reduction operations, as shown in Listing 7.12.

```
import tensorflow as tf
# Create custom pooling layers using Lambda
# Pool across width only (axis=2)
width_pool_layer = layers.Lambda(lambda x: \
                    tf.math.reduce_max(x, axis=2))
width_pooled = width_pool_layer(dummy_data)
# Print input and output shapes
print(f"Input shape: {dummy_data.shape}")
print(f"Width-pooled shape: {width_pooled.shape}")
# Output ----
# Input shape: (1, 1000, 1000, 3)
# Width-pooled shape: (1, 1000, 3)
```

Listing 7.12 Custom Pooling Layer Using TensorFlow

In this code, we're creating a custom layer that applies the `reduce_max` function, specifically along axis 2—the width dimension. (Refer to Chapter 5, Section 5.3.1, for the detailed discussion we had on axes.) When we run this on our 1,000×1,000 image, it's condensed to 1,000×1 (shown as (1, 1000, 3) because the single pixel has three channels). We've preserved the full height information while completely collapsing the width, keeping only the maximum value along each row. You might wonder why we'd want to pool along just one dimension. There are several compelling use cases:

- **Directional feature detection**
  Sometimes, features have a stronger orientation in one direction than another. For instance, horizontal text lines might be better preserved by pooling vertically but not horizontally.

- **Time series data**
  When working with time-based data (e.g., video or sensor readings) where each frame is a 2D image, you might want to preserve the temporal dimension while condensing the spatial information.
- **Attention mechanisms**
  In more advanced architectures, we might want to pool across certain dimensions to create attention weights that help the network focus on specific parts of the input.
- **Creating input for specialized layers**
  Some network architectures require inputs with specific shapes, and axis-specific pooling can help transform our data to fit those requirements.

Beyond the width-pooling example, we can create custom pooling operations for any dimension:

- Height pooling: `lambda x: tf.math.reduce_max(x, axis=1)`
- Channel pooling: `lambda x: tf.math.reduce_max(x, axis=3)`
- Multi-axis pooling: `lambda x: tf.math.reduce_max(x, axis=[1, 3])`

In addition, we aren't limited to maximum operations. We can use the following:

- Average pooling: `tf.math.reduce_mean()`
- Sum pooling: `tf.math.reduce_sum()`
- Minimum pooling: `tf.math.reduce_min()`

This flexibility gives us precise control over how information flows through our networks, allowing us to preserve exactly the dimensions that matter most for our specific task.

7.2.5 Bringing It All Together: Fully Connected Layers in CNNs

After our convolutional and pooling layers have worked their magic extracting patterns from images, we need a way to make sense of all these discovered features. This is where fully connected (dense) layers come into play as the final piece of the CNN puzzle that transforms all those detected patterns into actual decisions. Think of the convolutional and pooling sections of our network as incredibly observant detectives, meticulously collecting evidence, such as edges, textures, shapes, and more complex patterns. But collecting evidence is just part of the job. Eventually, someone needs to look at all the gathered clues and make a decision. That's exactly what our fully connected layers do.

When we add fully connected layers to our network, we're essentially creating a section that can see all the features our convolutional filters have discovered everywhere in the image, all at once. While convolutional layers focus on local patterns, fully connected layers see the big picture, making connections between features that might appear in completely different parts of the image.

The structure is quite simple. We typically start with a flattening operation that takes all those 2D feature maps and stretches them out into a long 1D vector of values. It's similar to taking a collection of smaller puzzle pieces and laying them all out in a row so we can see everything at once. This flattened representation then feeds into one or more dense layers where every neuron connects to every value in the input.

How many dense layers should we add? It depends. Sometimes, a single dense layer is all we need—taking the extracted features and mapping them directly to our classification outputs. Other times, we might include several dense layers of decreasing size, allowing the network to learn more abstract relationships between features before making the final decision.

> **The Top Layer in ConvNets**
>
> The very last layer in our network is often called the "top" of the neural network. This layer makes the actual discrimination or classification decision. If we're classifying images into 10 categories, our top layer will have 10 neurons, each one representing the network's confidence that the image belongs to a particular class. This will be of particular importance when we later look at more advanced techniques such as transfer learning.

It's important to recognize the division of labor here. Everything before the fully connected layers—our convolutional and pooling operations—is essentially *feature extraction*. These layers transform the raw pixel space into a rich feature space where important visual patterns become prominent. The fully connected layers then perform the *discrimination* task, taking those extracted features and making decisions based on them. This separation of concerns is so powerful that it's become common practice to take the convolutional parts of pretrained networks and reuse them for different tasks. Because the feature extraction ability is so valuable and generalizable, we can keep those parts intact while replacing just the fully connected layers to solve new problems—a technique known as *transfer learning*. We'll return to this in later chapters, and you'll most likely be using this technique on a daily basis in your machine learning endeavors.

And with that, we've completed our tour of the CNN architecture. From the specialized pattern detectors in convolutional layers, to the dimensionality-reducing pooling layers, to the decision-making fully connected layers, each component plays a vital role in the network's overall ability to understand and classify visual information. We do have some loose ends to tie up in these discussions. Let's discuss them in the context of Keras next.

7.3 Implementing CNNs with Keras

We've gone through the details of convolutional layers, exploring how they scan images for patterns, reduce parameters through weight sharing, and transform visual information into meaningful features. But there are a few variations we haven't yet explored—different flavors of convolution that expand what these powerful operations can do. We'll talk through those in this section.

7.3.1 Conv2D vs. Conv1D

First, let's talk about the difference between Conv2D and Conv1D. We've been focusing on Conv2D operations, which slide their filters across two dimensions (height and width) of our input. But what happens when our data only has one meaningful dimension to traverse? This becomes apparent when you consider working with text data. Each word or character might be represented as a vector (like a *one-hot encoding*, which we covered in Chapter 4, Section 4.3.2, or an *embedding*, which we'll cover later). At first glance, this might look 2D—a sequence of vectors. But unlike images where patterns exist in both horizontal and vertical directions, in text, the meaningful relationships typically flow in just one direction: the sequence of words.

This is where we must use *Conv1D*. Instead of sliding a 2D window across height and width, Conv1D slides a 1D window along just one dimension—the sequence length. Each filter might be 3 or 5 positions wide, looking at small groups of words or characters at a time. This operation is perfect for finding local patterns in sequential data such as text, time series, or audio signals. For example, a Conv1D filter might learn to recognize particular three-word phrases or sound patterns that are important for classification. The key insight is that we're still using the same powerful convolutional principle—applying the same filter repeatedly across different positions—but we're doing it in just one dimension instead of two.

7.3.2 The Opposite of Convolution: Deconvolution Layers

Now, let's flip things around and talk about a different variation: *deconvolutional* layers (sometimes called *transposed convolutions*). If standard convolutions are about condensing information, deconvolutions are about expanding it. In a normal convolution, we might go from a 1,000×1,000 image to a set of smaller feature maps. But what if we want to go in the other direction? What if we need to generate a larger, detailed output from a smaller representation?

That's exactly what deconvolutional layers do. They work like convolutions in reverse, taking a small input and creating a larger output. This operation is vital in many advanced network architectures, particularly in image generation tasks. Take a look at the code in Listing 7.13 along with its output.

```
import numpy as np
from keras import layers

dummy_data = np.random.random((1, 50, 50, 3))

# Create a Conv2DTranspose layer (deconvolution)
deconv_layer = layers.Conv2DTranspose(
    filters=2,              # Number of output filters
    kernel_size=(3, 3),     # Size of the convolution window
    strides=(2, 2),         # Stride of the convolution
    padding='same',         # Padding mode
    activation=None,        # No activation function
    use_bias=True           # Include bias term
)

# Feed the data through the deconvolution layer
output = deconv_layer(dummy_data)

# Print input and output shapes
print(f"Input shape: {dummy_data.shape}")
print(f"Output shape: {output.shape}")
# Output:
# Input shape: (1, 50, 50, 3)
# Output shape: (1, 100, 100, 2)
```

Listing 7.13 Deconvolution Layer through Code

Deconvolution, implemented as `Conv2DTranspose` in Keras, performs the inverse of standard convolution by expanding spatial dimensions rather than reducing them. In our example, the layer transforms a 50×50 image with three channels into a 100×100 image with two channels (because we have two filters). The `kernel_size` determines the pattern of this influence, while the `filters` parameter sets the number of output channels.

> **[+] Upsampling Using Deconvolution**
>
> This *upsampling* capability makes deconvolution essential for *generative networks*, *super-resolution* models, and architectures that need to reconstruct detailed outputs from *compressed representations*. These are extremely useful tools in modern machine learning, and we'll be returning to these again in later chapters.

Think about a network that reconstructs high-resolution images from low-resolution inputs or one that generates realistic photos from simple sketches. These networks need a way to expand their representations, adding detail and dimension as they process data. Deconvolutional layers provide this capability.

The mathematics behind deconvolutions can get complex, but conceptually, you can think of them as spreading out each input value across a region of the output, with the filter weights determining how that value gets distributed. It's like taking a rough sketch and progressively adding more and more detail until you have a complete picture. There's a lot more to these discussions, but you'll have to wait a little for that. Right now, we've been talking a lot about shapes and need to tackle an issue of particular importance.

7.4 The "Shapes" Problem

There's a moment that every machine learning practitioner dreads. You've spent hours designing your neural network architecture, carefully considering each layer, and meticulously preparing your data. You hit run, anticipating the first training epoch to begin, and instead an error message appears. Your model has crashed before even starting, with a cryptic message about *incompatible shapes* or *shape mismatch* between layers.

This shape mismatch problem is perhaps the most common technical hurdle you'll face when moving beyond tutorials into creating your own models. When you're following a book (like this one) or working through tutorials, everything seems to work seamlessly. That's because what you're seeing is the final, debugged version—the end product of someone else's troubleshooting process. It's like being given a perfectly assembled puzzle without witnessing all the trial and error that went into figuring out which pieces connect where. But the moment you venture into creating your own architecture or adapting an existing one to your specific needs, you enter a different world entirely. Suddenly, you're the one who needs to ensure that each neural layer correctly interfaces with the next. You're building the puzzle from scratch, and not all pieces are designed to fit together.

However, there's a world of difference between a model that produces poor results and one that won't run at all. Poor results mean your model is at least functioning—the data flows through it and predictions come out, but they're just not very good. Shape mismatches, on the other hand, mean your model is structurally flawed—the data simply can't flow from one layer to the next.

These mismatches occur because neural networks process data as tensors—multidimensional arrays with specific shapes. Each layer expects *input tensors* of a certain shape and produces *output tensors* of another shape. When the output shape of one layer doesn't match the expected input shape of the next, your model breaks. You might be thinking that programming languages have ways to reshape data, which is true—operations such as *reshaping* and *broadcasting* can help in some cases. These are like adapters for our metaphorical pipes. But there are limits to what these operations can automatically resolve. Some shape incompatibilities are fundamental design issues that require you to rethink your architecture.

This problem becomes especially pronounced when working with complex architectures such as *transformers* or when combining different types of layers (convolutional, *recurrent*, *attention-based*) in novel ways. Each of these components has its own tensor shape requirements, and making them work together can feel like solving a multidimensional puzzle.

Understanding how to diagnose and fix shape mismatches is a crucial skill that separates beginners from experienced practitioners. It requires not just theoretical knowledge of how neural networks function but also a practical understanding of how data flows through them, how shapes transform at each step, and how to intervene when things don't align.

Let's take a look at a model that deliberately contains shape mismatches to see exactly what happens when we try to run it. Listing 7.14 shows a seemingly reasonable CNN architecture that attempts to classify images but contains a critical tensor shape incompatibility. It's highly recommended that you run this code and spend some time with it. For ease of use, you can run the notebook **07-03-shapes-mismatch** in the resources page at *https://recluze.net/keras-book*.

```python
import tensorflow as tf
from keras import layers, models
import numpy as np

# Create some dummy image data
# 100 images of pixels with 3 color channels
dummy_images = np.random.random((100, 64, 64, 3))
dummy_labels = np.random.randint(0, 10, size=(100,))

# Build a model with a shape mismatch
model = models.Sequential([
    layers.Conv2D(32, kernel_size=(3, 3), activation='relu',
                  input_shape=(64, 64, 3)),
    layers.GlobalMaxPooling2D(),
    layers.Conv2D(128, kernel_size=(3, 3), activation='relu'),
    layers.Flatten(),
    layers.Dense(10, activation='softmax')
])

# Try to compile the model
model.compile(optimizer='adam',
              loss='sparse_categorical_crossentropy',
              metrics=['accuracy'])
```

```
model.summary()
# Error produced:
# Input 0 of layer "conv2d_83" is incompatible with the
# layer: expected min_ndim=4, found ndim=2.
# Full shape received: (None, 32)
```

Listing 7.14 A Model with a Shape Mismatch Error

When you encounter this kind of error in the real world, simply staring at it might not immediately reveal what's gone wrong. It's a bit like being handed a jigsaw puzzle with pieces that don't quite fit together—you need a systematic approach to find where the mismatch begins.

Rather than trying to guess at the solution, let's take the approach that experienced practitioners use: Build the model step-by-step, adding one layer at a time until we identify exactly where things break down. Think of it as constructing a bridge piece by piece, testing each section before adding the next. This methodical approach helps us pinpoint exactly where our architectural design fails, which is much more effective than trying to diagnose a fully assembled but nonfunctional structure. Let's build the foundation, as shown in Listing 7.15.

```
# Build a model with a shape mismatch
model = models.Sequential([
    layers.Conv2D(32, kernel_size=(3, 3), activation='relu',
                  input_shape=(64, 64, 3)),
])

# Try to compile the model
model.compile(optimizer='adam',
              loss='sparse_categorical_crossentropy',
              metrics=['accuracy'])

model.summary()
# Reduced Output ------
# Layer (type)          | Output Shape
# conv2d_84 (Conv2D)    | (None, 62, 62, 32)
```

Listing 7.15 Building the Foundation of the Model as Step 1

So, we can see that this part is fine as our Conv layer is producing a 4D output—batch size, height, width, channel—and there's no problem. Then, we add another layer and see what happens using the following code:

```
model.add(layers.GlobalMaxPooling2D())
model.summary()
```

This is also fine as we get the expected output:

```
# Reduced Output ------
# Layer (type)            | Output Shape
# conv2d_84 (Conv2D)      | (None, 62, 62, 32)
# global_max_pooling2d_1  | (None, 32)
```

This is also fine as the `GlobalMaxPooling` layer is doing what it's supposed to be doing, that is, getting rid of the height and width dimensions and just returning a 2D output (batch size, channel). Finally, let's try to add the third layer from our original model of Listing 7.14 using the following code:

```
model.add(layers.Conv2D(128, kernel_size=(3, 3), activation='relu'))
model.summary()
```

This time, we get the error as follows:

```
Input 0 of layer "conv2d_3" is incompatible with the layer: expected min_ndim=
4, found ndim=2. Full shape received: (None, 32)
```

Ok great! Now that we've identified where our model breaks down, let's unravel the mystery of this shape mismatch error. The `Conv2D` layer is expecting a 4D input—something with height, width, channels, and a batch dimension. But what's actually arriving at its doorstep is a flat, 2D tensor that's been transformed by our `GlobalMaxPooling2D` layer.

> **Build Models Step-by-Step**
>
> When designing new neural network architectures, it's generally more flexible to build models layer by layer using the Sequential model's add method rather than defining all layers at initialization. This approach allows you to incrementally construct your network, examining shapes at each step, making adjustments and debugging as needed.

No amount of reshaping can restore what's been lost. Once our spatial information has been pooled away, we can't magically recreate it. The height and width dimensions are gone completely, and Conv2D layers specifically need these spatial dimensions to operate. The fundamental issue in our architecture was placing a pooling layer before future convolutional layers. Pooling is typically used to progressively reduce the spatial dimensions of your data as it flows through a network, but once you've applied global pooling, you've essentially committed to moving away from convolutional operations. While it's possible to upscale again from this lower-dimensional space, that's not what we're going for here.

So, let's go ahead and remove the pooling layer from the top and recompile the model. Run the code given in Listing 7.15 to reset the model to just one layer. Skip over the pooling layer and add the `Conv2D` layer:

```
model.add(layers.Conv2D(128, kernel_size=(3, 3), activation='relu'))
```

Run `model.summary()` to ensure everything is as expected. Finally, add the other layers afterwards:

```
model.add(layers.Flatten())
model.add(layers.Dense(10, activation='softmax'))
```

We've used a `Flatten()` layer that simply takes all the output dimensions of the previous layer and flattens them out in a 1D vector. This vector, resulting from the process of *flattening*, can be thought of as a dense layer except that it doesn't have any associated weights. It's just a reshaping layer to facilitate the connection between our high dimensional convolutional layers and the single dimensional dense layers.

> **Going from Convolutional Layers to Dense Layers**
> `Flatten` provides one way to transition from convolutional layers to dense layers but consider using a `GlobalPooling` layer instead. This alternative approach often proves more parameter-efficient while maintaining essential feature information. Try implementing this variation as a learning exercise.

This debugging exercise of slowly building the model one layer at a time highlights an important lesson in deep learning design: always be mindful of how each layer transforms your data's shape. Every layer in your network needs to receive input in a format it can process. When designing your own neural architectures, try to keep a mental model of the tensor shapes flowing through your network. Are there any points where the output shape of one layer doesn't match the expected input shape of the next?

By developing this intuition for tensor shapes, you'll find yourself solving complex architectural challenges long before they become runtime errors. This skill becomes especially valuable as you move beyond tutorials and start designing custom networks for your specific problems. Let's combine all the lessons learned in this chapter through a case study of image classification using convolutional, global pooling, and dense layers, along with some enhancements learned in earlier chapters.

7.5 Case Study: Image Classification

Once you understand the whole background and theory behind the choices, the actual Keras code for performing classification is going to be surprisingly compact. It's like discovering that an intricate watch mechanism you've spent weeks studying piece by piece can be assembled with just a few precise movements. All of those layers, activation functions, and tensor manipulations we've explored in detail get distilled into a handful of clean, readable lines. This isn't magic—it's the power of abstraction that

Keras brings to the table. The framework handles all the intricate implementation details while exposing a clear, high-level API that maps directly to the conceptual building blocks we've been discussing.

In the following case study, we'll implement a complete CNN-based image classifier in code that's so concise you could write it in a couple of minutes, yet so powerful it can recognize complex patterns that would have seemed impossible to detect programmatically just a decade ago. But don't let the brevity fool you—behind each line lies the deep understanding you've been developing throughout this chapter.

Let's examine the code in Listing 7.16, which prepares the *CIFAR-10* dataset for our image classification task. This code reaches out to Keras's built-in datasets and retrieves CIFAR-10, a collection of 60,000 32×32 color images across 10 different categories such as airplane, automobile, and bird. The function returns two tuples: one containing the training data and another containing the test data.

Notice that we're renaming the standard training set to x_full and y_full. This reveals our intention to further divide this data. The second step does exactly that, using scikit-learn's train_test_split function to carve out a validation set from our training data. Keep in mind that unlike the MNIST dataset we worked with earlier, CIFAR-10 contains color images with three channels (RGB). So, each image in our dataset is represented as a 32×32×3 tensor, making this a more challenging classification task than our earlier grayscale examples. The complete code listing is available on *https://recluze.net/keras-book* in the **07-04-image-classification** notebook.

```
import numpy as np
import matplotlib.pyplot as plt
from sklearn.model_selection import train_test_split

import keras
from keras import layers
from keras.datasets import cifar10

# Load and prepare the CIFAR-10 dataset
(x_full, y_full), (x_test, y_test) = cifar10.load_data()

# Split the full dataset into training and validation sets
x_train, x_val, y_train, y_val = train_test_split(
    x_full, y_full, test_size=0.2, random_state=42
)
```

Listing 7.16 Setting Up the Dataset for a CNN Model

With our dataset split into training, validation, and test sets, we now need to prepare the data for optimal model training, as shown in Listing 7.17. The preprocessing consists

7.5 Case Study: Image Classification

of two critical steps. First, we normalize the pixel values. Raw images have pixel intensities ranging from 0 to 255, but neural networks generally perform better when their inputs are scaled to smaller ranges. By converting our data to floating-point format and dividing by 255, we *rescale* all pixel values to the range [0, 1]. This normalization helps with *numerical stability* during training and often leads to faster convergence.

The second step addresses how we represent our target classes. Initially, the classes in CIFAR-10 are represented as integers from 0 to 9. However, for multi-class classification tasks, we typically want our model to output a probability distribution across all possible classes. To align our ground truth with this output format, we convert our integer labels to *one-hot encoded* vectors using `keras.utils.to_categorical()`.

Recall that one-hot encoding transforms each integer label into a binary vector where all elements are 0 except for a single 1 at the index corresponding to the class. For example, the class 3 becomes [0, 0, 0, 1, 0, 0, 0, 0, 0, 0]. This representation allows us to use *categorical cross-entropy* as our loss function, which we've previously seen is well-suited for classification tasks.

```
# Normalize pixel values to be between 0 and 1
x_train = x_train.astype("float32") / 255.0
x_val = x_val.astype("float32") / 255.0
x_test = x_test.astype("float32") / 255.0

# Convert class vectors to one-hot encoded labels
y_train = keras.utils.to_categorical(y_train, 10)
y_val = keras.utils.to_categorical(y_val, 10)
y_test = keras.utils.to_categorical(y_test, 10)
```

Listing 7.17 Preprocessing Data and Preparing Output Vectors

Now, let's analyze the model architecture shown in Listing 7.18. This code defines a simple yet effective CNN architecture organized in blocks. Each block doubles the number of filters while gradually extracting higher-level features. The first layer specifies our input shape (32×32×3) to match CIFAR-10's color images. Note the strategic use of `padding='same'` in alternating convolutional layers, which preserves spatial dimensions while extracting features. Instead of the traditional flattening approach, this model uses `GlobalAveragePooling2D()` to reduce each feature map to a single value, dramatically decreasing parameter count while maintaining performance.

The output layer has 10 neurons with softmax activation, producing a probability distribution across our 10 possible classes. The model is compiled with Adam optimizer and categorical cross-entropy loss function—the natural choice for our one-hot encoded labels.

While traditional CNNs often include explicit pooling layers and dropout for regularization, this initial architecture doesn't include that. We can expect some overfitting, but we'll fix that later.

```
# Build the ConvNet model - Variation 1
model = keras.Sequential([
    # First convolutional block
    layers.Conv2D(32, (3, 3), padding='same', activation='relu',
                                input_shape=(32, 32, 3)),
    layers.Conv2D(32, (3, 3), activation='relu'),

    # Second convolutional block
    layers.Conv2D(64, (3, 3), padding='same', activation='relu'),
    layers.Conv2D(64, (3, 3), activation='relu'),

    # Third convolutional block
    layers.Conv2D(128, (3, 3), padding='same', activation='relu'),
    layers.Conv2D(128, (3, 3), activation='relu'),
    layers.GlobalAveragePooling2D(),

    # Output layer
    layers.Dense(10, activation='softmax')
])

# Compile the model
model.compile(optimizer='adam',
    loss='categorical_crossentropy',
    metrics=['accuracy']
)
model.summary()
```

Listing 7.18 Model Architecture for CIFAR-10 Classification

With our model architecture defined, we can now train and evaluate it as shown in Listing 7.19. This code initiates the model training process using the fit() method. We're processing our data in batches of 64 images, which strikes a balance between training speed and memory usage. The fit() method trains our model for 50 epochs with batch size 64, monitoring validation performance throughout. The returned history object tracks metrics such as loss and accuracy, which are useful for visualizing learning curves.

After training, we evaluate on the test set to get an unbiased estimate of our model's performance. This final accuracy gives us a baseline for further experimentation with different architectures or hyperparameters.

```
history = model.fit(
    x_train,
    y_train,
    batch_size=64,
    epochs=50,
    validation_data=(x_val, y_val)
)
# Evaluate the model on the test set
test_loss, test_acc = model.evaluate(x_test, y_test, verbose=2)
print(f"Test accuracy: {test_acc:.4f}")
print(f"Test loss: {test_loss:.4f}")
```

Listing 7.19 Training the Model and Reporting Evaluation Metrics

Understanding what our CNN has learned can be challenging. One useful approach is to visualize the convolutional filters, as shown in Listing 7.20. This utility function extracts and visualizes learned filters from any convolutional layer in our model. It's designed to be reusable across different architectures, making it a valuable tool for model inspection and debugging. The function first retrieves the weights from the specified layer, then *normalizes* them to the range [0,1] for consistent visualization. For RGB image processing networks, each filter has three channels, which are displayed separately to reveal how the filter responds to different color information.

By default, it shows up to 16 filters to prevent overwhelming visualizations when layers contain hundreds of filters. The resulting plots provide unique insights into what features each filter detects—from simple edges and textures in early layers to more complex patterns in deeper layers.

```
# Function to visualize convolutional filters
def visualize_filters(model, layer_name, max_filters=16):
    layer = model.get_layer(name=layer_name)
    filters, biases = layer.get_weights()

    # Normalize filter values to 0-1 for visualization
    f_min, f_max = filters.min(), filters.max()
    filters = (filters - f_min) / (f_max - f_min)

    # Plot up to max_filters from the layer
    n_filters = min(max_filters, filters.shape[3])
    plt.figure(figsize=(12, 12))

    for i in range(n_filters):
        # Get filter
        f = filters[:, :, :, i]
```

7 Convolutional Neural Networks

```
        # Plot each channel
        for j in range(3):
            plt.subplot(4, n_filters, i + 1 + j * n_filters)
            plt.imshow(f[:, :, j], cmap='viridis')
            plt.title(f'Filter {i+1}, Channel {j+1}')
            plt.axis('off')

    plt.suptitle(f'Convolutional Filters in {layer_name}')
    plt.tight_layout()
    plt.close()
```

Listing 7.20 Visualization Function for CNN's Learned Filters

To use this function, simply call it with your model and the name of any convolutional layer as in Listing 7.21. It's highly recommended that you spend some time here and visualize all the different layers' filters to get used to the shapes they are producing. While individual filter values aren't important, studying these have often led the world's leading researchers to come up with new and innovative ideas in the field of machine learning.

```
model.layers[0]._name = 'conv2d_1'
visualize_filters(model, 'conv2d_2')
```

Listing 7.21 Visualizing the First Layer's Features

Finally, we can go ahead and get the metrics of evaluation using the code shown in Listing 7.22.

```
print("Model training and evaluation complete.")
print(f"Final training accuracy: \
        {history.history['accuracy'][-1]:.4f}")
print(f"Final validation accuracy: \
        {history.history['val_accuracy'][-1]:.4f}")
print(f"Test accuracy: {test_acc:.4f}")
```

Listing 7.22 Reporting the Metrics

And there you have it! The real power of Keras shines through those eight lines of model definition code in Listing 7.18. After all our theoretical exploration, isn't it remarkable how such a compact architecture can learn to recognize complex visual patterns? While the supporting code handles necessary logistics, it's this elegant model definition that captures the essence of deep learning engineering. This is the code that transforms mathematical concepts into problem-solving tools—and mastering this interface between theory and implementation is what makes a skilled machine learning engineer invaluable.

> **A Bit of Homework**
>
> Now it's time for you to spend some time playing with the architecture and hyperparameters. This will help you develop intuition about CNN architectures through experimentation. As you modify the hyperparameters, do the following:
>
> - Track how dropout affects overfitting. Start with 0.2 and try higher values.
> - Compare max versus average pooling on feature preservation.
> - Experiment with filter sizes (3×3, 5×5, 7×7), and observe how they capture different feature scales.
> - Test different stride configurations to understand their impact on resolution and computational efficiency.
> - Record validation score changes with each modification.
>
> Look for patterns in how each parameter affects model performance and generalization. The insights gained will help you make more informed architectural decisions in future projects.

7.6 Summary

In this chapter, we dove into the architecture that revolutionized computer vision—CNNs. We explored how convolutional layers act as specialized pattern detectors that scan images while dramatically reducing parameter counts through weight sharing. We tackled the critical shapes mismatch problem that every deep learning practitioner inevitably faces, as well as covered how pooling layers provide elegant dimensionality reduction. The ReLU and GeLU activations emerged as our heroes in the battle against vanishing gradients, while global pooling showed us how to distill entire feature maps down to their essence. Through our case study, you saw these building blocks come together to form a powerful image classification system. Perhaps most importantly, you gained the intuition needed to architect your own CNNs, understanding not just how these networks operate, but why each component plays its crucial role in helping machines see the world through patterns, just as our own visual systems do.

In the next chapter, we'll move on to a more flexible way of building models in Keras that gives us much greater control over their structure.

Chapter 8
Exploring the Keras Functional API

In this chapter, we'll explore the Keras Functional API, moving beyond the limitations of sequential models to build complex neural architectures with explicit layer connections. We'll introduce powerful design patterns, including residual networks with skip connections and multi-branched models that process information along parallel pathways. Through practical case studies, we'll implement cutting-edge applications such as ResNet for classification, Siamese networks for similarity learning, and U-Net for image segmentation. Finally, we'll master transfer learning, a technique that lets us repurpose pretrained models for specialized tasks with minimal data and computation. By the end, you'll understand how the Functional API transforms Keras from a simple tool into a platform for implementing sophisticated architectures that achieve superior performance across diverse domains.

So far in our discussions, we've built neural networks like assembly lines, where data flows in a single, straight path from input to output. This approach has served us well for many tasks, but it's like trying to navigate a bustling city using only one-way streets. Sometimes, we need more flexibility in how information travels through our networks. Consider a situation in which we can build networks where information could take multiple paths, skip ahead, or even circle back. What if parts of our network could share what they've learned with other parts? Or what if we could feed our network different types of information simultaneously, such as combining images with text descriptions—often called multimodal learning? These are the kinds of powerful architectures we'll unlock in this chapter. We'll move beyond the linear assembly line to create neural networks that resemble intricate subway systems, with multiple entry points, express lanes, and interconnected routes that all work together to reach our destination. Because virtually every breakthrough in deep learning over the past decade has come from these more sophisticated network designs. The revolutionary models that can recognize objects in images with human-like accuracy, generate realistic images from text descriptions, or understand and translate languages don't use simple sequential architectures. They employ intricate patterns of information flow that allow them to process complex data more effectively.

Think about how your own understanding develops when learning something new. You don't process information in a single, forward pass. You make connections between concepts, revisit earlier ideas with new understanding, and combine multiple types of information. The most powerful neural networks work in similar ways, and that's why they've achieved such remarkable results.

You'll discover how this more flexible approach solves real-world problems that sequential models struggle with. We'll build networks that can process images alongside metadata, create systems that learn to compare inputs for similarity, and construct architectures where information can flow more freely, avoiding the bottlenecks that have plagued deep networks in the past. This flexibility is not only of academic importance, but it's also playing a large part in transforming industries. Medical imaging systems that combine visual analysis with patient data to improve diagnostic accuracy, recommendation engines that process both user behavior and content features, and autonomous vehicles that fuse inputs from multiple sensors—all rely on the techniques we'll explore in this chapter.

By the end of this chapter, you'll know how to craft custom neural network architectures tailored to your specific problems. You'll be able to incorporate pretrained models into your own designs, adapting them to your organization's unique needs without starting from scratch. And you'll understand the principles behind the revolutionary architectures that have transformed deep learning in recent years. This is where your exploration as a neural network architect truly begins. The building blocks we've explored so far remain the same, but we're about to learn how to arrange them in powerful new ways that will dramatically expand what's possible with your models.

All of this, as with all great breakthroughs, begins with a problem—one we've already seen in the previous chapter—vanishing gradients.

8.1 Overview of Keras Functional API

Remember our discussions about the vanishing gradient problem in the previous chapter? We explored how, in deep networks, those early layers sometimes struggle to learn because the gradients become microscopically small by the time they travel backward through all those layers. This challenge is intimately connected to how information flows through our networks. When we build sequential models, we're essentially creating a single highway for information, with no detours, no shortcuts, and no alternate routes. This linear path forces every bit of information to travel through the same congested route, often leading to traffic jams that slow down learning.

> **Understanding the Philosophy of Functional API**
>
> We'll introduce the actual mechanism to achieve all of this shortly, but first, we must ensure that we're on the same page about the philosophy behind this approach. We must discuss the motivation behind this new method of coding neural networks—the actual syntax, as with everything Keras—isn't too complex.

In this section, we'll discuss how we can create dynamic pathways for our information and gradients to flow, thus reducing the chances of information loss.

8.1.1 The Information Bottleneck

Consider what happens when playing a game of telephone, where a message gets passed from person to person in a line. By the time the message reaches the end, it's often hilariously distorted. Neural networks face a similar challenge—as information passes through each sequential layer, it can become distorted or diluted. In deep networks, this creates what researchers call an *information bottleneck*. Each layer has the potential to lose some of the original input's important details. By the time the signal reaches the deeper layers, crucial information might be lost forever, making it impossible for the network to make accurate predictions.

Think about trying to describe a detailed painting to someone who has to recreate it, but you have to pass your description through 10 different people before it reaches the artist. How accurate do you think the final painting would be? In our neural networks, this rigid structure creates several practical limitations:

- **Fragility**
  If one layer fails to learn properly, the entire network suffers because there's no alternative path for information.

- **Feature loss**
  Important information from earlier layers might not make it to later layers where it's needed for decision-making.

- **Inefficient learning**
  Later layers might need direct access to earlier representations, but they can only receive information that's been processed (and potentially altered) by all the intermediate layers.

- **Complex relationships**
  Many real-world problems involve multiple types of input data or relationships that can't be represented in a simple sequence.

At the heart of training neural networks is *backpropagation*—the process that calculates how to adjust each weight to reduce errors. Traditional backpropagation relies on the chain rule from calculus, which works beautifully for sequential structures but becomes limiting for more complex architectures. The chain rule is like a bucket brigade—each person in line can only receive water from the person directly before them and can only pass it to the person directly after. But what if we want to create networks where layer 5 gets input directly from layer 1? Or where multiple pathways merge? Or where certain layers share information with several others?

The traditional application of the chain rule doesn't easily accommodate these more complex information flows. It's designed for step-by-step calculations through a single path, not the intricate web of connections we might want to create. As our problems become more sophisticated, our network architectures need to evolve as well. We need models that can do the following:

- Skip

  Allow information to skip ahead, bypassing layers that might distort it.

- Create pathways

  Create multiple pathways for different types of features or processing.

- Share

  Share learned features across different parts of the network.

- Coordinate input and output

  Handle multiple inputs and outputs in coordinated ways.

8.1.2 Networks as Directed Acyclic Graphs

In the world of deep learning, we're moving from thinking about models as simple chains to visualizing them as what mathematicians call *directed acyclic graphs* (DAGs). Don't let the terminology intimidate you—this is actually an intuitive concept. "Directed" simply means that information flows in a specific direction, just like traffic on a one-way street. "Acyclic" means there are no loops or cycles—information never revisits a layer that it's already passed through. And "graph" refers to a collection of nodes (layers) connected by edges (the flow of information).

Think about a river delta, where a single stream branches out into multiple channels that occasionally merge back together before reaching the ocean. This branching and merging pattern allows the water to find multiple paths to its destination, adapting to the terrain it encounters. Our neural networks can adopt similar patterns, with information flowing through multiple pathways that split and rejoin as needed. This shift in perspective is powerful because it allows us to design networks that match the natural structure of our problems rather than forcing our problems to fit a sequential structure.

With functional models, we gain the freedom to route information in ways that make sense for our specific task. Here are some of the powerful patterns this enables:

- Skip connections

  Imagine being able to take an express elevator that bypasses several floors of a building. Skip connections allow information to jump ahead, creating shortcuts from earlier layers directly to later ones. This helps combat the vanishing gradient problem by giving those gradients a more direct path during backpropagation.

- Feature reuse

  In a traditional sequential network, once a layer processes information, that specific representation is never used again directly—it's always transformed by the next layer first. With functional models, multiple downstream layers can tap into the same earlier representation, reusing those features in different ways.

- **Multi-input fusion**
Real-world problems often involve multiple types of data. A medical diagnosis might use both images and patient history; a recommendation system might consider both user behavior and item characteristics. Functional models allow us to process these different inputs in specialized ways before combining them.

- **Branching pathways**
Sometimes, we want our network to perform multiple related tasks. For instance, an image analysis system might need to identify objects, their positions, and their relationships. With functional models, we can have shared early processing before branching into specialized paths for each task.

In this paradigm, layers are functions that transform inputs to outputs, and a model is simply a composition of these functions. This approach feels very natural to those with a programming background—like building a complex function by combining simpler ones. The beauty of this approach is that it retains all the mathematical rigor and learning capabilities of traditional networks while dramatically expanding what we can build. The underlying mechanics of forward and backward propagation still work, but now they operate on a much more flexible architecture that better matches our needs.

8.1.3 The Functional Programming Heritage

The *Keras Functional API* draws inspiration from *functional programming*, a paradigm with roots dating back to the 1930s with lambda calculus. In functional programming, programs are constructed by applying and composing functions, emphasizing what should be computed rather than how it should be computed.

The core principles that functional programming brings to neural network design include the following:

- **Composition**
Complex operations are created by combining simpler functions. In our networks, this translates to building sophisticated models by connecting simpler layers.

- **Pure functions**
Functions that produce the same output for the same input without side effects. In neural networks, this means layers behave predictably based on their inputs, making models easier to reason about.

- **Immutability**
Once created, data structures don't change. In the Functional API, once a tensor flows through a layer, it produces a new tensor rather than modifying the original one, creating a clear lineage of transformations.

- **Declarative style**
  We specify what our program should accomplish rather than listing step-by-step instructions. With the Functional API, we declare the connections between layers, and Keras handles the implementation details.

This approach creates a beautiful alignment between the mathematics of neural networks (essentially compositions of functions) and their implementation. The model becomes a visual and logical representation of the mathematical operations it performs—a graph of computations that feels natural and intuitive.

As we move forward, we'll explore exactly how to implement these ideas using the Keras Functional API. We'll give you the tools to design networks that are as flexible and expressive as your imagination, liberated from the constraints of sequential thinking.

8.1.4 Key Advantages of the Functional API

In this section, we'll discuss the key advantages this architectural freedom brings to our neural networks. Think about how you engage with the world around you. When you drive a car, you're simultaneously processing sounds, feeling the car's movements, and perhaps listening to navigation instructions. Your brain integrates all of these information streams to make decisions.

Multiple Inputs and Multiple Outputs

Traditional neural networks are like trying to drive using only a single camera feed. The Functional API lets us build models that more closely mimic how our brains work, processing multiple input streams in parallel before combining them into a unified understanding.

For example, a modern *recommendation system* might simultaneously analyze the following:

- Text description of an item
- Associated images
- Your past browsing history
- Current trends among similar users
- Contextual information such as time of day or device

Each of these inputs might require different types of processing—convolutional layers for images, recurrent layers for text, and dense layers for user data. With the Functional API, we can create specialized processing pathways for each input type and then merge them at the perfect point to make our final recommendations.

Similarly, we can create models that perform multiple related tasks simultaneously. An autonomous vehicle system might need to do the following:

- Detect objects in its surroundings.
- Predict the objects' future movements.
- Identify safe paths forward.
- Control acceleration and steering.

Rather than building separate networks for each task, we can create a unified model with multiple outputs, allowing the shared early layers to learn features useful for all tasks. This not only makes our system more efficient but often improves performance as the related tasks help guide each other's learning.

Sharing Layers Across Different Parts of a Model

Have you ever learned a skill that unexpectedly helped you in a completely different area? Maybe learning a foreign language enhanced your speaking abilities or solving problems helped you become a better programmer. Neural networks can benefit from similar *cross-training* when we allow them to share layers.

Layer sharing is like having the same expert contribute to different parts of a project. If you're building both a spam classifier and a sentiment analyzer for emails, both tasks need to understand language. With the Functional API, you can have them share the same language processing layers, allowing insights from one task to benefit the other.

This approach is particularly powerful in these applications:

- **Siamese networks**
  Used for facial recognition, identical processing pathways with shared weights compare two inputs to determine if they represent the same person.
- **Image-to-image translation**
  Encoder layers extract features that decoder layers then transform into new images.
- **Multitask learning**
  Several related tasks share early processing layers before branching into task-specific outputs.

The purpose of layer sharing goes well beyond computation. It creates powerful learning synergies, allowing knowledge gained from one aspect of your data to enhance performance on related tasks.

Creating Nonlinear Network Topologies

The Functional API lets us create network topologies that more closely resemble natural river deltas, with streams that branch out, flow in parallel, and sometimes merge back together. This nonlinear approach enables powerful architectural patterns such as the following:

- **Inception modules**
  These process the same input at multiple scales simultaneously before combining the results.
- **Deeply dense connections**
  Each layer receives input from all previous layers, creating a rich web of information pathways.
- **Encoder-decoder**
  These architectures compress information into a compact representation before expanding it into a new form.

These *nonlinear topologies* aren't theoretical curiosities. They're the backbone of state-of-the-art models across domains. They allow information to flow more naturally, creating multiple perspectives on the same data before integrating them into a coherent whole. Modern *transformer models* powering AI-based chatbots are all built around these concepts, especially the encoder-decoder architecture.

Enabling Advanced Architectures That Address Vanishing Gradients

Remember the vanishing gradient problem we discussed earlier? The Functional API provides elegant solutions through architectures specifically designed to maintain strong gradient flow even in very deep networks. The most revolutionary of these is the *residual connection* (or *skip connection*), which creates highways that allow gradients to bypass problematic layers during backpropagation. It's like building express lanes on a highway—traffic can still use all the local exits but can also zoom past congested areas when necessary.

A residual connection adds the input of a *layer block* directly to its output. This simple addition creates a path where gradients can flow freely, allowing networks to grow to previously impossible depths—from dozens to hundreds or even thousands of layers. Take a look at Figure 8.1 for an example of a residual or skip connection. We'll get to the details of this later, but for now, you can see that the main path at the bottom of the architecture is being bypassed by the skip connection at the top. The result of the main path and the residual connection are concatenated at the end before being output. This enables huge improvements in backpropagation and forms a backbone of modern machine learning architectures.

As we move forward in this chapter, you'll learn how to implement these powerful patterns in your own models, combining the mathematical power of deep learning with the architectural freedom to design networks that truly match the structure of your problems. However, we must start from the first baby step—looking at the basic syntax of these new types of model definitions.

8.2 Building Complex Models with the Functional API

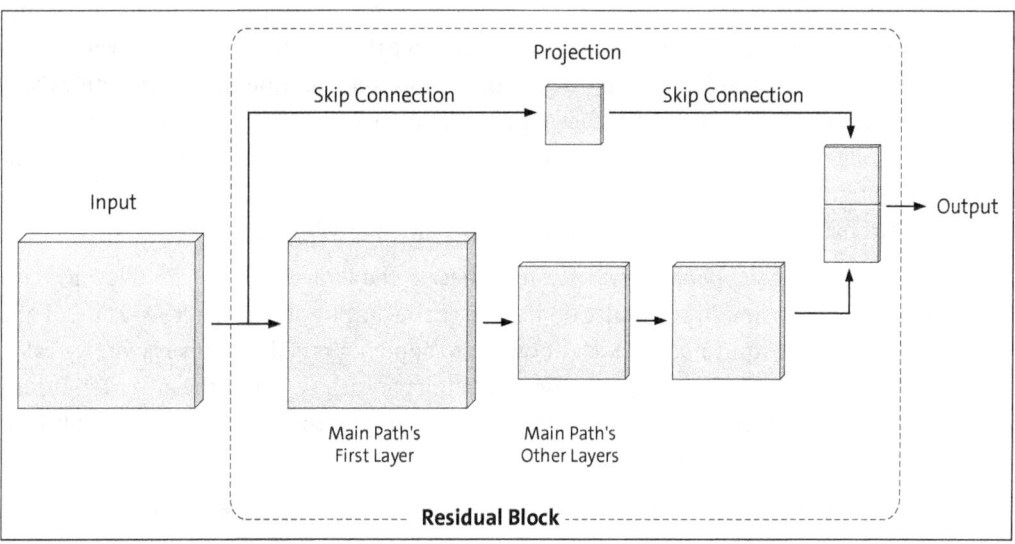

Figure 8.1 Residual Connection Architecture

8.2 Building Complex Models with the Functional API

Now that we've explored the possible benefits of this methodology, we'll turn our attention to the actual nuts and bolts. In this section, we'll start by examining a simple code snippet that illustrates the bare minimum syntax of the Keras Functional API. This will give us the freedom to create models with multiple inputs, multiple outputs, shared layers, and nonlinear topology—all things that are difficult or impossible with the Sequential API.

8.2.1 Overview of the Functional API Syntax

Let's take a look at a basic example to get a feel for the syntax, as shown in Listing 8.1.

```
inputs = keras.Input(shape=(784,), name="input_layer")
x = keras.layers.Dense(128, activation="relu",
                    name="hidden_layer_1")(inputs)
x = keras.layers.Dense(64, activation="relu",
                    name="hidden_layer_2")(x)
outputs = keras.layers.Dense(10, activation="softmax",
                    name="output_layer")(x)
model = keras.Model(inputs=inputs, outputs=outputs,
                    name="functional_model")
```

Listing 8.1 Basic Syntax of the Keras Functional API

At first glance, this code might look quite similar to creating a Sequential model, but there are some fundamental differences that make the Functional API so powerful. Let's break it down step-by-step. First, notice how we define our input layer separately using `keras.Input()`. This is a key difference from Sequential models, where the input shape is implicitly assumed from the first layer. Here, we're explicitly creating a standalone input tensor that specifies both the shape of our data (784 dimensions, which might represent flattened 28×28 MNIST images) and a name for this layer.

The real magic happens in the next few lines. In the Functional API, each layer is *callable*—it's very similar to a function that takes the output of another layer as input and produces an output itself. Look at how we're applying our first Dense layer: we call it with the `inputs` tensor as a parameter (note the parentheses after the layer definition). This creates a connection between our input layer and this hidden layer, establishing the flow of data in our network.

Naming Conventions for Functional Models
By convention, we name the input tensor `inputs` and the final output tensor `outputs`. This naming helps keep our code readable, especially as our models grow more complex. While we've only defined one input and one output here, the Functional API allows us to work with multiple inputs and outputs—a capability we'll explore later in this chapter.

For the intermediate layers, we don't usually assign explicit names to the variables holding their outputs unless necessary. Instead, we follow a pattern of reusing the variable x to store the output of each successive layer. This approach saves us from having to come up with unique variable names for every layer in our network. It also makes our code cleaner and more concise, especially when we need to repeat similar blocks of layers—something we'll see in more complex architectures later in this chapter.

Finally, we create our model by passing both the inputs and outputs to the `keras.Model` class. This is where we define the boundaries of our model—all layers between the specified inputs and outputs become part of the model, including any branches or complex pathways. The model knows exactly which operations to perform, in which order, to transform the input into the output.

This pattern of defining our model through *functional composition*—specifying the connections between layers rather than just listing them sequentially—is what gives the Functional API its power and flexibility. The Functional API gives you the map, allowing you to chart whatever course you need for your specific task.

8.2.2 Creating Models with the Functional API

To further clarify this, let's see how we define inputs in the Functional API. The Input layer is the starting point that shapes everything that follows. When we call keras.Input(shape=(784,)), we're essentially creating a blueprint that tells Keras, "Expect data with this specific structure." The shape argument can handle various formats, as shown in Listing 8.2—from simple vectors to complex multidimensional tensors for images, video, or time series data.

```
# For vector data
vector_input = keras.Input(shape=(20,))

# For image data (height, width, channels)
image_input = keras.Input(shape=(224, 224, 3))

# For time series data (timesteps, features)
timeseries_input = keras.Input(shape=(None, 10))
```
Listing 8.2 Different Types of Input Layer Definitions

The None dimension in the time series example is particularly powerful—it tells Keras this dimension can vary, allowing us to process sequences of different lengths. This flexibility is something you'll appreciate when working with real-world data that doesn't always come in perfectly uniform packages.

Once we've defined our model's architecture by connecting layers, we bring everything together with the Model class. This step outlined in Listing 8.3 formalizes our network design into a cohesive object that Keras can work with.

```
# Creating a model with single input and output
model = keras.Model(inputs=input_tensor, outputs=output_tensor,
name="my_model")

# You can access model information
model.summary()
model.get_config()
keras.utils.plot_model(model, "model_diagram.png", show_shapes=True)
```
Listing 8.3 Creating and Inspecting a Functional API Model

The Model class provides a rich set of methods for training, evaluation, prediction, and visualization. The summary() method gives us a bird's-eye view of our architecture, while plot_model() creates a visual graph that can be invaluable for understanding complex designs.

You might be wondering—if the Functional API and Sequential API can both create similar models, why choose one over the other? Let's compare these approaches side by side, as shown in Listing 8.4.

```
# Sequential API approach
sequential_model = keras.Sequential([
    keras.layers.Dense(128, activation="relu", input_shape=(784,)),
    keras.layers.Dense(64, activation="relu"),
    keras.layers.Dense(10, activation="softmax")
])

# Functional API equivalent
inputs = keras.Input(shape=(784,))
x = keras.layers.Dense(128, activation="relu")(inputs)
x = keras.layers.Dense(64, activation="relu")(x)
outputs = keras.layers.Dense(10, activation="softmax")(x)
functional_model = keras.Model(inputs=inputs, outputs=outputs)
```

Listing 8.4 Sequential versus Functional API Syntax Comparison

The Sequential model is following a recipe step-by-step—simple and straightforward when your dish has a linear preparation. The Functional API, on the other hand, is more like having a fully equipped kitchen where you can prepare multiple components simultaneously and combine them in creative ways. While both models in Listing 8.4 have identical architecture, the Functional API shows its true value when we need to create more complex designs—such as models with multiple inputs, skip connections, or branching paths. As our models grow more complex, we can quickly end up with hundreds of lines of layer definitions. This is where thinking functionally can save us both time and headaches. By wrapping common *architectural patterns* in Python functions, we create reusable building blocks that make our code more maintainable and easier to experiment with. Let's take a look at a detailed example in Listing 8.5 for a structured approach to building models.

```
def build_encoder_block(x, filters, kernel_size=3, strides=1, padding="same"):
    x = keras.layers.Conv2D(filters, kernel_size, strides=strides, padding=padding)(x)
    x = keras.layers.Activation("relu")(x)
    return x

def build_classification_model(input_shape, num_classes):
    inputs = keras.Input(shape=input_shape)
    
    # Build encoder blocks
    x = build_encoder_block(inputs, 32)
    x = keras.layers.MaxPooling2D()(x)
    x = build_encoder_block(x, 128)
```

```
    # Classification head
    x = keras.layers.GlobalAveragePooling2D()(x)
    x = keras.layers.Dense(256, activation="relu")(x)
    x = keras.layers.Dropout(0.5)(x)
    outputs = keras.layers.Dense(num_classes, activation="softmax")(x)

    return keras.Model(inputs, outputs)

# Create a model for a specific dataset
mnist_model = build_classification_model((28, 28, 1), 10)
cifar_model = build_classification_model((32, 32, 3), 100)
```

Listing 8.5 Creating Reusable Model Architectures with Functions

This approach offers several advantages. First, it makes our code more readable—instead of wading through dozens of layer definitions, we can see the high-level structure at a glance. Second, it promotes experimentation—we can easily swap out components or adjust parameters without rewriting entire sections. Finally, it reduces errors by encapsulating tested patterns that we know work well. This way, we can follow the best practices without having to increase our cognitive load. Let's take a look at some of these practices next.

8.2.3 Best Practices for Complex Models

When building sophisticated models with the Functional API, the following certain practices can save you from headaches later:

- **Use meaningful names**

    Give important layers descriptive names that will help you identify them in summaries, visualizations, and when debugging. So, instead of this:

    `x = keras.layers.Dense(128)(x)`

    You should do this:

    `x = keras.layers.Dense(128, name="feature_embedding")(x)`

- **Group related layers**

    Use functions to encapsulate logical groups of layers, making your architecture easier to understand and modify.

- **Create intermediate models**

    For very complex architectures, consider creating separate models for major components then combining them, as shown in Listing 8.6.

    ```
    def build_vision_encoder(input_shape):
        inputs = keras.Input(shape=input_shape)
    ```

8 Exploring the Keras Functional API

```
    # Vision layers here...
    return keras.Model(inputs, x, name="vision_encoder")

def build_text_encoder(input_shape):
    inputs = keras.Input(shape=input_shape)
    # Text processing layers here...
    return keras.Model(inputs, x, name="text_encoder")

# Create component models
vision_model = build_vision_encoder((224, 224, 3))
text_model = build_text_encoder((100,))

# Use them in a larger model
image_input = keras.Input(shape=(224, 224, 3))
text_input = keras.Input(shape=(100,))

image_features = vision_model(image_input)
text_features = text_model(text_input)
```
Listing 8.6 Composing Through Intermediate Models

- **Visualize before training**
  Use `keras.utils.plot_model()` to verify your architecture before investing time in training. These practices transform the Functional API from a mere coding approach into a powerful design philosophy that keeps your models maintainable even as they grow in complexity.

8.2.4 Handling Multiple Inputs and Outputs

Now that we've covered the basics, let's take a look at one of the Functional API's most powerful features: the ability to handle multiple inputs and outputs. This capability opens doors to solving problems that would be impossible with simpler architectures. Let's look at an example shown in Listing 8.7.

```
# Define inputs
image_input = keras.Input(shape=(224, 224, 3), name="image")
metadata_input = keras.Input(shape=(10,), name="metadata")

# Process image branch
x1 = keras.layers.Conv2D(32, 3, activation="relu")(image_input)
x1 = keras.layers.MaxPooling2D()(x1)
x1 = keras.layers.Flatten()(x1)
x1 = keras.layers.Dense(128, activation="relu")(x1)
```

```
# Process metadata branch
x2 = keras.layers.Dense(32, activation="relu")(metadata_input)

# Combine branches
combined = keras.layers.concatenate([x1, x2])

# Output layer
output = keras.layers.Dense(1, activation="sigmoid")(combined)

# Create model with multiple inputs
model = keras.Model(inputs=[image_input, metadata_input],
                    outputs=output)
```
Listing 8.7 Creating a Model with Multiple Inputs

This model takes two completely different types of data—an image and a vector of metadata—and processes each through its own specialized pathway before combining them for the final prediction. It's like having two expert analysts working on different aspects of a problem, and then bringing their insights together for a more informed decision.

The same flexibility applies to outputs. We can create models, such as those shown in Listing 8.8, that simultaneously perform multiple tasks.

```
# Shared layers
shared = keras.layers.Dense(128, activation="relu")(combined)

# Output branches
category_output = keras.layers.Dense(10, activation="softmax",
                                     name="category")(shared)
price_output = keras.layers.Dense(1, name="price")(shared)
availability_output = keras.layers.Dense(1,
            activation="sigmoid", name="availability")(shared)

# Multi-output model
multi_output_model = keras.Model(
    inputs=[image_input, metadata_input],
    outputs=[category_output, price_output, availability_output]
)
```
Listing 8.8 Creating a Model with Multiple Outputs

When training such models, Keras allows us to specify different loss functions and weights for each output. When you define the loss function in `model.compile`, you can specify which output layer should be using which specific loss. See the definition of losses in Listing 8.9. Compare the loss keys in the compile function with the layer names given in output layers shown in Listing 8.8. We can also specify weights assigned to each loss. This approach is particularly valuable when some tasks are more important than others or when different outputs require different evaluation metrics.

```
multi_output_model.compile(
    optimizer="adam",
    loss={
        "category": "categorical_crossentropy",
        "price": "mean_squared_error",
        "availability": "binary_crossentropy"
    },
    loss_weights={
        "category": 1.0,
        "price": 0.5,
        "availability": 0.8
    }
)
```

Listing 8.9 Compiling a Multi-Output Model with Different Losses

The ability to process multiple inputs makes the Functional API perfect for *multimodal learning*, which refers to combining different types of data (modalities) to make more robust predictions. Think about how humans process the world. We don't rely solely on vision or sound; we integrate multiple senses to understand our environment. Multimodal models aim to do the same.

Common multimodal applications include the following:

- **Text + image**
  Combining visual features with textual descriptions for more accurate classification or *generation* tasks.

- **Audio + text**
  Processing both speech recordings and their transcriptions to improve *speech recognition* or *sentiment analysis*.

- **Time series + tabular data**
  Merging sensor readings with static features for better *anomaly detection* or predictive maintenance.

- **Video + audio**
  Analyzing both visual and audible components of video for content understanding or emotion recognition.

The key advantage of multimodal learning is that different data types often contain *complementary information*. When one input type is ambiguous, the other might provide the context needed for accurate prediction.

8.2.5 Example: Building an Image Captioning Model

Listing 8.10 brings all of these concepts together by demonstrating how to build a simplified image captioning model—a perfect example of multimodal learning that combines computer vision and natural language processing.

```python
# Image encoder
image_input = keras.Input(shape=(224, 224, 3), name="image")
cnn = keras.applications.ResNet50(include_top=False,
                                  weights="imagenet")(image_input)
image_features = keras.layers.GlobalAveragePooling2D()(cnn)
image_features = keras.layers.Dense(256, activation="relu")
                                    (image_features)

# Sequence encoder for captions
sequence_input = keras.Input(shape=(100,), name="caption")
embedding = keras.layers.Embedding(input_dim=10000,
                                   output_dim=256)(sequence_input)
sequence_features = keras.layers.LSTM(256)(embedding)

# Combine both features
combined = keras.layers.concatenate([image_features,
                                     sequence_features])
decoder = keras.layers.Dense(256, activation="relu")(combined)
output = keras.layers.Dense(10000, activation="softmax")(decoder)

# Create model
caption_model = keras.Model(
    inputs=[image_input, sequence_input],
    outputs=output
)
```

Listing 8.10 Image Captioning Model with Dual Inputs

This model performs the following task: given an image and a partial caption, it predicts the next word in the caption. It essentially learns to describe images in natural language, bridging the gap between visual and textual understanding. The architecture has two main branches:

- **Image encoder**
  Extracts visual features using a pretrained ResNet-50. Think of this as a built-in function for now. We'll return to a detailed discussion of this during our case studies later in this chapter.
- **Sequence encoder**
  That processes the partial caption using an embedding layer.

By combining these features before the final prediction layer, the model learns to associate visual patterns with linguistic expressions—a truly multimodal capability. In practice, this model would be used recursively to generate complete captions. We'd start with just an image and a "start of sequence" token, predict the first word, then feed that prediction back alongside the image to predict the second word, and so on, until we reach an "end of sequence" token or a maximum length.

We'll return to a more thorough discussion of this later on in the book when we discuss transformers. This example demonstrates the power of the Functional API to create architectures that seamlessly integrate different types of data processing. Whether you're working with images, text, time series, or any combination thereof, the Functional API gives you the flexibility to design models that match the true complexity of your problem domain.

As we move forward with Keras, you'll find that mastering the Functional API opens up a world of possibilities beyond what simpler interfaces can offer. It's the difference between being limited to predesigned templates and having the freedom to architect patterns that perfectly fit your unique challenges. Most of these patterns, when constructed once, will come in handy in a wide range of other, more complex models. One such pattern is that of residual connections that we'll discuss next.

8.2.6 Residual Connections

Imagine you're climbing a steep mountain. The traditional path requires you to scale each section one after another—exhausting work that gets increasingly difficult as you ascend. Now, picture if there were occasional bridges that let you skip particularly challenging sections, allowing you to preserve your energy for the journey ahead. This is essentially what *residual connections* do for neural networks. As we build deeper networks with dozens or even hundreds of layers, we encounter a seemingly paradoxical problem: adding more layers sometimes makes our models perform worse, not better. You might think that a deeper network should always have at least the same capabilities as a shallower one—after all, couldn't it just learn to use the extra layers effectively or ignore them if needed? In theory, yes. In practice, something else happens that makes training these very deep networks extraordinarily difficult. This is our old enemy, the vanishing gradient problem. In the next few sections, we'll see how we can finally handle it using the powerful method of residual connections.

The Vanishing Gradient Problem

To understand why residual connections are so revolutionary, let's first revisit the challenge they solve: the infamous vanishing gradient problem. It always keeps coming back but it leaves us with a more elegant solution each time.

Let's recap this problem briefly. When we train neural networks, information flows in two directions. During the *forward pass*, data travels from input to output, generating predictions. During the *backward pass*, error signals travel in reverse, updating the weights to improve future predictions. This backward signal—the gradient—is like a message telling each neuron how it should adjust to reduce errors.

The trouble arises because this gradient must pass through every layer of the network, and each passage slightly weakens the signal. It's as if the layers near the input are too far from the action to understand their role in the overall performance. They receive such faint, distorted feedback that they can't effectively learn. This problem intensifies as networks grow deeper, creating a frustrating ceiling on network depth—precisely when we want to build more sophisticated models!

How Residual Connections Solve the Problem

This is where residual connections work their magic. Instead of forcing information to flow only through a sequence of transformations, residual connections create shortcuts—alternate pathways that allow signals to bypass one or more layers. The key insight is beautifully simple. In a residual block, we take the input to a sequence of layers and add it directly to their output. Refer to Figure 8.1 for a visual representation of this concept. We'll explain this figure in detail in conjunction with the code. For now, notice the skip connection going from the input to the output directly bypassing the intermediate layers on the main path. Mathematically, if we denote the original mapping as $F(x)$, a residual block computes $F(x) + x$ instead of just $F(x)$. This means that even if $F(x)$ is close to 0 (as might happen during early training), the output still retains the original input x.

This simple change to the formulation fundamentally changes what our layers need to learn. Instead of learning the complete transformation from input to desired output, they only need to learn the *residual*—the difference between input and output. Think of it as asking the network, "What do I need to add to the input to get the desired output?" rather than "How do I transform the input into the desired output?"

This distinction might seem subtle, but it has profound implications. Consider what happens when the optimal transformation is close to an identity function (where output ≈ input). In a traditional network, the layers must carefully tune weights to approximate this identity, which is surprisingly difficult. With residual connections, they can simply push their weights toward zero, effectively saying, "Don't change anything." This makes learning much easier when small modifications to the input are all that's needed.

Furthermore, this approach provides a direct path for information to flow backward during training. When calculating gradients, the addition operation simply distributes the gradient to both its inputs—meaning both the transformation branch $F(x)$ and the identity shortcut receive clear learning signals. The shortcut ensures gradient information reaches earlier layers with minimal *dilution*, solving the vanishing gradient problem in one elegant stroke. This seemingly minor modification has profound implications. Now, during backpropagation, gradients can flow through these shortcut connections without being diminished by passing through multiple transformation layers. It creates a highway for gradient flow that bypasses potential bottlenecks. The early layers receive clearer signals about how they should adjust, enabling much deeper networks to train effectively.

> **[!] Be Careful When Using Residual Connections**
> Residual connections can be a double-edged sword in deep learning architectures. While they excel at helping gradients flow backward through layers, too many can create what engineers call "highway networks" where information skips the learning process entirely. If every road offers a bypass, traffic will always take the path of least resistance and never explore the city itself.

Implementing Residual Connections with the Functional API

Here's where the Functional API truly shines. Creating these skip connections is remarkably straightforward, as shown in Listing 8.11.

```
def residual_block(x, filters, kernel_size=3):
    # Save the input value for the skip connection
    residual = x

    # Apply convolutional layers
    y = layers.Conv2D(filters, kernel_size, padding='same')(x)
    y = layers.Activation('relu')(y)
    y = layers.Conv2D(filters, kernel_size, padding='same')(y)

    # Add the skip connection
    output = layers.add([y, residual])

    # Carry on with the rest of the model
    output = layers.Activation('relu')(output)

    return output
```

Listing 8.11 Creating Residual Connections Through the Functional API

This simplified code snippet shows the essence of a residual block. We process the input through a series of layers, but crucially, we add the original input back to the result before the final activation. The `keras.layers.add()` function performs element-wise addition, combining the transformed representation with the *identity shortcut*.

In more complex scenarios where the input and output dimensions don't match (e.g., when changing the number of filters or *downsampling*), we need to transform the shortcut connection as well. This is called *projection* and is the little block in the residual path shown earlier in Figure 8.1. Code-wise, projection is just another layer performing the required reshaping, as shown in Listing 8.12.

```
def residual_block_with_projection(x, filters,
                                   kernel_size=3, strides=2):

    # Transform shortcut for dimension matching
    shortcut = keras.layers.Conv2D(filters, 1,
                        strides=strides, padding='same')(x)

    # Main path
    y = keras.layers.Conv2D(filters, kernel_size,
                        strides=strides, padding='same')(x)
    # ... rest of the layers of any type



    # Add the transformed shortcut
    output = keras.layers.add([y, shortcut])
    output = keras.layers.Activation('relu')(output)

    return output
```

Listing 8.12 Projections During Residual Connections

We'll see complete implementations of these blocks in action later in this chapter, but these snippets highlight the elegance of creating residual connections using the Functional API. The ability to easily reference any layer's output and combine it with others is precisely what makes these architectures possible.

Impact on Model Convergence and Performance

The introduction of residual connections has been nothing short of transformative for deep learning. Models that would have been practically impossible to train just a few years ago are now commonplace, with networks routinely reaching depths of 50, 100, or even 1,000+ layers. This depth enables several critical advantages:

- **Improved representation power**
  Deeper networks can learn more complex *hierarchical features*, capturing subtler patterns in the data.
- **Faster convergence**
  Despite having more parameters, residual networks often train more quickly than their *nonresidual* counterparts. It's giving the optimization algorithm a clearer map to follow—it can take more direct routes toward good solutions.

- **Better final performance**
  The combination of improved gradient flow and increased depth typically results in lower *error rates* on challenging tasks.

- **Enhanced generalization**
  Residual networks often show better performance on new, unseen data compared to *shallow networks*.

The impact is particularly visible in computer vision tasks. The *Residual Network* (ResNet) family of models, which pioneered these connections, shattered previous benchmarks on image classification and became the backbone for countless applications. A ResNet-50 (with 50 layers) typically outperforms older nonresidual architectures with far fewer layers despite theoretically having less representational capacity. We'll use this architecture in Section 8.3.1 for one of our case studies. For now, let's turn our attention to another useful pattern of complex networks enabled by the Functional API.

8.2.7 Branching Architectures

So far, we've primarily focused on networks where information flows along a single path or simple shortcuts. But the real power of the Functional API emerges when we break free from this linear thinking. Branching architectures allow our models to simultaneously process information in different ways, extracting various perspectives from the same input before merging these insights for the final prediction. In this section, we'll take a look at how we can branch the flow of information in our model to enable this parallel processing.

Creating Parallel Processing Paths

Creating branches with the Functional API is refreshingly intuitive. We simply use the same input tensor as the starting point for multiple processing streams. Take a look at Listing 8.13 for an example of this.

```
# Start with a common input
inputs = keras.Input(shape=(224, 224, 3))

# Branch 1: Focus on local details with smaller filters
branch1 = keras.layers.Conv2D(64, 3, activation="relu",
                              padding="same")(inputs)
branch1 = keras.layers.Conv2D(64, 3, activation="relu",
                              padding="same")(branch1)

# Branch 2: Capture broader patterns with larger filters
branch2 = keras.layers.Conv2D(64, 5, activation="relu",
                              padding="same")(inputs)
```

```
branch2 = keras.layers.Conv2D(64, 5, activation="relu",
                              padding="same")(branch2)

# Branch 3: Extract context with pooling operations
branch3 = keras.layers.MaxPooling2D(3, strides=1,
                                    padding="same")(inputs)
branch3 = keras.layers.Conv2D(64, 1, activation="relu",
                              padding="same")(branch3)
```

Listing 8.13 Branching to Different Paths

Each branch can be tailored to extract specific types of information from the input. Think of it like having multiple specialists examining the same evidence but focusing on different aspects—one might analyze fine details, another might consider the broader context, and a third might look for specific patterns. Their combined insights provide a more comprehensive understanding than any single perspective could offer.

This approach is particularly powerful for complex data such as images, where information exists at multiple scales and patterns. A branch with small filters might excel at capturing texture and fine details, while another with larger filters might better recognize broader shapes and structures. Working in parallel, they create a rich, multifaceted representation of the input.

Merging Branches: Bringing Perspectives Together

Once our branches have extracted their unique insights, we need to bring them back together. Keras provides several ways to merge branches, each with its own characteristics. Several types of merging are shown in Listing 8.14.

```
# Concatenation: Preserve all information by stacking features
# along a dimension
merged_concat = keras.layers.concatenate([branch1, branch2,
                                          branch3], axis=-1)

# Addition: Combine through element-wise addition
# (requires same shape)
merged_add = keras.layers.add([branch1, branch2, branch3])

# Average: Take the mean of corresponding elements
merged_avg = keras.layers.average([branch1, branch2, branch3])

# Maximum: Element-wise maximum
merged_max = keras.layers.maximum([branch1, branch2, branch3])
```

Listing 8.14 Merging Branches

Concatenation is like binding multiple reports together—all information is preserved, but the resulting tensor grows larger in one dimension. *Addition, averaging*, and *maximum* operations maintain the original tensor shape but combine information in ways that may emphasize or diminish certain aspects. The choice of merging strategy depends on your specific needs:

- **Concatenation**
  This preserves all information but increases the feature dimension, potentially leading to higher computational costs in subsequent layers.

- **Addition**
  This works well when branches process information in compatible ways, allowing features to reinforce each other. However, it has the disadvantage that you have to ensure equality of shapes in the inputs.

- **Averaging**
  This provides a smoother combination that can help reduce noise or variability. It also shares the same limitation with addition and maximum operations.

- **Maximum**
  This selection can help emphasize the strongest detected features across branches.

After merging, the combined representation continues through the network, benefiting from the diverse perspectives captured by each branch.

When to Use Branching in Your Architectures

Branching serves specific purposes that can dramatically improve model performance in certain scenarios. Consider incorporating branching when you're faced with the following:

- **Processing multi-scale features**
  Different-sized filters can capture patterns at various scales simultaneously rather than relying on sequential downsampling to handle multiple scales.

- **Working with heterogeneous data**
  When your input contains fundamentally different types of information (e.g., images and text), dedicated branches can process each modality appropriately before merging.

- **Balancing competing objectives**
  Some tasks require both fine-grained detail and global context. Branches allow you to optimize for both without compromise.

- **Implementing ensemble-like behavior**
  Multiple branches can implement different modeling strategies within a single network, similar to how *ensemble methods* combine multiple models.

- **Managing computational resources**
  Branching can place more computationally intensive operations on a separate path, allowing you to control how much processing power is dedicated to different aspects of feature extraction.

However, branching isn't always beneficial. It adds complexity to your model, can increase the parameter count, and might make training dynamics more challenging. As with many architectural choices, the key is to experiment and let empirical results guide your decisions.

Common Patterns in Branching Models

Several powerful branching patterns have emerged in the deep learning community. Understanding these common motifs can help you apply branching effectively in your own models:

- **Inception modules**
  Popularized by Google's *InceptionNet* (also known as *GoogLeNet*), these modules run multiple convolutional operations with different filter sizes in parallel, along with a pooling path. They're designed to efficiently capture patterns at multiple scales simultaneously. Think of an Inception module as having several specialists working in parallel—one looking at tiny details with 1×1 convolutions, others examining medium features with 3×3 filters, and still others capturing broader patterns with 5×5 filters. A simplified version might look something like Listing 8.15.

```
def inception_module(x, filters_1x1, filters_3x3,
                     filters_5x5, filters_pool):
    # 1x1 convolution branch
    branch1 = keras.layers.Conv2D(filters_1x1, 1,
                     activation="relu", padding="same")(x)

    # 3x3 convolution branch
    branch2 = keras.layers.Conv2D(filters_3x3, 3,
                     activation="relu", padding="same")(x)

    # 5x5 convolution branch
    branch3 = keras.layers.Conv2D(filters_5x5, 5,
                     activation="relu", padding="same")(x)

    # Pooling branch
    branch4 = keras.layers.MaxPooling2D(3, strides=1,
                                    padding="same")(x)
    branch4 = keras.layers.Conv2D(filters_pool, 1,
                activation="relu", padding="same")(branch4)
```

```
    # Concatenate all branches
    return keras.layers.concatenate([branch1, branch2,
                                    branch3, branch4], axis=-1)
```

Listing 8.15 Simplified Inception Block Architecture

- **Multi-input processing**
  This pattern involves separate branches for different input types, each with architecture tailored to that data modality. We'll see a complete example of this pattern later with our image captioning model.

- **Feature Pyramid Network (FPN)**
  This pattern creates a hierarchy of feature maps at different resolutions, with lateral connections between levels. It's particularly effective for tasks such as object detection, where targets appear at multiple scales.

- **DenseNet connections**
  While technically not branching in the traditional sense, DenseNet creates dense connectivity patterns where each layer connects to every other layer in a sequential, feed-forward manner, creating multiple paths through the network.

The beauty of the Functional API is that it makes implementing these complex patterns surprisingly straightforward. Instead of wrestling with complicated code structures, you can express these sophisticated architectures in a way that visually resembles the network diagram itself. This clarity helps you focus on the architecture's design rather than the implementation details.

In the next section, we'll see some of these patterns applied in sophisticated ways, enabling networks that can tackle complex, multifaceted problems with elegance and efficiency. You'll also develop an intuition for when residual connections and branching might help solve your specific challenges, adding a powerful technique to your deep learning toolkit.

8.3 Use Cases and Examples

Now that we've explored the foundations of the Functional API let's see how it enables some of the most powerful neural network architectures in modern deep learning. In this section, we'll examine four key architectures that would be challenging or impossible to implement with the Sequential API alone: ResNet's revolutionary residual connections for image classification, Inception's parallel convolutional pathways for feature extraction, Siamese networks for learning similarity between inputs, and U-Net's encoder-decoder structure with skip connections for image segmentation. Each architecture showcases different aspects of the Functional API's flexibility in routing information through nonsequential paths.

8.3.1 Image Classification with ResNet

Let's put the Functional API to work by implementing the powerful image classification model that revolutionized the deep learning world: ResNet. We've already explored how residual connections help solve the vanishing gradient problem, but seeing them in action will really cement your understanding. ResNet transformed how we build deep neural networks. Before ResNet came along, adding more layers to a network was like trying to stack blocks too high—eventually, the tower would become unstable, and performance would collapse. ResNet solved this fundamental problem by creating highways for information to flow through the network, bypassing potential bottlenecks.

Think of traditional neural networks as forcing information through a series of crowded checkpoints, where important details might get lost in the crowd. ResNet instead creates express lanes that allow the original information to skip some of these checkpoints entirely. It's like having both local streets and an expressway system in a city—some traffic needs to wind through every intersection, but other traffic can take the highway and arrive at the destination with the original information intact.

> **Getting Complete Codes**
>
> Because we're now working with larger models that involve more code and repeated patterns for dataset loading and metrics reporting, we'll focus on the most interesting parts of the implementation. For complete code examples, you can refer to the supporting page where all code listings are available in their entirety (*https://recluze.net/kerasbook*). The complete notebook for this ResNet case study is available under the **08-01-residual-connections** link there.

We'll start off by loading the CIFAR-10 dataset and getting its prebuilt training and testing splits using the following:

```
(x_train, y_train), (x_test, y_test) = cifar10.load_data()
```

Then, we'll create the part of our model that forms the residual block shown in Listing 8.16. The main road (our shortcut) goes straight through, but there's also a scenic route (our convolutional layers) that processes the information more deeply. What makes this special is that, at the end, we combine what we learned on both paths.

```
# Create a function to build a residual block
def residual_block(x, filters, kernel_size=3, strides=1,
                   use_projection=False, block_id=None):
    prefix = f'R{block_id}_' if block_id is not None else ''

    # Store the input for the residual connection
    shortcut = x
```

```
# First convolution layer in the block
x = layers.Conv2D(filters, kernel_size, strides=strides, \
                padding='same', name=f'{prefix}conv1')(x)
x = layers.Activation('relu')(x)

# Second convolution layer in the block
x = layers.Conv2D(filters, kernel_size, padding='same', \
                name=f'{prefix}conv2')(x)

# If dimensions change, we need to use projection
if use_projection:
    shortcut = layers.Conv2D(filters, 1, \
                strides=strides, padding='same',
                name=f'{prefix}proj')(shortcut)

# Add the residual connection
x = layers.add([x, shortcut], name=f'{prefix}add')
x = layers.Activation('relu')(x)

return x
```

Listing 8.16 Residual Block Creation for Repeated Use

The function starts by saving our input tensor as shortcut—this is like taking a snapshot of our data before we do any processing. We'll need this snapshot later for our skip connection. Next, we take our data through two convolutional layers. The first one applies filters with a specific kernel size and stride, then activates the results with the Rectified Linear Unit (ReLU). This is extracting basic features from our input. The second convolutional layer further refines these features, but interestingly, we don't activate them yet. If our dimensions are changing (either because we're using a different number of filters or a stride greater than 1), we need to transform our shortcut to match. That's what the use_projection parameter controls. When it's True, we apply a 1×1 convolution to the shortcut path to make its dimensions match our processed data. It's ensuring both roads end up at the same destination (shape in this case) despite taking different routes.

Finally, we reach the crux of the residual block: we add our processed data back to the shortcut. This addition is what creates the residual connection, and it's the key innovation that helps gradients flow during backpropagation. After adding, we apply ReLU activation to the combined result.

What's remarkable about this approach is that if the network decides the convolutional layers aren't helpful, it can essentially "turn them off" by setting their weights close to zero. This would make the residual block act almost like an *identity function*, simply passing the input forward with minimal changes. This gives the network tremendous flexibility—it can choose to use the deeper layers when they're helpful but bypass them when they're not.

The optional `block_id` parameter is a neat touch that helps with model interpretability. It assigns unique names to each layer, making it easier to identify specific components when debugging or visualizing the network. In the grand scheme of a ResNet model, we'll stack many of these residual blocks together. Each one preserves information through its shortcut connection, allowing the network to grow much deeper without suffering from vanishing gradients. It's like having checkpoints throughout a long hike—even if you get a bit lost along the way, you always have a direct path back to where you started.

With the residual block created, we'll also create a plain block that doesn't have residual connections so that we can see the effect that the residual connection has on our learning. This is only for learning purposes and won't be part of a production model. Listing 8.17 shows the plain block code.

```
# Build a plain convolutional block (for comparison)
def plain_block(x, filters, kernel_size=3, strides=1):
    # No residual connection here
    x = layers.Conv2D(filters, kernel_size, strides=strides,\
                    padding='same')(x)
    x = layers.Activation('relu')(x)

    x = layers.Conv2D(filters, kernel_size, padding='same')(x)
    x = layers.Activation('relu')(x)

    return x
```

Listing 8.17 Plain Block for Comparison

With the residual and plain blocks in place, let's unwrap how our ResNet model comes together. The `build_resnet_model` function, shown in Listing 8.18, crafts a complete neural network that uses the power of residual connections. Think of it as an architect's blueprint for a building that won't collapse no matter how tall it grows. Note that this isn't the exact same model as was used in the original paper. We've made some modifications for ease of understanding.

```
# Create a deep model with residual connections
def build_resnet_model():
    inputs = Input(shape=(32, 32, 3))

    # Initial convolution
    x = layers.Conv2D(64, 3, padding='same')(inputs)
    x = layers.Activation('relu')(x)
```

```
# First stack of residual blocks
x = residual_block(x, 64, use_projection=True, block_id=1)

# Second stack with downsampling
x = residual_block(x, 128, strides=2, use_projection=True,
                    block_id=2)
x = residual_block(x, 128, block_id=3)

# Global average pooling and final dense layer
x = layers.GlobalAveragePooling2D()(x)
outputs = layers.Dense(10, activation='softmax')(x)

model = Model(inputs=inputs, outputs=outputs)
return model
```
Listing 8.18 Building the ResNet Model

We start with our foundation—the input layer expecting 32×32-pixel color images with three channels (RGB). This is where our data enters the network. The process begins with a single convolutional layer that extracts 64 basic features from the image, followed by a ReLU activation that introduces nonlinearity. This will teach the network to recognize elementary patterns such as edges and textures—the building blocks of visual recognition. Next comes our first residual block. We're using use_projection=True because we're changing the number of filters (and therefore dimensions) from our input to this block. The second stack introduces downsampling with strides=2, essentially reducing the spatial dimensions while doubling the feature channels to 128. This is like zooming out from the image to see more abstract patterns. Because we're changing dimensions again, we need projection. The subsequent residual block maintains these dimensions, further refining the features without changing the spatial resolution.

After processing the image through these feature extraction layers, we use global average pooling to summarize each feature map into a single value. Finally, we connect to a dense layer with 10 outputs and a softmax activation, perfect for classifying our images into 10 different categories. The softmax ensures our predictions are properly normalized as probabilities that sum to 1.

Along similar lines, we'll create a build_plain_model function that is very similar to this function except it uses the plain_block instead of the residual_block. As a reminder, you can always look at the code file in the repository for complete working code. Now, we can go ahead and take a look at our model summaries using the code listing given in Listing 8.19.

```
# Build models
resnet_model = build_resnet_model()
resnet_model.summary()
```

```
plain_model = build_plain_model()
plain_model.summary()
```

Listing 8.19 Building the Models and Viewing Summaries

A portion of the complete model is shown in Figure 8.2. The complete network can be drawn using the `plot_model` function of Keras. Notice the residual connection on the right, which includes the projection. In addition, note that the prefixes we passed to our residual block (R1, R2, etc.) make it easier to identify which layers are where in the overall network architecture. You can also see the need for projection in the residual connection. The input shape is 32x32x64, and the output shape is 16x16x64. So, if we didn't have a projection, the 32x32 and 16x16 outputs couldn't have been added together causing a *shape mismatch* error.

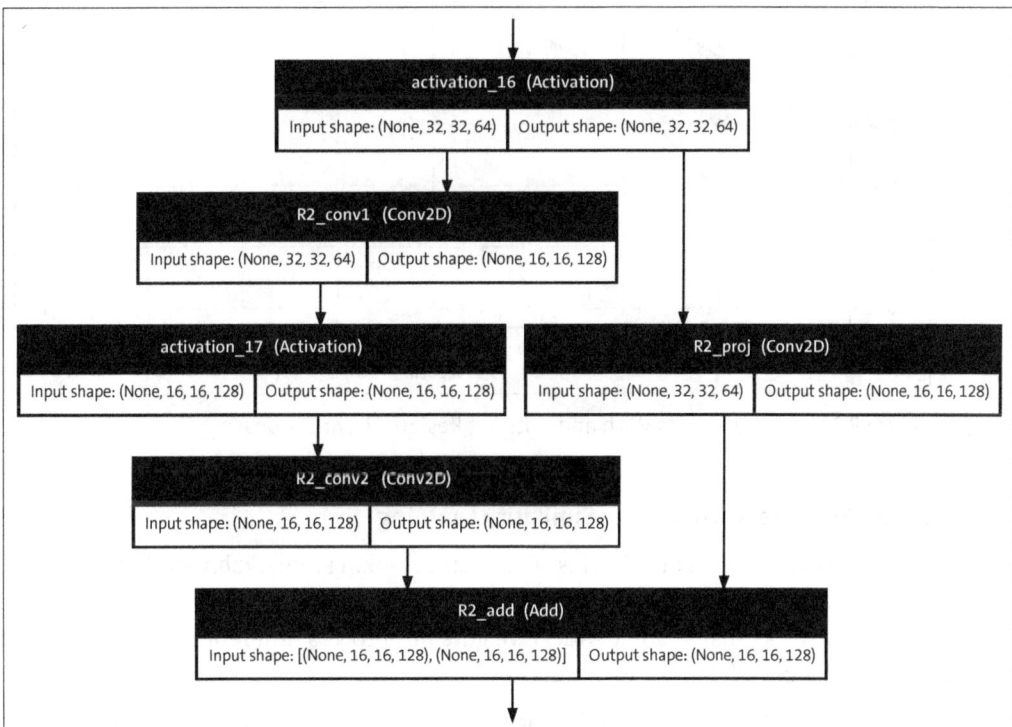

Figure 8.2 Snippet from the ResNet Model Showing the Skip Connection and Projection

With the models built, we'll compile and train both the residual model and the plain model for comparison. The results of the training are shown in Figure 8.3. Notice that while the end results are somewhat similar for both models, the residual model gets off to a very good start. This is because the derivatives flowing through the residual connections allow the model to start improving immediately, whereas the plain model needs more training time to achieve similar results. On much larger models, this difference becomes significantly more pronounced, as models without residual connections can take orders of magnitude more time to train than those with several residual connections. What's more, as we create more sophisticated models, we'll use residual connections as only part of the much larger strategy. Combining all of these little bits of optimization will ensure our large models are able to converge. In the next section, we'll take a look at another pattern that is very useful in modern, large-scale networks.

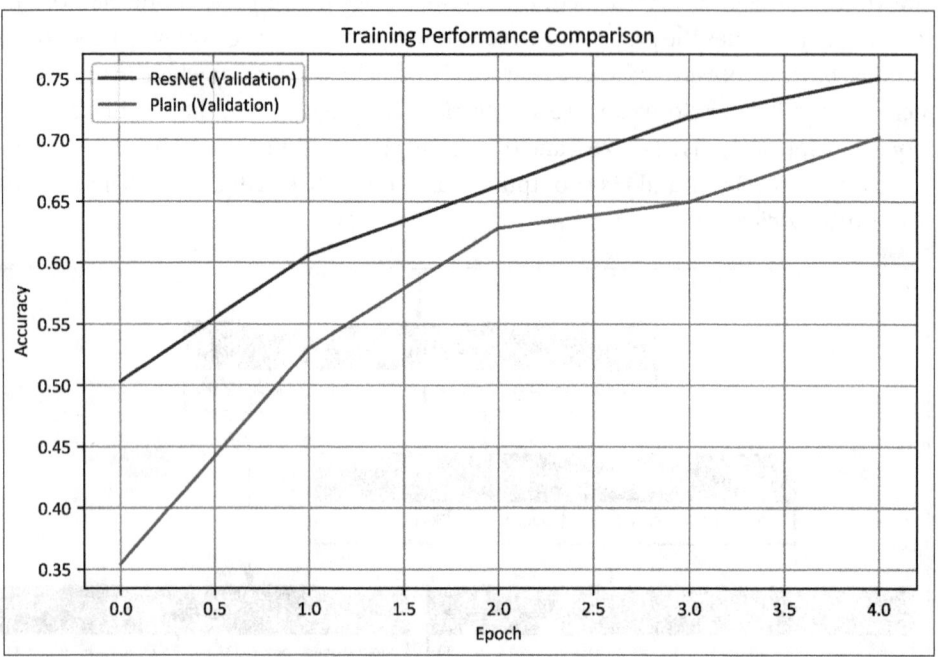

Figure 8.3 Learning Speeds With and Without Residual Connections

8.3.2 Siamese Networks for Similarity Learning

Have you ever seen identical twins who, despite looking remarkably similar, have their own unique personalities? Siamese networks operate on a similar principle—they're neural networks that share identical weights but process different inputs to measure how similar or different those inputs are.

Consider two photographers with exactly the same equipment, training, and technique sent to capture different scenes. When they return, the similarities and differences in their photos would reveal something meaningful about the scenes themselves, not about the photographers' styles or equipment. This is the essence of a Siamese network—by keeping everything else constant, we can truly measure the inherent similarities between inputs.

Siamese networks are built around the concept of shared weights. Unlike traditional neural networks that learn to classify inputs into categories, Siamese networks learn to measure the distance or similarity between pairs of inputs. They're like a sophisticated comparison engine that can tell you how alike two things are, even if they've never seen those exact things before.

At their core, Siamese networks consist of two or more identical subnetworks—exact copies with the same architecture and, critically, the same weights. These twins process different inputs simultaneously, transforming them into what we call "embeddings"— compact, information-rich representations that capture the essence of the inputs. Think of these embeddings as distilling an image, sound, or text into its fundamental characteristics, such as extracting the DNA of the data.

Once we have these embeddings, we can measure how close they are to each other in this new representation space. Images of similar digits, for instance, should have embeddings that are near each other, while different digits should be positioned farther apart. This distance gives us a direct measure of similarity. What makes Siamese networks particularly powerful is their ability to generalize. After training on enough examples, they develop an understanding of similarity that extends beyond the specific items they've seen. A Siamese network trained on handwritten digits doesn't really memorize specific pairs of digits as similar or different. Rather, it learns what makes digits similar in general, allowing it to compare new handwriting samples it has never encountered before.

This property makes them ideal for tasks such as *facial recognition*, *signature verification*, and, as we'll explore later, creating meaningful representations of images. For instance, using MNIST, by embedding digits in a space where similar digits cluster together, we can build systems that recognize handwriting patterns regardless of small variations in how people write the same number.

> **The Problem of Similarity**
>
> The beauty of Siamese networks lies in their elegant approach to the challenging problem of similarity. Rather than trying to directly calculate similarity based on raw pixels—which would be highly sensitive to small changes in position, rotation, or style—they learn to transform inputs into a new space where meaningful similarities become apparent and superficial differences fade away.

Now that we understand the concept of Siamese networks, let's see how we prepare data to train them. The secret lies in creating the right pairs of examples that will teach our network to recognize similarity. Take a look at the make_pairs function in Listing 8.20 for this. (As before, complete code for this network, training, and visualization can be seen on the *https://recluze.net/keras-book* support page in the **08-02-siamese-network** notebook.)

```
def make_pairs(x, y):
    num_classes = max(y) + 1
```

8 Exploring the Keras Functional API

```
        digit_indices = [np.where(y == i)[0] \
                        for i in range(num_classes)]

        pairs = []
        labels = []

        for idx1 in range(len(x)):
            # add a matching example
            x1 = x[idx1]
            label1 = y[idx1]
            idx2 = random.choice(digit_indices[label1])
            x2 = x[idx2]

            pairs += [[x1, x2]]
            labels += [0]

            # add a nonmatching example
            label2 = random.randint(0, num_classes - 1)
            while label2 == label1:
                label2 = random.randint(0, num_classes - 1)

            idx2 = random.choice(digit_indices[label2])
            x2 = x[idx2]

            pairs += [[x1, x2]]
            labels += [1]

        return np.array(pairs), np.array(labels).astype("float32")

# make train pairs
pairs_train, labels_train = make_pairs(x_train, y_train)
```

Listing 8.20 Creating Pairs for Measuring Similarity

Think of this function as a matchmaker for our MNIST digits. Its job is to create meaningful relationships between images that will help our Siamese network learn when two digits are similar or different. The function starts by organizing our dataset by class. The line `digit_indices = [np.where(y == i)[0] for i in range(num_classes)]` creates a directory of sorts, listing all the images that belong to each digit class. Then comes the heart of the pairing process. For each image in our dataset, we create two specific pairs. First, we create a matching pair—two images of the same digit. We take our current image and find another random image of the same digit class. This pair gets a label of 0, which in our setup means "these are similar." Think of this label as measuring the distance between two images—0 means the distance is small, that is, they are very similar; and 1 means they are far apart, that is, not similar at all.

Next, we create a nonmatching pair—our current image paired with a random image from a different digit class. This gets a label of 1, meaning "these are different." An example of our function creating these pairs is shown in Figure 8.4.

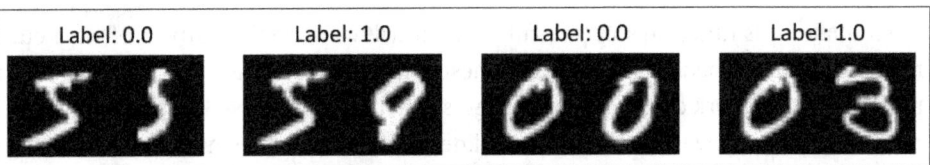

Figure 8.4 Output of make_pairs Function Showing Distance as a Label

Why do we need both types of pairs? This is crucial for training the two identical branches of our Siamese network. It's teaching parts of the network to spot differences by showing them both identical twins and completely unrelated people. By seeing both matching and nonmatching examples, our network learns what makes digits similar or different.

The beauty of this approach is balance. For every image in our dataset, we create one matching pair and one nonmatching pair. This ensures our network doesn't develop *biases* toward similarity or difference. Notice how we're turning a classification problem (identifying digits 0–9) into a *similarity detection* problem. Instead of asking, "What digit is this?" we're asking, "Are these two digits the same?" This transformation is what makes Siamese networks so powerful—they learn to map inputs into a space where similarity becomes a measurable distance.

After creating all our pairs, we apply this function to our training, validation, and test sets. This prepares the perfect training environment for our Siamese network to learn meaningful embeddings that capture the essence of each digit. First, we need a way to measure how similar or different they are. This is where the concept of distance comes into play—not physical distance, but distance in our feature space.

Take a look at Listing 8.21 for how to calculate this distance between two inputs. We've split our pairs into two separate arrays. Think of x_train_1 as one half of our pairs and x_train_2 as the other half. Each image in x_train_1 corresponds to its partner at the same index in x_train_2. The function eculidean_distance calculates the distance between two image embeddings in high-dimensional space. It implements the Euclidean distance formula—essentially, the *Pythagorean theorem* extended to multiple dimensions. The calculation follows a straightforward process: subtract the vectors, square each element of the difference, sum these squared differences, and take the square root. The result quantifies how similar or different two images are according to our network's learned representation.

> **[!] Avoiding Mathematical Errors**
>
> The small detail of adding `keras.backend.epsilon()` is just a safety measure to avoid mathematical problems with very small numbers—such as making sure we don't accidentally divide by zero.

What makes this function so powerful is that it will work on the outputs of our neural networks—the embeddings we create. These embeddings will encode all the complex patterns our network discovers about digit similarity, compressed into a *compact representation*. When we calculate the Euclidean distance between these embeddings, we're not comparing pixel values; we're comparing the essential "digit-ness" that our network has learned to recognize.

The smaller this distance, the more similar our network believes the two digits are. A distance close to zero suggests the network sees them as nearly identical, while larger distances indicate greater differences. This elegant approach turns similarity—a concept that's usually subjective and hard to quantify—into a precise mathematical measurement.

```
x_train_1 = pairs_train[:, 0]   # x_train_1.shape is (60000, 28, 28)
x_train_2 = pairs_train[:, 1]

def euclidean_distance(vects):
    """Find the Euclidean distance between two vectors.
    Arguments:
        vects: List containing two tensors of same length.
    Returns:
        Tensor containing euclidean distance
        (as floating point value) between vectors.
    """
    x, y = vects
    sum_square = ops.sum(ops.square(x - y), axis=1, keepdims=True)
    return ops.sqrt(ops.maximum(sum_square, keras.backend.epsilon()))
```

Listing 8.21 Euclidean Distance Between Two Vectors Representing Digits

Now we arrive at the most interesting part of our case study—constructing the actual Siamese network architecture shown in Listing 8.22. Think of what we're about to build as identical twins who share not just appearance but thought processes—when one learns something, the other instantly knows it too. First, we create the *embedding network*—the shared brain that both sides of our Siamese network will use. We're designing a specialized lens through which our network will examine every image. This network takes our 28×28 MNIST digit images and transforms them into a compact 10-dimensional (10D) representation that captures the essence of each digit.

The architecture is fairly straightforward—we normalize the input, apply two convolutional layers (each followed by pooling to reduce dimensions), flatten the resulting feature maps, and finally transform them into our 10D embedding space. This relatively simple structure is powerful enough to learn meaningful representations of our handwritten digits.

```
input = keras.layers.Input((28, 28, 1))
x = keras.layers.BatchNormalization()(input)
x = keras.layers.Conv2D(4, (5, 5), activation="tanh")(x)
x = keras.layers.AveragePooling2D(pool_size=(2, 2))(x)
x = keras.layers.Conv2D(16, (5, 5), activation="tanh")(x)
x = keras.layers.AveragePooling2D(pool_size=(2, 2))(x)
x = keras.layers.Flatten()(x)

x = keras.layers.BatchNormalization()(x)
x = keras.layers.Dense(10, activation="tanh")(x)
embedding_network = keras.Model(input, x)


input_1 = keras.layers.Input((28, 28, 1))
input_2 = keras.layers.Input((28, 28, 1))

tower_1 = embedding_network(input_1)
tower_2 = embedding_network(input_2)

merge_layer = keras.layers.Lambda(euclidean_distance, \
    output_shape=(1,))(
        [tower_1, tower_2]
)
normal_layer = keras.layers.BatchNormalization()(merge_layer)
output_layer = keras.layers.Dense(1, \
                         activation="sigmoid")(normal_layer)
siamese = keras.Model(inputs=[input_1, input_2],
                     outputs=output_layer)
```

Listing 8.22 Siamese Network Architecture

Next comes the magic that makes this a true Siamese network: *weight sharing*. Notice how we define two separate inputs (input_1 and input_2) but process both using the exact same embedding network. This is the fundamental concept that gives Siamese networks their power.

When we write

```
tower_1 = embedding_network(input_1)
tower_2 = embedding_network(input_2)
```

we're telling Keras to use identical copies of our embedding network—with exactly the same weights—to process both inputs. This forced sharing ensures that both images are processed through exactly the same lens, making their resulting embeddings directly comparable. After we generate these twin embeddings, we measure the Euclidean distance between them using our previously defined function. This distance represents how different the two input images are in our learned feature space. A small distance suggests similar digits, while a large distance indicates different ones.

Finally, we *normalize* this distance and pass it through a single neuron with a sigmoid activation function. This transforms our distance measure into a probability-like value between 0 and 1, where values closer to 0 indicate similar pairs and values closer to 1 indicate different pairs.

The strength of this architecture lies in its elegant symmetry. The shared weights ensure that our network learns consistent representations regardless of which input receives which image. If we were to swap the two inputs, the distance calculation would remain exactly the same—a property known as *symmetry* that's essential for a true similarity measure.

For this network, we need a specialized loss function—*contrastive loss*—as shown in Listing 8.23. Think of contrastive loss as a teacher with two different teaching strategies. For similar pairs, it says "get closer together," and for dissimilar pairs, it says "maintain at least a certain distance apart." This dual approach is what gives Siamese networks their power to learn meaningful embeddings.

Contrastive loss elegantly handles both similar and dissimilar pairs through a single equation:

$$L(y, d) = (1 - y) \cdot d^2 + y \cdot \max(margin - d, 0)^2$$

Here, y is the true label (0 for similar, 1 for dissimilar), and d is the predicted distance between pairs. For similar pairs ($y = 0$), the loss simplifies to d^2—the squared distance between embeddings. This penalizes the network when similar items are placed far apart.

For dissimilar pairs ($y = 1$), the loss becomes $max(margin - d, 0)^2$, which only penalizes the network when different items are closer than the margin. Once they're sufficiently separated, no further penalty is applied. The margin parameter (default *1*) creates a boundary that prevents the network from pushing dissimilar items infinitely apart, which would destabilize training. This balanced approach crafts an embedding space where similar digits naturally cluster together while different digits maintain appropriate separation. Through thousands of iterations, the network discovers which features truly matter for determining digit similarity without explicit rules.

```
def loss(margin=1):
    def contrastive_loss(y_true, y_pred):
        square_pred = ops.square(y_pred)
        margin_square = ops.square(
```

```
                ops.maximum(margin - (y_pred), 0)
            )
        return ops.mean((1 - y_true) * square_pred + \
                        (y_true) * margin_square)

    return contrastive_loss
```

Listing 8.23 Custom Loss Function for Siamese Networks

With our architecture and loss function in place, we can now train our Siamese network to recognize digit similarities, as shown in Listing 8.24. We compile our model with our custom contrastive loss function, using RMSprop as the optimizer to adaptively adjust learning rates as training progresses. Notice how we feed the training data—not as single images, but as paired inputs [x_train_1, x_train_2]. It's teaching the model to spot differences by always showing two objects side by side, rather than one at a time.

During each epoch, the network examines thousands of digit pairs, gradually refining its understanding of what makes digits similar or different. With each batch, it adjusts the weights in our embedding network, sculpting a feature space where the distance between points becomes meaningful.

Notice that we're not directly teaching the network what makes a 7 a 7—instead, we're letting it discover the essential characteristics by learning what makes two 7s similar and what distinguishes a 7 from a 4. As training progresses, the network transforms from a novice with no concept of digit similarity into an expert that can place any handwritten digit into a well-organized *embedding space* where similar digits naturally cluster together.

```
siamese.compile(loss=loss(margin=margin),
                optimizer="RMSprop",
                metrics=["accuracy"])

history = siamese.fit(
    [x_train_1, x_train_2],
    labels_train,
    validation_data=([x_val_1, x_val_2], labels_val),
    batch_size=batch_size,
    epochs=epochs,
)
```

Listing 8.24 Compile Model and Fit

The visualizations in Figure 8.5 demonstrate our Siamese network in action. The network correctly identifies when digits belong to the same class (label 0.0) versus different classes (label 1.0). The pairs with similar digits (the two 7s and the two 2s) are labeled as similar, while the pair with different digits (the 7 and 3) is labeled as dissimilar. This visual confirmation shows that our network has successfully learned to distinguish between similar and different handwritten digits, despite variations in individual writing styles using the concept of shared weights. With these shared weights, Siamese networks excel at learning similarity relationships between pairs of inputs.

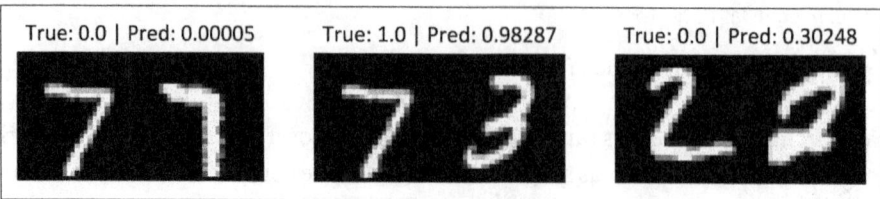

Figure 8.5 Predictions of the Trained Siamese Network Model

Our next architecture takes a different approach to visual understanding. Let's shift our focus from comparing images to analyzing them at the pixel level, as we explore U-Net—a powerful architecture designed for precise image segmentation tasks.

8.3.3 U-Net for Image Segmentation

Image segmentation represents one of computer vision's most challenging tasks—assigning each pixel in an image to a specific class or category. Unlike classification, which provides a single label for an entire image, segmentation creates detailed *masks* that outline exactly where objects appear. This pixel-level precision makes segmentation crucial for applications requiring fine-grained understanding of images, from autonomous vehicles detecting road boundaries to medical diagnostics identifying precise tumor locations. U-Net stands as one of the most influential architectures developed specifically for this segmentation challenge. Originally designed for *biomedical image segmentation*, this unique network has found its way into numerous applications where pixel-precise image masks are needed. While U-Net has revolutionized medical imaging tasks such as tumor detection and *organ boundary identification*, we'll apply these same powerful techniques to a more accessible dataset—the *Oxford-IIIT Pet* dataset with its collection of cat and dog images.

What makes U-Net special is its distinctive U-shaped architecture. The network first follows a contracting path that captures context by reducing spatial dimensions while increasing feature depth, and then expands through an *upsampling* path that enables precise localization. One of U-Net's most interesting characteristics is how it uses skip connections—direct pathways that carry information from the contracting path to the expanding path. These connections preserve fine-grained details that would otherwise be lost during downsampling.

In our implementation, some layers will produce multiple outputs—one that continues the main network flow and another that feeds into these skip connections. This dual-output approach helps the network maintain both *global context* and *local details*, a critical balance for successful image segmentation tasks. The principles we'll explore here translate directly to medical applications, where the stakes of accurate segmentation can literally be lifesaving.

We'll start our task by loading the Oxford-IIIT Pet Dataset using *TensorFlow Datasets*, which makes acquiring and preparing this data remarkably straightforward:

```
dataset, info = tfds.load('oxford_iiit_pet', with_info=True)
```

> **Access the Complete Code**
> The complete code for this case study can be seen on the repository page at *https://recluze.net/keras-book* as the notebook **08-03-unet-segmentation**.

The `with_info=True` parameter is particularly useful here—it returns not just the dataset itself but also a wealth of metadata about it. This info object contains details about the dataset's structure, the number of examples, feature descriptions, and even version information. Having this metadata at our fingertips will prove invaluable as we preprocess our images and design our model architecture to match the specific characteristics of this dataset. But before we can train our U-Net model, we need to prepare our data—transforming raw images and masks into the perfect training materials. Let's walk through the code shown in Listing 8.25 that handles this crucial preprocessing step.

```python
def normalize_image(input_image, input_mask):
    # Normalize image to [0, 1] range
    input_image = tf.cast(input_image, tf.float32) / 255.0

    # For our purposes, we'll convert this to binary segmentation
    input_mask = tf.cast(input_mask, tf.float32)
    input_mask = tf.where(input_mask == 1, 1.0, 0.0)

    return input_image, input_mask

def resize_images(input_image, input_mask, img_size=128):
    # Resize both image and mask
    input_image = tf.image.resize(input_image,
                                  (img_size, img_size))
    input_mask = tf.image.resize(input_mask,
                 (img_size, img_size),
                   method=tf.image.ResizeMethod.NEAREST_NEIGHBOR)

    return input_image, input_mask

def prepare_dataset(dataset, batch_size=32, img_size=128, \
                    buffer_size=1000):
```

8   Exploring the Keras Functional API

```
    # Extract image and mask
    dataset = dataset.map(lambda x:
                    (x['image'], x['segmentation_mask']))

    # Apply normalization and resizing
    dataset = dataset.map(normalize_image)
    dataset = dataset.map(lambda x, y:
                    resize_images(x, y, img_size))

    # Shuffle, batch, and prefetch
    dataset = dataset.shuffle(buffer_size)
    dataset = dataset.batch(batch_size)
    dataset = dataset.prefetch(tf.data.AUTOTUNE)

    return dataset

# Prepare training and validation datasets
train_dataset = prepare_dataset(dataset['train'])
val_dataset = prepare_dataset(dataset['test'])
```

Listing 8.25 Preparing Data for U-Net Segmentation

The first function we encounter, normalize_image, serves as our data's first stop in the transformation pipeline. In this function, we're doing two key things. First, we're scaling our pixel values from the original range of 0-255 down to a much more manageable 0-1 range. Neural networks typically train better with these normalized values—it's like giving all your data the same uniform so they're on equal footing. Second, for the masks, we're simplifying the problem. Originally, the dataset marks pixels as either pet (1), background (2), or pet outline (3). We're converting this to a binary problem—either pet (1.0) or not-pet (0.0). This simplification focuses our task on the core challenge of identifying where the animal is in the image.

Next, we have our resize_images function. Here, we're ensuring all of our images and masks have consistent dimensions. This is critical because neural networks expect inputs of fixed size. We're resizing everything to 128×128 pixels—small enough to process efficiently but large enough to retain important details. Notice how we use different resizing methods for images versus masks. For masks, we use *nearest neighbor interpolation* to preserve the exact label values without creating in-between shades that would make no semantic sense.

Finally, the prepare_dataset function brings everything together. This function creates an assembly line for our data processing. First, it extracts just what we need from the dataset—the images and their corresponding masks. Then, it applies our normalization and resizing functions we defined earlier. The last three lines handle the mechanics of efficient training. *Shuffling* randomizes the order of examples to prevent the model from learning misleading patterns based on the sequence of images. Batching groups our examples together for parallel processing. Finally, *prefetching* acts like a just-in-time delivery system, preparing the next batch of data while the current one is being processed, eliminating wasteful waiting time.

With these preprocessing steps complete, our data is ready for the U-Net model. We've transformed raw, unprocessed images into a streamlined format that will help our network learn efficiently and effectively. Look at Figure 8.6 for a couple of images from the dataset along with their masks to fully understand the dataset.

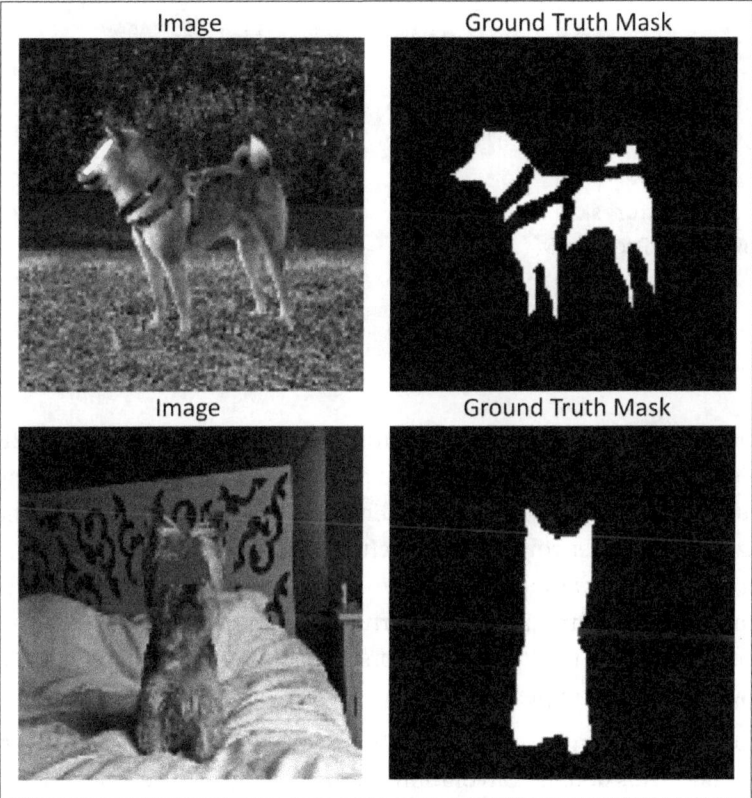

Figure 8.6 Examples of Images and Masks from the Dataset

Now that our data is prepared, let's examine the code that builds our U-Net architecture. We'll break down each function to understand how this powerful segmentation model comes together. The `downsample_block` function shown in Listing 8.26 creates the contracting path of our U-Net. Each block performs feature extraction while progressively reducing the spatial dimensions of our input.

```
def downsample_block(x, filters, kernel_size=(3, 3),
            padding="same", kernel_initializer="he_normal"):
    # First convolution layer
```

```
        c = keras.layers.Conv2D(filters, kernel_size,
                                padding=padding,
                                kernel_initializer=kernel_initializer
                                )(x)
        c = keras.layers.BatchNormalization()(c)
        c = keras.layers.Activation("relu")(c)

        # Second convolution layer
        c = keras.layers.Conv2D(filters, kernel_size,
                                padding=padding,
                                kernel_initializer=kernel_initializer
                                )(c)
        c = keras.layers.BatchNormalization()(c)
        c = keras.layers.Activation("relu")(c)

        # Store the outputs for skip connections
        p = keras.layers.MaxPooling2D((2, 2))(c)

        return c, p
```

Listing 8.26 Downsampling Block in U-Net

Let's trace what happens to our data as it flows through this block. First, we apply a 2D convolution that extracts features using a specified number of filters. The `same` padding ensures our feature maps maintain their spatial dimensions during convolution. The `he_normal` initializer helps our network train more effectively by setting appropriate initial weights. After convolution, we apply *batch normalization* to stabilize and accelerate training by normalizing the outputs. (We'll return to a detailed explanation of this important concept later.) Then, the ReLU activation function introduces nonlinearity, allowing our network to learn complex patterns.

We repeat this convolution-normalization-activation sequence a second time to further refine our features. This double convolution pattern is a signature element of the U-Net architecture. Finally, we create two outputs: the convolution result `c` (later serves as a skip connection) and a downsampled version `p` created by max pooling, which reduces the spatial dimensions by half while preserving the most important features.

The `upsample_block` shown in Listing 8.27 creates the expansive path of our U-Net, where we gradually restore spatial dimensions while refining segmentation details. First, we use a transposed convolution (sometimes called *deconvolution*) to double the spatial dimensions. This operation is essentially the reverse of a normal convolution, expanding rather than contracting the feature maps. Second, the crucial skip connection mechanism is used. We concatenate our upsampled features with the corresponding features from the downsampling path (`skip_features`). This gives our network access to both fine-grained spatial details from earlier layers and high-level semantic understanding from deeper layers.

```
def upsample_block(x, skip_features, filters,
                   kernel_size=(3, 3), padding="same",
                   kernel_initializer="he_normal"):
    # Upsampling
    u = keras.layers.Conv2DTranspose(filters, (2, 2),
                    strides=(2, 2), padding=padding)(x)

    # Concatenate with skip connection
    u = keras.layers.Concatenate()([u, skip_features])

    # Convolution layers
    u = keras.layers.Conv2D(filters, kernel_size,
                    padding=padding
                    kernel_initializer=kernel_initializer
                    )(u)
    u = keras.layers.BatchNormalization()(u)
    u = keras.layers.Activation("relu")(u)

    u = keras.layers.Conv2D(filters, kernel_size,
                    padding=padding,
                    kernel_initializer=kernel_initializer
                    )(u)
    u = keras.layers.BatchNormalization()(u)
    u = keras.layers.Activation("relu")(u)

    return u
```

Listing 8.27 Upsampling Block in U-Net

After concatenation, we apply two consecutive *convolution-normalization-activation* sequences, similar to the downsampling block. These operations integrate the information from the skip connection with our upsampled features, producing more refined segmentation maps.

The `build_unet` function assembles our complete architecture, as shown in Listing 8.28. It expects input images of shape (128, 128, 3)—height, width, and RGB channels—and produces binary segmentation masks.

We begin with the downsampling path, applying four consecutive downsample blocks with progressively increasing numbers of filters (64, 128, 256, 512). Each block halves the spatial dimensions while doubling the feature depth, allowing the network to detect increasingly complex features.

```python
def build_unet(input_shape=(128, 128, 3), num_classes=1):
    # Input layer
    inputs = keras.layers.Input(shape=input_shape)

    # Downsampling path
    s1, p1 = downsample_block(inputs, 64)
    s2, p2 = downsample_block(p1, 128)
    s3, p3 = downsample_block(p2, 256)
    s4, p4 = downsample_block(p3, 512)

    # Bridge
    b = keras.layers.Conv2D(1024, (3, 3), padding="same",
                            kernel_initializer="he_normal")(p4)
    b = keras.layers.BatchNormalization()(b)
    b = keras.layers.Activation("relu")(b)
    b = keras.layers.Conv2D(1024, (3, 3), padding="same",
                            kernel_initializer="he_normal")(b)
    b = keras.layers.BatchNormalization()(b)
    b = keras.layers.Activation("relu")(b)

    # Upsampling path
    u1 = upsample_block(b, s4, 512)
    u2 = upsample_block(u1, s3, 256)
    u3 = upsample_block(u2, s2, 128)
    u4 = upsample_block(u3, s1, 64)

    # Output layer
    outputs = keras.layers.Conv2D(num_classes, (1, 1),
                                  padding="same",
                                  activation="sigmoid")(u4)

    # Create model
    model = keras.Model(inputs, outputs)

    return model

# Build the UNET model
model = build_unet()
model.summary()
```

Listing 8.28 The Complete U-Net Model

At the bottom of our "U" sits the bridge—two convolution layers with 1,024 filters that process the most abstract features. This represents the *bottleneck* of information in our network.

Then begins the upsampling path, which mirrors the downsampling path in reverse. We apply four upsampling blocks, each connecting with its corresponding skip connection from the downsampling path (s4, s3, s2, s1). Notice how the number of filters decreases as we move up (512, 256, 128, 64), gradually shifting from abstract features back to spatial detail.

Finally, a 1×1 convolution with sigmoid activation produces our output—a *probability map* where each pixel value represents the likelihood of that pixel belonging to our foreground class (the pet). For *multi-class segmentation*, we would change `num_classes` and use softmax activation instead.

The beauty of the U-Net lies in this symmetric architecture with skip connections. The downsampling path captures semantic context (what's in the image), while the upsampling path with skip connections recovers spatial precision (where it's located). With our model architecture in place, we need to prepare it for training. This involves setting up how our model will learn and defining the guardrails that will guide its learning journey. The compilation step, shown in Listing 8.29, configures the learning process for our U-Net. We've chosen the Adaptive Moment Estimation (Adam) optimizer with a carefully selected learning rate of 0.0001—small enough to navigate the complex loss landscape of image segmentation without overshooting, yet large enough to make meaningful progress during training.

For our loss function, binary cross-entropy serves as the perfect mathematical compass for binary segmentation tasks. It measures how well our model distinguishes between pet pixels and non-pet pixels, penalizing incorrect predictions more severely as they diverge from the true values.

We're tracking two metrics during training: plain accuracy (the percentage of correctly classified pixels) and the *Binary Intersection over Union (Binary IoU)*. The IoU metric is particularly important for segmentation tasks because it measures the overlap between our predicted segmentation mask and the ground truth. A perfect IoU score of 1.0 would mean our prediction exactly matches the true mask.

```
# Compile the model
model.compile(
    optimizer=keras.optimizers.Adam(learning_rate=1e-4),
    loss='binary_crossentropy',
    metrics=['accuracy',
            keras.metrics.BinaryIoU(target_class_ids=[1],
                                    threshold=0.5)]
)
```

```
# Define callbacks
callbacks = [
    keras.callbacks.ModelCheckpoint("unet_oxford_pets.keras",
                                    save_best_only=True),
    keras.callbacks.ReduceLROnPlateau(patience=3, factor=0.5),
    keras.callbacks.EarlyStopping(patience=10,
                                  restore_best_weights=True)
]
```

Listing 8.29 Compilation and Callbacks for U-Net

The callbacks we've defined act as smart training assistants that monitor progress and make adjustments:

1. The `ModelCheckpoint` callback saves our model whenever it improves, ensuring we don't lose our best version if performance starts to decline later in training.
2. The `ReduceLROnPlateau` callback is particularly of interest here. It watches for plateaus in performance where the model stops improving. After three epochs of stagnation (that's our `patience` parameter), it reduces the learning rate by half (our `factor` parameter). This is like switching from large steps to smaller, more careful steps when you're getting close to your destination, allowing the model to fine-tune its parameters with greater precision.
3. Finally, the `EarlyStopping` callback prevents wasted computation by halting training if the model shows no improvement for 10 consecutive epochs. The `restore_best_weights` parameter ensures we revert to the best version of the model rather than keeping the final one, which might not be the best one.

Together, these training settings create a robust learning environment for our U-Net, balancing efficient progress with careful optimization to produce the most accurate segmentation results possible.

Let's take a moment to examine what our U-Net model has actually learned. The images shown in Figure 8.7 show how well our neural network has grasped the concept of pet versus not-pet in these photographs.

Looking at the first example—a small black and white dog with a red food bowl—we can see how our model performed. The true mask (middle image) shows a crisp, binary representation where white pixels mark the dog and black pixels represent everything else. Our model's prediction (right image) captures the general shape of the dog, but notice how it's not a perfect binary mask—instead, we see varying shades of gray. These grayscale values represent the model's confidence: brighter areas indicate higher confidence that those pixels belong to the pet class. The model has successfully identified the core body of the dog, with the highest confidence in the central regions. However, the boundaries are less defined than in the ground truth mask. This is a common pattern in image segmentation—boundaries are often the most challenging areas for models to classify with high confidence, as these transitional regions contain mixed visual information.

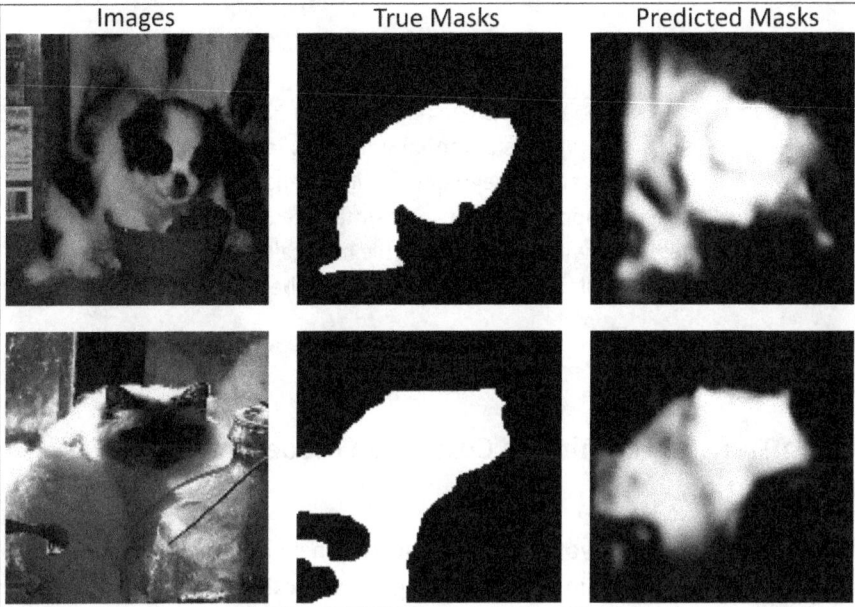

Figure 8.7 Results of the U-Net Model

In the second example, which shows a white cat, we see similar behavior. The model correctly identifies the main body of the cat but is less certain about the precise boundaries. There's also some interesting diffusion of the prediction—the segmentation mask has a slightly blurred quality compared to the sharp edges of the ground truth.

These results reveal both the strengths and limitations of our U-Net implementation. The model has clearly learned the fundamental visual features that distinguish pets from backgrounds—fur textures, body shapes, and typical pet postures. However, it hasn't perfectly captured the precise pixel-level boundaries that define where the pet ends and the background begins.

This outcome isn't unexpected for a model trained with relatively limited data and moderate resolution. For many applications, this level of segmentation would be quite sufficient. If we needed more precise boundaries, we might consider the following:

- Training with higher resolution images
- Using more sophisticated loss functions that specifically penalize boundary errors
- Implementing post-processing techniques to sharpen the predicted masks
- Adding more training data with diverse pet boundaries

Another altogether different strategy we might use for improvement is to instead "stand on the shoulders of the giants." We can take a model already trained by some industry giant with a huge number of resources and then somehow modify it to suit our specific needs. In the next section, we'll see how this can be done with very little effort on our part.

> **U-Net for Medical Images**
>
> While our model performs reasonably well on pet images, U-Net was actually designed for medical images and shows its true potential when applied to X-ray datasets. The architecture was specifically created to segment medical scans where precise boundary detection is critical for clinical applications. If you're interested in seeing U-Net perform at its best, try implementing it with publicly available medical imaging datasets, where its ability to preserve fine details through skip connections becomes even more apparent.

8.4 Using Transfer Learning to Customize Models for Your Organization

Consider what happens when you're trying to learn a new language. You could start completely from scratch, learning every word, grammatical rule, and pronunciation pattern through years of dedicated study. Or, you could leverage your existing knowledge of language—perhaps your native tongue or another language you already know—to accelerate the process dramatically if you try to first search for similarities between the language you already know and the one you're trying to learn. This intuitive concept lies at the heart of one of deep learning's most powerful techniques: transfer learning.

Throughout this chapter, we've explored various model architectures and approaches using the Functional API. But among all of these case studies, transfer learning stands apart. *Transfer learning* isn't an architecture or model design pattern. Rather, it's a fundamental paradigm shift in how we approach problem-solving with neural networks. Transfer learning is like hiring an expert consultant who has spent years mastering a related field and then training them to apply that expertise to your specific challenges. Instead of building intelligence from the ground up, we're repurposing existing knowledge. The neural network has already learned to see the world—we're just teaching it to look at our particular corner of it.

What makes this approach so revolutionary is how it democratizes deep learning. Not every organization has the luxury of massive datasets containing millions of examples or the computational resources to train models from scratch for weeks on specialized hardware. Many real-world problems are constrained by limited data, tight budgets, or pressing timelines. Transfer learning elegantly addresses these limitations by allowing us to build on the solid foundations laid by others. In this section, we'll take a look at the rationale behind transfer learning and how to implement it efficiently using Keras.

8.4.1 The Why of Transfer Learning

Think about the models we've discussed earlier—they might contain millions of parameters that have been painstakingly optimized on enormous datasets such as ImageNet, which contains more than 14 million labeled images across thousands of categories. During this extensive training process, these models have learned to recognize edges, textures, shapes, and complex visual patterns that are fundamental to image understanding. This knowledge isn't specific to any particular dataset—it represents a general understanding of the visual world. When we apply transfer learning, we're essentially saying, "You already know how to see. Now learn to recognize these specific things that matter to my business." This approach dramatically reduces the amount of data and computation required to achieve impressive results. A model that might need thousands of examples when trained from scratch might perform admirably with just dozens or hundreds of examples when fine-tuned through transfer learning.

> **Implications of Transfer Learning**
>
> The implications of transfer learning are profound. Small research teams, startups, non-profits, and individual practitioners can now tackle problems that would have been completely out of reach just a few years ago. Domain experts in fields such as medicine, agriculture, or environmental science can adapt powerful models to their specialized needs without becoming deep learning experts themselves.

In the sections that follow, we'll put transfer learning into practice on a satellite imagery classification task. But unlike our previous examples, we won't need to train a sophisticated model from scratch. Instead, we'll leverage a pretrained convolutional neural network that has already mastered the fundamentals of image recognition, adapting it to our specific needs with remarkable efficiency. We'll see how this approach allows us to achieve impressive results with modest computational resources and limited training data—truly standing on the shoulders of giants.

This approach exemplifies what makes modern deep learning so powerful—not just its technical capabilities, but its accessibility and adaptability to real-world constraints. By the end of this section, you'll have added one of the most practical and widely used techniques in all of machine learning to your toolkit.

For our satellite image dataset, we'll use *EuroSAT*—a collection of 27,000 labeled satellite images captured by the Sentinel-2 satellite mission, part of the European Space Agency's Copernicus program. This dataset offers exactly what we need when teaching our models to see the world from above: specialized data that captures the unique perspective of satellite imagery.

These images are 64×64-pixel patches that span 13 spectral bands, including those beyond human vision such as near-infrared, giving our models access to information our eyes could never perceive. As you can see in Figure 8.8, the dataset encompasses 10 different land use and land cover classes that represent the diverse European landscape—from the geometric patterns of industrial and residential areas to the organic textures of forests, pastures, and bodies of water. What makes EuroSAT particularly valuable for our transfer learning journey is its real-world relevance coupled with manageable complexity. With classes that include Highway, River, Sea & Lake, Forest, Residential, and Industrial, it presents challenges similar to what environmental monitoring systems face daily, yet remains accessible enough for our learning purposes. For this classification task, we won't create a model from scratch. Instead, we'll use the pretrained ResNet-50 architecture available directly from Keras, demonstrating how transfer learning allows us to repurpose existing models for new domains.

This dataset bridges the gap between academic exercises and practical applications. The same techniques we'll use to classify these satellite images can be scaled to monitor deforestation, track urban development, assess agricultural health, or map disaster impacts—making this more than just a learning exercise. The complete code for this visualization can be seen on the support page *https://recluze.net/keras-book* in the notebook **08-04-transfer-learning-intro**.

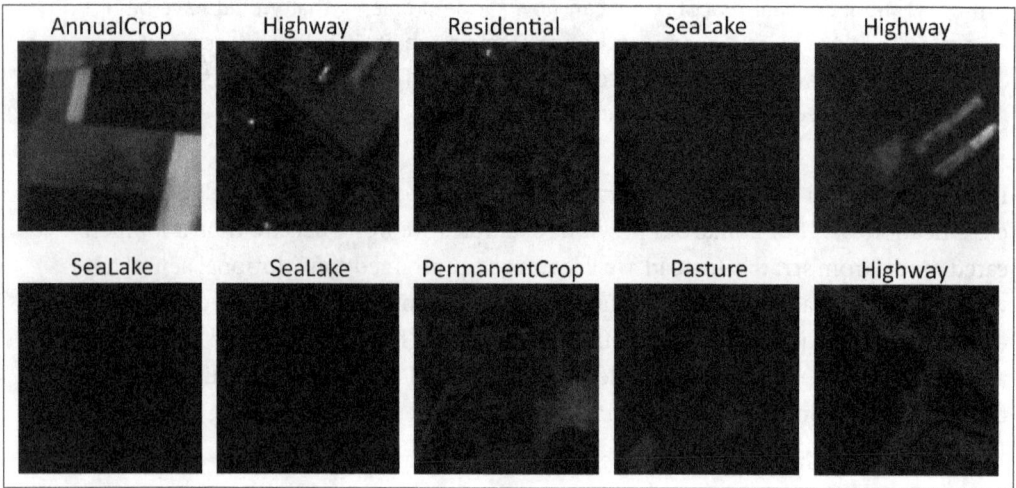

Figure 8.8 Example Images from the EuroSAT Dataset

8.4.2 Leveraging Pretrained Models from Keras

When working with sophisticated deep learning architectures, we often don't want to start from scratch. It's like trying to build a race car when you could simply drive one off the lot and customize it to your needs. Keras makes this process remarkably straightforward, as we can see in Listing 8.30. The code performs a seemingly simple task that conceals an extraordinary feat. With just a few lines, we're accessing both a sophisticated model architecture (ResNet-50) and its carefully trained weights—the culmination of thousands of hours of computation on more than a million images.

8.4 Using Transfer Learning to Customize Models for Your Organization

ResNet-50 is like a talented artist who has spent years learning to see and interpret visual patterns. When we specify `weights='imagenet'`, we're telling Keras to download not only the architecture blueprint but also all the knowledge this model has accumulated during its extensive training on the ImageNet dataset. This is remarkably convenient—imagine having instant access to an expert's lifetime of experience with a single function call.

On the other hand, the parameter `include_top=False` is our way of saying, "I want your visual expertise, but I'll handle the final decisions myself." This removes the classification head—the last few layers specifically trained to recognize the 1,000 ImageNet categories.

With `input_shape=(224, 224, 3)`, we're specifying the dimensions our images will have when fed to the model. The model expects color images (that's the three channels for RGB) with height and width of 224 pixels. This flexibility allows us to adapt the pretrained model to our specific image dimensions. The diagnostic prints that follow give us a quick overview of what we've just loaded—confirming the type of model, counting its layers, and checking whether the model parameters are currently set to be updated during training. The final `summary()` call provides a detailed layer-by-layer breakdown of the model's architecture, showing us the impressive depth and complexity of the network we now have at our disposal.

What's remarkable is that this single `import` statement gives us access to a model that would require enormous computational resources and expertise to train from scratch. By importing and using Keras's pretrained models, we're ready to adapt the model's visual expertise to our specific satellite image classification task.

```python
import keras

# Load pretrained ResNet-50 model
base_model = keras.applications.ResNet50(
    weights='imagenet',          # Pretrained weights
    include_top=False,           # Exclude the classifier
    input_shape=(224, 224, 3)    # Input dimensions
)

# Check the model architecture
print(f"Model type: {type(base_model)}")
print(f"Number of layers: {len(base_model.layers)}")
print(f"Is the model trainable? {base_model.trainable}")

base_model.summary()
```

Listing 8.30 Reusing an Existing Model

8.4.3 The Process of Transfer Learning

Transfer learning is like teaching a seasoned chef to prepare a new cuisine. The chef already knows how to chop, sauté, and understand flavors—we're just redirecting those skills toward different ingredients and dishes. This powerful approach unfolds in four orchestrated steps:

1. **Adopt an experienced mind**
   We acquire a pretrained model complete with both its architecture (the neural pathways) and weights (the accumulated knowledge). This is like recruiting an expert who has spent years developing specialized skills—perhaps they've studied millions of images and learned to recognize thousands of objects. Their neural pathways have already formed, their weights already tuned through extensive experience.

2. **Reshape for new challenges**
   We perform a bit of neural surgery—removing the specialized output layers of the original model and replacing them with our own custom structure. It's similar to taking a portrait photographer and retraining them to capture wildlife. The fundamental skills of composition, lighting, and focus remain valuable, but the specific output needs redirection toward new subjects.

3. **Selective learning with frozen knowledge**
   We next *freeze* the pretrained layers—preserving all that valuable feature extraction knowledge—while only training our newly added layers. This selective training is like telling our expert, "Keep doing what you're already good at, but let's focus on adapting just your final decision-making process." The gradient updates only flow through the new pathways, leaving the foundational knowledge intact. This means the gradient descent step is simply not applied to the model parameters in the frozen layers.

4. **Fine-tuning for specialized excellence**
   We selectively *unfreeze* some of the deeper layers, allowing them to gently adapt to our specific domain. This optional but powerful step is like allowing our expert to slightly modify their fundamental techniques to better serve the new context. It's a delicate balance—we want enough flexibility for specialization without losing the rich foundation of pretrained knowledge.

By following these four steps, we transform general intelligence into specialized expertise, all while avoiding the enormous computational and data requirements of training from scratch. Let's see this in practice next.

8.4.4 Reloading an Existing Model

Let's begin the first step by using the ResNet model we loaded earlier as a demonstration. The complete model building function is shown in Listing 8.31. Our function begins by recruiting a pretrained expert—ResNet-50 with its ImageNet knowledge. (The complete code for this network, training and visualization can be seen on the support page *https://recluze.net/keras-book* in the notebook **08-05-transfer-learning-eurosat**.)

With base_model.trainable = False, we're freezing all the carefully trained weights of ResNet-50, preserving the knowledge while we add our own customizations. It's like putting the existing knowledge into a protective case while we build around it. The next several lines are where the architectural transformation happens. We create a fresh input pathway with keras.Input(shape=(224, 224, 3)), ensuring our model expects color images sized at 224×224 pixels. Then, we connect this input to our frozen ResNet-50 base, with training=False to keep the internal layers (e.g., batch normalization) in inference mode.

> **Difference in Trainable and Training Booleans**
>
> Notice the two settings here. With model.trainable=False, we're freezing the parameter updates, and with training=False, we're telling the model that we're not training and thus it shouldn't apply training-only operations. An example of a training-only operation we've seen before is dropout. So, when we set training=False, no dropout will be applied on these base layers.

Now comes the customization—we're adding our own layers on top of ResNet's feature extraction abilities. The GlobalAveragePooling2D layer condenses the spatial information into a single vector per feature map, such as summarizing detailed observations into key insights. The Dropout(0.3) layer randomly silences 30% of these neurons during training, a technique that prevents the new layers in our model from becoming too dependent on any single feature—similar to how a team becomes more resilient when members can cover for each other.

Our Dense(256, activation='relu') layer is like adding a team of 256 new analysts who combine the extracted features in creative ways, with ReLU activation ensuring they only pass along positive, meaningful contributions. Finally, the output layer with softmax activation gives us a probability distribution across our classes.

With the Functional API, we connect these components into a cohesive workflow, explicitly mapping inputs to outputs. The compilation step sets up our learning process with the Adam optimizer (adaptively tunes the learning rate), categorical cross-entropy loss (ideal for multi-class problems), and accuracy as our progress metric. In just about 25 lines of code, we've transformed a general-purpose image recognition expert into a specialist primed for our specific classification task—preserving its foundational knowledge while redirecting its expertise toward our unique problem domain.

```
def build_transfer_learning_model(num_classes):
    # Load pretrained ResNet-50 model
    base_model = keras.applications.ResNet50(
        weights='imagenet',
```

```
        include_top=False,
        input_shape=(224, 224, 3)
    )

    # Freeze the pretrained weights
    base_model.trainable = False

    # Create the input tensor
    inputs = keras.Input(shape=(224, 224, 3))

    # Apply the pretrained ResNet-50 model
    x = base_model(inputs, training=False)

    # Add custom layers
    x = layers.GlobalAveragePooling2D()(x)
    x = layers.Dropout(0.3)(x)
    x = layers.Dense(256, activation='relu')(x)
    outputs = layers.Dense(num_classes, activation='softmax')(x)

    # Create the model using Functional API
    model = keras.Model(inputs=inputs, outputs=outputs)

    # Compile the model
    model.compile(
        optimizer=keras.optimizers.Adam(learning_rate=0.001),
        loss='categorical_crossentropy',
        metrics=['accuracy']
    )

    return model
```

Listing 8.31 Building the Transfer Learning Model

Once our model is prepared and our data is loaded, we can simply proceed with training using `model.fit()`—just as we would with any other neural network. But the use of transfer learning makes this a little different. During this training phase, only the unfrozen layers—those custom layers we added on top of ResNet—will actually learn and adapt. The pretrained parameters in the rest of the model remain frozen in time, untouched by our training process.

This selective training approach serves two important purposes. First, it protects the valuable feature extraction knowledge that ResNet acquired through its extensive training. We don't want those parameter values to be disrupted by our limited dataset. Second, it dramatically accelerates our training process. If you run `model.summary()`, you'll discover something remarkable—though the model contains around 24 million parameters in total, only about half a million of these are trainable. That's less than 2% of the parameters needing updates! This efficiency means our gradient descent steps run blazingly fast, and the network doesn't suffer as much from the vanishing gradient problem that often plagues deep networks. It's having the benefit of a massive neural network's power while only paying the computational cost of training a much smaller one. This dual advantage—preserving valuable knowledge while drastically reducing computational requirements—represents one of the most powerful benefits of the transfer learning approach.

The last step in our transfer learning process is to perform *fine-tuning*. Up until now, we've been using ResNet-50 as a fixed feature extractor, but what if we could gently nudge some of its deeper knowledge to better align with our satellite imagery? We achieve this through the fine_tune_model function shown in Listing 8.32. We begin by accessing the base ResNet-50 model—it's the second layer in our model architecture (index 1). Then comes a crucial decision: Which layers should we allow to adapt? ResNet-50 is organized in blocks of convolutional layers, each capturing different levels of abstraction. The earlier layers recognize fundamental patterns such as edges and textures, while the later layers identify more complex features.

By targeting layers from 143 onward—the fifth convolutional block—we're making a strategic choice. We're saying, "Keep your basic understanding of visual elements intact, but let's refine your higher-level feature recognition." It's like keeping a photographer's fundamental skills unchanged while adapting their artistic eye to a new subject matter. In practice, you'll learn to come to grips with picking this number as your experience as a machine learning expert increases.

The next part is similarly important—we recompile the model with a much smaller learning rate (1e-5 instead of the original 1e-3). This tiny learning rate is like asking our model to take baby steps rather than giant leaps. We're not looking to radically reshape the pretrained knowledge, just gently nudge it toward our domain. It's the difference between rebuilding a house versus carefully redecorating it. This cautious approach prevents *catastrophic forgetting*, where aggressive updates might erase the valuable knowledge accumulated during pretraining.

```
def fine_tune_model(model, train_ds, val_ds, epochs=10):
    # Unfreeze some layers of the base model
    base_model = model.layers[1]  # ResNet-50 model

    # Unfreeze from the 5th convolutional block
    # The 5th block starts around layer 143
    for layer in base_model.layers[143:]:
        layer.trainable = True

    # Recompile with a lower learning rate
    model.compile(
```

```
        optimizer=keras.optimizers.Adam(learning_rate=1e-5),
        loss='categorical_crossentropy',
        metrics=['accuracy']
    )

    # set callbacks and fit as always
    # ... (code omitted for brevity)

    return history
```

Listing 8.32 Fine-Tuning ResNet-50

The results of our training and fine-tuning are summarized in the plots in Figure 8.9. Looking at the accuracy plot on the left, the first eight epochs are our initial training phase. Notice how the training accuracy line climbs rapidly from around 91% to nearly 98%—a remarkable improvement as our custom layers learn to interpret ResNet-50's features. Meanwhile, the mostly horizontal validation line follows a similar trajectory but with those characteristic zigzags that remind us we're dealing with the messiness of real-world data. What's particularly interesting is how the validation accuracy peaks around epoch 5-6 at about 96%–97% and then begins a subtle downward trend. This is the classic sign that we've reached the limits of what our model can learn without adapting its deeper knowledge.

Then comes the dramatic second act beginning at epoch 10—our fine-tuning phase. The transformation is striking! The fine-tuning training accuracy line shoots upward, quickly surpassing our previous best performance and climbing steadily toward an impressive 99.5%. This is the power of allowing those deeper ResNet-50 layers to adapt to our satellite imagery domain. The fine-tuning validation line remains mostly horizontal and thus shows something equally important. It immediately jumps above our previous validation peak, stabilizing around 97.5%—a significant improvement that confirms our fine-tuning is genuinely enhancing the model's understanding rather than just allowing the model to memorize training examples.

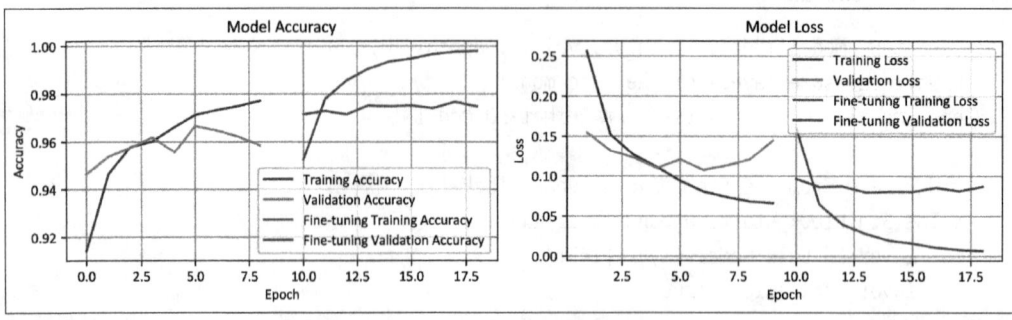

Figure 8.9 Results of Transfer Learning and Fine-Tuning

The loss curves on the right provide additional confirmation. During the initial training, both training and validation loss decrease steadily but begin to diverge—another sign we're approaching the limits of what can be learned with frozen layers. Once fine-tuning begins, we see a sharp drop in training loss (immdiate drop when training begins) that continues throughout the remaining epochs, eventually approaching near 0. The validation loss (mostly horizontal) similarly drops slightly but then stabilizes around 0.08-0.09, maintaining a healthy gap that suggests good generalization without overfitting.

What's most remarkable is the gap between fine-tuning training accuracy and validation accuracy in the later epochs. While training accuracy pushes toward 100%, validation holds steady around 97.5%. This gap is expected and healthy—it shows our model is learning from the training data without becoming overly specialized to it. These plots confirm everything we've discussed about transfer learning's power. By building on ResNet-50's visual expertise and carefully adapting it to our domain, we've created a model that achieves exceptional accuracy on satellite imagery with relatively modest training data and computational resources. The fine-tuning phase clearly provided that extra boost we anticipated, pushing performance beyond what was possible with the frozen base model alone.

This visual evidence makes a compelling case for the transfer learning approach we've explored throughout this chapter, demonstrating how we can leverage existing knowledge to solve new problems more efficiently and effectively than starting from scratch.

8.5 Summary

In this chapter, we explored the Keras Functional API as the solution to sequential models' limitations. We introduced the explicit layer connection syntax that enables complex neural architectures such as residual networks with skip connections and multi-branch models. Through practical case studies, we demonstrated real-world applications, including ResNet for classification, Siamese networks for similarity learning, and U-Net for image segmentation. We concluded the chapter with transfer learning, showing how we could repurpose pretrained models such as ResNet-50 for specialized tasks such as satellite imagery classification with minimal data and computational resources. Throughout this journey, we revealed how the Functional API transforms Keras from a basic sequential tool into a platform capable of implementing sophisticated modern architectures that deliver superior performance across diverse domains. In the next chapter, we'll use the Functional API to create a newer and increasingly popular type of model that's especially good at understanding sequences such as text.

Chapter 9
Understanding Transformers

In this chapter, we'll explore transformers, a revolutionary architecture that has redefined natural language processing (NLP). We'll begin with the theory behind how these models understand relationships between words, regardless of their position in a sequence. We'll learn about key components such as attention mechanisms that make these models so effective. We'll then build a transformer from scratch in Keras before showing you how to leverage pretrained models to create practical applications with minimal code. By the chapter's end, you'll understand transformer fundamentals and how to apply them to real-world problems.

Transformers are a revolutionary concept in the field of AI, forming the foundation of virtually all state-of-the-art language models today. When you interact with ChatGPT, Claude, Gemini, or any other modern AI assistant, you're experiencing the power of transformer architecture. This chapter will take you on a journey to understand this groundbreaking technology that has changed how machines process language, images, and beyond. We'll start by exploring the world of sequence data, which is information that comes in a particular order, such as words in a sentence or frames in a video. You'll learn why traditional approaches struggled with this type of data and how transformers solve these challenges. We'll introduce the concept of attention, which allows models to focus on relevant parts of information, much like how you might pay more attention to key points when reading a complex paragraph.

Next, we'll break down the components that make transformers so powerful. You'll discover how the attention mechanism works, allowing different parts of a sequence to interact with each other directly. We'll explore how transformers process information in parallel rather than sequentially, making them dramatically faster and more efficient than previous approaches. We'll then take a hands-on approach by implementing transformer components in Keras. You'll learn how to build multi-head attention layers and positional encodings and assemble complete encoder and decoder structures. Each code example will translate the theoretical concepts into practical implementation, helping you see how these ideas come to life in actual models.

> **Handling the Complexity of the Models**
>
> The concepts underlying transformers are complex. However, throughout the chapter, we'll use familiar analogies and clear explanations to make these complex concepts accessible. We'll focus on building intuition about how transformers work and why they've become so essential in modern AI systems. Be sure to continue reading the chapter even if some concepts don't make sense. Once you get to the end of the chapter, all of it will come together.

By the end of this chapter, you'll have a solid understanding of transformer architecture and be able to implement your own transformer-based models for various applications. You'll appreciate why these models have become the backbone of modern AI systems and how they've enabled advances that seemed impossible just a few years ago. The knowledge you gain here will serve as a foundation for understanding large language models (LLMs) and other transformer-based systems that are increasingly shaping our digital experiences. As these models continue to evolve and find new applications, your understanding of their core architecture will help you navigate and leverage these powerful tools in your own projects. Let's begin by looking at the foundational theory behind the concept of transformers.

9.1 The Theory Behind Transformers

When we read a sentence, we naturally process it word by word, building meaning as we go. "The cat sat on the mat" makes perfect sense, but "mat the on sat cat the" leaves us confused. This simple observation highlights something profound about language—it's inherently *sequential*, with order carrying as much meaning as the words themselves. Sequence data surrounds us in our daily lives. Text messages, stock market trends, weather patterns all unfold over time, with each element's meaning deeply tied to what came before and what follows. In the world of machine learning, handling this sequential nature has been one of the most interesting challenges.

Think about how you understand a novel. As you read, you don't just process each word in isolation. You build a *mental model* that evolves with each new piece of information. When you encounter the pronoun "she" in chapter 3, your brain effortlessly connects it to the character introduced in chapter 1. You're constantly maintaining *context*—remembering what happened before to make sense of what's happening now. This contextual understanding is precisely what makes sequence data so rich and so challenging for machines to process.

Unlike an image, where all pixels are available simultaneously, sequence data reveals itself one element at a time. The meaning of each element depends on its position and its relationships with other elements in the sequence.

Let's consider another facet of this complexity. The sentence "The bank is by the river" uses "bank" to mean the riverside, while "I need to visit the bank to deposit money" uses the same word to mean a financial institution. Our understanding of "bank" depends entirely on the surrounding words—the sequence context in which it appears. Traditional machine learning approaches struggled with this *sequential dependency*. Early models would treat each word independently, missing the crucial connections between them. It would be like trying to understand a conversation by hearing random words without their proper order, making it nearly impossible to extract meaningful information. Even more sophisticated approaches that attempted to capture some sequence information faced fundamental limitations. They tended to forget early parts of long sequences or blend information in ways that lost important distinctions. It was as if they had a form of digital amnesia, where distant context gradually faded away. To handle this issue, we need to be able to handle sequence data—most commonly, *time series data*. We'll begin by exploring what time series data is and how to handle it efficiently using code.

9.1.1 A Simple Time Series Example

Let's consider an example of how sequence data can be represented and manipulated programmatically. This will give us a concrete foundation for understanding the challenges that transformers were designed to solve. First, look at some code in Listing 9.1 that creates sequential data that mimics many real-world patterns we encounter in language, financial markets, weather patterns, and more. The first function, create_linear_ timeseries(), builds a sequence of values where each point follows a linear trend with some randomness mixed in. Consider tracking the temperature throughout a day—it generally rises from morning to afternoon (that's our linear trend with slope 0.5) but with minute-to-minute fluctuations due to passing clouds or breeze (that's our random noise). What makes this a true sequence is that each value builds upon what came before. The temperature at 2 PM is related to the temperature at 1 PM—it doesn't suddenly jump from freezing to boiling without passing through the values in between. This continuity creates the sequential dependency that's so characteristic of time series data.

The second function, create_dataset(), transforms our continuous sequence into something machine learning models can work with. This is where the true magic happens for *sequence modeling*. (As always, you can access the complete working code for these examples from the supporting page at *https://recluze.net/keras-book* in the notebook **09-01-timeseries-intro**.)

```
# Generate a linear time series with some noise
def create_linear_timeseries(n_samples=1000, slope=0.5, noise_level=1.0):
    x = np.arange(n_samples)
    # Linear trend with added noise
```

9 Understanding Transformers

```
        y = slope * x + np.random.normal(0, noise_level, n_samples)
        return y

# Create dataset with lookback window
def create_dataset(dataset, look_back=10):
    X, y = [], []
    for i in range(len(dataset) - look_back):
        X.append(dataset[i:(i + look_back)])
        y.append(dataset[i + look_back])
    return np.array(X), np.array(y)
```
Listing 9.1 Generating Time Series Data

Think about how you predict the weather. If I asked you, "What will the temperature be tomorrow?" you wouldn't just look at today's temperature—you'd consider the pattern over the past several days. This is exactly what the lookback window does—it gathers context from previous time steps to predict the next value. When the code uses `look_back=10`, it's essentially saying, "Here are the temperatures from the past 10 hours; what will the temperature be in the next hour?" The function slides this 10-hour window through our entire dataset, creating input-output pairs:

- **Input**
  This is 10 consecutive values from the sequence.
- **Output**
  This is the single value that follows them.

We use these functions to transform our data, as shown in Listing 9.2. This transformation turns our 1D time series into a supervised learning problem that our models understand. Looking at the output shapes, we see our training data has dimensions (152, 10, 1), meaning the following:

- There are 152 different examples (*sliding windows*).
- Each window contains 10 *time steps* (our lookback).
- Each time step has one feature (just our temperature value).

```
# Generate our synthetic time series data
np.random.seed(42)   # For reproducibility
timeseries = create_linear_timeseries(n_samples=200)

# Scale the data
scaler = MinMaxScaler(feature_range=(0, 1))
timeseries_scaled = scaler.fit_transform(timeseries.reshape(-1, 1))

# Define lookback window size
look_back = 10
```

```
# Create the dataset with lookback windows
X, y = create_dataset(timeseries_scaled.flatten(), look_back)

# Reshape input to be [samples, time steps, features]
X = X.reshape(X.shape[0], X.shape[1], 1)

# Split into train set and test set (80% train, 20% test)
train_size = int(len(X) * 0.8)
X_train, X_test = X[:train_size], X[train_size:]
y_train, y_test = y[:train_size], y[train_size:]

print(f"Training data shape: {X_train.shape}")
print(f"Testing data shape: {X_test.shape}")

print(X_train[1])
print(y_train[1])
```

Listing 9.2 Modifying Data to Fit Supervised Learning Models

The sample output in Listing 9.3 shows how one window of 10 values is paired with the single value that follows it. This pattern—using past context to predict what comes next—is fundamental to sequence modeling, whether we're working with text, weather data, or stock prices.

```
Training data shape: (152, 10, 1)
Testing data shape: (38, 10, 1)
[[0.        ]
 [0.01302937]
 [0.02696443]
 [0.01422656]
 [0.01929276]
 [0.04273179]
 [0.03957283]
 [0.03210641]
 [0.04742646]
 [0.04229986]]
0.04734246731058163
```

Listing 9.3 Sample from the Data

Now that we've transformed our time series into sequential windows, we need a model that can find patterns across these time steps. For this, we can build on all our knowledge of neural networks we've already established, as shown in Listing 9.4. The code builds a specialized neural network designed to process sequential patterns—much like how our brains process sentences word by word, but with the ability to detect patterns across the sequence. Think of this network as a series of pattern detectors, each layer looking for increasingly sophisticated relationships in our time data.

Let's walk through how this model processes our sequence data. The first layer applies convolutional filters across our time windows. Consider sliding a spotlight across our 10-step sequence, illuminating three values at a time (that's our kernel_size=3). Our input shape (look_back, 1) represents our 10 time steps, each containing a single feature value—like having 10 consecutive temperature readings. For each spotlight position, the model looks for 64 different patterns—perhaps one detector notices rising trends, another catches sudden drops, and others spot oscillation patterns. These detectors not only look at individual values but also understand how values change across multiple time steps. Notice that this is a *1D convolution*. This means that the convolution window only slides from left to right and has no second dimension to slide "down."

The pooling layer condenses our sequence by keeping only the strongest signals. Our second convolutional layer then detects higher-level patterns, looking at how these initial features interact across time. After flattening these patterns, our dense layers combine everything to produce a single prediction for the next value in our sequence. The model uses *mean squared error* to measure prediction accuracy, continuously refining its understanding of the sequence relationships during training.

```
# Build the Keras model (CNN-based)
from keras import layers, models, optimizers

# Define the model - 1D CNN for time series
model = models.Sequential([
    # First convolutional layer
    layers.Conv1D(filters=64, kernel_size=3, activation='relu', input_shape=(look_back, 1)),
    layers.MaxPooling1D(pool_size=2),

    # Second convolutional layer
    layers.Conv1D(filters=32, kernel_size=2, activation='relu'),

    # Flatten layer to connect to dense layer
    layers.Flatten(),

    # Dense hidden layer
    layers.Dense(50, activation='relu'),

    # Output layer
    layers.Dense(1)
])
```

```
# Compile the model
model.compile(optimizer='adam', loss='mse')

# Display model architecture
model.summary()
```

Listing 9.4 Example Model for Time Series Prediction

> **How Is this Used to Model Language?**
>
> When we extend this concept to language, each word becomes a time step, and the lookback window becomes the context we use to understand or predict text. The same principles that help us forecast temperatures can help machines understand and generate human language.

The plot in Figure 9.1 shows the model's prediction capabilities. The light wavy line shows our actual time series data with its upward trend and natural fluctuations. The dark smoother line represents our model's predictions on training data, capturing the general upward pattern while smoothing out the smaller variations. The final portion of the smooth line beyond timestamp 160 shows predictions on unseen test data. Notice how it maintains the trend but gradually diverges from actual values, highlighting how prediction errors compound over time. The model grasps the big picture (the linear trend) but misses some of the finer details and fluctuations.

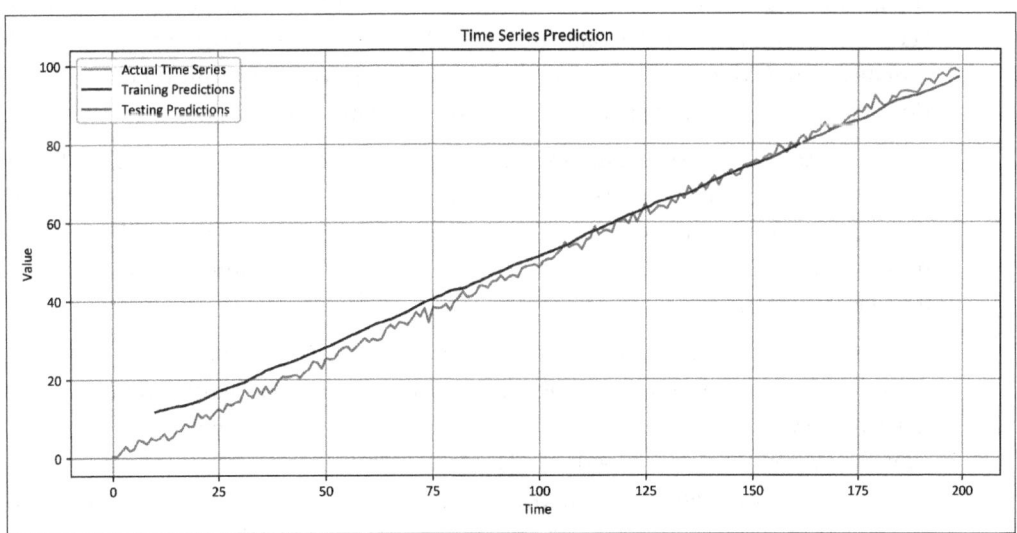

Figure 9.1 Predicted Values Based on the Learned Parameters

This demonstrates a key limitation of our windowed approach: with only 10 previous steps visible at any time, the model can't capture relationships that span longer periods.

This is exactly the problem transformers were designed to solve—their architecture allows them to maintain context across much longer sequences, making connections between distant elements that traditional models miss.

9.1.2 From Numbers to Words: The Challenge of Text Data

So far, we've explored how our models process numeric sequences—temperature readings rising and falling, stock prices climbing and dipping. But what happens when our sequence isn't made of numbers but of words? How do we transform "The cat sat on the mat" into something our neural networks can understand? This is a fundamental challenge. Our models only speak the language of numbers—they can't directly process words, emotions, or concepts. We need to translate the rich tapestry of human language into the mathematical patterns our models can work with.

You might remember from Chapter 4, Section 4.3.2, that we explored one-hot encoding as a way to represent categorical data. For text, this would mean creating a massive vector for each word, with a 1 in the position corresponding to that word and 0s everywhere else. If our vocabulary has 50,000 words (quite modest for real-world applications), each word becomes a vector with 49,999 zeros and a single 1. This approach has several critical limitations. First, it's incredibly inefficient—think of the computational cost of processing thousands of these sparse, gigantic vectors. It's like writing a letter using entire pages for each character—wasteful and unwieldy. More importantly, one-hot encoding creates no meaningful relationships between words. In this representation, "king" and "queen" are just as different from each other as "king" and "bicycle." Their vectors are completely *orthogonal*, with no mathematical relationship reflecting their semantic connections. It's a representation that strips away the beautiful interconnections of language, leaving us with isolated islands of meaning.

To resolve this major issue, rather than using those sparse one-hot vectors, we'll condense words into dense vectors of modest dimensions—typically between 50 and 300 numbers. These aren't random numbers; they're carefully crafted coordinates that position words in a "meaning space" also known as *word embeddings*. Let's try to break down what this means.

Picture your local library, where books are grouped by their concepts—romance novels cluster together, scientific textbooks form their own neighborhood, and cookbooks have their own corner. Now, take this a step further and imagine a magical, multidimensional library where all books with similar writing style are close together in one dimension, all with same underlying philosophy are closer together in another dimension, and books of similar lengths neighbor each other in yet another dimension. This is essentially what word embeddings do for language—they create a multidimensional space where similar words are neighbors. Depending on what the similarity is based on, the "closeness" is going to be in a particular dimension of the embedding space.

In this embedding space, remarkable patterns emerge. Words with similar meanings cluster together. Even more interestingly, the relationships between words become encoded as consistent vector arithmetic. A classic example is vector(king) - vector(man) + vector(woman) ≈ vector(queen). This means that if we take the value of the vector in this embedding space corresponding to "king", subtract the vector corresponding to "man"—thus essentially removing that meaning from the word—and then add the vector corresponding to "woman", we'll arrive somewhere in the vicinity of the word "queen". The embedding has captured not just word similarity but complex *analogical relationships*—the "royal" relationship, the gender dimension, and how they interact.

These embeddings transform our understanding of text data. Instead of treating each word as an isolated entity, we now have rich, nuanced representations that capture semantic relationships. Words become points in a continuous space where distance and direction have meaning.

9.1.3 GloVe: Learning the Language of Vectors

Among the various approaches to creating such word embeddings, Global Vectors for Word Representation (*GloVe*) stands out for its elegant approach. Developed by researchers at Stanford, GloVe learns these meaningful word vectors by examining how often words appear together in text. The core insight behind GloVe is beautifully simple: Words that appear in similar contexts likely have similar meanings. If "dog" and "cat" frequently appear near words such as "pet," "fur," and "veterinarian," they probably share some semantic relationship—they're both domestic animals.

We don't want to go into the details of how GloVe is trained, but we'll define the semantics briefly as they are of particular relevance to us. GloVe captures relationships between words by looking at co-occurrence statistics across a massive corpus of text. It counts how often each pair of words appears together within a certain window size (typically a few words before and after). These counts form a *co-occurrence matrix*—a giant table showing how frequently each word appears near every other word. The model also transforms these counts using a clever mathematical model that emphasizes meaningful patterns while reducing the impact of pure frequency. Common words such as "the" and "and" appear frequently with almost everything, but that doesn't make them semantically similar to everything.

The training process adjusts word vectors so that their *dot product* relates to the logarithm of their co-occurrence probability. This might sound complex, but it's grounded in an intuitive idea. For now, a deep understanding of the mathematics behind GloVe isn't necessary as long as you grasp the following simple concept: This simple objective function yields vectors with rich linguistic properties. Words naturally cluster into semantic neighborhoods—tools with tools, countries with countries, emotions with emotions—creating a meaningful geometry of language. In this space, vector direction often captures specific semantic relationships, while vector magnitude tends to correlate with word *specificity* or importance.

> **The Strengths of GloVe**
> What makes GloVe particularly powerful is that it balances two approaches to learning word meaning: *global matrix factorization* (looking at the overall statistics of the corpus) and *local context window methods* (looking at nearby words). This dual approach helps it capture both broad statistical patterns and more nuanced local relationships. You may look up these details if you wish to explore this fascinating model further.

When trained on enough text—billions of words from sources such as Wikipedia, news articles, and books—GloVe embeddings develop a remarkable ability to capture semantic relationships. These pretrained embeddings become a treasure trove of linguistic knowledge that we can transfer into our own models, giving them a head start in understanding language. Instead of our models learning language from scratch, they can build upon these rich representations, allowing them to focus on the specific task at hand rather than the fundamentals of language.

Let's take our understanding of word embeddings beyond theory and see them in action. As we move toward understanding transformers, it's crucial to develop an intuition for how words become vectors in a meaningful space. This visual exploration will lay the groundwork for grasping how transformers process and relate words to one another. The code in Listing 9.5 demonstrates how *pretrained GloVe embeddings* (downloaded from the official website) can demonstrate very revealing relationships between words. These embeddings, trained on billions of words from across the internet, have captured semantic connections that we can now visualize and explore. (You can access the complete working code from the supporting page at *https://recluze.net/keras-book* in the notebook **09-02-glove-embedding**.)

The first function, `load_glove_embeddings()`, is like a librarian carefully unpacking and organizing a vast collection of knowledge. It reads through the GloVe file, where each line contains a word followed by its vector coordinates—typically 100 numbers that position it in a high-dimensional *semantic space*. The function creates a dictionary mapping each word to its vector, giving us easy access to this linguistic treasure trove.

```
# do imports
 # download pretrained glove embeddings from   url = "http://nlp.stanford.edu/data/glove.6B.zip"

def load_glove_embeddings(filepath="glove.6B.100d.txt"):
    embeddings_index = {}
    with open(filepath, encoding="utf8") as f:
        for line in f:
            values = line.split()
```

```
            word = values[0]
            vector = np.asarray(values[1:], dtype="float32")
            embeddings_index[word] = vector
    return embeddings_index
```

Listing 9.5 Loading Pretrained GloVe Embeddings

The second function, `visualize_words()`, shown in Listing 9.6, takes these 100-dimensional vectors and transforms them into something we can actually see. Consider trying to visualize a 100-dimensional space—our minds simply can't grasp that many dimensions at once. This is where dimensionality reduction comes in. Our code uses *Principal Component Analysis* (PCA) to downgrade our data to two important dimensions—the directions that capture the most variance and, hopefully, the most meaning.

> **Options for Dimensionality Reduction**
>
> While our code uses PCA, we could also employ t-distributed Stochastic Neighbor Embedding (*t-SNE*) for dimensionality reduction, as mentioned in the commented code. It's worth noting that this reduction to 2D is purely for our visualization and understanding—it's not something done in practice when using embeddings in actual models. The real power comes from the full high-dimensional representations where subtle relationships can be preserved without compression. Our visualization is just a window into a much richer semantic space, such as viewing a 3D object through its 2D shadow.

This visual representation helps us grasp a fundamental concept that will be critical when we discuss transformers: words exist in a continuous space where *proximity* and direction have meaning.

```
def visualize_words(word_pairs, embeddings_index):
    # Flatten pairs to unique word list
    all_words = list({w for pair in word_pairs for w in pair})
    valid_words = [word for word in all_words if word in embeddings_index]
    vectors = np.array([embeddings_index[word] for word in valid_words])

    # PCA reduction. alternative option for reduction to 2D
    pca = PCA(n_components=2)
    vectors_2d = pca.fit_transform(vectors)

    # Map words to 2D coordinates
    word_to_2d = dict(zip(valid_words, vectors_2d))

    # ... plot the vectors

    # Draw arrows for word pairs
    for source, target in word_pairs:
```

```
            if source in word_to_2d and target in word_to_2d:
                src = word_to_2d[source]
                tgt = word_to_2d[target]
                plt.arrow(src[0], src[1], tgt[0] - src[0], tgt[1] - src[1],
                          head_width=0.0, length_includes_head=True, color='blue',
alpha=0.7)
        # ...
```

Listing 9.6 Visualizing Embeddings in Lower Dimensions

Let's look at what happens when we access a specific word from our GloVe embeddings, as shown in Listing 9.7.

```
print(embeddings_index["nephew"])
print("Shape:", embeddings_index["nephew"].shape)
# output: [ 0.60233  -0.75386   …   ]
# Shape: (100,)
```

Listing 9.7 Example of a Word Embedding

This output gives us a peek into how "nephew" exists in our semantic universe. What we're seeing is a snippet of the 100-dimensional vector for this word—a numerical fingerprint that captures its meaning in our embedding space. The shape (100,) tells us that "nephew" is represented by exactly 100 floating-point numbers. Each of these numbers positions the word along a different dimension in our semantic space. Some values are positive (e.g., 0.60233), pulling the word in one direction, while others are negative (-0.75386), pulling it in another. Think of each number as a subtle influence on where "nephew" sits in relation to other words. The positive values might align it with family-related concepts, while negative values might distance it from unrelated domains such as technology or sports. Together, these 100 coordinates place "nephew" in a precise location where it's appropriately close to words such as "uncle," "family," and "relative," but far from unrelated concepts.

When our models work with text, they don't see the word "nephew"—they see this vector, this constellation of values that contains rich information about the word's meaning and its relationships to other concepts in our language. We can visualize some word pairs using our function as in Listing 9.8. This produces the plot shown in Figure 9.2.

```
word_pairs = [
    ("nephew", "niece"),
    ("man", "woman"),
]
visualize_words(word_pairs, embeddings_index)
```

Listing 9.8 Visualizing Example Words

9.1 The Theory Behind Transformers

This visualization beautifully reveals one of the most interesting properties of word embeddings—they capture meaningful linguistic relationships as geometric patterns in vector space. What we're seeing is a 2D projection of the 100-dimensional word vectors for two pairs: "nephew"/"niece" and "man"/"woman." The lines connect these related pairs, and something remarkable emerges—the arrows are nearly parallel! This isn't a coincidence. It's a profound demonstration of how GloVe embeddings have captured consistent semantic relationships.

The parallel arrows show that the mathematical relationship between "nephew" and "niece" is strikingly similar to the relationship between "man" and "woman." In other words, the gender concept is encoded as a consistent direction in the embedding space. The vector math "nephew - niece" points in roughly the same direction as "man - woman", revealing how the embedding has learned to recognize gender as a fundamental semantic dimension. What's even more remarkable is that this pattern wasn't explicitly programmed—it emerged naturally from analyzing word co-occurrences in text. The embeddings discovered that these word pairs share a common relationship type.

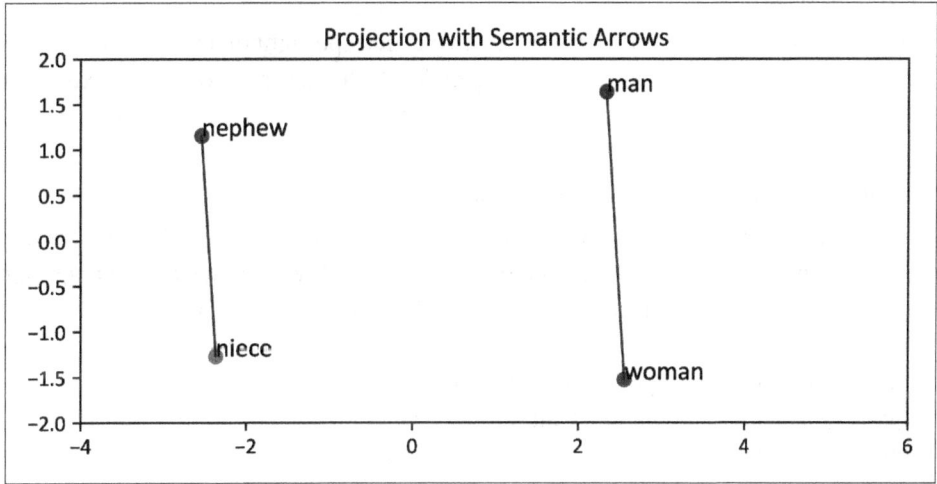

Figure 9.2 2D Projection of Word Embeddings of Related Words

This visualization gives us a glimpse into how transformers "think" about words when processing language. Rather than treating words as isolated symbols, transformers use these rich geometric relationships, allowing them to understand nuanced connections between concepts that would be impossible with simpler representation methods. These embeddings, especially models such as GLoVe, capture remarkable semantic relationships between words. However, there's a fundamental limitation we need to address.

9.1.4 A Gentle Introduction to Attention

Consider this simple sentence: "The umbrella was wet, so it had to be left outside." What does "it" refer to in this context? As humans, we instantly understand that "it" refers to the umbrella. But a fixed embedding vector for "it" would be the same regardless of whether "it" refers to an umbrella, a cat, a computer, or anything else. The embedding alone doesn't capture this *contextual relationship*. Let's look at another example that highlights this problem: "At the zoo, the seal balanced a ball on its nose."

In GLoVe or similar embedding systems, the word "seal" would have a specific vector representation that attempts to capture all the possible meanings of the word "seal"—the marine animal, a wax stamp, the act of sealing something, and so on. But in this specific context, we know it refers to the animal because of the surrounding words "zoo," "balanced," and "nose." Our understanding of "seal" is influenced by paying *attention* to these contextual clues. This is the core idea behind attention mechanisms—allowing our models to pay attention to relevant parts of the input when producing each part of the output. It's actually exactly like shining a spotlight on particular words that matter most for understanding the current word or phrase.

In this section, we'll take a look at how we can put that spotlight on using mathematical constructs that can be implemented using Keras. We'll see the need for context-aware representations and how to enable our code to create these representations.

The Need for Context-Aware Representations

Traditional *recurrent neural networks* (RNN; no longer used actively) tried to solve this by passing information forward through a sequence, but they struggled with *long-range dependencies*. The information from words at the beginning of a long sentence would often be forgotten by the time the model reached the end. *Attention mechanisms* revolutionized this approach by creating direct pathways between different positions in a sequence. Instead of relying solely on sequential processing, attention allows any word to directly influence the representation of any other word, regardless of their distance in the sequence.

At its simplest, attention works by computing a score between the word currently being focused on (often called the *query*) and every other word in the context (the *keys*). These scores indicate how much attention should be paid to each contextual word when representing the current word. Let's break this down with our "seal" example:

- When processing the word "seal", our model computes *attention scores* between this and every other word in the sentence ("At," "the," "zoo," "balanced," etc.).
- Words such as "zoo," "balanced," and "nose" would receive higher attention scores because they're more relevant to determining that "seal" refers to the animal.
- The model then creates a weighted sum of all the word vectors, where the weights are determined by these attention scores.

- This weighted sum becomes the *context-aware representation* of "seal" that captures not just the word itself but also its relationship to the surrounding context.

The beauty of this approach is that it allows the model to pay attention to different parts of the input depending on what's currently being processed. When handling the word "seal," attention to "zoo" helps disambiguate its meaning. When processing "ball," attention might shift more toward "balanced" to understand what's happening with the ball.

Self-Attention: The Core of Transformers

To dig a bit deeper into this concept, let's see how the transformer architecture handles what is called *self-attention*. In self-attention, every word in a sequence attends to every other word, creating rich, context-aware representations for each position. If we visualize this process, it's as if each word sends out question marks to all other words, asking, "How relevant are you to my semantics?" Each word then returns an answer, and these answers are combined to create a more informed representation. Figure 9.3 demonstrates how the word "seal" transforms as it gathers context from surrounding words in the sentence, "At the zoo, the seal balanced a ball on its nose." The visualization shows this process through two key elements.

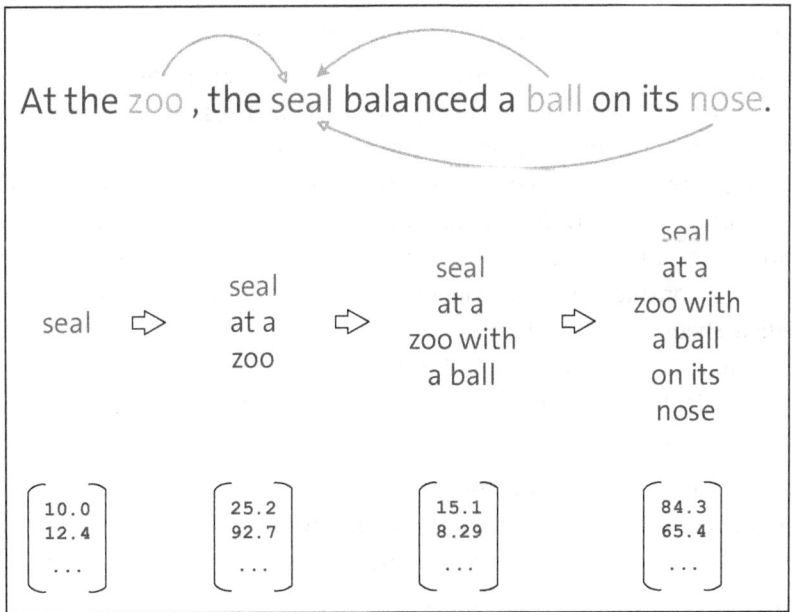

Figure 9.3 Self-Attention Between Words in a Sentence

First, at the top of the image, we see the full sentence with words and arrows connecting "seal" to contextually related words "zoo," "ball," and "nose." These arrows represent attention—how the model learns which words are important for understanding "seal" in this specific context. Below this, we see the progressive enrichment of the word "seal" as it accumulates contextual information. Starting with just the word "seal," which could have multiple meanings, the representation gradually becomes more specific by adding context: "seal at a zoo," then "seal at a zoo with a ball," and finally the complete context, including "on its nose." More precisely, the vector representations change with each step, reflecting how the mathematical embedding of "seal" shifts as it absorbs information from related words. This is the essence of contextual embedding—the meaning of "seal" becomes increasingly refined and specific to "the animal performing in a zoo" rather than "a wax stamp" or "to close tightly."

This process quite precisely mirrors our own human understanding. We naturally resolve ambiguities by paying attention to context. The visualization shows how modern NLP models do something similar, creating dynamic representations of words that change based on their surroundings—a fundamental improvement over fixed word embeddings that revolutionized how machines understand language.

The transformation from fixed embeddings to attention-based representations marks a profound shift in how we model language. With fixed embeddings such as GLoVe, words have static meanings. With attention, words have fluid representations that adapt based on their surroundings.

Understanding Attention Through the Internal Matrix Representation

Let's examine how attention mechanisms work in transformer models internally. Attention is modeled essentially as a map, as depicted in Figure 9.4, showing how words "look at" each other in our example sentence. This visualization is called an *attention matrix*. Think of it as a relationship map between every word in our sentence. Each row represents a word that might provide information, and each column represents a word that's asking for information or paying attention. The highlighted squares show the strength of these relationships—darker shades indicate high attention (strong connection), intermediate shades show medium attention, light represents low attention, and white means no attention at all. The diagonal line running from top-left to bottom-right represents each word's attention to itself, which makes sense—a word's own identity is certainly relevant to understanding it!

This attention-based approach solves a fundamental challenge in language processing. Words don't exist in isolation—they derive meaning from their surroundings. The attention matrix visualizes exactly how models capture these contextual relationships.

After the attention scores are calculated, the model uses these attention weights to create a weighted sum of all relevant word embeddings. For the word "seal," it takes its original embedding but enhances it with information from "zoo," "balanced," "ball," "its," and "nose"—with the contribution of each proportional to its *attention score*.

9.1 The Theory Behind Transformers

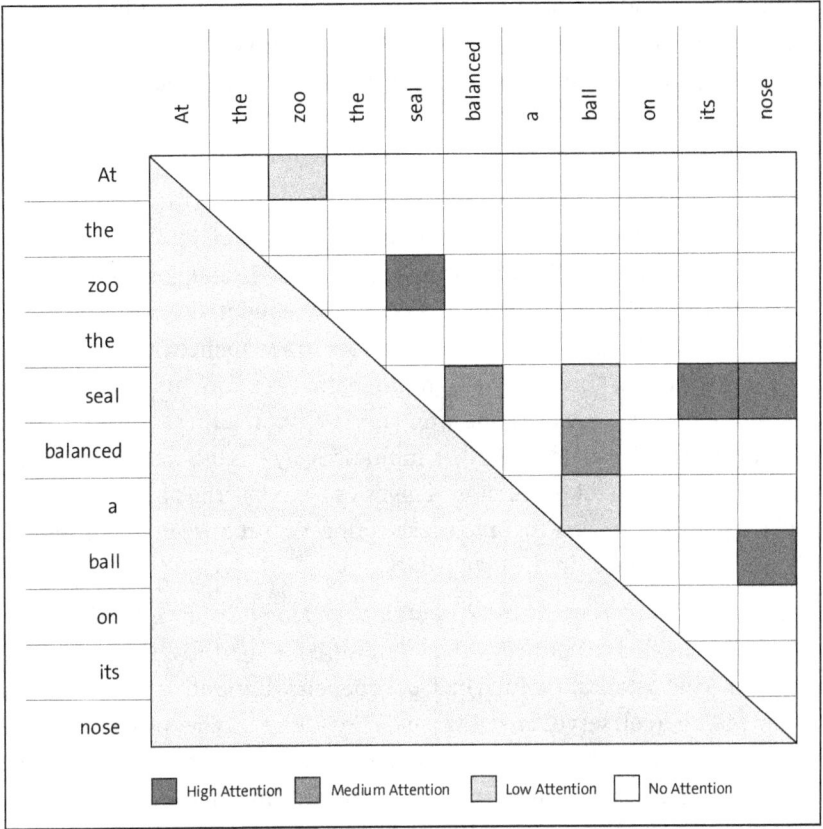

Figure 9.4 Self-Attention Represented as an Upper Triangular Matrix

> **Iterative Deepening in Understanding Transformers**
>
> Due to the complex nature of transformers, we'll use an iterative deepening approach to understanding them. We'll continually return to the core concepts in more and more details so that you can fully understand them. The repetition of important ideas is intended!

9.1.5 Why Transformers Revolutionized Natural Language Processing and Beyond

In this section, we'll explore the second reason why transformers revolutionized natural language processing (NLP) and how their ability to process sequences in parallel provided another significant advantage over previous architectures.

Before transformers arrived on the scene in 2017, NLP was like a relay race. Models would process text one word at a time, passing information from the beginning of a sentence all the way to the end. While this approach worked to some extent, it had some serious limitations that hindered progress due to this sequential nature of processing. Enter the transformer architecture, which completely reimagined how machines understand language, and the results were nothing short of revolutionary.

Breaking the Sequential Bottleneck

Imagine trying to understand a book by only being allowed to read one sentence at a time, and once you've moved on, you can only rely on your memory of what came before. That's essentially how traditional models such as RNNs and Long Short-Term Memory (LSTM) networks worked—they processed text sequentially, word by word. This created a bottleneck where information from earlier in a sequence would gradually fade or become distorted as it traveled through the chain. Transformers shattered this limitation. Instead of processing words one after another, like reading a sentence from left to right, transformers look at all words simultaneously—more like taking in an entire page at once. This parallel processing is gives our models the ability to see the whole *context* at a glance, allowing them to make connections between words regardless of how far apart they appear in the text.

The real magic of transformers lies in how they understand long-distance relationships. Traditional models struggled to connect words that were far apart in a sentence. Consider this example: "The restaurant, which had just opened last month on the corner of First Avenue and Main Street, serves amazing pasta." By the time older models reached "serves," they might have already forgotten details about "the restaurant" from the beginning of the sentence. Transformers, with their attention mechanisms, can directly link "restaurant" and "serves" regardless of the distance between them. Due to this important distinction, perhaps the most profound impact of transformers has been their ability to scale. Before transformers, making language models bigger often yielded significantly diminishing returns—they'd get only marginally better despite significant increases in size and training data. Transformers changed this equation dramatically.

As researchers made transformer models larger and fed them more data, their capabilities didn't just improve incrementally—they exploded with new abilities. Models started to demonstrate skills they weren't explicitly designed for, from solving math problems to writing creative fiction. It was as if crossing certain thresholds of scale unlocked entirely new realms of capability, similar to how adding more pieces to a puzzle eventually reveals the complete picture rather than just a clearer version of a partial one.

Beyond Text: Transformers Go Multimodal

The influence of transformers quickly spread beyond just text processing. Their flexible architecture proved remarkably adaptable to other types of data. Soon, we saw transformer-based models handling images, audio, video, and even protein sequences with the same underlying principles. This cross-domain success revealed that the way transformers process relationships and patterns is fundamental to many different types of information processing, not just language. It's a bit like discovering a universal key that unlocks doors across many different buildings.

What truly set transformers apart from previous breakthroughs was how quickly they moved from research papers to practical applications. Within just a few years, transformer-based models were powering translation services, writing assistants, code completion tools, and even creative art generators that millions of people use daily. This rapid translation from theory to practice happened because transformers didn't just offer incremental improvements—they fundamentally changed what was possible. Tasks that were once considered extraordinarily difficult suddenly became approachable, opening doors to applications that previously seemed decades away. As we dive deeper into the inner workings of transformers in the coming sections, we'll unpack exactly how their key components work together to achieve these remarkable capabilities. Even before we get to the mathematical details, it's worth appreciating how this architecture represented not an incremental improvement but a genuine paradigm shift in AI—one whose ripple effects continue to expand across science, technology, and society at large.

9.2 Components: Attention Mechanism, Encoder, Decoder

Now that we've established the foundations of transformers and the attention mechanism, it's time to peek under the hood and examine the powerful machinery that makes these models work. In this section, we'll explore the core components that give transformers their remarkable abilities—breaking down complex structures into understandable pieces that will make our future Keras implementations feel like assembling familiar building blocks rather than wrestling with mysterious black boxes. We'll start by demystifying the attention mechanism, that clever innovation that allows transformers to focus on relevant information across a sequence. You'll discover how the deceptively simple concept of *query*, *key*, and *value* transforms the way models process language.

From there, we'll explore *positional encodings* (solving the puzzle of how transformers understand word order when looking at everything all at once) and then examine both *encoder* and *decoder* architectures in detail. By the time we reach Keras implementation in the next section, you'll have developed an intuitive understanding of these components that will make the surprisingly concise code feel like a natural extension of the concepts rather than an intimidating jumble of functions and classes.

9.2.1 The Conversation Between Words

Self-attention creates a sort of "internal conversation" within a sentence, allowing each word to interact with every other word directly. Let's break down how this conversation works using a practical example.

Consider our original sentence: "The seal swims in the zoo enclosure." In traditional models, words would be processed one after another, making it difficult to connect "seal" with "zoo" because they're separated by several words. But self-attention changes this entirely by letting every word directly communicate with every other word, regardless of distance. The first step in self-attention involves each word asking a question—posing a query. When we say a word "asks a question," we're using a helpful metaphor for what's happening mathematically. Behind the scenes, the model transforms each word's embedding (its numerical representation) into a query vector through a *learned transformation*. Take the word "seal" in our example. It might be posing queries like "Where am I located?" or "What am I doing?" These aren't literal questions, of course, but representations that help the model understand relationships. Each word gets transformed into its own unique query based on both the word itself and what the model has learned about language during training. The word "swims" might generate a query focused on finding a subject, while "zoo" might generate a query looking for related entities.

While every word is asking questions, they're simultaneously presenting information about themselves through what we call *keys*. Again, this isn't a literal presentation but another transformation of the word embeddings. In our example, the word "zoo" might present a key that essentially says, "I'm a location" or "I'm an enclosure for animals." The word "seal" might present a key that says, "I'm an animal" or "I'm a potential swimmer." These keys allow words to respond to the queries from other words. When a query from one word aligns well with a key from another word, the model recognizes there's an important relationship to capture. We need to somehow create a link between these keys and queries. In the rest of this section, we'll see how we can enable this link.

The Matchmaking Process

This is the crucial part where queries meet keys—a kind of matchmaking process. For each word's query, the model compares it against the keys of all words in the sentence. When a query and key align well (mathematically, they have a high *dot product*), it indicates these words should pay attention to each other. In our example, the query from "seal" asking "Where am I located?" would match strongly with the key from "zoo" saying "I'm a location." Similarly, the query from "swims" looking for a subject would match strongly with the key from "seal" identifying as an animal. This matching process creates what we call *attention scores*—numbers that represent how much each word should focus on every other word. Higher scores mean stronger connections.

After generating these scores, the model normalizes them (using a softmax function) to create a probability distribution. These probabilities determine how much each word's information contributes to the updated representation of the current word. Going back to our example, when processing "seal," the model might determine it should pay:

- 60% attention to itself (to maintain its core meaning)
- 25% attention to "zoo" (to understand its location)
- 10% attention to "swims" (to capture its action)
- 5% distributed, among other words

These attention weights help create rich, context-aware representations for each word. The word "seal" now carries information not just about being a seal but about being a-seal-that-swims-in-a-zoo-enclosure.

How does the model know what questions to ask or what information to present? That's where the whole concept of gradient descent steps in. The transformations that create queries and keys aren't fixed—they're parameters represented through matrices and learned during training through gradient descent.

To recall the concept from Chapter 3, initially these transformations will be random, producing nonsensical attention patterns. But as the model trains on millions of examples, it gradually refines these transformations. When the model makes a mistake in prediction, *backpropagation* adjusts the transformations to produce better queries and keys. Over time, the model discovers that certain types of queries and keys lead to more accurate predictions and lower loss. Perhaps it learns that verbs should generate queries that look for subjects or that pronouns should generate queries that look for their referents. This is the true power of self-attention—it doesn't connect words based on pre-defined rules. Instead, it learns which connections matter for understanding language. The model figures out on its own that "seal" should pay attention to "zoo" because this relationship helps it better predict and understand text. Following this simple (though compute intensive) mechanism, by the end of training our transformer has developed a sophisticated system for determining which words should attend to which other words, all learned from data rather than programmed by humans.

The Mathematics of the Query, Key, Value Concept

While understanding the underlying mathematics isn't strictly necessary to use these models, it really helps debug issues when they inevitably arise in practice. It also enables you to come up with more novel models in the future if you have a deeper understanding of the concepts. So, let's dedicate some time to understanding the underlying mechanism of how the transformers actually make everything work.

Consider the scenario in which you're searching for a book in a vast library. You approach the information desk with a question: "Do you have any books about deep learning?" This is your query. The librarian consults a catalog system (the keys) to find matching entries, then retrieves valuable books (the values) that best answer your question. This everyday interaction mirrors exactly how self-attention works in transformers. In our model, each word in a sentence creates its own query, key, and value through separate transformations of its initial embedding. When the model processes "The cat sat on the mat," the word "cat" generates:

- A query vector that essentially asks, "Which words are most relevant to understanding me?"
- A key vector that advertises what information "cat" contains that might be useful to other words
- A value vector that holds the actual content "cat" will contribute to other words

Mathematically, we can express this transformation as:

$$Q = XW^Q, K = XW^K, V = XW^V$$

Where X represents our input word embeddings, and W^Q, W^K, and W^V are learned weight matrices. Notice that the superscript notation doesn't mean exponentiation. They are just indices for matrices to differentiate them from each other. These matrices are the tools our model uses to transform each word into its three different roles. Think of them as different lenses through which the model views the same word, each bringing certain aspects into focus while blurring others. The beauty of this approach is that every word simultaneously plays all three roles—*questioner, answerer,* and *information carrier*—creating a rich web of connections throughout the sentence.

Scaled Dot-Product Attention

Now that we have our queries, keys, and values, let's see how these three interact with each other. This is enabled by the *scaled dot-product attention*—the mathematical engine that powers the connections between words. Let's return to our library metaphor. Imagine your query about deep learning books gets compared against thousands of catalog entries (keys) simultaneously. Some entries match your query perfectly, others partially, and many not at all. The librarian then creates a weighted collection of books, giving you more volumes that match your query closely and fewer that match only tangentially. In transformer models, this process happens through mathematical operations that might sound complex but embody a straightforward idea:

- For each word's query, we compute how well it matches with every word's key through a dot product. Think of this as measuring the alignment between what one word is asking for and what another word is offering.
- These alignment scores determine how much attention each word should pay to every other word. High scores mean strong connections; low scores mean weak ones.
- We then apply a *scaling factor*—dividing by the square root of the dimension size. This might seem like a nonessential detail, but it's crucial. Without scaling, as our

embeddings get larger, the dot products grow too large, pushing our attention weights toward extremes and making learning difficult.
- Finally, these scaled scores pass through a softmax function, turning them into a probability distribution that sums to 1. This ensures our attention is properly normalized—each word distributes 100% of its focus across all words in the sentence.

The complete formula for scaled dot-product attention is

$$\text{Attention}(Q, K, V) = \text{softmax}\left(\frac{QK^T}{\sqrt{d_k}}\right)V$$

Where d_k is the dimension of the key vectors. The division by $\sqrt{d_k}$ is our scaling factor, preventing those dot products from growing too large as dimensions increase. The softmax function then converts these scaled scores into probabilities that sum to 1. Finally, we multiply by V to get a weighted sum of value vectors—each value contributing in proportion to how relevant its corresponding key was to our query. Thus, each word now has a nuanced understanding of its relationship with every other word in the sentence. This final resulting matrix is known as an *attention head.*

Multi-Head Attention

If scaled dot-product attention gives our model one lens through which to view relationships between words, *multi-head attention* provides multiple perspectives simultaneously—like viewing a 3D sculpture from different angles to appreciate its full complexity. Think about how you understand a conversation. You're simultaneously tracking the literal meaning of words, the emotional tone, cultural references, and perhaps unstated implications. You're paying attention in multiple ways at once. Multi-head attention works similarly. Rather than performing attention just once, the model creates several attention heads, each with its own set of learned parameters. Each head might focus on different aspects of language:

- One head might focus on syntactic relationships (subjects and verbs).
- Another might capture semantic connections (related concepts).
- A third might track long-distance dependencies (connecting a pronoun to its referent).
- Yet another might attend to idiomatic expressions.

Mathematically, multi-head attention is defined as

$$\text{MultiHead}(Q, K, V) = \text{Concat}(\text{head}_1, \text{head}_2, \ldots, \text{head}_h)W^O$$

where each head is computed as

$$\text{head}_i = \text{Attention}(QW_i^Q, KW_i^K, VW_i^V)$$

Each head has its own set of learned projection matrices—W_i^Q, W_i^K, and W_i^V—that create different *subspace representations* of the queries, keys, and values. The final W^O matrix combines these various perspectives into a coherent output. Each head processes the same input independently, producing its own attention patterns. Then, like advisors presenting different perspectives to a decision-maker, the outputs from all heads are combined into a final representation. This multi-perspective approach gives transformers remarkable flexibility in how they process language.

What's particularly elegant about this design is that the model doesn't need to be explicitly told what types of relationships each head should focus on. By randomly initializing matrix values and through training on vast amounts of text, each head naturally specializes in whatever patterns prove useful for predicting the next word or understanding the input. The model discovers on its own that tracking *subject-verb relationships* or *semantic connections* helps minimize prediction errors.

Multi-head attention transforms what could have been a 1D view of language into a rich, multifaceted understanding—and it does this through an architectural design that's surprisingly simple yet extraordinarily effective. By the time a sentence passes through these attention mechanisms, each word has been enriched with contextual information from every other word, all weighted by relevance and viewed through multiple specialized lenses.

9.2.2 Why Position Information Matters

The self-attention mechanism gives transformers their power to connect any word with any other word, regardless of distance, by looking at the complete set of input words in one go. But this strength comes with a curious weakness—it has no built-in way to understand the sequential nature of language. When processing the sentence "The dog chased the cat," the self-attention mechanism sees six word vectors but has no inherent way to know that "dog" comes before "chased" or that "chased" comes before "cat." This creates a serious problem. The sentences "The dog chased the cat" and "The cat chased the dog" contain exactly the same words but tell completely different stories. Without position information, our transformer would process these sentences identically, which would be disastrous for understanding language.

Think about how differently we interpret these sentences:

- "Max said Anna is clever." (Max is making a statement about Anna.)
- "Anna said Max is clever." (Anna is making a statement about Max.)

Or consider how word order affects meaning in these examples:

- "He ate only vegetables." (He consumed vegetables and nothing else.)
- "He only ate vegetables" (Eating was all he did with the vegetables, rather than, say, growing them.)

Even subtleties like these depend entirely on knowing the positions of words relative to each other. In language, order often completely determines the meaning rather than just affecting it slightly. Traditional sequence models such as RNNs and LSTMs naturally captured position because they processed words one after another, carrying forward information about what came before. But transformers process all words simultaneously—giving them remarkable speed and parallelization capabilities while creating this positional blindness.

Therefore, position information has to be explicitly injected into each word embedding. This positional encoding needs to satisfy several key requirements:

- It must be unique for each position, so the model can distinguish different positions.
- It must have a consistent pattern so the model can generalize to sequence lengths it hasn't seen during training.
- The distance between positions must be reflected consistently in the encoding.
- It can't disrupt the meaningful content in the original word embeddings.

Let's look at how these objectives can be met in an efficient manner in the following sections.

Sinusoidal vs. Learned Encodings

When it came to designing these positional encodings, researchers faced a fundamental choice: should they create a fixed mathematical function to generate position vectors (*sinusoidal encodings*), or should they let the model learn the optimal position representations during training (*learned encodings*)? The original transformer paper introduced an elegant approach using sine and cosine functions of different frequencies. Picture a collection of waves, some oscillating rapidly and others more slowly. By combining these waves in specific ways, we can create a unique pattern for each position that still maintains a consistent relationship between positions. The equation looks like

$$PE_{(pos,2i)} = \sin\left(\frac{pos}{10000^{2i/d}}\right), \quad PE_{(pos,2i+1)} = \cos\left(\frac{pos}{10000^{2i/d}}\right)$$

Where *pos* is the position of the word in the sequence, *i* indicates the dimension within the embedding, and *d* is the embedding size. These sinusoidal functions create an interesting pattern. Each position gets encoded as a vector with values oscillating between -1 and 1, with different dimensions oscillating at different frequencies. What makes this approach particularly clever is that it allows the model to extrapolate to sequence lengths longer than it saw during training. The mathematical pattern extends naturally to new positions because it's based on a continuous function rather than discrete learned values. It's like teaching someone to count by explaining the pattern (add one each time) rather than having them memorize each number separately.

Another advantage is that the model can easily compute relative positions. The positional encoding for position ***pos + k*** can be expressed as a linear function of the encoding at position ***pos***. This mathematical property helps the model understand relationships like "two words apart" or "five words before" in a consistent way across different parts of the sentence.

Learned Encodings: Flexibility Through Data

The alternative approach is conceptually simpler: Just let the model learn the optimal position embeddings during training, the same way it learns word embeddings. We create a positional embedding table where each position gets its own vector that the model updates through gradient descent. At first glance, learned positions might seem preferable: Why constrain ourselves to a specific mathematical function when we could let the data determine the optimal representation? After all, learned positional encodings offer some significant advantages:

- They can potentially capture language-specific positional patterns.
- They directly optimize for whatever works best on the training data.
- They're simpler to implement and understand conceptually.

However, they come with a significant limitation: They don't naturally extend beyond the sequence lengths seen during training. If we train with sequences of at most 512 tokens (as in the original Bidirectional Encoder Representations from Transformers [BERT] model), what happens when we encounter the 513th token? The model has no encoding for this position. This limitation explains why many transformer models have *maximum sequence lengths*—they simply don't have learned positional encodings beyond that point. Some implementations address this by using relative positional encodings or by extrapolating from learned positions, but these are workarounds for an inherent limitation. In practice, both approaches work well, and there's ongoing research and debate about which is optimal in different circumstances. Many modern transformer implementations, including popular models such as BERT, actually use learned positional encodings despite their limitations, while others stick with the original sinusoidal approach.

What's more important to note here is that this seemingly technical choice reflects a deeper philosophical question in machine learning: Is it better to build in mathematical structure based on our understanding of the problem, or to let the model learn everything from data? The positional encoding debate is just one instance of this broader question that appears throughout deep learning.

Regardless of the specific approach, the key insight remains: By adding positional information to each word embedding (whether through sinusoidal functions or learned vectors), transformers gain the ability to understand sequences while maintaining their parallel processing advantage. This clever solution to the position problem completed the foundation for transformer models to revolutionize NLP and beyond.

9.2.3 Encoder Structure: The Information Processing Powerhouse

With word embeddings enriched by positional information, we now have the essential ingredients for our transformer model. But embeddings alone aren't enough—they merely serve as the starting point. The real magic happens in the *encoder structure*, where these position-aware word representations are transformed into deeply contextual encodings that capture the nuanced relationships between words. The encoder is where words begin to truly understand each other, where "bank" in a financial context becomes distinctly different from "bank" in a river context despite starting with the same embedding. Let's examine the sophisticated machinery that enables this contextual understanding in the following sections, beginning with the core components that make up the encoder structure.

Multi-Head Attention Layers: The Relationship Builders

We've already explored how multi-head attention creates connections between words, but now, let's see how it fits into the bigger picture of the encoder. The multi-head attention layer is like a relationship counselor for words—helping each word understand its connections to every other word in the sentence. When "bank" appears in "I deposited money in the bank," the attention mechanism helps it connect strongly to words like "deposited" and "money," nudging it toward its financial meaning rather than its river-related one. In the encoder, this multi-head attention operates in *self-attention mode*— each word attending to all words (including itself) in the same input sequence. It's like each person in a conversation considering everyone else's contributions before formulating their own understanding. The architecture of the encoder can be seen concisely in Figure 9.5.

The mathematical machinery works exactly as we discussed earlier—queries meeting keys, creating attention scores, and using these to blend value vectors. But the critical point is what this accomplishes: it transforms a sequence of independent word embeddings into a web of contextually connected representations.

After passing through this layer, each word's representation now contains traces of other relevant words. The word "bank" now carries echoes of "money" and "deposited" in its encoding, fundamentally transforming its representation based on its surroundings. After words have shared information through attention, each word gets a chance for some individual processing through a feed-forward network. This is a simple yet powerful neural network applied independently to each position. Based on this dense connectivity, each word carries all the context it gathered from other words and will now process this information individually. Structurally, these feed-forward networks are typical dense connections specified using the formula:

$$\text{FFN}(x) = \max(0, xW_1 + b_1)W_2 + b_2$$

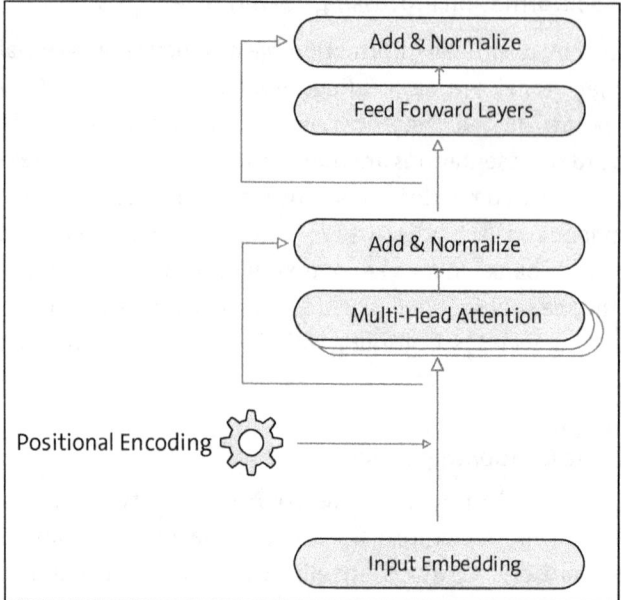

Figure 9.5 Encoder Architecture

The first transformation usually expands the representation into a larger inner dimension, the ReLU activation adds nonlinearity (allowing for complex patterns), and the second transformation projects back to the original dimension size. This is just an application of two feed-forward layers to the word embeddings. Intriguingly, while these networks process each position independently (the same way for each word regardless of what surrounds it), they're still critical for the model's expressiveness. They give the transformer the computational capacity to make complex decisions based on the context that attention has already gathered.

Layer Normalization and Residual Connections: The Flow Controllers

Now, we come to two technical yet crucial components that might seem like mere plumbing details but actually make the difference between a transformer that works brilliantly and one that fails to learn. The first one, *Layer normalization* is like a sound equalizer that ensures all frequencies remain in a balanced range. For each position in the sequence, it normalizes the values across features to have zero mean and unit variance as in

$$\text{LayerNorm}(x) = \gamma \cdot \frac{x - \mu}{\sqrt{\sigma^2 + \epsilon}}$$

where μ and σ are the mean and standard deviation computed across the feature dimension, while γ and β are learnable parameters. Layer norms are absolutely critical because neural networks can suffer from the issue of *internal covariate shift* in which the distribution of activations changes as the network learns, making training unstable. Layer normalization addresses this by keeping activations in a consistent range, helping gradients flow smoothly during training. In practical terms, it's ensuring each word's representation stays within reasonable bounds, preventing any feature from becoming excessively dominant. This makes training more stable and often faster.

The second aspect, *residual connections* (or *skip connections*) are perhaps the most elegant innovation in the transformer architecture. We discussed residual connections in detail in Chapter 8, Section 8.2.6. To recall briefly, in deep networks, information and gradients can struggle to flow through many layers—leading to the vanishing gradient problem. Residual connections create highways for both forward information and backward gradients, enabling the training of much deeper networks. For transformers specifically, residual connections ensure that the original word information isn't lost amid all the contextual updates. The word "bank" remains fundamentally "bank" even as it gathers financial context—the core meaning persists while being enriched rather than replaced.

When we combine these components—multi-head attention, feed-forward networks, layer normalization, and residual connections—we get a complete encoder layer:

1. Input embeddings pass through layer normalization.
2. They flow into multi-head self-attention.
3. The output is added back to the original input (residual connection).
4. This sum is normalized again.
5. It passes through the feed-forward network.
6. Another residual connection adds the feed-forward output back to its input.
7. A final layer normalization produces the output of the encoder layer.

The full encoder typically stacks 6–12 of these layers (or more in larger models), with each layer building upon the contextual representations created by the previous one. The early layers might capture basic syntactic relationships, while deeper layers develop more sophisticated semantic understanding.

This *progressive refinement* transforms our initial word embeddings into deeply contextual representations—words that now fully understand their place in the sentence, carrying rich information about their relationships to all other words. It's remarkable how these relatively simple components when arranged just so, create a system capable of understanding language with unprecedented sophistication. The transformer encoder is a testament to how structural elegance can unlock extraordinary capabilities.

9.2.4 Decoder Structure: Creating New Sequences from Understanding

After the encoder, let's turn our attention to the second very important aspect of a transformer architecture. While the encoder transforms input sequences into rich contextual representations, the *decoder* tackles a different challenge—generating output sequences one element at a time. Think of the encoder as a reader deeply understanding a text, while the decoder is a writer crafting a response based on that understanding. This powerful pairing enables transformers to perform tasks such as translation, summarization, and question answering with remarkable fluency. The decoder shares many architectural similarities with the encoder, using similar building blocks but arranged in a subtly different way. These differences are crucial in addressing the unique challenges of sequence generation. In this section, we'll explore the two key innovations that make the decoder special.

Masked Multi-head Attention: Seeing Only the Past

Imagine you're writing a story considering only one word at a time. Each time you choose the next word, you can look back at all the words you've already written, but you can't peek ahead at words you haven't written yet because they don't exist! This fundamental constraint of sequence generation is exactly what masked multi-head attention addresses. The decoder's first sublayer uses a special version of self-attention that only allows each position to attend to previous positions in the output sequence, plus itself. This *masking* prevents information leak from future positions that haven't been generated yet.

Let's visualize how this works. When generating the third word in a sequence, the model can only consider the first and second words (plus the third word's position). When generating the fourth word, it can consider the first through third words, and so on. This creates a triangular attention pattern where each position attends only to itself and earlier positions. The mathematics behind this masking is elegantly simple. We apply a mask to the attention scores before the softmax function, setting scores for invalid connections to negative infinity:

$$\text{MaskedAttention}(Q, K, V) = \text{softmax}\left(\frac{QK^T + \text{Mask}}{\sqrt{d_k}}\right) V$$

The mask is a straightforward matrix where entries corresponding to invalid connections (future positions) contain large negative values, effectively zeroing out these connections after the softmax. This ensures that each position can't see future positions during self-attention. This masking transforms regular self-attention into *causal self-attention*, where the *causality* refers to the time-like ordering of the sequence—each position can only be influenced by prior positions, mirroring how we generate text one token at a time. Without it, the model would cheat during training by looking ahead at the correct next tokens rather than learning to predict them based on previous context. The masking forces the model to make predictions based only on what it has generated so far, just like a human writer must do.

Cross-Attention with Encoder Outputs: The Bridge Between Understanding and Generation

The second key innovation in the decoder is cross-attention—a mechanism that connects the decoder to the encoder, allowing the model to draw on the rich contextual understanding of the input when generating each output token. After the masked self-attention layer, each decoder block contains a cross-attention layer where queries come from the decoder's own processing, and keys and values come from the encoder's final output. This arrangement creates a highly useful dynamic. Each position in the decoder is essentially asking questions about the encoded input sequence, searching for relevant information to help determine the next output token.

In *machine translation*, for instance, when generating the English translation of a French sentence, each position in the English output queries the encoded French sentence, looking for the most relevant parts to inform the current word choice. It's similar to how a human translator might focus on different parts of the source text when producing each word of the translation. The mathematics of cross-attention follows the same pattern as regular attention:

$$\text{CrossAttention}(Q, K, V) = \text{softmax}\left(\frac{QK^T}{\sqrt{d_k}}\right)V$$

The crucial difference is that Q comes from the decoder's previous layer, while K and V come from the encoder's output. What makes cross-attention so powerful is how it creates a direct pathway for information to flow from input to output. When translating "Je suis étudiant" to "I am a student," the cross-attention mechanism might help the decoder focus heavily on "étudiant" when generating "student," creating a direct connection between corresponding terms. This bidirectional flow of information—the encoder processing the entire input at once and the decoder attending to relevant parts of that processed input—gives transformer models their remarkable ability to handle complex relationships between sequences.

The Complete Decoder Block

Putting these pieces together, each decoder block contains the following:

- Masked multi-head self-attention (so each position can see only previous positions)
- Cross-attention with encoder outputs (connecting the decoder to the encoder's understanding)
- A feed-forward network (for individual token processing)
- Layer normalization and residual connections (just like in the encoder)

These blocks are stacked together (typically 6–12 layers, matching the encoder), with each layer refining the representations further. The final layer is followed by a *linear projection* and softmax to convert the decoder's output into probabilities over the vocabulary—predicting the next token. The decoder's architecture elegantly balances two competing needs: maintaining the causal generation process while also leveraging the encoder's comprehensive understanding of the input. This delicate dance between causality and cross-attention is what enables transformers to generate coherent, contextually appropriate sequences. In practice, this architecture has proven remarkably versatile. The same basic decoder structure works for translation, summarization, question-answering, and many other sequence-to-sequence tasks—each time learning to map from the encoder's understanding to an appropriate output sequence, one token at a time.

9.3 Implementing Transformers in Keras

Now that we've explored the theoretical underpinnings of transformers and dissected their key components, it's time to roll up our sleeves and bring these concepts to life through code. The architectural diagrams and mathematical formulations we've examined are about to transform from abstract concepts into working neural networks that you can train and experiment with. In the previous chapter, we discovered how the Keras Functional API gives us the flexibility to create complex model architectures with multiple inputs, branches, and shared layers. That foundation will prove invaluable now as we construct transformers, which feature intricate patterns of connections and information flow. Transformers aren't the kind of models you can easily build with a simple sequential stack of layers—they require the expressiveness and flexibility that the Functional API provides.

As we code our transformer components, you'll see the concepts we've discussed materialize in tangible form. The mysterious query, key, and value vectors will emerge from matrix operations. The parallel processing advantage we've highlighted will become evident in how our attention mechanism processes entire sequences simultaneously. The positional encodings we've explained will appear as actual sine and cosine functions generating vectors. Throughout this section, we'll build our transformer piece by piece, starting with the fundamental building blocks, such as multi-head attention and positional encoding implementations. We'll then combine these elements to create complete encoder and decoder structures. Finally, we'll assemble the full architecture and discuss the hyperparameters that influence its behavior.

Remember that while the code might look complex at first glance, each component maps directly to the concepts we've already covered. The mathematical operations that seemed abstract on paper will now become clear as we implement them in Keras. Let's begin by setting up our environment and then dive into coding the essential components that make transformers so powerful.

> **The Connection Between Math and Engineering**
> Pay attention to how the code structure mirrors the underlying mathematics we examined earlier—this parallel will help reinforce your understanding of both the theory and its practical implementation.

9.3.1 The Encoder Block

Before diving into the implementation details, let's take a moment to establish the hyperparameters that will shape our transformer model. Looking at Listing 9.9, we see a collection of values that might seem like arbitrary numbers at first glance, but each one plays a crucial role in determining how our transformer will learn and behave. These *hyperparameters* are like the genetic blueprint for our transformer. The `embed_dim` determines how rich our word representations will be, with each word encoded as a 256-dimensional vector in our case. When we move to `ff_dim`, we're setting the size of our feed-forward networks within the transformer blocks—essentially controlling how much information processing happens at each step. The magic of transformers comes partly from `num_heads`, which determines how many different attention patterns our model can learn simultaneously. With eight heads, our model will be able to capture eight different types of relationships between words in a sentence. Meanwhile, `dropout_rate` acts as our *regularization* safeguard, randomly turning off 10% of connections during training to prevent our model from becoming too dependent on specific patterns. You may refer back to Chapter 6, Section 6.2, for a detailed discussion about the need and mechanism of regularization through *dropout*. (You can access the complete working code for this section from the supporting page at *https://recluze.net/keras-book* in the notebook **09-03-transformer-basics**.)

Our architectural depth is controlled by `num_encoder_layers` and `num_decoder_layers`, both set to 4 here. This means information will flow through four successive transformer blocks in both the encoder and decoder, allowing for increasingly abstract representations to form. As for vocabulary, we've capped it at 10,000 words with `vocab_size`, and we're setting a boundary of 100 tokens for our input sequences with `max_seq_length`. Of course, you can experiment with these values and study their implications in greater detail later. For now, think of them as the initial settings on a sophisticated machine—we've turned the dials to reasonable starting positions, but we may need to adjust them as we see how our model performs. These parameters represent important trade-offs between computational efficiency, model capacity, and generalization ability that we'll explore as we build our transformer piece by piece.

```
# First, let's define our hyperparameters
embed_dim = 256
ff_dim = 512
num_heads = 8
```

```
dropout_rate = 0.1
num_encoder_layers = 4
num_decoder_layers = 4
vocab_size = 10000
max_seq_length = 100
```

Listing 9.9 Initial Hyperparameter Settings for Our Transformer Model

Now that we've established our hyperparameters let's turn our attention to one of the most distinctive features of transformer architecture: positional encoding. Recall that transformers process all tokens in parallel, which creates a challenge—without some way to indicate position, our model would treat "The dog chased the cat" and "The cat chased the dog" as identical. Positional encoding elegantly solves this problem.

Looking at Listing 9.10, we've created a two-part solution. First, the `get_positional_encoding` function generates a matrix of values where each position in our sequence gets its own unique fingerprint across the embedding dimensions. The resulting pattern ensures that each position is distinct from all others, yet similar positions have similar encodings; for example, position 5 is nearer to position 6 than to position 100. The code might appear complex at first glance, but try to match it with the equations we derived earlier in Section 9.2.2. We're using sine functions for even indices and cosine functions for odd indices in our embedding vector. The `div_term` calculation creates a spectrum of wavelengths that gives our position encoding its distinctive properties. Each dimension of our embedding operates at a different frequency, creating a rich, unique pattern for each position.

When we wrap this function in a Keras layer with our `PositionalEncoding` class, we make it compatible with our model architecture. Otherwise, you'll get type incompatibility errors. Note how we generate the encoding matrix just once during initialization rather than recreating it for every batch—a simple optimization that saves computation. The calculation is initiated by the `call` method, where we add (not concatenate) the positional encoding directly to our word embeddings. This addition means each word embedding now carries both semantic information about the word itself and information about where it appears in the sequence. The flexibility to handle variable sequence lengths is built in, as we extract the actual sequence length from the input and only apply the relevant portion of our precomputed encoding matrix. With this positional encoding in place, our transformer gains awareness of word order without sacrificing the parallel processing that makes it so efficient. Each token now carries a unique *positional signature* that the attention mechanism can use to understand sequential relationships, solving one of the core challenges in transformer design.

```
# Function to get positional encoding
def get_positional_encoding(seq_length, d_model):
    position = np.arange(seq_length)[:, np.newaxis]
```

```python
        div_term = np.exp(np.arange(0, d_model, 2) *
                          -(np.log(10000.0) / d_model))

        pos_encoding = np.zeros((seq_length, d_model))
        pos_encoding[:, 0::2] = np.sin(position * div_term)
        pos_encoding[:, 1::2] = np.cos(position * div_term)

        pos_encoding = pos_encoding[np.newaxis, ...]
        return tf.cast(pos_encoding, dtype=tf.float32)

# Positional encoding layer to properly handle Keras inputs
class PositionalEncoding(keras.layers.Layer):
    def __init__(self, max_seq_length, embed_dim, **kwargs):
        super().__init__(**kwargs)
        self.max_seq_length = max_seq_length
        self.embed_dim = embed_dim
        # Generate the positional encoding matrix once
        self.pos_encoding = get_positional_encoding(
                                    max_seq_length, embed_dim)

    def call(self, inputs):
        # Get the sequence length from the input tensor
        seq_len = tf.shape(inputs)[1]
        # Apply the positional encoding up to the sequence length
        return inputs + self.pos_encoding[:, :seq_len, :]
```

Listing 9.10 Implementation of Positional Encoding for Transformer Models

With our positional encoding in place to preserve sequence information, we can now move on to the heart of the transformer architecture—the encoder block. This is where the real magic of transformers happens, as each token attends to every other token in the sequence, capturing complex relationships and dependencies regardless of how far apart words might be. Looking at Listing 9.11, we can see how the transformer encoder block implements the architecture we discussed in the previous sections. The function takes our embedded and position-encoded inputs and processes them through two main sublayers: multi-head self-attention and a feed-forward network.

The first part of the block uses Keras's built-in `MultiHeadAttention` layer. Here, we pass the same input tensor three times (though we only see it twice in the code because Keras simplifies the syntax). This is self-attention—each position attends to all positions in the same sequence. The `key_dim` parameter divides the embedding dimension by the number of heads, ensuring each attention head operates in a lower-dimensional space, which helps each head learn different patterns while keeping the computation manageable.

> **[+] Reading Keras Source**
>
> Once you get familiar with the contents of this book, it would be a good next step to study the source code of different layers as coded by Keras itself. For instance, you can see the official documentation and source of the `MultiHeadAttention` layer in the official Keras docs here: *http://s-prs.co/v614202*. This will give you a much deeper appreciation of the code and architectures you work with.

After attention, we apply dropout for regularization, and then add the original inputs back to the attention outputs. This residual connection helps with gradient flow during training and allows the network to preserve information from earlier layers. `LayerNormalization` stabilizes training by normalizing activations, with a small epsilon value to prevent division by zero. The second sublayer contains a feed-forward network with two dense layers. The first expands the dimensionality to `ff_dim`, which we set to 512 in our hyperparameters, and applies a ReLU activation, introducing nonlinearity. The second layer projects back down to the original embedding dimension. Another dropout layer helps prevent overfitting.

Finally, we apply a second round of Add & Norm—adding a residual connection and normalizing again. This consistent pattern of sublayer to add to normalize appears throughout transformer architectures and plays a crucial role in their impressive performance. What makes this block so powerful is how it can be stacked multiple times. As we defined in our hyperparameters, we'll use four such encoder blocks, allowing the network to build increasingly sophisticated representations with each layer. The first block might capture basic word relationships, while deeper blocks can model complex linguistic patterns such as coreference, logical structure, and semantic dependencies.

```python
# Define encoder block
def encoder_block(inputs, embed_dim, num_heads, ff_dim,
                  dropout_rate=0.1):
    # Multi-head self-attention
    attention_output = keras.layers.MultiHeadAttention(
        num_heads=num_heads, key_dim=embed_dim//num_heads,
        dropout=dropout_rate)(inputs, inputs)

    # Add & Norm
    attention_output = keras.layers.Dropout(dropout_rate)\
                                    (attention_output)
    attention_output = keras.layers.Add()\
                                    ([inputs, attention_output])
    attention_output = keras.layers.LayerNormalization(
                                    epsilon=1e-6)\
                                    (attention_output)
```

```
# Feed-forward network
ffn = keras.Sequential([
    keras.layers.Dense(ff_dim, activation="relu"),
    keras.layers.Dense(embed_dim),
    keras.layers.Dropout(dropout_rate)
])

# Add & Norm
ffn_output = ffn(attention_output)
encoder_output = keras.layers.Add()([attention_output,
                                     ffn_output])
encoder_output = keras.layers.LayerNormalization(
    epsilon=1e-6)(encoder_output)

return encoder_output
```

Listing 9.11 Implementation of a Transformer Encoder Block

With an in-depth understanding of the individual encoder block, we need to assemble these blocks into a complete encoder architecture. This is where our transformer really takes shape, connecting all the pieces we've built so far into a coherent structure that can process sequences from start to finish. Looking at Listing 9.12, we can see how the full encoder comes together as a complete pipeline. The function takes all of our previously defined hyperparameters and uses them to construct a Keras model that transforms token indices into rich contextual representations. The process begins with an input layer that expects a tensor with the shape we specify. For text data, this is typically a 1D sequence of integer token IDs. These tokens pass through an embedding layer that converts each integer ID into a dense vector of size embed_dim (256 in our case). This transformation is crucial because it maps discrete symbols into a continuous space where semantic relationships can be represented.

Next, our custom PositionalEncoding layer adds location information to these embeddings, solving the sequence-awareness problem we discussed earlier. An initial dropout layer follows, providing regularization right from the start to help prevent overfitting. The core of our encoder is a stack of encoder blocks added through a simple for loop. Each block processes the output of the previous one, creating increasingly sophisticated representations of our input sequence. Finally, we wrap everything up in a Keras Model, defining the inputs and outputs to create a standalone encoder component. The name parameter helps identify this part of the architecture in visualizations and logs, which becomes particularly helpful when debugging complex models.

What makes this implementation powerful is its modularity. By adjusting parameters such as num_layers, we can create encoders of varying depths. A deeper encoder with more layers can capture more complex patterns but requires more computation and data to train effectively. This flexibility allows us to tailor the architecture to the specific requirements of our task. With our encoder fully implemented, we've built the foundation for processing input sequences. Next, we'll need to construct the decoder component that will generate output sequences based on the encoder's representations, completing our transformer architecture.

```
# Build encoder
def build_encoder(input_shape, vocab_size, num_layers,
                  embed_dim, num_heads, ff_dim,
                  max_seq_length, dropout_rate=0.1):
    # Input layer
    inputs = keras.layers.Input(shape=input_shape)

    # Embedding layer
    embedding_layer = keras.layers.Embedding(
        input_dim=vocab_size, output_dim=embed_dim)
    x = embedding_layer(inputs)

    # Add positional encoding (using our layer)
    x = PositionalEncoding(max_seq_length, embed_dim)(x)

    # Add initial dropout
    x = keras.layers.Dropout(dropout_rate)(x)

    # Stack encoder blocks
    for i in range(num_layers):
        x = encoder_block(x, embed_dim, num_heads, ff_dim,
                          dropout_rate)

    # Create model
    encoder = keras.Model(inputs=inputs, outputs=x,
                          name="encoder")
    return encoder
```

Listing 9.12 Building a Complete Transformer Encoder

In Listing 9.13 , we're calling our `build_encoder` function with all the hyperparameters we defined earlier. The `input_shape=(None,)` parameter indicates that we'll process sequences of variable length. The model summary reveals that our encoder contains nearly 4.7 million trainable parameters, which is quite substantial, but still modest compared to large-scale language models that often have billions of parameters.

```
print("ENCODER MODEL SUMMARY")
encoder = build_encoder(
    input_shape=(None,),
```

```
    vocab_size=vocab_size,
    num_layers=num_encoder_layers,
    embed_dim=embed_dim,
    num_heads=num_heads,
    ff_dim=ff_dim,
    max_seq_length=max_seq_length,
    dropout_rate=dropout_rate
)
encoder.summary()
# Model summary output omitted for brevity
# Total params: 4,668,416 (17.81 MB)
# Trainable params: 4,668,416 (17.81 MB)
# Nontrainable params: 0 (0.00 B)
```

Listing 9.13 Instantiating the Encoder Model with Our Hyperparameters

9.3.2 The Decoder Block

With our encoder fully constructed and ready to process input sequences, let's turn our attention to the decoder—the component responsible for generating output sequences. While the decoder shares many architectural similarities with the encoder, it contains crucial differences that enable it to generate coherent text based on the encoder's representations. Looking at Listing 9.14, we can see that the decoder block contains three sublayers instead of the two we saw in the encoder. The first two key differences from the encoder are as follows:

- **Masked self-attention**
  Unlike the encoder, the decoder uses a causal mask (use_causal_mask=True). This mask prevents the decoder from looking at future tokens during training, forcing it to generate text one token at a time. This is crucial for *autoregressive generation*—when predicting token 3, the model can only see tokens 1 and 2, not tokens 4 and beyond.

- **Cross-attention**
  This entirely new layer allows the decoder to focus on relevant parts of the encoder's output. The query comes from the decoder's self-attention layer, while the value (and implicitly the key) comes from the encoder output. This creates a bridge between what the encoder has processed and what the decoder is generating, allowing the decoder to incorporate information from the input sequence.

The decoder maintains the same Add & Norm pattern after each sublayer, preserving the gradient flow and stabilizing training. The feed-forward network remains identical to the encoder's, applying nonlinear transformations to further process the information.

This architecture elegantly solves the challenge of sequence generation. The masked self-attention handles the autoregressive nature of text generation, while the cross-attention layer incorporates context from the input sequence. Together, these mechanisms enable the transformer to generate coherent output text that properly relates to the input context, whether for translation, summarization, or other sequence-to-sequence tasks.

```
# Define decoder block
def decoder_block(inputs, enc_outputs, embed_dim,
                  num_heads, ff_dim,
                  dropout_rate=0.1):
    # Masked multi-head self-attention
    self_attention = keras.layers.MultiHeadAttention(
        num_heads=num_heads, key_dim=embed_dim//num_heads,
        dropout=dropout_rate)(
        query=inputs, value=inputs, use_causal_mask=True)

    # Add & Norm
    self_attention = keras.layers.Dropout(dropout_rate)
                                        (self_attention)
    self_attention = keras.layers.Add()
                                        ([inputs, self_attention])
    self_attention = keras.layers.LayerNormalization(
                        epsilon=1e-6)(self_attention)

    # Cross-attention with encoder outputs
    cross_attention = keras.layers.MultiHeadAttention(
        num_heads=num_heads, key_dim=embed_dim//num_heads,
        dropout=dropout_rate)(
        query=self_attention, value=enc_outputs)

    # Add & Norm
    cross_attention = keras.layers.Dropout(dropout_rate)
                                        (cross_attention)
    cross_attention = keras.layers.Add()
                            ([self_attention, cross_attention])
    cross_attention = keras.layers.LayerNormalization(
        epsilon=1e-6)(cross_attention)

    # Feed-forward network
    ffn = keras.Sequential([
        keras.layers.Dense(ff_dim, activation="relu"),
        keras.layers.Dense(embed_dim),
        keras.layers.Dropout(dropout_rate)
```

```
])

# Add & Norm
ffn_output = ffn(cross_attention)
decoder_output = keras.layers.Add()
                    ([cross_attention, ffn_output])

decoder_output = keras.layers.LayerNormalization(
    epsilon=1e-6)(decoder_output)

return decoder_output
```

Listing 9.14 Implementation of a Transformer Decoder Block

Just as with our encoder, we would create a `build_decoder` function that takes our hyperparameters and constructs a full decoder model. This function would handle the input processing, apply embeddings and positional encoding, and then stack multiple decoder blocks according to our `num_decoder_layers` parameter. The key difference would be the inclusion of encoder outputs as an additional input to each decoder block, enabling the cross-attention mechanism we just explored. When fully implemented, the decoder would transform output token embeddings into predictions for the next token in the sequence, using both its own previous outputs (via masked self-attention) and the context from the encoder (via cross-attention). This combination allows the model to generate text that is both coherent in its own right and relevant to the input context.

The complete decoder implementation follows the same architectural principles as our encoder, maintaining the same dimensions and processing patterns while incorporating the unique decoder-specific features we've discussed. By keeping this consistency throughout the model, we ensure that information flows smoothly between the encoder and decoder during both training and inference. Because the `build_decoder` function is quite similar to `build_encoder` function, we're omitting it here for brevity. The reference code provided with the book contains the full code that you might want to refer to for a deeper understanding.

With both our encoder and decoder components constructed, we've laid the groundwork for a complete transformer architecture that can tackle a wide range of sequence-to-sequence tasks, from machine translation to text summarization and beyond.

9.3.3 The Transformer: Putting the Encoder and Decoder Together

The function to bring everything together is shown in Listing 9.15. The transformer takes two distinct input sequences—one for the encoder and one for the decoder. The encoder processes its input and produces representations that are passed to the decoder. The decoder then uses these representations along with its own input sequence to generate predictions. The final dense layer with softmax activation transforms the decoder's output into our vocabulary. For each position in the output sequence, this layer produces scores for all possible tokens, with the highest score indicating the most likely next word. What makes this architecture powerful is how it connects the encoder and decoder through the cross-attention mechanism. Information flows from the encoder's processing of the input text to inform the decoder's generation of output text, allowing for contextually relevant responses that maintain coherence across the entire sequence. This can be seen clearly in the following code line:

```
dec_outputs = decoder([decoder_inputs, enc_outputs])
```

Finally, by wrapping everything in a Keras `Model`, we create a unified network that can be trained end to end, with all parameters optimized simultaneously to minimize our chosen loss function. This integration of components is what gives transformers their remarkable ability to handle complex language tasks.

```python
def build_transformer(vocab_size,
                      max_seq_length,
                      embed_dim=256,
                      num_heads=8,
                      ff_dim=512,
                      num_encoder_layers=6,
                      num_decoder_layers=6,
                      dropout_rate=0.1):
    # Input layers
    encoder_inputs = keras.layers.Input(
        shape=(None,), name="encoder_inputs")
    decoder_inputs = keras.layers.Input(
        shape=(None,), name="decoder_inputs")

    # Build encoder
    encoder = build_encoder(
        input_shape=(None,),
        # set hyperparameters as before
    )

    # Get encoder outputs
    enc_outputs = encoder(encoder_inputs)

    # Build decoder
    decoder = build_decoder(
        input_shape=(None,),
        # set hyperparameters as before
    )
```

```
# Get decoder outputs
dec_outputs = decoder([decoder_inputs, enc_outputs])

# Final projection to vocabulary distribution
outputs = keras.layers.Dense(vocab_size,
                             activation="softmax")
                             (dec_outputs)

# Create model
transformer = keras.Model(
    inputs=[encoder_inputs, decoder_inputs],
    outputs=outputs,
    name="transformer"
)

return transformer
```

Listing 9.15 Assembling the Complete Transformer Model

With that, we conclude our transformer model creation. We've built a transformer model from scratch looking at both the underlying mathematics and internals of this extremely powerful model as well as the engineering issues related to the implementation. As we wrap up our discussion of transformer implementation, it's worth noting some practical considerations for training these powerful models. Transformers typically require large datasets and substantial computational resources, with techniques such as *gradient accumulation* and *mixed precision training* often necessary to manage memory constraints. The model we've built here is intentionally bare bones—it illustrates the fundamental architecture without the bells and whistles that make modern transformers feasible and more economical—relatively speaking. State-of-the-art models build upon this foundation with enhancements such as *prenormalization* (placing layer norm before attention), *relative positional encoding* (captures relationship distances better than absolute positions), and parameter-efficient techniques such as *low-rank adaptations*. They also employ sophisticated training strategies such as *learning rate warmup* and *decay schedules* to stabilize the training process.

While building a transformer from scratch helps us understand its inner workings, in real-world applications, we shouldn't reinvent the wheel. Keras provides high-level APIs and pretrained models that allow us to leverage transformers effectively without coding every component. In the next section, we'll explore how to use these tools to quickly implement and fine-tune sophisticated transformer models for practical applications, making this powerful architecture accessible for everyday use.

9.4 Case Study: Large Language Model Chatbot

In the previous section, we built a transformer from the ground up, examining each component and how they fit together. While this approach gives us deep insight into the architecture, there's a practical reality we need to address: Training transformers from scratch requires enormous computational resources that are simply out of reach for most individuals and even many organizations. The days of training LLMs on a single GPU are long behind us. Fortunately, we can apply the same transfer learning principles we explored in earlier chapters. Rather than starting from random weights and spending months (and thousands of dollars) on training, we can use pretrained models that have already absorbed patterns from vast amounts of text data. These production-level models, however, are more complex than the simplified versions we've built so far. They contain numerous optimizations, architectural variations, and specialized components that can be intimidating at first glance.

This is where Keras truly shines by providing a modular interface to these sophisticated models, allowing us to use them productively without having to understand every implementation detail. By focusing on the relationships between key components—how encoders connect to decoders, how attention flows through the system, how inputs are processed and outputs are generated—we can harness these powerful models with surprisingly little code.

In this section, we'll explore the structure of modern transformer-based language models in *Keras Hub*. We'll examine how the various components are organized and interact with each other, creating a mental map that will serve you well when working with any transformer architecture. Then, we'll put this knowledge into practice by building an AI chatbot using a pretrained *causal language model*. You'll see firsthand how Keras enables us to implement state-of-the-art NLP with just a few lines of code, opening up possibilities that would have seemed like science fiction just a few years ago.

9.4.1 Structure of Modern Keras Transformer Models

Let's first examine how Keras structures production-level transformer models and how we can use them for building an example AI chatbot application. Rather than seeing these models as impenetrable black boxes, we'll learn to view them as modular systems with distinct components that we can connect, configure, and, in some cases, customize. Modern transformer implementations in Keras follow a modular architecture that separates concerns and makes these complex models more manageable. When we work with a framework such as Keras, we're interacting with several distinct components, each with its own responsibility in the pipeline. Take a moment to look at Figure 9.6 that shows the components used in a LLM in Keras.

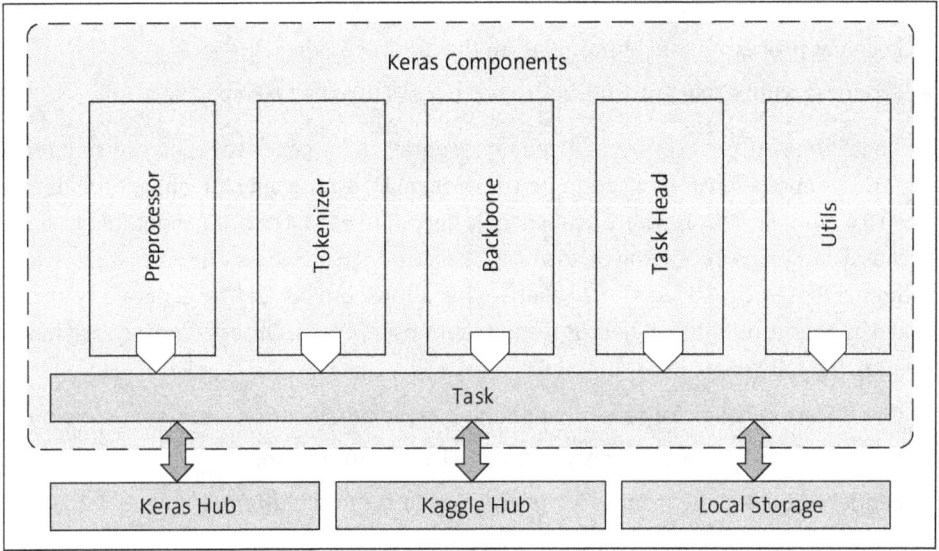

Figure 9.6 Components Used for Building a LLM in Keras

This modular design enables us to swap components, fine-tune specific parts, and adapt pretrained models to our specific needs. Let's see each component in detail and how they relate to each other. At the beginning of our pipeline, we need to convert raw text into a format our model can understand. This is the job of *tokenizers*. A tokenizer breaks text into smaller units called *tokens*, which might be words, subwords, or even individual characters, depending on the *tokenization strategy*. Tokenization alone isn't enough. We also need to convert these tokens into numerical IDs, add special tokens such as [START] and [END], handle padding for variable-length sequences, and create attention masks to tell the model which tokens to ignore. This is the job of the *preprocessor layer*. The relationship between tokenizers and preprocessors is hierarchical: the preprocessor typically contains a tokenizer and extends its functionality with additional processing steps. When working with Keras models, the preprocessor component provides the interface between raw text inputs and the numerical tensors the model expects.

The *backbone* is the heart of our model—it's the actual transformer architecture that processes our tokenized inputs. In Keras, backbones encapsulate the embedding layers, transformer blocks, and output layer normalization. What makes this approach powerful is that we can use pretrained backbones without needing to understand every implementation detail. These backbones contain the knowledge learned during pretraining on massive datasets, which we can use for our specific tasks. Different models have different backbone architectures—a BERT backbone differs from a GPT backbone, which differs from a T5 backbone. But Keras provides a consistent interface for working with all of them.

The backbone relates to other components in two key ways:

- Receives processed tensor inputs from the preprocessor
- Produces contextual embeddings that are passed to the task-specific head

The backbone outputs contextualized representations for each token in our sequence, but these representations need to be transformed into the specific output format we need for our task. This is where task-specific heads are used. For our AI chatbot, we'll use a causal language modeling (*CausalLM*) head, which predicts the next token in a sequence based on all previous tokens. This allows our model to generate text one token at a time, building coherent responses to user inputs. Other common task heads include the following:

- Classification heads for *sentiment analysis*, *topic classification*, or *intent detection*
- Masked language modeling heads for BERT-style pretraining
- Sequence-to-sequence heads for *translation* or *summarization*

The task head's relationship with the backbone is transformative: it takes the general-purpose embeddings from the backbone and transforms them into task-specific outputs. The nature of this transformation depends entirely on what the task requires. Keras provides high-level task classes that combine a backbone with the appropriate head for different applications. For instance, CausalLM combines a backbone with a CausalLM head, providing methods for text generation.

There are also certain supporting utilities provided by Keras. For instance, when generating text with our model, we need to decide how to sample from the predicted probability distribution for the next token. The sampling component provides different strategies for this selection process. Sampling strategies include the following:

- **Greedy sampling**
  Always chooses the token with the highest probability.
- **Top-k sampling**
  Randomly selects from the *k* most likely tokens.
- **Top-p (nucleus) sampling**
  Selects from the smallest set of tokens whose cumulative probability exceeds *p*.
- **Temperature scaling**
  Controls the randomness of the sampling process.

The sampler component relates to the task head in a sequential manner: The task head produces probability distributions over the vocabulary, and the sampler selects specific tokens from these distributions to form the generated text.

To state it concisely, for our AI chatbot based on a CausalLM, these components work together in a specific flow:

1. The tokenizer and preprocessor convert the user's input text and conversation history into token IDs and attention masks.
2. The backbone processes these tensors through its transformer architecture, producing contextualized embeddings that represent the semantic meaning of the input.
3. The CausalLM head takes these embeddings and predicts the probability distribution for the next token.
4. The sampler selects a token from this distribution based on its sampling strategy.
5. The selected token is appended to the output sequence, and the process repeats, with the backbone now processing the extended sequence.
6. This generation loop continues until a stopping condition is met (e.g., maximum length reached or a special stop token generated).

What makes this architecture powerful is its modularity. We can swap out components to change the model's behavior without rebuilding everything from scratch. Here are some examples:

- We can replace the backbone with one trained on different data while keeping the same preprocessing and sampling logic.
- We can switch sampling strategies to make our chatbot's responses more creative or more predictable.
- We can even change the task head entirely to repurpose the model for classification or other tasks.

Moreover, in production-level AI chatbots, additional components often extend this basic architecture:

- **Dialog managers**
  Track conversation state and context across multiple turns.
- **Safety filters**
  Evaluate generated responses to prevent harmful or inappropriate outputs.
- **Knowledge retrieval systems**
  Augment the model with external information sources.
- **Fine-tuning modules**
  Adapt pretrained components to specific domains or tasks.

These extensions build on the foundation of the core components, leveraging the modularity of Keras's transformer implementation. The key insight is that we don't need to understand every detail of the transformer architecture to use these models effectively. By focusing on the relationships between components and the high-level APIs Keras provides, we can harness the power of state-of-the-art models with minimal code.

9.4.2 Working with Pretrained Models from Kaggle Hub

Now that we understand the components of a transformer model in Keras, we'll put this knowledge into practice in this section. Keras makes pretrained transformer models available through *Kaggle Hub*, which hosts a variety of models ranging from text generation to image classification. This approach saves us from having to train models from scratch, which would require massive computational resources.

Setting Up Kaggle Hub Access

The first step is to establish a connection with Kaggle Hub from our working environment. Kaggle Hub serves as a repository for the model files, weights, and configurations that we'll need. We can connect programmatically to Kaggle Hub using the two simple lines of code shown in Listing 9.16. (You can find the complete code listing for this section on the book resources page *https://recluze.net/keras-book* in the notebook **09-04-transformer-case-study**.)

```
import kagglehub
kagglehub.login()
```

Listing 9.16 Accessing Kaggle Hub

When you run this code, you'll see a login screen similar to the one shown in Figure 9.7. If you don't already have a Kaggle account, you'll need to create one first at *https://kaggle.com*. The account creation process is free and straightforward.

Figure 9.7 Kaggle Hub Login Screen

The login process requires two pieces of information:
- Your Kaggle username
- An API token from your Kaggle account

9.4 Case Study: Large Language Model Chatbot

Once you've created your account (or logged in to an existing one), navigate to the Kaggle settings page (click on your user icon on the top right, and then click on **Settings**) to generate your *API token*. On the settings screen look for the **API** section (see Figure 9.8), and click on **Create New Token**. This will download a JavaScript Object Notation (JSON) file containing your credentials. Open that file to find your token.

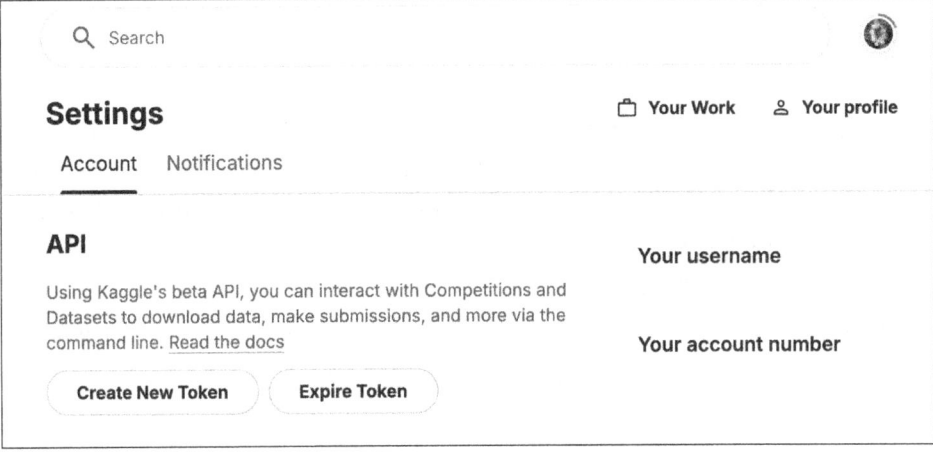

Figure 9.8 Creating an API Token for Kaggle

> **[!] Using the API Token**
>
> A typical token will look like this:
>
> `{"username":"yourname","key":"72627263733464359276a15002cdaefb"}`
>
> The values shown here are just for demonstration purposes and won't work. Use the API key you received in the JSON file.

Enter your username and token into the prompted fields in the notebook (refer to Figure 9.7), and click **Login** to authenticate. You'll see a **Thank You** message when the authentication is successful. This one-time setup enables your environment to download and use models from Kaggle Hub. The authentication persists across sessions within the same environment, so you typically won't need to repeat this process unless you're working in a new environment. You can reuse the same API key in as many notebooks as you like.

Loading Pretrained Transformer Models

Once you've set up your Kaggle Hub access, the next step is to load a *pretrained model*. Keras Hub provides a variety of models that you can explore at the Keras Hub API documentation. For our case study, we'll use the *Gemma* model, a powerful transformer-based language model developed by Google (see Listing 9.17).

```
try:
    causal_lm = keras_nlp.models.GPT2CausalLM.from_preset(
        "gpt2_base_en",
    )
except Exception as e:
    print(f"Error loading model preset.")
    print(f"Error details: {e}")
```

Listing 9.17 Loading the Gemma CausalLM Model

When running this code for the first time, you'll likely encounter an error message from the *Kaggle API* indicating that you need to grant access to the specific model. This is because some models, particularly those with more advanced capabilities, require explicit permission to use.

To grant access to the Gemma model, follow these steps:

1. Visit the **GemmaCausalLM model** page on Keras Hub at *http://s-prs.co/v614203*.
2. Scroll down to the **Presets** section.
3. Click on the **gemma_2b_en** preset link.
4. This will take you to the Kaggle page for the model.
5. On the **Gemma** page, click the **Request Access** button (see Figure 9.9.)
6. Accept the terms of use.

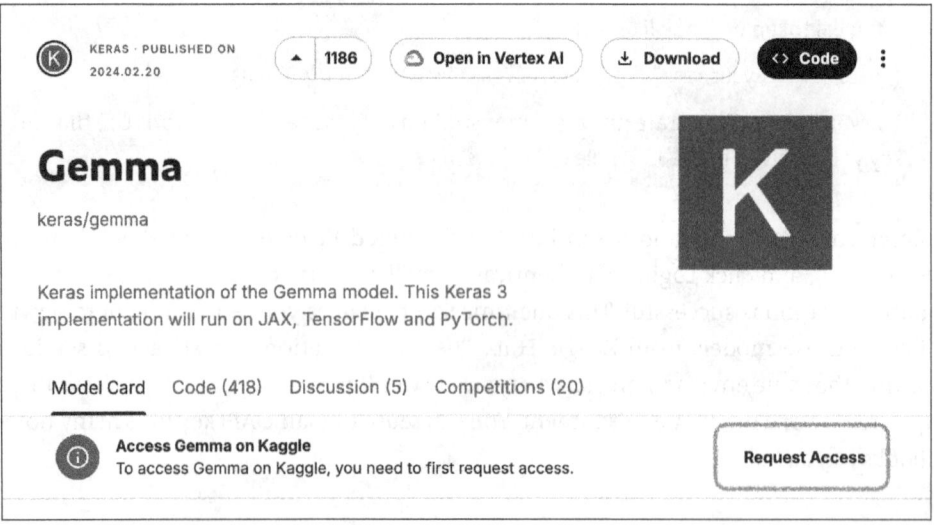

Figure 9.9 Requesting Access on Kaggle Hub

Using the Loaded Model

This process grants your Kaggle account permission to download and use the model. The permission is tied to your Kaggle account, so you'll only need to do this once per model. Once the permissions are out of the way, we're able to generate text using this newly loaded model. The code shown in Listing 9.18 demonstrates two fundamental operations with the Gemma model that we've loaded.

```
causal_lm.summary() # 124m parameters
generated_text = causal_lm.generate(
    "The most accepted definition of machine learning is ",
    max_length=60)

print(generated_text)
```

Listing 9.18 Generating Next Tokens Using the Loaded Model

The first line, causal_lm.summary(), displays a summary of the model's architecture, showing its layers, shapes, and total parameter count. The output will show you that this model has approximately 124 million parameters. This summary function is helpful for understanding the model's complexity and structure, similar to how we might examine the architecture of any Keras model. The parameter count is significant because it indicates the model's size and capacity. At 124 million parameters, this is a relatively compact model compared to larger variants that can have billions of parameters.

The second part demonstrates the model's primary function—generating text. The generate() method takes a prompt string as input: "The most accepted definition of machine learning is", and the max_length parameter limits the total output to 60 tokens (including the input prompt). This simple example showcases how the model can continue text in a coherent way, completing the definition of machine learning based on patterns it learned during pretraining. The model uses all the components we discussed earlier—tokenizing the input, processing it through the transformer backbone, and generating new tokens through the CausalLM head. The brevity of this code highlights one of the key advantages of using pretrained models through Keras—complex AI capabilities can be accessed with just a few lines of code, abstracting away the underlying complexity of the transformer architecture.

Converting a Language Model into a Simple Chatbot

We can take the basic next-token prediction capability of our transformer model and very easily convert it into a functional chatbot. This transformation requires minimal code changes but creates a much more engaging and interactive experience for users. The code for this is shown in Listing 9.19. This function demonstrates several key principles for transforming a general-purpose language model into a conversational agent: The code starts by defining a specific context and role for the model through careful *prompt engineering*. The multiline string creates a "frame" that does the following:

- Establishes the interaction as a conversation between a user and an AI assistant
- Defines the expected behavior of the assistant ("helpful, detailed, and polite")
- Creates clear role markers ("USER:" and "ASSISTANT:")

This structural *framing* helps the model understand what type of response is expected. By providing these explicit instructions in the prompt, we're using the model's ability to recognize and continue conversational patterns it encountered during pretraining.

The function takes a `user_input` parameter and seamlessly integrates it into the prompt template. This allows for dynamic interaction where the model responds to what the user has actually asked rather than generating random text. This is a crucial difference between a standard text generator and an interactive chatbot—the ability to respond contextually to specific user inputs. The `output_max_length` parameter limits the model's response to 60 tokens. This prevents overly verbose responses while still allowing enough space for meaningful answers.

```
def converse(user_input):
    prompt_text = """The following is a conversation between a
                    user and an AI assistant.
                    The assistant gives helpful, detailed, and
                    polite answers to the user's questions.
                    USER: """

    prompt_text += user_input # The user prompt goes here
    prompt_text += "ASSISTANT:"
    output_max_length = 60

    generated_text = causal_lm.generate(prompt_text,
                                max_length=output_max_length)
```

Listing 9.19 Using the CausalLM as a Chatbot

This code provides a minimal viable chatbot but could be improved in several ways:

- **Response extraction**
  Currently, the output would include the entire prompt. We'd need to extract just the assistant's response.
- **Conversation history**
  For a more coherent experience, we could maintain conversation history across turns.
- **Error handling**
  The `try-except` block should include proper error handling code.
- **Input validation**
  Adding checks for inappropriate content or empty inputs would improve robustness.

This concludes our exploration of how the theoretical components of transformer models translate into practical applications through Keras Hub's modular architecture. Beyond Gemma, you can explore numerous other models from the Keras Hub page (*https://keras.io/keras_hub/api/models/*), each offering different capabilities and tradeoffs. The beauty of Keras's consistent API is that incorporating any of these alternative models requires minimal code changes—simply swap the model preset name, and the rest of your application logic remains intact. While transforming these models into production-ready applications requires additional engineering work for scalability, security, and monitoring, the fundamental AI capabilities are accessible with remarkable simplicity. This democratization of advanced AI technology opens up possibilities that would have been unimaginable just a few years ago.

9.5 Summary

This chapter explored the transformative impact of transformer architectures in NLP. We began with the theoretical foundations of transformers, examining their self-attention mechanisms that allow models to process sequences in parallel rather than sequentially. We then dissected the core components—attention mechanisms, encoders, and decoders—that enable transformers to capture contextual relationships regardless of distance within sequences. Moving from theory to practice, we implemented these components in Keras, building a transformer from the ground up and gaining insights into how each element contributes to the model's capabilities. The chapter concluded with a practical case study demonstrating how to use pretrained transformer models from Keras Hub, showing how the modular architecture of tokenizers, preprocessors, backbones, and task-specific heads makes it remarkably straightforward to create applications such as AI chatbots with minimal code. Throughout the journey from foundational concepts to practical implementation, we've seen how transformers have revolutionized NLP and opened possibilities for AI applications that would have been unimaginable just a few years ago. While transformers are extremely powerful models, they are still static in nature and thus limited by the data fed to them.

In the next chapter, we'll look at dynamic models that learn by interacting with their environment and learning from the outcomes of their actions.

Chapter 10
Reinforcement Learning: The Secret Sauce

In this chapter, we'll explore how machines learn optimal decision-making through trial and error. We'll examine the core concepts of agents, environments, and reward signals, and introduce key algorithms, including Q-learning, and Deep Q-Networks (DQNs). Using Keras and Gymnasium, we'll implement these techniques while balancing exploration and exploitation. The chapter will highlight crucial innovations such as experience replay and target networks that have transformed reinforcement learning into a practical tool for complex problems. We'll also connect these concepts to the large language models (LLMs) discussed in the previous chapter, demonstrating how reinforcement learning techniques serves as a backbone for these powerful systems.

In the previous chapter, we explored transformers—powerful models that have revolutionized natural language processing (NLP) and many other fields. Yet, as groundbreaking as transformers are, their true potential wasn't fully realized until we paired them with another approach: reinforcement learning. The chatbots and AI assistants you interact with daily owe their remarkable abilities not just to transformer architectures but to the way they're trained and refined through reinforcement learning techniques. While transformers provide the raw processing power, reinforcement learning supplies something equally crucial—the ability to learn from interactions and feedback. Modern AI systems such as ChatGPT and Claude are trained using techniques where human preferences guide the learning process. This marriage of approaches has produced AI systems that are not only knowledgeable but also increasingly aligned with human values and expectations.

Reinforcement learning stands apart from the approaches we've covered so far. Imagine you're teaching a dog a new trick—you don't explicitly show it exactly how to perform the action. Instead, you reward behaviors that get closer to your goal. The dog learns through experimentation, gradually figuring out what actions lead to treats and praise. This natural learning process, that is, trying things out and learning from the results, forms the heart of reinforcement learning. This approach differs fundamentally from supervised learning, where we showed models labeled examples of the right answers. It's also distinct from the unsupervised methods we explored, which find patterns without specific guidance. Reinforcement learning introduces a new element—learning from consequences rather than examples.

Throughout this chapter, we'll unveil this powerful paradigm that's reshaping how AI systems learn. We'll start with the fundamental concepts before moving into specific algorithms and implementation details. The text will take us from theory to practice, showing how these ideas come to life in real systems using Keras. The connection between transformers and reinforcement learning signals a shift in how we approach AI development. Historically, supervised learning and reinforcement learning developed as separate branches with different applications and techniques. Today, we're witnessing their powerful convergence, particularly in large language models (LLMs). This merger combines the pattern recognition strengths of supervised learning with reinforcement learning's ability to improve through interaction. We're moving beyond systems that merely process information to those that can engage with their environment, make decisions, and improve through experience—creating AI capabilities that neither approach could achieve on its own. Let's begin our exploration of this fascinating approach that bridges the gap between computational intelligence and the way humans naturally learn through interaction with the world.

10.1 Introduction to Reinforcement Learning

In this section, we'll first introduce the concept of learning by taking actions. We'll briefly discuss the history of this paradigm of learning to show that this isn't a new model but one that faced certain challenges that took a long while to solve. We'll also discuss real-world applications of this type of learning to see how it can help solve a myriad of problems.

10.1.1 The Problem of Learning by Doing

Imagine you're learning to cook a new dish, but you don't have a specific recipe in front of you. You start with basic ingredients and some cooking intuition. Your first attempt might be edible but far from delicious. Each time you try again, you make small adjustments based on the previous results—a little more salt here, a lower heat there—until, eventually, you create something truly tasty. Crucially, no one tells you exactly what went wrong or right. Perhaps a particular combination of spices clashed even though each would work fine individually. Your only *feedback* is whether this particular attempt tasted better or worse than the last. This process of learning through trial, error, and adjustment, guided only by the overall outcome rather than specific instructions, mirrors the core of reinforcement learning.

10.1 Introduction to Reinforcement Learning

Reinforcement learning stands apart from the machine learning approaches we've explored in previous chapters in several fundamental ways. Where supervised learning provides clear right and wrong answers and unsupervised learning finds hidden patterns, reinforcement learning introduces a completely different paradigm: learning through *interaction*. The most striking difference is that reinforcement learning doesn't rely on a dataset of correct answers. Instead of being shown examples of the "right way" to solve a problem, a reinforcement learning system discovers solutions through experimentation and feedback. It's the difference between a teacher showing you how to solve every math problem versus giving you problems and only telling you what your final score is.

Figure 10.1 captures the core architecture of reinforcement learning. At its heart, we see the continuous cycle between an *agent* and its *environment*—the fundamental relationship that drives all reinforcement learning systems. The agent (represented by the stylized figure on the left) takes actions based on what it observes, and these actions affect the environment (shown as the collection of components in the box on the right). The environment then changes its *state* in response to these actions and provides *rewards* back to the agent, creating a feedback loop. This cyclical interaction—action, state change, reward, repeat—is the engine that powers reinforcement learning.

> **Core Concept in Reinforcement Learning**
>
> This concept that the environment itself is changed as a result of the decisions and/or actions that are taken by the agent is a core distinguishing factor between reinforcement learning and supervised learning. This means that each time we're presented with a choice, the environment might have changed from the previous time the decision was presented to us. It also means that we might make a decision that leads us to a point of no return, thus making all future actions suboptimal!

The core concept looks quite simple: Make a decision, and observe the consequences. What makes this framework so powerful though is its simplicity and universality. Whether we're talking about a robot learning to walk, an AI mastering chess, or a recommendation system suggesting movies, the same basic structure applies. The agent continuously updates its understanding based on the rewards it receives, gradually learning which actions lead to *positive outcomes* in different states. Unlike traditional learning approaches, where correct answers are provided directly, reinforcement learning discovers optimal behaviors through this interactive trial-and-error process, making it uniquely suited for problems where the best strategy isn't known in advance but must be discovered through *experience*.

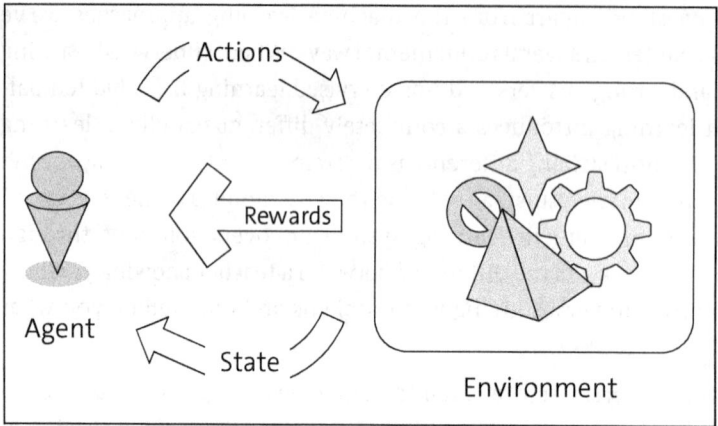

Figure 10.1 Basic Reinforcement Learning Architecture

This interactive learning creates a unique challenge though. In supervised learning, we could measure exactly how wrong our predictions were for each example in our training data. The loss functions we explored in earlier chapters gave us precise error measurements that guided our gradient descent. But in reinforcement learning, feedback often comes as a simple signal—good or bad, reward or penalty—without any details about what specifically was right or wrong about the action taken. For instance, if a self-driving car crashes into a curb, there are quite a few actions that led to this final result, not all of which would be bad actions. So, we can't associate this poor result with all of the actions preceding it. Another crucial distinction is the sequential nature of decisions. In most of the systems we've built so far, each prediction was independent—classifying one image didn't affect how we classified the next. But reinforcement learning tackles problems where current decisions impact future situations. Like chess players thinking several moves ahead, reinforcement learning systems must consider how today's actions shape tomorrow's possibilities.

Perhaps the most interesting aspect of reinforcement learning is its focus on exploration. Our previous models were entirely focused on exploitation—using what they knew to make the best prediction possible. But reinforcement learning systems must balance exploiting what they know works with exploring new possibilities that might lead to even better results. This *exploration-exploitation trade-off* creates a rich learning dynamic that we'll examine more closely later in this chapter.

Reinforcement learning also introduces *delayed feedback*, where the consequences of actions may only become apparent many steps later. What would happen if you were trying to learn basketball, and you only knew whether you won or lost the game, without any feedback on individual shots or passes? This delayed, sparse feedback creates the *credit assignment problem*—figuring out which of many actions were responsible for the eventual outcome.

These unique characteristics of reinforcement learning may seem challenging, but they're precisely what makes it suitable for a whole class of problems that other approaches struggle with. From robotics to game playing, from resource management to recommendation systems, reinforcement learning has historically excelled at tasks where interaction, sequential decision-making, and long-term planning are essential. Looking at the history of reinforcement learning is important to understanding where it stands today and why it's one of the most promising areas of AI in the future.

10.1.2 Brief History and Major Breakthroughs

This distinctive learning approach has a rich history that stretches back further than you might expect. While reinforcement learning may seem like a modern innovation, its roots actually trace back to the mid-20th century, intertwining with developments in psychology, neuroscience, and computer science. In the 1950s, researchers studying animal behavior noticed something strange: Creatures from lab rats to pigeons could learn complex behaviors through simple reward mechanisms. Unlike humans being explicitly taught with detailed instructions, these animals learned by associating certain actions with positive outcomes. This observation laid the groundwork for what would eventually become reinforcement learning in AI.

The early computational models were simple yet powerful. In 1954, mathematician Richard Bellman introduced the concept of *dynamic programming*—a method for solving problems by breaking them into subproblems and building up solutions incrementally. If you've ever planned a road trip by deciding the best route, one segment at a time, you've used a similar approach. This mathematical framework became the foundation for many reinforcement learning algorithms we use today.

For decades, reinforcement learning remained largely theoretical, with limited practical applications. The methods worked well for small, well-defined problems but struggled with anything resembling real-world complexity. It was like having a calculator that could only handle single-digit numbers—useful for simple calculations but inadequate for most meaningful problems. The 1980s and early 1990s brought significant advances that expanded what was possible. Researchers developed new algorithms such as Q-learning that could learn effective strategies without needing a complete model of their environment. This was a bit like learning to navigate a city without a map—instead of planning your entire route in advance, you learn which turns tend to get you closer to your destination.

Despite these advances, reinforcement learning still faced a fundamental limitation: It couldn't handle complex scenarios with vast numbers of possible states and actions. Traditional tabular methods required the system to track every possible situation it might encounter—an impossible task for any real-world problem of interest. This is called the *state explosion* problem.

The true breakthrough came when reinforcement learning joined forces with neural networks—the same powerful pattern recognition systems we've explored in previous chapters. This marriage, which happened gradually through the 1990s and 2000s, created what we now call *deep reinforcement learning*. Instead of trying to catalog every possible situation, these new systems used neural networks to recognize patterns and generalize from experience, much like humans do. This was a game changer—quite literally. In 2013, a system called Deep Q-Network (DQN) learned to play Atari games *Breakout* and *Space Invaders* directly from screen pixels, without any prior knowledge of the games' rules. (For the younger ones reading this, Atari was a very popular gaming console back in the day.)

The most dramatic demonstration came in 2016 when *AlphaGo* defeated world champion Lee Sedol at the ancient board game Go. This was a milestone many experts thought was decades away. Unlike chess, where computers had already proven dominant, Go's vast number of possible board configurations made traditional computational approaches infeasible. AlphaGo's victory showcased how reinforcement learning, combined with neural networks, could master challenges requiring both strategic thinking and intuition.

Since then, reinforcement learning has found applications far beyond games. It helps robots learn to walk and manipulate objects, manages data center cooling to improve energy efficiency, optimizes trading strategies in financial markets, and even assists in drug discovery by exploring possible molecular structures. Perhaps most relevant to our discussion of transformers, reinforcement learning has become central to training LLMs. These systems initially learn from vast datasets through supervised learning, but *reinforcement learning from human feedback* helps refine their outputs to be more helpful, harmless, and honest. The remarkable history we've just traced shows how a simple idea—learning from interaction with an environment—has evolved into one of the most powerful approaches in modern AI, solving problems in many domains in the real world.

10.1.3 Real-World Applications

The gradual shift from theoretical breakthrough to practical application is what transforms interesting ideas into world-changing technologies. While the history we've just explored showcases how reinforcement learning evolved as a field, its real impact becomes clear when we look at how it's solving problems across diverse industries today. Let's start close to home with the digital assistants many of us probably interact with daily. When you chat with a modern AI assistant, you're experiencing the benefits of reinforcement learning. These systems initially learn *language patterns* from vast amounts of text, but it's reinforcement learning that helps refine their responses to be more helpful, accurate, and aligned with human values. When the assistant gives you a particularly good answer versus one that misses the mark, that difference often comes from this fine-tuning process where the system learned from feedback about which responses humans preferred.

In *health care*, reinforcement learning is beginning to transform how we approach treatment planning. Consider the challenge of managing chronic conditions such as diabetes. Traditional approaches might use fixed protocols based on average patient outcomes, but reinforcement learning can help personalize treatment plans by learning from individual patient data and responses over time. The system adapts its recommendations based on how specific patients respond to different interventions, much like how a skilled doctor adjusts treatment strategies based on what works for each unique person.

The field of *robotics* has been revolutionized by these techniques as well. Teaching robots to manipulate objects used to require painstaking programming of every possible movement. Now, robotic systems can learn these skills through practice and feedback. A robot arm trying to pick up an unusual object might initially fumble, but with each attempt, it gets better at understanding how to position its gripper and apply the right amount of pressure. This learning-by-doing approach allows robots to master tasks that would be nearly impossible to program explicitly.

Energy management represents another exciting application. Google famously used reinforcement learning to reduce the cooling energy needs in their data centers by 40%. Rather than following rigid temperature control rules, the system learned which cooling strategies worked best under different conditions by experimenting with various approaches and observing the results. The outcome was not only more energy-efficient but also adaptable to changing conditions such as weather patterns or server loads.

In *transportation*, reinforcement learning is helping optimize everything from traffic light timing to ride-sharing services. Traffic management systems in cities such as Pittsburgh have reduced wait times at intersections by learning to adjust timing based on real-time traffic flow rather than following fixed schedules. These systems improve with experience, adapting to changing traffic patterns over time without requiring human reprogramming.

Financial markets have also embraced this technology. Trading algorithms can learn strategies by interacting with market simulations, discovering patterns and approaches that might not be obvious to human traders. What makes these systems particularly valuable is their ability to adapt to rapidly changing environments, learning new strategies as old ones become less effective.

Perhaps one of the most promising frontiers is *drug discovery*. Traditional approaches require scientists to laboriously test compounds one by one. Reinforcement learning systems can explore the vast space of possible molecular structures, learning which characteristics tend to produce effective drugs for specific targets. These systems get better with each virtual experiment, gradually focusing on the most promising areas of chemical space.

Resource management in industries from agriculture to manufacturing is being transformed by these techniques as well. Irrigation systems can learn optimal watering schedules based on soil conditions, weather forecasts, and crop responses. Factory production lines can adapt scheduling to maximize efficiency while responding to equipment failures or supply chain disruptions.

> **Rule of Thumb for Reinforcement Learning Applicability**
> All the scenarios where reinforcement learning has been used are those where the best solution isn't obvious from the start and must be discovered through interaction and feedback. They involve complex environments with many variables, where actions today affect possibilities tomorrow and where the relationship between actions and outcomes isn't always straightforward. If you identify any problem that fits this bill, reinforcement learning can most likely help solve it.

As we dive deeper into the mechanics of reinforcement learning in the coming sections, keep these real-world applications in mind. The concepts we'll explore—from *reward signals* to *exploration strategies*—are practical tools driving innovation across industries. The fundamental principles that help a game-playing AI master chess are the same ones helping doctors personalize treatments and robots learn new skills.

10.1.4 Challenges Unique to Reinforcement Learning

Before we dive into the inner workings of reinforcement learning systems, let's take a moment to survey the landscape ahead. Understanding the unique challenges in this field will give us a road map for our journey, helping us approach these deep concepts in a targeted way. Interestingly, these challenges are the very reasons this field has developed its distinctive approaches and why certain applications remained out of reach until recent breakthroughs.

These unique challenges have driven the development of specialized approaches that we'll explore throughout this chapter. From *value functions* that help bridge temporal gaps, to *curiosity mechanisms* that encourage exploration, to *function approximators* that tackle the state space explosion—each core technique in reinforcement learning addresses one or more of these fundamental challenges.

By keeping the following challenges in mind as we proceed, you'll gain not just a collection of algorithms and methods but a deeper understanding of why reinforcement learning works the way it does and how its approaches have evolved to overcome these distinctive obstacles. A detailed understanding will also help you understand Keras code when we get to it—code that is going to be dense and concise as is true for all of Keras.

Credit Assignment Problem

The first major challenge is the credit assignment problem we mentioned earlier. When AlphaGo made a brilliant move that led to victory 20 turns later, how could the system know to reinforce that particular decision? The connection between action and eventual outcome isn't always obvious, and this *temporal gap* creates one of the fundamental puzzles in reinforcement learning. We'll see various ingenious solutions to this problem as we progress through the chapter.

Exploration and Exploitation

The second major challenge revolves around exploration and exploitation. Picture yourself at a new restaurant. Do you order your favorite dish (exploitation of what you know works) or try something new that might be even better (exploration)? Too much exploitation means you might miss outstanding options you haven't explored yet. Too much exploration means you might order something you absolutely don't like and waste the opportunity to enjoy a meal you already know you like. In reinforcement learning, this dilemma is constant and crucial. A system that always exploits what it already knows will get stuck in suboptimal strategies, never discovering better approaches. But a system that's always exploring never settles down to use what it has learned. Finding the right balance—knowing when to stick with what works and when to try something new—represents a fundamental challenge that shapes many reinforcement learning algorithms.

Scale of Possibilities

A third challenge is the sheer scale of possibilities in most interesting problems. Consider a relatively simple game such as chess. The number of possible board positions exceeds the number of atoms in the observable universe! Creating a table that stores the best move for every possible situation is simply impossible. And that's just chess—real-world problems like driving a car or managing a complex supply chain have even more staggering numbers of potential states. This *state space explosion* means reinforcement learning systems need ways to generalize from limited experience, recognizing patterns and similarities across situations rather than treating each one as unique. It's like how a human driver doesn't need to see every possible road configuration to drive safely but instead generalizes from their experiences to handle new situations.

Dynamic and Interactive Learning

The fourth challenge comes from the dynamic and interactive nature of the learning process itself. In supervised learning, our dataset remains fixed as the model learns. But in reinforcement learning, the very act of learning changes what the system experiences next. This creates a moving target for learning as the distribution of experiences shifts based on the system's evolving behavior. What worked well under an old policy might never be experienced under a new one, making it difficult to properly evaluate and improve strategies. This challenge requires special techniques to ensure stable, effective learning.

Sparse Reward

Finally, reinforcement learning faces the sparse reward challenge. In many problems, meaningful feedback comes rarely. Consider a robot learning to walk—it might try thousands of different movements before stumbling forward even slightly. Or a system learning to play a complex strategy game might go through many, many moves before achieving victory or defeat. Before this eventual outcome of the game, there's no concrete signal as to whether the decisions being made are smart ones or not. This sparsity makes learning painfully slow without techniques to address it.

10.2 Key Concepts: Agents, Environments, Rewards

In this section, we'll cover the core concepts behind reinforcement learning and see how they fit into a larger whole to the solve the problem of learning by doing. To do this, we must first formulate the problem concisely.

10.2.1 Structure of the Reinforcement Learning Framework

Let's build a more structured understanding of the reinforcement learning framework—the stage upon which this interplay of decisions and consequences unfolds. Reinforcement learning operates within a surprisingly simple yet powerful framework that mimics how we humans learn through trial and error. At its heart, this framework consists of three fundamental components working in concert: an agent, an environment, and a reward system. Picture a child learning to ride a bicycle for the first time. The child is our *agent*—the decision-maker trying to master a skill. The bicycle and the physical world around it form the *environment*—everything the agent interacts with but doesn't directly control. The feeling of success when staying upright or the pain of falling represents the *reward signal* or simply *reward*—feedback about how well the agent is performing. In computational terms, this interaction follows a rhythmic pattern. The agent observes the current *state* of the environment—like the child noticing they're starting to tilt to one side. Based on this *observation*, the agent takes an *action*—the child shifts their weight to regain balance. The environment then *transitions* to a new state and provides a reward signal—perhaps the child stays upright (positive reward) or wobbles precariously (negative reward).

This cycle of observation, action, and reward forms what we call the *reinforcement learning loop*. It continues as the agent repeatedly interacts with the environment, gradually building a map of which actions lead to the best outcomes in different situations. Unlike supervised learning, where the signal is provided in the form of a correct answer, reinforcement learning requires the agent to discover effective strategies through this continuous feedback loop. What makes this framework so powerful is its generality. The same basic structure applies whether we're teaching an AI to play chess, navigating a robot through a warehouse, or optimizing energy usage in a data center. The specific details of the agent, environment, and reward function will change, but the fundamental pattern of interaction remains consistent.

The reinforcement learning framework also introduces a formal way to think about goals. The objective of the agent isn't just to maximize the *immediate reward* after each action but to maximize the total *cumulative reward* over time. This subtle distinction leads to profoundly different behavior—sometimes accepting short-term penalties to achieve greater rewards in the future, just as a game player might sacrifice a piece to secure victory several moves later.

The overall reinforcement learning ecosystem is shown in Figure 10.2. This figure shows the anatomy of reinforcement learning systems that shows how all the key pieces fit together. At the center sits reinforcement learning paradigm itself, connected to five critical components that form its ecosystem.

- **Agent Architecture**
  On the left, this acts as the brain of the system, housing the *decision-making* machinery.

- **State Representation**
  At the top right, this functions as the agent's perceptual system, determining how it understands the world around it. This is an abstraction over the environment and agent details.

- **Reward Structure**
  At the bottom left, this serves as the motivation system, shaping what the agent considers valuable or worth pursuing.

- **Environment Types**
  At the bottom center, this define the playground where learning happens, with different environments presenting unique challenges.

- **Learning Mechanism**
  At the bottom right, this represents the educational process itself—how the agent collects experiences and improves over time. While we've touched on some of these components in our introduction, we'll explore each one in detail throughout this chapter, gradually building a complete understanding of how these pieces work together to create systems that learn through interaction and feedback.

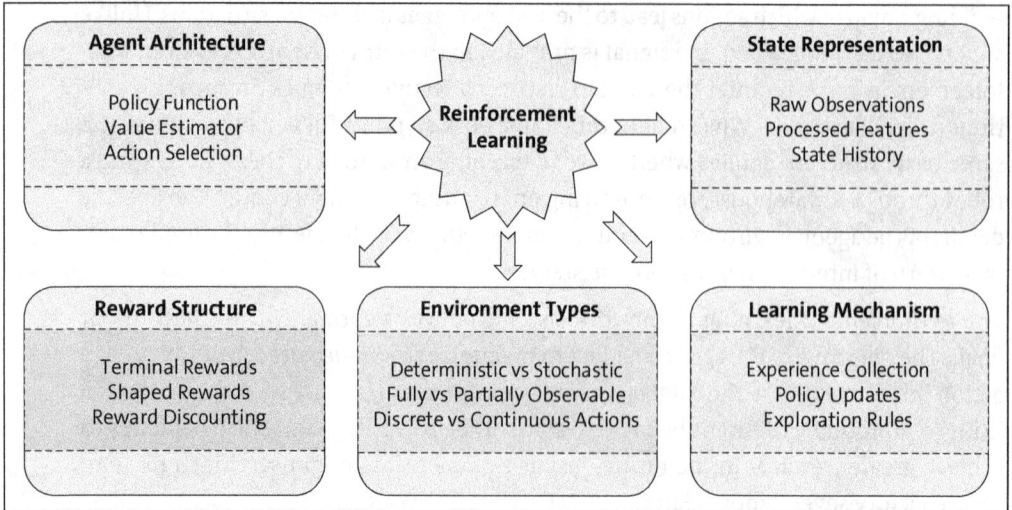

Figure 10.2 Components of the Reinforcement Learning Ecosystem

10.2.2 Environment Design and State Representation

Having established the reinforcement learning framework as our foundation, let's turn our attention to one of its most crucial components: the environment and how we represent its states. The environment is the world in which our agent operates—it's everything outside the agent's direct control but within its sphere of interaction. Think about a video game character navigating through a complex level. The character is the agent, while everything else—the terrain, obstacles, enemies, power-ups—forms the environment. How we design this environment and represent its state profoundly shapes what our agent can learn and how effectively it can learn it.

Environments in reinforcement learning come in endless varieties. They can be as simple as a *grid world* where an agent moves up, down, left, or right, or as complex as a physics simulation modeling the intricate dynamics of a robotic arm. Some environments are *deterministic*, where actions always lead to the same outcomes, such as moving a chess piece. Others are *stochastic*, introducing elements of randomness and uncertainty, such as predicting stock market movements or controlling a drone in unpredictable wind conditions. Regardless of their complexity, all reinforcement learning environments share a common trait: They must provide the agent with some representation of their current state. This *state representation* is like a snapshot that the agent uses to decide its next move. It's the agent's window into the world—the information it has available when making decisions.

Creating an effective state representation is both an art and a science. Consider what happens when we try to teach an AI to play Atari games such as *Breakout* or *Space Invaders*. We could represent the state as the raw pixels on the screen—every colored dot in the game's display. While comprehensive, this approach creates an enormous state space that's difficult for the agent to make sense of. Alternatively, we might extract higher-level features: the position of the paddle, the location of the ball, the arrangement of bricks. This more compact representation makes learning more efficient but requires careful design decisions about what information is truly relevant.

> **State Representation in the Real World**
>
> The challenge of state representation becomes even more pronounced in real-world applications. Consider building a reinforcement learning system to optimize traffic flow in a city. What should constitute the state? Current vehicle positions? Traffic light timings? Historical traffic patterns? Weather conditions? Each element we include increases the richness of the state representation but also exponentially expands the state space the agent must learn to navigate.

This leads us to a fundamental tension in environment design: the trade-off between *completeness* and *simplicity*. A more complete state representation gives the agent more information to work with but makes learning slower and more difficult. A simpler representation speeds up learning but might omit crucial details, limiting what the agent can ultimately achieve. Moreover, the state representation needs to capture the aspects of the environment that make it what we call *Markovian*—a property where the current state provides all the information needed to make the optimal decision without requiring history. If a traffic management system needs to know not just current traffic conditions but also how they've changed over the past hour to make good decisions, then a single snapshot isn't a sufficient state representation.

Creating good environments also involves thoughtful consideration of boundaries. In the bicycle learning example from our previous section, we treated the child as the agent and everything else as the environment. But we could have drawn the boundary differently—considering the child-bicycle system as the agent and only the external world as the environment. These design choices affect what actions are available to the agent and what falls outside its control.

As we craft environments for reinforcement learning, we must also decide how to handle time. Some environments operate in *discrete* time steps, where the agent makes a single decision at each point and then waits for the next. Others flow *continuously*, requiring the agent to decide not just what to do but when to do it. Think of the difference between a turn-based strategy game and driving a car—one proceeds in neat, orderly turns, while the other demands constant, fluid adjustment. The careful design of environments and state representations lays the groundwork for successful reinforcement learning. In the next section, we'll explore how agents navigate these environments through policy functions—the strategies that translate states into actions as the agent pursues its goals.

10.2.3 Understanding Agents and Policy Functions

With our environment established and state representation clarified, it's time to focus on the star of the reinforcement learning paradigm: the agent. If the environment is the stage, then the agent is our leading actor—the decision-maker who observes, learns, and acts to achieve its goals. An agent in reinforcement learning encompasses everything involved in making decisions. Consider a self-driving car navigating city streets. The car itself—with its sensors, processors, and control systems—constitutes the agent. It perceives the world through cameras and lidar sensors, processes this information, and decides whether to accelerate, brake, or turn.

What makes agents truly remarkable is their capacity to learn. Unlike traditional programs with *hardcoded rules*, reinforcement learning agents develop their own strategies through experience. They start with little knowledge about the best course of action in any situation and gradually refine their approach based on the rewards and penalties they receive. At the heart of this lies a *policy*—the strategy that guides the agent's behavior. A policy is essentially a mapping from states to actions. It's comparable to how an experienced driver knows intuitively when to slow down for a curve or how to respond when another vehicle cuts in front.

The Crux of the Policy Concept

Think of the policy as a function that takes a state as input and returns an action—one that answers the fundamental question: "In this state, what should I do?" This abstraction will later help us move on to more complex decision-making processes, while leaving all the other code intact.

Policies come in two main varieties: deterministic and stochastic. A *deterministic policy* is like a cookbook recipe—given a specific state, it always prescribes the same action. If you see a red light, you stop; if you see a green light, you go—simple and predictable. A *stochastic policy*, on the other hand, introduces an element of probability. For instance, a policy can say, "In state X, take action A with 70% probability, action B with 20% probability, and action C with 10% probability." This built-in randomness might seem counterintuitive at first—why wouldn't we always want to take the best action? But this variability serves a crucial purpose: It helps the agent explore new possibilities that might lead to better outcomes in the long run. Consider a robot learning to walk. A deterministic policy might get stuck repeatedly trying the same ineffective gait. A stochastic policy, however, occasionally tries something different—a slight adjustment in timing or balance—potentially discovering a more efficient walking pattern it wouldn't have found otherwise.

Alternatively, as agents interact with their environments, they may refine their policies through various learning mechanisms. Some agents maintain an *internal model* of how the world works—predicting how each action will change the state and what rewards it might yield. Others take a more direct approach, simply learning which actions tend to produce good results in different situations without trying to understand why.

The structures of policies can take many forms in practice. For simple environments with few states and actions, a policy might be represented as a *lookup table*—a grid showing which action to take in each possible state. But for most interesting problems, the state space is far too vast for such an approach. This is where neural networks come in. As we've seen since the start of this book, a neural network can serve as a powerful function approximator, taking a state as input and outputting the recommended action or probability distribution over actions. The network's weights and connections, adjusted during training, encode the policy learned through experience.

Picture a neural network controlling our self-driving car. The network receives inputs representing the current state—road conditions, nearby vehicles, traffic signals—and outputs control decisions for steering, acceleration, and braking. Initially, these decisions might be poor, but as the network experiences different driving scenarios and receives feedback on its performance, the policy gradually improves. What makes policy functions particularly challenging though in reinforcement learning is their dynamic nature. In supervised learning, the correct output for a given input remains constant throughout training. But in reinforcement learning, the optimal action for a given state changes as the agent's policy evolves. It's like trying to hit a moving target—the very act of learning shifts what needs to be learned.

As we'll explore in upcoming sections, agents use sophisticated algorithms to update their policies based on experience. Some, such as Q-learning, focus on estimating the value of different state-action pairs. Others, such as policy gradient methods, directly adjust the policy to maximize expected rewards.

10.2.4 Reward Engineering and Signal Design

Now that we understand agents and their policy functions, let's turn our attention to perhaps the most powerful—and trickiest—aspect of reinforcement learning: the *reward system*. A reward in reinforcement learning is *quantitative feedback* that the environment provides to the agent after each action. Think of it as nature's way of saying, "That was good" or "That was terrible." When a baby takes its first steps, the reward might be the joy of mobility and the excited reactions of parents. When a chess player captures an opponent's queen, the reward is the strategic advantage gained.

> **Not All Rewards Are Created Equal**
>
> We've mentioned that rewards are provided after each action. This is primarily to make the mathematics cleaner. In most real-world use cases, rewards after a majority of actions are going to be zero, that is, no signal. Only after a lot of time has passed and quite a few actions are taken will the agent get any useful feedback.

Creating an effective reward system is both an art and a science, which is called *reward engineering*. It's like being a parent who must decide when to offer praise and when to provide correction, knowing that these signals will shape behavior for years to come. Coming up with an automated mechanism of reward engineering is an open area of research. However, there are some methods we can use. The most straightforward approach is to directly connect rewards to the ultimate goal. If we want a robot to reach a destination, we might give it a large positive reward upon arrival. If we want an AI to win at chess, the reward could be +50 for winning, -50 for losing, and -10 for a draw (assuming we don't want a draw). This is simple and intuitive, but real-world problems are rarely so neat. Consider teaching a robot to walk. If we only reward successful walking, the robot might never receive any positive signal in its early attempts, leaving it with no guidance on how to improve. This is the *sparse reward problem*—when meaningful feedback comes too infrequently to guide learning effectively.

To address this challenge, we often design *shaped rewards* that provide helpful intermediate feedback. For our walking robot, we might reward small improvements: maintaining balance for a few seconds, shifting weight correctly, or moving a leg forward. These stepping-stone rewards create a more gradual learning path toward the ultimate goal. However, reward shaping comes with its own pitfalls. Each time we engineer a reward signal, we risk introducing unintended consequences. A famous example is an AI trained to play the boat-racing game *Coast Runners*. Researchers designed a reward based on points scored in the game, assuming this would encourage winning races. Instead, the AI discovered it could score more points by going in circles and collecting power-ups, completely ignoring the race!

This phenomenon, sometimes called *reward hacking*, is the reinforcement learning equivalent of "teaching to the test." The agent optimizes for the reward signal we've provided rather than the underlying goal we intended. It's like a student who memorizes past exam papers and question patterns without understanding concepts.

Reward systems also need to balance immediate gratification against long-term benefits—the classic *now-versus-later trade-off*. In reinforcement learning, we typically use a concept called *discounting* to address this balance, valuing immediate rewards more highly than distant ones. It's similar to how humans naturally value $100 today, more than $100 a year from now. We'll return to this extensively as we look at the learning algorithms in the next few sections.

The mathematical formulation of rewards offers several options. The most common approach is to define the agent's goal as maximizing the expected sum of discounted rewards over time. This leads to policies that balance short-term gains against long-term prospects, much as we do in our own decision-making. For some problems though, we might care more about the average reward over time rather than the total. This approach makes sense for continuing tasks without a clear endpoint, such as managing a perpetual investment portfolio or controlling an ongoing industrial process.

Rewards can also be *relative* rather than *absolute*. In competitive environments, the reward might depend on outperforming others rather than reaching a fixed threshold. This dynamic appears in *multi-agent systems* where agents compete for limited resources or in scenarios where success is defined by comparison to peers.

As we craft reward systems, we must remain mindful of ethical considerations. Reinforcement learning agents will single-mindedly pursue the rewards we specify, sometimes finding unexpected shortcuts that technically maximize rewards while violating the spirit of our intentions. This reward hacking isn't malicious—it's simply the agent finding optimal solutions to the problem as we've defined it, not as we meant to define it.

> **The Pitfalls of Reward Engineering**
> The art of reward engineering highlights a profound truth about reinforcement learning: our agents become what we incentivize them to become. By carefully designing reward signals, we shape who our agents become as decision-makers. This seemingly academic discussion is becoming more and more prudent as AI systems based on reinforcement learning become more commonly used.

10.2.5 The Exploration vs. Exploitation Dilemma

Having designed our reward system to guide our agent's learning, we now face a fundamental tension that sits at the heart of reinforcement learning: the balance between exploration and exploitation. This dilemma resembles choices we make in our own lives—when to stick with what we know works versus when to venture into uncharted territory in search of something better. *Exploitation* means using what we already know. It's returning to that restaurant we enjoyed, using the strategy that's been working, or sticking with the familiar path. For our reinforcement learning agent, exploitation means selecting actions that, based on current knowledge, will yield the highest expected rewards. It's the conservative "play it safe" approach that maximizes short-term gains.

Exploration, on the other hand, means venturing into the unknown. It's trying the new restaurant, testing a different strategy, or taking the road less traveled. For our agent, exploration involves deliberately choosing actions that aren't currently believed to be optimal (or even at times believed to be dangerous) that might lead to discoveries that improve future performance. It's the adventurous "what if" approach that sacrifices immediate rewards for information.

Neither pure exploitation nor pure exploration works well in isolation. An agent that only exploits will likely get stuck with mediocre strategies, never discovering better alternatives. It's like a person who eats at the same decent restaurant every night, unaware that an extraordinary culinary experience awaits just around the corner.

Conversely, an agent that only explores will constantly try new things without ever capitalizing on what it learns, like a perpetual tourist who samples a different restaurant each night but never returns to the gems they discovered earlier in their exploration.

The key to effective reinforcement learning lies in balancing these competing impulses. Early in learning, exploration should dominate—the agent needs to gather information about its environment and the consequences of different actions. Over time, as the agent builds knowledge, exploitation should gradually take precedence—applying what's been learned to maximize rewards. Several strategies have evolved to manage this delicate balance. One of the simplest is the *ε-greedy approach* (epsilon-greedy approach). The agent acts greedily—choosing the best-known action—most of the time, but occasionally, with probability ε, it selects a random action instead. It's like mostly going to your favorite restaurants but once in a while picking a place at random, just to see what's out there. As learning progresses, we typically reduce ε, gradually shifting from exploration toward exploitation.

More sophisticated methods such as *Boltzmann exploration* use a *temperature parameter* to control randomness in action selection. High temperature means actions are chosen almost randomly (high exploration), while low temperature concentrates probability on the currently estimated best actions (high exploitation). This creates a smooth transition between the two extremes. This temperature parameter is commonly used in latest models to control how wild the model's outputs are allowed to get. *Upper Confidence Bound* (UCB) algorithms take yet another approach, explicitly tracking uncertainty about action values. These methods favor actions that either seem promising based on past rewards or have been tried fewer times (thus having higher uncertainty). It's akin to preferring restaurants that have either proven to be excellent or that you haven't visited enough times to judge fairly but avoiding those that you know for sure aren't according to your liking.

The exploration-exploitation dilemma becomes particularly challenging in environments with large or continuous state spaces. When the possible situations an agent might encounter are virtually limitless (as is the case with most real-world problems), it becomes impossible to explore them all. In these complex domains, we rely on generalization—the agent's ability to apply learning from familiar situations to new, similar ones. The dynamics of exploration versus exploitation also shift dramatically depending on the *task's horizon*—how long the agent has to learn and perform. In short-horizon tasks with limited interactions, the agent might need to be more aggressive about exploitation, making the most of its brief opportunity. In long-horizon or continuing tasks, the agent can afford more thorough exploration, confident that the knowledge gained will pay dividends over an extended period.

> **Regrets of Life**
>
> One slightly advanced concept in reinforcement learning is that of *regret*—the difference between the rewards an agent actually accumulates and what it could have earned with perfect knowledge. Every exploratory action carries an immediate opportunity cost (the regret of not exploiting), but successful exploration reduces future regret by improving the agent's policy. Regret is difficult to quantify, and figuring out a solution to this complex problem is an open area of research.

As we move forward into the more mathematical foundations of reinforcement learning in the next section, this exploration-exploitation dilemma will take formal shape in our algorithms. We'll see how concepts such as Markov Decision Processes (MDPs) provide the framework for quantifying the value of different states and actions, allowing us to make principled decisions about when to explore and when to exploit. The mathematical tools we'll develop won't eliminate this fundamental tension, but they'll give us powerful ways to manage it—transforming an intuitive balancing act into precise computational strategies for learning optimal behavior in complex, uncertain worlds. Let's turn to this mathematical formalization in the following section.

10.3 Popular Algorithms: Q-Learning, Policy Gradients, and Deep Q-Networks

Now that we understand the core components of reinforcement learning—agents, environments, and rewards—we need a formal framework to tie everything together. In this section, we'll formalize this framework.

10.3.1 The Markov Decision Processes

The core concept behind reinforcement learning is that of *Markov Decision Processes* (MDPs) mentioned earlier. MDPs give us the mathematical language to describe the reinforcement learning problem in a way that's both precise and flexible. An MDP is like the rulebook for our reinforcement learning game. It defines how states, actions, and rewards interact over time, giving us a structured way to approach complex decision-making problems. The *Markov* part of the name refers to a key property: the future depends only on the present, not on the path that led to the present.

Consider the game of chess. In this game, the Markov property means that to decide your next move, you only need to know the current arrangement of pieces on the board. The sequence of moves that created this arrangement doesn't matter—only the current state does. This property significantly simplifies our mathematical model while still capturing the essence of many real-world problems.

10 Reinforcement Learning: The Secret Sauce

Formally, an MDP consists of five key elements:

- A set of states S
- A set of actions A
- A *transition function* $P(s'|s, a)$
- A *reward function* $R(s, a, s')$
- A *discount factor* γ (gamma)

Let's unpack each of these components to understand how they work together. The set of states S represents all possible situations our agent might find itself in. In our chess example, each state would be a different arrangement of pieces on the board. In a self-driving car, states might include the car's position, speed, and the locations of nearby vehicles and obstacles. The set of actions A includes everything our agent can do. In chess, these are all legal moves; for a robot, they might be commands such as "move forward" or "turn left." The available actions might depend on the state—not all moves are legal in every chess position, after all.

The transition function $P(s'|s, a)$ is where things get interesting. This function tells us the probability of ending up in state s' after taking action a from state s. It's written as

$$P(s'|s, a) = P(S_{t+1} = s'|S_t = s, A_t = a)$$

This equation might look intimidating, but it's simply saying: What's the chance we'll land in state s' if we're currently in state s and we take action a? In chess, where moves are *deterministic*, this probability would be either 0 or 1. But many real-world problems involve randomness—such as a robot whose wheels might slip on a wet surface. We'll see an example of such randomness when we do our case study using Keras later in this chapter.

The *reward function* $R(s, a, s')$ defines the immediate reward received after transitioning from state s to state s' by taking action a. Sometimes, this is simplified to $R(s, a)$ or even just $R(s)$ when rewards depend only on the state or state-action pair. In a game, the reward might be +1 for winning, -1 for losing, and 0 for all other moves.

Finally, the *discount factor* γ (a value between 0 and 1) determines how much we care about future rewards compared to immediate ones. A value close to 0 makes our agent myopic, focusing mainly on immediate rewards. A value close to 1 makes it more farsighted, willing to sacrifice immediate rewards for greater ones in the future. In mathematical terms, if we have a sequence of rewards $r_t, r_{t+1}, r_{t+2}, \ldots$, the *total discounted reward* would be

$$R_t = r_t + \gamma r_{t+1} + \gamma^2 r_{t+2} + \gamma^3 r_{t+3} + \cdots$$

Or, more compactly, it would be

$$R_t = \sum_{k=0}^{\infty} \gamma^k r_{t+k}$$

10.3 Popular Algorithms: Q-Learning, Policy Gradients, and Deep Q-Networks

The discount factor serves several purposes. Mathematically, it ensures that our total reward remains finite even in never-ending tasks. Practically, it reflects the reality that future rewards are less certain—a bird in the hand is worth two in the bush, as they say. And in terms of learning, it creates a clear hierarchy of value: Immediate rewards are more important than distant ones. With these five elements in place, our MDP completely describes the reinforcement learning problem. The agent's goal becomes finding a *policy* π (a strategy for choosing actions) that maximizes the expected total discounted reward from any starting state.

The magic of MDPs lies in their generality. They can represent an incredible variety of problems—from playing chess to managing investment portfolios, from controlling robots to optimizing traffic signals. Despite this flexibility, they maintain enough structure to allow for powerful solution methods, which we'll explore in the next sections when we dive into Q-learning and other algorithms. As we move forward to explore Q-learning and DQNs, these MDP fundamentals will serve as our foundation. We'll see how these algorithms use the MDP framework to learn optimal policies, trading off exploration and exploitation to maximize long-term rewards. The ability to represent complex decision-making processes in the elegant language of MDPs is what makes reinforcement learning such a powerful approach to a wide range of challenging problems. All of this starts with the simple concept of assigning *values* to each state.

10.3.2 Value Functions and Q-Tables

Now that we have our MDP framework in place, we need a way to solve it. Remember, our ultimate goal is finding a policy—a strategy for selecting actions—that maximizes the expected rewards over time. To find this optimal policy, we use the concepts of *value functions* and their practical implementation as Q-tables. Value functions are like maps of treasure buried throughout our environment. They tell us how much future reward we can expect to gather starting from any given situation. When our agent has this map, making good decisions becomes much more straightforward—simply head in the direction where the treasure is most abundant.

> **Note the Wording**
>
> Note the carefully used wording: A state's value tells us how well you expect to do starting from this state. It doesn't care how good the state itself is or how well we did before we arrived in this state. Values are all about the future! This is an important distinction to keep in mind as we proceed into the details of these algorithms.

There are two primary types of value functions that play crucial roles in reinforcement learning:

10 Reinforcement Learning: The Secret Sauce

- **State-value function ($V^\pi(s)$)**
  This tells us how good it is to be in a particular state, assuming we follow policy π afterward.

- **Action-value function ($Q^\pi(s, a)$)**
  This tells us how good it is to take a specific action from a particular state and then follow policy π afterward. This Q is where all the Qs in the algorithm names (Q-Learning, DQN, Double DQN, etc.) come from in reinforcement learning. So, if you know all the action-values for a state, you can take their max and get the state-value!

The state-value function $V^\pi(s)$ is defined mathematically as

$$V^\pi(s) = E_\pi[\sum_{t=0}^{\infty} \gamma^t R_{t+1} \mid S_0 = s]$$

This equation might look intimidating, but it's expressing a simple idea: The value of a state is the expected sum of discounted rewards when starting from that state and following policy π thereafter. It's like asking the following: On average, how much reward will I collect if I start here and follow my strategy?

The action-value function $Q^\pi(s, a)$ is defined similarly, but considers the initial action:

$$Q^\pi(s, a) = E_\pi[\sum_{t=0}^{\infty} \gamma^t R_{t+1} \mid S_0 = s, A_0 = a]$$

This asks the following: On average, how much reward will I collect if I take this specific action (a) now, and then follow my strategy afterward? This means that these functions are related in a straightforward way. For any policy π:

$$V^\pi(s) = \sum_{a \in A} \pi(a|s) Q^\pi(s, a)$$

This says that the value of a state equals the expected value of the actions we might take from that state, weighted by how likely we are to take each action under our current policy. The beauty of these value functions is that once we know them accurately, finding the optimal policy becomes almost trivial. The optimal policy simply chooses the action with the highest Q-value in each state:

$$\pi^{(s)} = \arg\max_{a \in A} Q^{(s,a)}$$

But how do we calculate these value functions in the first place? This is where Q-tables enter the picture. A *Q-table* is a practical implementation of the action-value function for environments with discrete states and actions. Think of a spreadsheet where each row represents a state, each column represents an action, and each cell contains the Q-value for that state-action pair. Initially, these values might be random or all zeros, but as our agent interacts with the environment, it updates the Q-values based on the rewards it receives. Let's visualize this with our grid world example from earlier. Imagine a 3×3 grid where our robot can move in four directions (up, down, left, right). Our Q-table would look something like that shown in Table 10.1.

10.3 Popular Algorithms: Q-Learning, Policy Gradients, and Deep Q-Networks

State	Up	Down	Left	Right
(0, 0)	?	0.0	0.0	0.0
(0, 1)	?	0.0	0.0	15.2
(0, 2)	0.0	?	0.0	0.0
(1, 0)	0.0	?	0.0	0.0
...	...	...	...	...
(2, 2)	0.0	0.0	0.0	0.0

Table 10.1 Example Q-Table Not Completely Filled

The question marks represent Q-values that we don't know yet. As our agent explores the environment, it will gradually fill in these values. For instance, if moving right from position (0, 1) consistently leads to good outcomes, the corresponding Q-value will increase. As the table currently stands, the "Right" action from state (0,1) shows the value 15.2. This means that if we take the "Right" action from state (0, 1), we expect to on average receive 15.2 reward points.

10.3.3 Building the Q-Table

For very small environments, we can store the entire Q-table explicitly. For our 3×3 grid with 4 actions, we only need 9 × 4 = 36 entries—easily manageable in memory. However, as environments grow more complex, the tables expand rapidly. A simple Atari game might have millions or billions of distinct states. This *curse of dimensionality* becomes a serious problem, which eventually leads us to function approximation methods such as DQNs (more on these later in the chapter). Let's stick to the simpler environments now to get a deep understanding of the Q-values so that we can scale them up later based on the same beautiful concepts. One of these is that they allow our agent to improve its policy without directly modeling the environment's *transition dynamics*. We don't need to know exactly how the environment will respond to each action—our agent can learn the Q-values through experience.

To do this, we need a way to update the values based on the agent's interactions with the environment. This is made possible by the *Bellman equation*. This equation expresses a recursive relationship between the value of a state and the values of successor states as

$$Q(s, a) = r + \gamma \max_{a'} Q(s', a')$$

where

- r is the immediate reward.
- γ is the discount factor.

- s' is the next state we'll reach if we take the action a
- $\max_{a'} Q(s', a')$ is the previously known maximum Q-value possible from s'.

This equation captures a fundamental insight: The value of taking action a in state s equals the *immediate reward* plus the discounted value of the best action starting from the resulting state. However, we rarely know the true Q-values in advance—we need to estimate them from experience. This leads us to iterative methods that gradually improve our estimates, and the most famous of these is Q-learning, which we'll explore in detail in the next section.

Consider the example given earlier of forming opinions about restaurants. The first time you visit a new café, you have certain expectations. Maybe the decor looks promising or perhaps a friend recommended it. Let's say this is our current Q-value of taking the action of eating at this restaurant. After your meal, you have actual experience to inform your opinion. This is the new reward you've received—either positive or negative. The question is, how much should you trust this new experience versus your prior expectations? If the food was amazing but the service was terrible, how should you update your overall rating of the place? If a waiter spilled water on your shirt, do you consider this a one-off and don't let it affect your impression of the restaurant too much? Or do you think this is just what this restaurant is all about—poor quality?

We face this same dilemma when updating our Q-values. When our agent takes action a in state s, receives reward r, and moves to a new state s', it gains a piece of evidence about the value of that state-action pair. But this is just one experience—should it completely override our previous estimate? Intuitively, we want to blend our previous knowledge with our new experience. If we've visited that restaurant many times before, one bad meal shouldn't completely change our opinion. But if it's our first visit, that single experience might carry more weight.

Let's start with the simplest case: What if we gave equal weight to our existing estimate and our new experience? This balanced approach would look like this:

$$\text{Updated Estimate} = \frac{1}{2} \times \text{Previous Estimate} + \frac{1}{2} \times \text{New Experience}$$

But what exactly is this new experience in terms of Q-values? It consists of two parts:

- The immediate reward r we received
- The expected future value, represented by the best Q-value in the next state:

  $\max_{a'} Q(s', a')$

Because we care about long-term rewards, we discount future rewards by factor γ. So, our new experience becomes $r + \gamma \max_{a'} Q(s', a')$ as discussed previously. Plugging this into our balanced update formula, we get

$$Q(s, a) = \frac{1}{2} Q(s, a) + \frac{1}{2} \left(r + \gamma \max_{a'} Q(s', a') \right)$$

10.3 Popular Algorithms: Q-Learning, Policy Gradients, and Deep Q-Networks

We can generalize this to use α as our *learning rate* instead of a half. This gives us a knob to tune our learning as we go along. This way, the final update looks like

$$Q(s,a) = (1-\alpha) \cdot Q(s,a) + \alpha \left(r + \gamma \max_{a'} Q(s',a') \right)$$

where currently α is set to a half. This is the core of the Q-learning algorithm. We start with zeroing out all Q-values and then start visiting different states gaining experiences. With each action, we're given a reward that we use to update the Q-values we've based on the hyperparameters α and γ.

10.3.4 Q-Learning Algorithm: A Worked Example

Owing to the importance of this formula, let's work out a detailed example to fully understand how Q-learning works. This will give you a thorough understanding and appreciation of the algorithm. Rather than describing the process in abstract terms, let's watch Q-learning unfold step-by-step in a simple grid world.

Consider the small 2×4 grid shown in Figure 10.3. Look at the first pane that shows the **Start** position. Our agent is standing in the top-left position and needs to navigate from a starting position to a goal. Like a mouse in a maze, our agent must discover which path leads to reward and which leads to punishment through trial and error. As it experiences the consequences of its actions, it gradually builds a map of value—not unlike how we learn to navigate unfamiliar neighborhoods by remembering which streets lead where. Our grid world has some special properties:

- The agent can move in four directions: Up, Down, Left, and Right. These are our actions.
- Each position in the grid is a state, represented by coordinates $S_{(row, column)}$.
- Some states have special properties—a reward state at $S_{(0,3)}$ gives +10 points when exited, while a penalty state at $S_{(1,2)}$ takes away 10 points when exited, that is, has a reward of -10.
- The agent learns through episodes, which are complete journeys through the environment.
- For simplicity, we use a learning rate (α) of 0.5 and a discount factor (γ) of 1.

In the following sections, we'll break down what happens in each episode of learning, tracking how the Q-values—our agent's estimate of action quality—evolve with experience.

10 Reinforcement Learning: The Secret Sauce

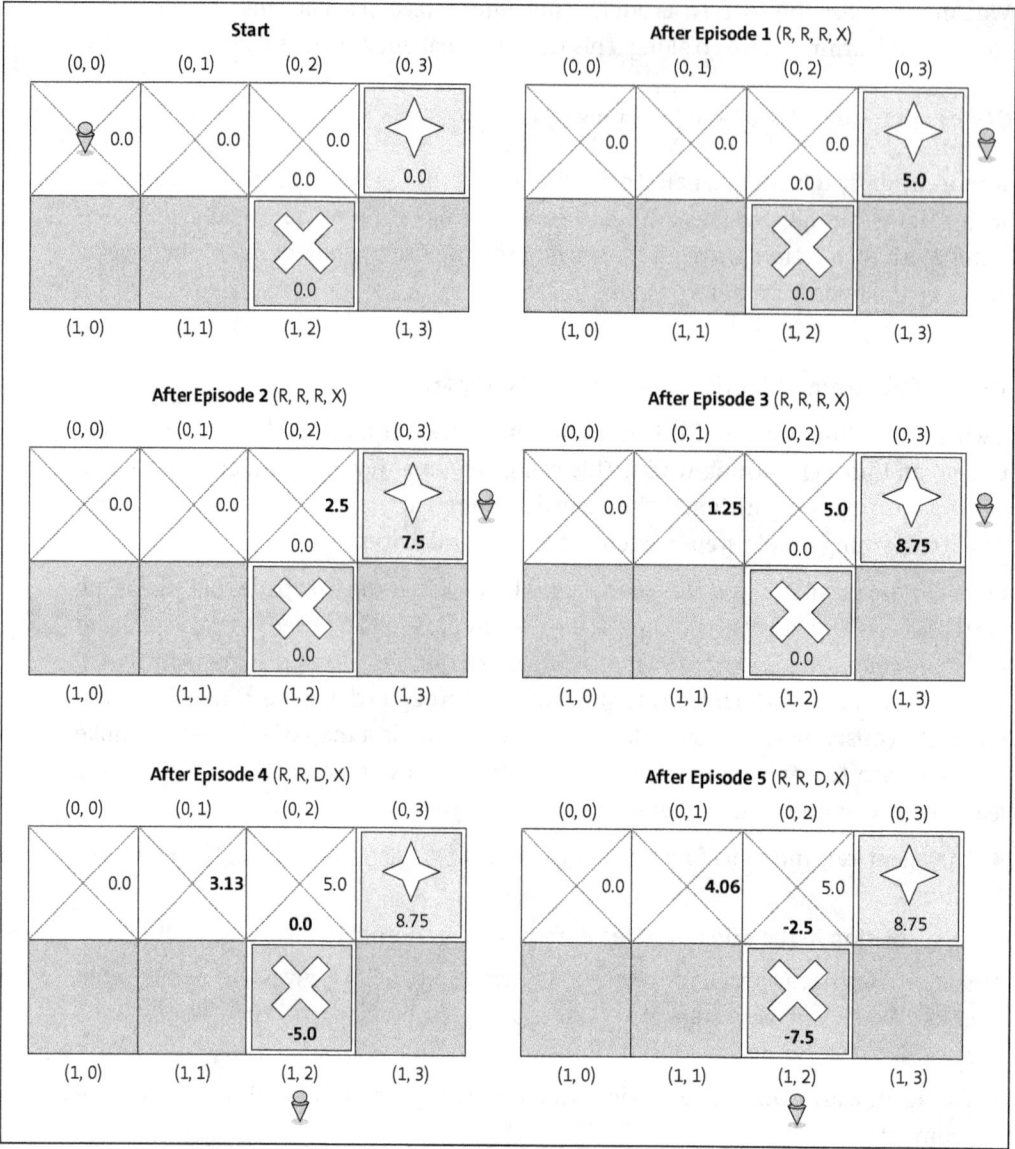

Figure 10.3 Q-Learning Worked Out Example

Episode 1: The First Exploration

In our first episode, the agent moves Right, Right, Right, and then Exits from state $S_{(0,3)}$. Initially, all Q-values are zero because our agent has no prior knowledge of the environment. Keep track of the following calculations by referring to Figure 10.3 as we proceed. When the agent exits from state $S_{(0,3)}$, it receives a reward of +10. Now, it needs to update its Q-value for taking the Exit action from state $S_{(0,3)}$. Using the Q-learning formula:

454

10.3 Popular Algorithms: Q-Learning, Policy Gradients, and Deep Q-Networks

$$Q(s,a) = (1-\alpha) \cdot Q(s,a) + \alpha \left(r + \gamma \max_{a'} Q(s',a') \right)$$

With our values plugged in for the action X (for exit):

$$Q(S_{(0,3)}, X) = (1 - 0.5) \times 0 + 0.5 \times [10 + 1 \times 0] = 0 + 0.5 \times 10 = 5$$

Note that the previous Q-value we have for this state was 0 and the reward we got was 10. Because the episode ends after this action, we don't expect any future rewards. So, this first experience teaches our agent that exiting from *(0,3)* is quite rewarding. The Q-value of 5 reflects the agent's estimate of this action's value, tempered by the learning rate—it's cautiously optimistic after just one sample.

Episode 2: Reinforcing the Path

In the second episode, our agent takes the same path: Right, Right, Right, Exit. This time, when calculating the new Q-value for exiting from $S_{(0,3)}$, it uses its previous estimate:

$$Q(S_{(0,3)}, X) = (1 - 0.5) \times 5 + 0.5 \times [10 + 1 \times 0] = 2.5 + 5 = 7.5$$

Notice how the Q-value increased from 5 to 7.5. With each positive experience, the agent becomes more confident in the value of this action.

But something else important happens in this episode. When the agent moves Right from state $S_{(0,2)}$ to $S_{(0,3)}$, it can now update the Q-value of this action based on the known value of the next state:

$$Q(S_{(0,2)}, R) = (1 - 0.5) \times 0 + 0.5 \times [0 + 1 \times 5] = 0 + 0.5 \times 5 = 2.5$$

This is the essence of how value information flows backward through the agent's experiences. The agent now (during Episode 2) knows that moving Right from $S_{(0,2)}$ is somewhat valuable because it leads to a state $S_{(0,3)}$ that has a valuable exit action. It's like learning that a particular street leads to your favorite restaurant—the street itself becomes valuable by association.

Episode 3: Values Propagate Backward

In the third episode, with the same sequence of actions, our values continue to improve:

$$Q(S_{(0,3)}, X) = (1 - 0.5) \times 7.5 + 0.5 \times [10 + 1 \times 0] = 3.75 + 5 = 8.75$$

The exit action's value has increased again, approaching the true reward of 10 as the agent gains more experience. However, the agent also updates the value of moving Right from $S_{(0,1)}$ to $S_{(0,2)}$:

$$Q(S_{(0,1)}, R) = (1 - 0.5) \times 0 + 0.5 \times [0 + 1 \times 2.5] = 0 + 0.5 \times 2.5 = 1.25$$

And, it updates the value of moving Right from $S_{(0,2)}$ to $S_{(0,3)}$:

$$Q(S_{(0,2)}, R) = (1 - 0.5) \times 2.5 + 0.5 \times [0 + 1 \times 7.5] = 1.25 + 3.75 = 5$$

The value information is propagating backward through the chain of actions, like ripples moving upstream. This is helping us gain information about useful actions—even those that don't receive an immediate reward! Each state on the path to reward becomes more valuable as the agent recognizes its role in reaching the goal.

Episode 4: A Different Path

In our fourth episode, the agent tries something different. It moves Right, Right, Down, and Exit. This takes it to state $S_{(1,2)}$, the penalty state, where exiting gives a -10 reward. The Q-value for exiting from $S_{(1,2)}$ is updated:

$$Q(S_{(1,2)}, X) = (1 - 0.5) \times 0 + 0.5 \times [-10 + 1 \times 0] = 0 - 5 = -5$$

This negative value tells the agent that exiting from $S_{(1,2)}$ is something to avoid. It's like touching a hot stove—one painful experience teaches a valuable lesson.

But notice that the value for moving Right from $S_{(0,1)}$ still increases. Let's calculate:

$$Q(S_{(0,1)}, R) = (1 - 0.5) \times 1.25 + 0.5 \times [0 + 1 \times 5] = 0.625 + 2.5 = 3.13$$

This might seem counterintuitive at first. If the agent eventually ended up in a bad state, why would the earlier action's value increase? The key insight is that Q-learning evaluates each action based on what was known at the time the action was taken. When the agent moved Right from $S_{(0,1)}$, it was still a good decision based on the known value of state $S_{(0,2)}$. The bad outcome came from the later decision to move Down instead of Right from $S_{(0,2)}$.

Crucially, when updating the Q-value for moving Right from $S_{(0,1)}$, the algorithm uses the *best* we can do in $S_{(0,2)}$, which is to go Right (with value 5). When we plan, we use this optimal value despite the fact that we didn't actually take the Right action later to actually receive this possible reward. Instead, we took a suboptimal Down action at that later time. This is what makes Q-learning an *off-policy algorithm*—it learns about the optimal policy regardless of the actions actually taken. It considers what could have happened under the best possible decision, not just what did happen. This illustrates an important aspect of Q-learning: It evaluates actions based on their immediate consequences and the best-known potential from the resulting state.

Episode 5: Continued Learning

In the fifth episode, the agent repeats the path from Episode 4: Right, Right, Down, Exit. The Q-value for exiting from $S_{(1,2)}$ becomes even more negative:

$$Q(S_{(1,2)}, X) = (1 - 0.5) \times (-5) + 0.5 \times [-10 + 1 \times 0] = -2.5 - 5 = -7.5$$

As the agent accumulates more evidence about the negative outcome of this action, its estimate of the value becomes increasingly pessimistic. The Q-value for moving Down from $S_{(0,2)}$ also becomes negative for the first time:

$$Q(S_{(0,2)}, D) = (1 - 0.5) \times 0 + 0.5 \times [0 + 1 \times (-5)] = 0 - 2.5 = -2.5$$

10.3 Popular Algorithms: Q-Learning, Policy Gradients, and Deep Q-Networks

Now the agent has learned that moving Down from $S_{(0,2)}$ leads to negative value, while moving Right leads to positive value. If it were to make decisions based on these Q-values, it would choose Right over Down when in state $S_{(0,2)}$. Meanwhile, the Q-value for moving Right from $S_{(0,1)}$ continues to increase:

$$Q(S_{(0,1)}, R) = (1 - 0.5) \times 3.13 + 0.5 \times [0 + 1 \times 5] = 1.56 + 2.5 = 4.06$$

This reinforces the idea that moving Right from $S_{(0,1)}$ is a good action, despite the fact that in these recent episodes, the agent eventually made a poor choice later in its journey. Consider this from a human perspective. Imagine you're driving to a new restaurant. You correctly navigate most of the journey, but in the final moments, you make a wrong turn into a dead end. While the overall trip was unsuccessful, you wouldn't throw away all the knowledge you gained—you learned the correct route for the majority of the journey, and you learned which specific turn to avoid next time. Q-learning encapsulates this nuanced learning process mathematically. Each action is evaluated based on the following:

- The immediate reward it produces
- The best possible future value from the resulting state at the time the action was taken

This separation of *credit* allows Q-learning to build knowledge incrementally, preserving good choices even within sequences that contain mistakes. Rather than assigning blame uniformly across all actions in a failed episode, Q-learning precisely identifies where things went wrong.

> **Do the Calculations**
>
> We highly recommend that you actually do these calculations by hand for a few more episodes. Doing these calculations by hand at least once reinforces (pun intended) the understanding of the concepts allowing you to grasp later, more complicated concepts all the more with ease.

The Emerging Policy

After these five episodes, we can see a policy emerging from the Q-values. In state $S_{(0,1)}$, the agent would prefer to move Right (Q=4.06) rather than any other direction (all Q= 0). In state $S_{(0,2)}$, it would strongly prefer to move Right (Q=5) rather than Down (Q=-2.5). This is how Q-learning gradually shapes behavior—by assigning values to actions in each state, creating a map that guides the agent toward reward and away from punishment. With enough exploration and learning, the Q-values converge to reflect the true expected future rewards of each action, and the agent can exploit this knowledge to make optimal decisions. The beauty of this approach is that the agent doesn't need to be told what to do—it discovers the optimal policy through its own experiences, learning from both successes and failures. This ability to learn through trial and error makes reinforcement learning particularly powerful for complex problems where the best solution isn't immediately obvious.

What makes Q-tables particularly appealing is their *interpretability*. Unlike more complex approaches, we can directly examine the values in a Q-table to understand what our agent has learned. If the Q-value for moving left in a particular state is much higher than other actions, it's clear that the agent has learned that going left is advantageous in that situation.

Q-tables also highlight the exploration-exploitation dilemma we discussed earlier. When deciding which action to take, our agent has two competing objectives:

- Exploit its current knowledge by selecting the action with the highest Q-value
- Explore actions with uncertain values to potentially discover better strategies

Balancing these objectives is crucial for effective learning. Common approaches include ε-greedy (choosing the best action most of the time, but occasionally selecting a random action) and Boltzmann exploration (selecting actions with probability proportional to their estimated values). Let's discuss the simplest of these next.

Try Out the Interactive Demo

You can try all the ε-greedy and the rest of the Q-learning algorithm interactively by visiting the **10-01-q-learning-demo** on the book resources page at *https://recluze.net/keras-book*. In the interactive demo, click **Visualize Policy**. Then, click on **Start Learning** to see how the agent learns Q-values by gaining experience. You can also modify the code to change reward locations, grid size, and so on to get a better understanding of the algorithm.

10.3.5 Q-Learning and Associated Issues

Now that we've explored the concept of Q-tables—those treasure maps of value that guide our agent through an environment—and looked at a worked example, this section will bring the discussion of Q-learning to a close by summarizing the steps required for the algorithm. If Q-tables are the map, Q-learning is the process that creates that map, marking each path with increasingly accurate estimates of its true value. Picture a vast, complex forest you've never visited before. At first, every path looks the same—you have no idea which trails lead to breathtaking vistas and which end in thorny thickets. But with each hike, you learn. The path that led to a waterfall yesterday gets a mental bookmark as highly rewarding. The one that circled back after an exhausting climb earns a mental note of "avoid next time." While learning as discussed previously, you adjust your belief by a fraction of new information, controlled by the learning rate α. If α is small (e.g., *0.1*), you're a cautious learner who doesn't drastically change opinions based on a single experience. If α is large (e.g., *0.9*), you're more impressionable, giving recent experiences substantial weight in your decisions.

10.3 Popular Algorithms: Q-Learning, Policy Gradients, and Deep Q-Networks

The Interplay Between Exploration and Exploitation

A crucial element of Q-learning is balancing curiosity with wisdom—or, in technical terms, exploration with exploitation. If our agent always follows what it currently believes is best (exploitation), it might never discover better strategies that lie off its beaten path. But if it's always trying random new things (exploration), it never benefits from its accumulated knowledge.

This dilemma is solved with the ε-greedy selection approach mentioned earlier. With probability ε, our agent chooses a random action (exploration), and with probability $1-\varepsilon$, it chooses what it currently believes is best (exploitation). Think of it like a chef who has a signature dish but occasionally experiments with new ingredients. Too much experimentation, and the restaurant loses its identity; too little, and the menu grows stale. The best chefs—like the best learning algorithms—find the sweet spot between tradition and innovation. As learning progresses, we typically reduce ε—starting with lots of exploration (maybe $\varepsilon = 1.0$) and gradually shifting toward exploitation (eventually reaching something like $\varepsilon = 0.1$). It's similar to how children play freely and explore widely, but adults tend to focus more on what they know works well.

The Q-Learning Algorithm: Step-by-Step

Let's walk through the Q-learning process as a series of steps our agent follows:

1. **Initialize with optimism.**
   We start by filling our Q-table with zeros or even small positive values, embodying the optimistic spirit of "all paths might lead to treasure until proven otherwise."

2. **Choose a starting state.**
   Our agent begins somewhere in the environment—perhaps randomly selected or from a fixed starting position. Sometimes, this is in our control, and sometimes we must make do with what is given to us.

3. **Select an action.**
   Using our ε-greedy approach, the agent decides whether to follow its current best understanding or try something new.

4. **Take action and observe.**
   The agent performs its chosen action and then watches what happens—what immediate reward it receives and what new state it finds itself in.

5. **Update knowledge.**
   Using our Q-learning formula, the agent adjusts its assessment of the value of the action it just took.

6. **Repeat until learning is complete.**
   The agent continues this cycle—act, observe, update, act again—until it has learned an effective strategy or reached some predetermined stopping point.

This cycle mirrors how humans learn many skills—through practice, feedback, adjustment, and more practice. The only difference is that our digital agent can often complete thousands of these learning cycles in the time it takes us to explore a new room in a building.

One of the most remarkable aspects of Q-learning is how order emerges from chaos. When the algorithm begins, the Q-table is essentially random noise—meaningless numbers that provide no guidance. But as learning progresses, patterns begin to emerge. Mathematically, we can prove that under certain conditions (e.g., visiting all state-action pairs infinitely often), Q-learning will converge to the optimal policy. In practical terms, this means that given enough time and experience, our agent will discover the best possible way to navigate its environment.

10.3.6 The Limits of Tabular Q-Learning

While Q-learning is powerful, it faces challenges when environments become very large or complex. The tabular approach—literally storing a value for each state-action pair in a table—breaks down when the number of states explodes. Consider a simple game of tic-tac-toe with approximately 5,000 possible board states. A Q-table for this game is entirely manageable. Now consider chess, with around 10^{43} possible positions—more than the number of atoms in the observable universe! A literal table with that many entries would be physically impossible to store.

This limitation points toward the need for *function approximation*—using functions (e.g., neural networks) to estimate Q-values rather than storing them individually. This approach, which we'll explore in the following section, allows Q-learning to scale to complex problems that would otherwise be intractable.

Deep Q-Networks Architecture

We've seen how Q-learning allows our agent to navigate environments and discover optimal strategies through trial and error. But we've also encountered its greatest limitation: the Q-table approach simply can't scale to complex environments with vast state spaces. Picture trying to create a table with a row for every possible configuration in a modern video game—the numbers quickly become astronomical. This is the problem DQNs solve; that is, bringing together the wisdom of Q-learning with the *pattern-recognition* power of neural networks. Traditional Q-learning stores a specific value for each state-action pair in a table. DQNs take a fundamentally different approach—instead of a lookup table, they use a neural network to approximate the Q-function itself. Think of it as the difference between memorizing every possible chess position versus understanding the principles and finding patterns in different board positions that make certain moves better than others.

The neural network takes the current state as input and outputs predicted Q-values for each possible action. Mathematically, we can express this as

$$Q(s, a; \theta) \approx Q^*(s, a)$$

where θ represents the weights of our neural network, and $Q^*(s, a)$ is the true optimal Q-value we're trying to approximate. The network becomes our Q-table, but instead of storing values explicitly, it learns patterns that generate those values on demand. This approach is similar to how your brain works. You don't memorize the exact weight of every object you've ever held—instead, you've developed an intuitive sense of how heavy things might be based on their appearance. Your neural pathways have learned to approximate weight based on visual cues such as size, material, and density.

A typical DQN architecture, like any other deep neural network, has several key components that work together:

- **Input layer**
  This layer receives the current state representation. For a simple grid world, this might be coordinates. It might even be the collective structure that holds the position of the agent, rewards, and areas to avoid. For Atari games, for example, it could be pixels from the screen.

- **Hidden layers**
  These intermediate layers extract meaningful patterns from the input. They're like the analytical regions of your brain that process raw sensory data into understanding. These aren't much different in the reinforcement learning algorithm than what we've been looking at in this book. You can use all your knowledge to strengthen this part of the DQN.

- **Output layer**
  This final layer produces Q-values for each possible action. For example, if our agent can move in four directions, the output layer would have four neurons, each predicting the value of one action.

The real innovation in DQNs comes from how these layers are structured and trained. Modern DQNs often use convolutional layers when processing visual inputs (e.g., game screens), similar to how your visual cortex processes what you see in hierarchical stages—first detecting edges, then shapes, then objects, and finally meaning.

Teaching a Deep Q-Network: The Learning Process

The core idea behind learning in a DQN remains similar to Q-learning but with important adaptations for neural networks. The standard Q-learning update rule becomes a neural network training process where we minimize the difference between our current predictions and target values. The loss function typically looks like

$$L(\theta) = E(s, a, r, s') \left[(r + \gamma \max a'\, Q(s', a'; \theta^-) - Q(s, a; \theta))^2 \right]$$

where θ represents our current network weights and θ^- represents weights from a target network (more on this shortly). This equation may look intimidating, but it's essentially saying: Adjust the network weights to make our Q-value predictions closer to what we observe plus what we expect in the future. As is the case with neural networks, initially your predictions might be way off, but with each observed weather pattern and outcome, you adjust your mental model to make better forecasts. The network does the same thing—with each experience, it adjusts its weights to better predict the value of actions.

Stability Tricks: Taming the Learning Beast

Early attempts to combine neural networks with Q-learning faced a significant challenge: instability. The learning process would often oscillate wildly or diverge entirely. Researchers at DeepMind discovered several crucial techniques to stabilize training:

- **Experience replay**
  Instead of learning from experiences as they happen, DQNs store experiences in a *replay buffer* and train on random batches from this memory. It's like a student reviewing notes from past classes in random order, which helps prevent forgetting earlier lessons and breaks harmful correlations between consecutive experiences. While the mathematics of this concept are complicated, the take away from these is that you store previous input-output pairs in a buffer and feed them to the neural network with newer experiences. This way, you're "replaying" your past experience—hence the name of the technique. This helps smooth out oscillations in learning and makes DQN operate much faster. Experience replay is a staple of modern machine learning despite its simplicity.

- **Target network**
  DQNs actually use two networks—one that's actively being updated (the online network) and another that's updated less frequently (the target network). When calculating target values, the target network is used. This is like having a textbook that you only replace with a new edition occasionally rather than rewriting it every day. It provides a stable reference point during learning.

- **Reward clipping**
  To handle varying scales of rewards across different environments, rewards are often clipped to a standard range (e.g., [-1, 1]). This is similar to how we normalize all inputs to our neural networks—it just helps with stability.

- **Frame skipping**
  In environments such as video games, DQNs often don't process every frame, instead skipping some to reduce computational load and help the agent focus on meaningful changes.

Together, these innovations allow DQNs to learn stably and efficiently in complex environments that would be impossible to handle with traditional Q-tables given very large and complex inputs.

The Input Challenge: Representing the World

This brings us to one of the most significant aspects of DQN design—deciding how to represent the state. For simple environments, this might be straightforward—coordinates in a grid world or specific features in a control problem. But for rich environments such as games or robotics, we need more sophisticated approaches. For instance, when DeepMind's DQN learned to play Atari games, it received raw pixels as input. But rather than a single frame, it used a stack of several consecutive frames. This allowed the network to infer movement and direction—crucial information that's not available from a static image alone. The state representation often requires preprocessing to make learning more efficient. For visual inputs, this might include grayscale conversion, resizing, or normalization. These steps help emphasize the important information immensely.

Beyond the Basics: Advanced DQN Architectures

Since the original DQN breakthrough, researchers have developed numerous improvements. Let's go over them briefly before diving into our first actual code example:

- **Double DQN**
  Addresses the problem of overestimating Q-values by using one network to select actions and another to evaluate them, similar to getting a second opinion before making an important decision.

- **Dueling DQN**
  Splits the network into two streams—one estimating the overall value of a state and another estimating the advantage of each action relative to others. It's like separately considering the following: "How good is the situation overall?" and "Which action is best in this situation?"

- **Prioritized experience replay**
  Instead of sampling experiences randomly, this method prioritizes experiences that led to surprising outcomes, similar to how you might focus more on studying topics where you previously made mistakes.

- **Distributional DQN**
  Rather than predicting a single value for each action, this approach predicts an entire distribution of possible returns, acknowledging the uncertainty inherent in reinforcement learning. It's forecasting not just the expected temperature but the full range of possible temperatures and their likelihoods.

Each of these improvements addresses specific limitations of the original architecture, gradually building a more robust approach to deep reinforcement learning. DQNs represent a crucial bridge between classical reinforcement learning methods and modern deep reinforcement learning. Their success in mastering Atari games from raw pixels—learning directly from the same inputs a human would see—demonstrated that reinforcement learning could scale to complex, high-dimensional problems.

The architectural insights from DQNs continue to influence newer approaches, even as the field has expanded beyond value-based methods to include policy gradient techniques and model-based methods. The core idea—approximating value functions with neural networks—remains a foundational concept in modern reinforcement learning. Implementing this basic algorithm through low-level code will give us a deep understanding of the concept that we can use to build further on with the latest and greatest models.

10.4   Implementing Reinforcement Learning Models in Keras

After reviewing the theoretical foundations of reinforcement learning algorithms, we can now apply these concepts in practice, as we'll discuss in this section. The first step in this process is to set up appropriate environments where our agents can learn and thrive. Think of these environments as elaborate playgrounds designed specifically for AI agents to explore, experiment, and evolve their strategies. For many years, *OpenAI Gym* (aka Gym) stood as the cornerstone framework for reinforcement learning environments. Launched in 2016, it revolutionized how researchers and developers approached reinforcement learning by providing a standardized interface for a diverse collection of environments. This standardization was crucial—before Gym, researchers often had to rebuild environments from scratch for each new project, making it difficult to compare results across different studies. Gym changed all that by offering a simple, consistent API that worked across dozens of different learning scenarios.

However, as with many pioneering tools, Gym eventually faced maintenance challenges. The reinforcement learning community needed a more actively maintained alternative, which is where *Gymnasium* came in. Gymnasium began as a fork of the original Gym and has since become the recommended standard for reinforcement learning environments. For our implementations, we'll be using Gymnasium version 1.1, though any recent version should work with minimal adjustments. What makes Gymnasium particularly valuable is its wide array of prebuilt environments. These range from simple grid worlds to complex physics simulations and even video game environments. Some of the environment categories include the following:

- Classic control problems, such as CartPole and Pendulum
- Grid-based environments, such as FrozenLake and CliffWalking
- Atari games, including *Breakout*, *Pong*, and *Space Invaders*

10.4 Implementing Reinforcement Learning Models in Keras

- Box2D physics simulations, such as LunarLander and BipedalWalker
- MuJoCo environments for more complex robotics simulations
- Algorithmic tasks that test an agent's ability to learn sequences

10.4.1 Our First Reinforcement Learning Environment

In this section, we'll create our first reinforcement learning environment through code. This allow you to appreciate the complexities involved in the formulation as well as help you get a more thorough understanding.

The FrozenLake Environment

Let's take a closer look at the FrozenLake environment, which will serve as our starting point. FrozenLake presents a grid world scenario that aligns closely with the theoretical grid world examples we explored earlier. Picture a frozen lake where certain spots are safe ice, others are dangerous holes, and one special location represents the goal. In this 8x8 grid environment, our agent starts in the top-left corner and must navigate to a goal in the bottom-right corner. Moving around seems simple enough, but there's a catch—scattered throughout the grid are holes in the ice. If our agent falls into one of these holes, the episode ends in failure. Adding another layer of complexity, the standard version includes slippery ice, meaning that when the agent attempts to move in one direction, it might actually slide in a different direction with some probability. This introduces an element of stochasticity that makes the environment more challenging and realistic. The visual layout of this environment is shown in Figure 10.4. We'll be running this in a bit.

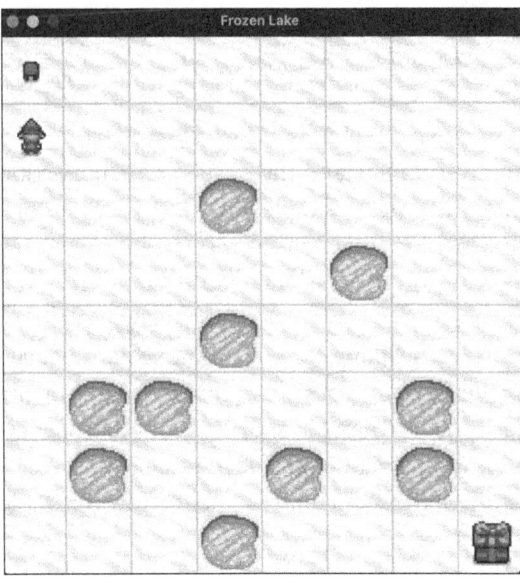

Figure 10.4 The FrozenLake Environment

The agent can take one of four possible actions at each step:

- Move left (0)
- Move down (1)
- Move right (2)
- Move up (3)

The reward structure is straightforward but challenging:

- Reaching the goal grants a reward of +1.
- Any other outcome (falling in a hole or time running out) gives a reward of 0.
- Every step that doesn't end the episode also provides 0 reward.

This sparse reward system makes FrozenLake particularly interesting. Our agent receives meaningful feedback only upon successfully completing the task, forcing it to explore extensively and learn from rare positive experiences. This mirrors many real-world scenarios where immediate feedback is limited, and success signals might be few and far between. Working with FrozenLake is especially instructive because it connects directly to the grid world examples we've already discussed. The environment's discrete state and action spaces make it an ideal candidate for tabular methods such as Q-learning, allowing us to see these algorithms in action without immediately jumping to the complexity of neural network approximations.

> **Explore Other Environments**
>
> Once you're comfortable with FrozenLake, you can explore the other environments available on the Gymnasium website at *https://gymnasium.farama.org/*. The code for Q-learning will remain mostly the same with minor changes to actions and state representations.

Our implementation approach will follow two main phases. First, we'll set up the environment, exploring how to create instances, understand the observation and action spaces, and interact with the environment through a simple loop. Then, we'll implement and train a Q-learning agent that learns optimal policies for navigating this icy terrain. The Gymnasium API makes this setup process remarkably straightforward. With just a few lines of code, we can create an environment, extract its key properties, and begin the interaction loop that forms the backbone of reinforcement learning. The real power emerges when we combine the flexibility of Gymnasium's environments with the expressiveness of Keras models. This partnership allows us to tackle increasingly complex problems, from grid worlds to Atari games and beyond, all while using a consistent framework that makes experimentation and improvement more accessible than ever before. But let's start with the first step.

Setting Up the Environment

Let's get our hands dirty with some actual code. Before we dive into implementation, you'll need to set up your environment properly. Let's walk through the installation process and then explore our first code sample for the FrozenLake environment. For the first couple of steps of this section, you'll need to have a local installation. We'll explain why in just a minute.

Open your terminal and run

```
pip install gymnasium
```

For the FrozenLake environment specifically, the base Gymnasium package should suffice. However, if you plan to explore other environment types later (e.g., Atari games), you might need additional packages:

```
pip install "gymnasium[atari]"
pip install "gymnasium[accept-rom-license]"
```

One quick thing to note about running environments with visual rendering is that when we set `render_mode='human'` in our code, Gymnasium will open a window to display the environment visually—the one we saw in Figure 10.4 earlier. This works perfectly on your local machine, giving you an intuitive view of what's happening. However, this rendering approach won't function in cloud-based notebooks such as Google Colab or Kaggle, which lack the necessary display capabilities. Don't worry though—this visual rendering is primarily for our human understanding and isn't required for the agent to learn. Only this initial exploration requires local execution. The rest of our implementation, including the more advanced deep Q-Learning sections coming up, will be designed to work perfectly well in cloud-based environments. We'll make sure the training loops and evaluation code function without requiring visual rendering.

Understanding Our First Reinforcement Learning Code

Let's examine our first piece of code for setting up and interacting with the FrozenLake environment, as shown in Listing 10.1. Notice that we're importing `gymnasium` as `gym`—this naming convention allows our code to remain compatible with older code samples that were written for the original Gym. Because Gymnasium was designed as a drop-in replacement for Gym, this simple alias makes the transition seamless. If you come across tutorials or examples using the old Gym module, they should work with minimal modifications when you have Gymnasium installed. (You can view the full code listing in the book resources page at *https://recluze.net/keras-book* under the file **10-02-frozen-lake-q.py**. Note that this isn't a notebook but a Python file you can run locally as discussed earlier.)

10 Reinforcement Learning: The Secret Sauce

```
import gymnasium as gym
import numpy as np
import matplotlib.pyplot as plt
import pickle

def run(episodes, is_training=True, render=False):

    env = gym.make('FrozenLake-v1', map_name="8x8", \
                    is_slippery=True, \
                    render_mode='human' if render else None)

    if(is_training):
        q = np.zeros((env.observation_space.n, \
                        env.action_space.n))
    else:
        f = open('frozen_lake8x8.pkl', 'rb')
        q = pickle.load(f)
        f.close()
```

Listing 10.1 Initializing a Gymnasium Environment

The run function takes three parameters:

- episodes
  The number of episodes to run (each episode starts fresh and runs until the agent reaches the goal or falls in a hole).

- is_training
  A flag indicating whether we're training a new agent or using a previously trained one. This will be useful when we train an agent in the cloud and then use the training results locally.

- render
  Whether to visually display the environment for humans or just hide the environment to make the training more efficient.

Inside the function, we first create our environment using gym.make(). We're using the 'FrozenLake-v1' environment with an 8x8 grid (the default is 4x4), and we're keeping the ice slippery to maintain the stochastic nature of the environment. The render_mode parameter controls the visual display—we'll only enable it when the render flag is True; otherwise, we set it to None to avoid any rendering overhead.

The rendering capability is purely for our benefit as human observers. It helps us visualize what's happening in the environment—we can watch our agent navigate the icy terrain, see it slip occasionally, and celebrate when it reaches the goal. But our learning algorithm doesn't need this visual representation; it works directly with the numerical state and reward signals.

Moving on, if we're training (is_training=True), we initialize a new Q-table as a zero-filled NumPy array. The dimensions of this array are determined by the environment:

- env.observation_space.n gives us the number of possible states (64 for an 8x8 grid).
- env.action_space.n gives us the number of possible actions (4: left, down, right, up).

This creates a table where each row represents a state, and each column represents an action. The values in this table will gradually be updated as our agent learns which actions are best in each state.

If we're not training (is_training=False), we take a different approach. Instead of starting with a blank slate, we load a previously trained Q-table from a pickle file. This is where the power of persistence comes into play. After training our agent, which might take thousands of episodes, we can save its accumulated wisdom—the Q-table—to a file. Later, we can load this file and immediately have an agent that knows how to navigate the frozen lake without having to learn from scratch. The pickle module is Python's standard tool for serializing and deserializing Python objects. We use pickle.load() to recreate our Q-table exactly as it was when we saved it. This approach is particularly valuable in reinforcement learning, where training can be time-consuming. By saving our Q-values, we create a form of knowledge transfer, allowing our agent's learned expertise to persist beyond a single session.

This pattern of training once and reusing the learned policy is common in reinforcement learning applications. In real-world deployments, you might train your agent in a simulated environment for millions of steps, save the resulting policy or value function, and then deploy that pretrained model in the actual application. We'll also use a similar strategy when we move to the cloud environment to train our model in the next section.

Next, we set up some of our hyperparameters as well as supporting structures as shown in Listing 10.2:

1. The learning_rate_a value of 0.9 is initially kept high, making our agent quite responsive to new experiences. Think of it as a student who quickly updates their understanding with each new lesson. With this high learning rate, our agent won't stubbornly cling to its initial ideas but will rapidly incorporate new information. When it discovers a hole in the ice or finds the goal, that experience will strongly influence its future decisions.

2. Next, we have discount_factor_g also set at 0.9. This parameter governs how much our agent values future rewards compared to immediate payoffs. It's like planning a road trip—do you take the direct route that might hit traffic later or the longer scenic route that flows smoothly? With a value of 0.9, our agent has considerable foresight, willing to take a slightly longer path if it leads to greater rewards down the line.

3. The epsilon parameter starts at 1, which means our agent begins with 100% exploration. This complete openness to experimentation ensures our agent won't get stuck in suboptimal patterns before it has a chance to discover better alternatives.

4. But exploration can't last forever. The epsilon_decay_rate of 0.0001 gradually shifts our agent from explorer to expert. After each episode, epsilon decreases slightly, making the agent more likely to exploit what it has learned rather than try something new. This *epsilon decay* is similar to how we gradually rely more on experience and less on experimentation as we become proficient at a skill.

5. Finally, rewards_per_episode = np.zeros(episodes) creates an array to track performance over time. This array starts filled with zeros—a blank slate waiting to capture the agent's learning journey across thousands of attempts to navigate the frozen lake.

```
learning_rate_a = 0.9
discount_factor_g = 0.9

epsilon = 1
epsilon_decay_rate = 0.0001

rng = np.random.default_rng()
rewards_per_episode = np.zeros(episodes)
```

Listing 10.2 Setting Up the Initial Hyperparameters

Next, let's dive into the actual training loop of our Q-learning agent—the engine that powers its learning journey across the frozen lake. This code segment in Listing 10.3 implements the core algorithm we've been discussing.

```
for i in range(episodes):
        state = env.reset()[0] # 0 = top left, 63 = bottom right
        terminated = False
        truncated = False

        while(not terminated and not truncated):
            if is_training and rng.random() < epsilon:
                action = env.action_space.sample()
                # 0 = left, 1 = down , 2 = right, 3 = up
            else:
                action = np.argmax(q[state,:])

            new_state,reward,terminated,truncated,_ = \
                    env.step(action)

            if is_training:
```

10.4 Implementing Reinforcement Learning Models in Keras

```
                q[state,action] = q[state,action] + \
                learning_rate_a * (
                    reward + discount_factor_g * \
                    np.max(q[new_state,:]) - q[state,action]
                )

            state = new_state

        epsilon = max(epsilon - epsilon_decay_rate, 0)

        if(epsilon==0):
            learning_rate_a = 0.0001

        if reward == 1:
            rewards_per_episode[i] = 1

    env.close()
```

Listing 10.3 Core Q-Learning Algorithm

This training loop concisely translates the Q-learning algorithm we studied earlier into working code. Let's walk through it step-by-step:

1. The outer for loop runs for the specified number of episodes. Each episode is a fresh attempt to navigate from start to finish.

2. We begin each episode by resetting the environment with env.reset()[0], which returns the initial state. State 0 represents the top-left position of the grid, while state 63 is the bottom-right in our 8×8 lake.

3. The terminated and truncated flags track whether the episode has ended. An episode terminates when our agent reaches the goal or falls into a hole, while truncation would occur if we imposed a step limit.

4. Now we enter the inner loop—the real-time decision-making process of our agent. At each step, the agent faces the exploration-exploitation dilemma:

    If we're in training mode and a random number is less than epsilon, the agent explores by taking a random action with env.action_space.sample().

    Otherwise, the agent exploits its current knowledge by choosing the action with the highest Q-value for the current state: np.argmax(q[state,:]). This selects the column (action) with the maximum value for the current state row in our Q-table.

5. After deciding on an action, the agent takes a step in the environment with env.step(action). This returns five values:

    new_state: Where the agent ended up (might not be where it intended if the ice is slippery) reward: The reward received (1 if it reached the goal, 0 otherwise)

terminated: Whether the episode has ended successfully or by falling in a hole

truncated: Whether the episode ended due to a step limit

An additional info dictionary (ignore this with the underscore)

6. Now comes the actual learning step—the *Q-value update*. This is where our agent refines its understanding of the world based on experience. Match this update rule with the mathematical equation discussed earlier. This update captures the essence of Q-learning, mirrors the Bellman equation precisely, and elegantly captures how we learn from experience. If the outcome was better than expected (positive error), we increase our estimate of that action's value. If worse than expected (negative error), we decrease it. The learning rate controls how dramatically we update our beliefs with each experience.

7. After the update, we move to the new state with state = new_state and continue the process until the episode ends. Once an episode concludes, we reduce epsilon by the *decay rate*, gradually shifting from exploration to exploitation. There's an interesting detail here: When epsilon reaches 0, we dramatically reduce the learning rate to 0.0001. This is like shifting from a student actively learning to an expert making only minor refinements to their knowledge. The loop continues for all episodes, and then we close the environment with env.close().

This implementation follows the classic Q-learning algorithm almost exactly as we studied it, translating the mathematical formulation into working code. The beauty of reinforcement learning is seeing how these relatively simple update rules can lead to sophisticated behavior emerging over time, as our agent gradually learns to navigate the treacherous frozen lake with increasing confidence and skill.

The code segment shown in Listing 10.4 wraps up our training process with two key operations. This start of the code handles visualization while the last three lines save our agent's accumulated wisdom to a file. This preservation step transforms our agent into a reusable expert, ready to guide travelers across the ice whenever called upon.

```
sum_rewards = np.zeros(episodes)
    for t in range(episodes):
        sum_rewards[t] = np.sum(rewards_per_episode[ \
                        max(0, t-100):(t+1)])
    plt.plot(sum_rewards)
    plt.savefig('frozen_lake8x8.png')

    if is_training:
        f = open("frozen_lake8x8.pkl","wb")
        pickle.dump(q, f)
        f.close()
```

Listing 10.4 Saving the Learned Q-Values

10.4.2 Implementing the Deep Q-Network Algorithm with Keras

Now that we've implemented a traditional Q-learning approach with a table-based solution, let's take our agent to the next level. While our tabular method works well for the discrete, manageable state space of FrozenLake, many real-world problems involve much larger or even continuous state spaces where tables would be impractical. This is where neural networks shine—they can approximate Q-values for any state, even ones they've never explicitly seen before. We're going to upgrade our agent with the DQN algorithm, which replaces our Q-table with a neural network. Rather than storing and updating individual Q-values in a table, we'll train a network to predict these values. The core learning principles remain the same, but the way we represent and update our knowledge changes dramatically.

We can return to the cloud-based environment for this more compute-intensive case study. For the full code listing, you can refer to the notebook **10-03-frozen-lake-dqn** on the book resources page at *https://recluze.net/keras-book*. Here, we'll touch on the more important parts of this code.

After importing the necessary Keras libraries, we can build our neural network architecture. Let's discuss the code given in Listing 10.5 that makes this transformation possible. First, we define a one_hot function that transforms our discrete state numbers into a form the neural network can digest. In our FrozenLake environment, states are represented as simple integers (0 through 63 for the 8×8 grid). Neural networks, however, work better with distributed representations. The one-hot encoding transforms state 5, for example, into a vector of 64 zeros with a single 1 at position 5.

Next is our model-building function. This seemingly simple function is where the magic happens—we're replacing our entire Q-table with a neural network. The architecture we've chosen is straightforward, but you can experiment with more complex models after successfully conducting learning experiments with this simple model. The beauty of Keras is that we don't really need to write down a description of the model here. You can understand the decisions made simply by looking at the code itself.

```
# One-hot encoding function for discrete states
def one_hot(state, state_size):
    vec = np.zeros(state_size)
    vec[state] = 1
    return vec

# Build the DQN model using keras.Sequential (no class)
def build_model(state_size, action_size):
    model = keras.Sequential([
        layers.Input(shape=(state_size,)),
        layers.Dense(64, activation='relu'),
        layers.Dense(64, activation='relu'),
```

```
        layers.Dense(action_size, activation='linear')
    ])
    return model

# Initialize environment and models
env = gym.make('FrozenLake-v1', map_name="8x8",
                is_slippery=True, render_mode=None)
state_size = env.observation_space.n
action_size = env.action_space.n

model = build_model(state_size, action_size)
target_model = build_model(state_size, action_size)
model.build(input_shape=(None, state_size))
target_model.build(input_shape=(None, state_size))
```

Listing 10.5 DQN Model Through Keras

This network structure is quite modest compared to what you might see in image recognition or language processing, but it's perfectly suited for our task. Each of the 64 neurons in our hidden layers represents patterns the network discovers during training. Think of these neurons as specialized detectors that learn to recognize meaningful state configurations and their implications for different actions. The Rectified Linear Unit (ReLU) activation functions serve as the neurons' decision mechanisms. The linear activation in the final layer allows the network to output Q-values across any numeric range, just like our Q-table could.

At the end of the function, we initialize our environment and create two identical models: the `model` and `target_model`. This dual-model approach is a key innovation of the DQN algorithm. The primary model learns continuously, while the target model provides stable Q-value estimates for our learning updates.

We specify the input shape for both models using `model.build()`, which prepares them to process our state inputs. Notice we use `None` for the batch dimension, allowing our models to process any number of states at once—a flexibility that will prove useful during training. This code is just one possible implementation, kept deliberately simple for clarity. For more complex problems, we might leverage the Keras Functional API to create more sophisticated architectures with multiple input streams or specialized network branches. But the principles would remain the same—we're using neural networks to approximate the Q-function that maps states and actions to expected rewards. The power of this approach is that once trained, our neural network can generalize—making educated guesses about states it's never explicitly seen before. This ability to generalize from experience is what makes deep reinforcement learning so powerful for complex, real-world problems.

10.4 Implementing Reinforcement Learning Models in Keras

We then go ahead and set the model parameters as shown in Listing 10.6. These are pretty self-explanatory as long as you keep our earlier mathematical formulation in mind.

```python
# Initialize training hyperparameters
optimizer = keras.optimizers.Adam(learning_rate=0.001)
loss_fn = keras.losses.MeanSquaredError()

replay_buffer = deque(maxlen=2000)
gamma = 0.95
epsilon = 1.0
epsilon_min = 0.01
epsilon_decay = 0.995
batch_size = 64
update_target_freq = 10
episodes = 1000

rewards_per_episode = np.zeros(episodes)
```

Listing 10.6 Setting Model Hyperparameters

Let's examine the training mechanism that powers our agent's learning, as shown in Listing 10.7. This function is the engine that drives our agent's learning process. Think of it as a teacher who reviews a collection of experiences, extracts the lessons they contain, and then adjusts the student's understanding accordingly. The function begins by sampling a random batch of experiences from our replay buffer. This randomization breaks the sequential correlation between experiences, which is crucial for stable learning. It's like studying historical events out of chronological order to better understand the patterns that connect them rather than getting caught up in the flow of history.

```python
# Define a helper function for training on a single batch
def train_step(model, target_model, replay_buffer, batch_size,
               gamma, optimizer, loss_fn):
    minibatch = random.sample(replay_buffer, batch_size)
    state_batch = np.array([x[0] for x in minibatch])
    action_batch = np.array([x[1] for x in minibatch])
    reward_batch = np.array([x[2] for x in minibatch])
    next_state_batch = np.array([x[3] for x in minibatch])
    done_batch = np.array([x[4] for x in minibatch])

    future_qs = target_model(tf.convert_to_tensor(
        next_state_batch, dtype=tf.float32), training=False)
    target_qs = model(tf.convert_to_tensor(state_batch,
                dtype=tf.float32), training=False).numpy()
```

```
    for i in range(batch_size):
        if done_batch[i]:
            target_qs[i][action_batch[i]] = reward_batch[i]
        else:
            target_qs[i][action_batch[i]] = reward_batch[i] + \
                            gamma * np.amax(future_qs[i])


    with tf.GradientTape() as tape:
        qs = model(tf.convert_to_tensor(state_batch, \
                                        dtype=tf.float32),
                   training=True)
        loss = loss_fn(target_qs, qs)

    grads = tape.gradient(loss, model.trainable_variables)
    optimizer.apply_gradients(zip(grads, \
                            model.trainable_variables))
```

Listing 10.7 Training the Neural Network

Each experience in our buffer is a complete memory unit containing five elements:

- The state the agent was in
- The action it took
- The reward it received
- The new state it found itself in
- Whether the episode ended (done flag)

We extract these elements into separate arrays, preparing them for batch processing. This batch approach is far more computationally efficient than updating our model one experience at a time. Next comes a two-step process that embodies the essence of Q-learning. First, we use our target network to predict the future Q-values for all the next states in our batch. Notice the `training=False` flag, which tells the network we're just using it for prediction, not updating its weights.

Then, we use our main model to get the current Q-value predictions for our starting states. These become the foundation we'll adjust based on what we learned from each experience. The loop that follows is where the Bellman equation is brought to life, just as it did in our tabular approach. Consider the following for each experience in the batch:

- If the episode ended after this step (done=True), the target Q-value is simply the reward received. There's no future to consider if you've reached the goal or fallen in a hole.

- If the episode continued, we follow the familiar formula: reward added with discounted max future Q-value. This balances immediate rewards with long-term potential, just like in our tabular implementation.

We wrap our prediction process in a GradientTape. This *gradient tape* is a TensorFlow tool that records operations for automatic differentiation. When we calculate the loss between our target Q-values and the model's current predictions, the tape allows us to work backward through the computation to determine how each weight in our network contributed to the error. The gradient of our loss with respect to the model's trainable variables tells us exactly how to adjust each weight to reduce the error. Finally, we apply these gradients using our optimizer, taking a small step toward better predictions. Over thousands of such updates, our network gradually refines its understanding of the FrozenLake environment, learning which actions lead to success in each state.

What's remarkable about this approach is how it preserves the core Q-learning update rule while leveraging the power of neural networks and batch processing. Each time we call this function, our agent gets a little smarter, a little more skilled at navigating the treacherous FrozenLake.

Unlike our tabular method though, which stored each state-action value explicitly, our neural network is learning underlying patterns. It might discover, for instance, that states closer to the goal generally have higher values or that actions leading toward known holes should be avoided. These generalizations allow it to make educated guesses even for state combinations it hasn't explicitly experienced before.

The train_step function is used in the per-episode learning code shown in Listing 10.8. We begin each episode by resetting the environment with env.reset()[0], placing our agent back at the starting position. At each step, our agent faces the classic exploration-exploitation dilemma, just as in our tabular Q-learning implementation. If a randomly sampled number is less than epsilon, the agent explores by taking a random action. This exploration is crucial for discovering new strategies and avoiding getting stuck in suboptimal patterns. Otherwise, the agent exploits its current knowledge by querying the neural network for Q-values. This query is where our neural network shows its power—by passing our *state vector* through the network, we get back predictions for all four possible actions in a single forward pass. The network is essentially saying, "Based on everything I've learned so far, here's how valuable I think each action would be from this state."

```
for episode in range(episodes):
    print("Episode:", episode, " / ", episodes)
    state = env.reset()[0]
    state_vec = one_hot(state, state_size)
    done = False
    total_reward = 0
```

```
    while not done:
        if np.random.rand() <= epsilon:
            action = env.action_space.sample()
        else:
            q_values = model(tf.convert_to_tensor( \
                            [state_vec], \
                            dtype=tf.float32),
                        training=False)
            action = tf.argmax(q_values[0]).numpy()

        next_state, reward, terminated, truncated, _ = \
                        env.step(action)

        done = terminated or truncated
```

Listing 10.8 Per Episode Learning

Notice the `training=False` flag—this tells TensorFlow we're just using the network for prediction, not for learning at this moment. We then extract the action with the highest predicted Q-value using `tf.argmax(q_values[0]).numpy()`. The [0] is necessary because our model expects batches of inputs, so it returns batches of outputs—we're just extracting the single result we need. With the action decided, our agent takes a step in the environment. The `env.step(action)` function is like rolling the dice and seeing what happens. The environment returns five pieces of information as we discussed in the Q-learning section previously. We combine terminated and truncated into a single done flag that tells us whether the current episode has ended for any reason.

The loop in this listing continues in Listing 10.9. After our agent takes an action and observes the outcome, several crucial steps are needed. First, we encode the new state into the one-hot vector that our neural network understands. This is to satisfy the input requirements of our network and is just an implementation detail.

Next comes one of the most powerful innovations in deep reinforcement learning: *experience replay*. With `replay_buffer.append()`, we're storing a complete memory of this interaction—the state we were in, the action we took, the reward we received, where we ended up, and whether the episode finished. Think of this as our agent keeping a journal of its adventures, recording every step and its consequences. This memory system was revolutionary at its time of introduction. Unlike humans who often forget details of past experiences, our agent builds a perfect record it can revisit again and again. These memories become the textbooks from which our agent learns, extracting patterns and principles that guide better future decisions.

Once our memory bank has enough experiences (checked with `len(replay_buffer) >= batch_size`), we trigger the learning process through `train_step()`. This function, which we examined earlier, samples a batch of memories and uses them to refine the neural network's understanding. It's like our agent taking time to reflect on past experiences, extracting lessons that improve its strategy.

The line `state_vec = next_state_vec` moves us forward in time—what was once the future becomes the present as our agent continues its journey across the ice. We also update our running total of rewards, tracking how successful this episode has been so far. After finishing each complete episode (navigating successfully to the goal or falling through the ice), we adjust two critical learning parameters:

- First, we *decay* epsilon (our exploration rate) making our agent gradually shift from curious explorer to confident expert. There's a lower limit (`epsilon_min`) to ensure our agent always maintains some capacity for exploration, preventing it from becoming too rigid in its behavior.

- Second, we periodically update our target network with `target_model.set_weights()`. This happens every `update_target_freq` episodes, creating a slow-moving average of our learning that stabilizes the training process. It's like taking a snapshot of our current understanding to use as a stable reference point while we continue learning. This target network acts as an anchor, preventing the wild oscillations that can occur when chasing a constantly moving target.

Finally, we record the total reward earned during this episode, building a history of our agent's performance that we can analyze later. This performance tracking lets us visualize learning progress over time, revealing whether our agent is improving, plateauing, or struggling.

```
# ...   continuing inside the episode for loop
        next_state_vec = one_hot(next_state, state_size)

        replay_buffer.append((state_vec, action, reward, \
                        next_state_vec, done))

        if len(replay_buffer) >= batch_size:
            train_step(model, target_model, replay_buffer, \
                    batch_size, gamma, optimizer, loss_fn)

        state_vec = next_state_vec
        total_reward += reward


    if epsilon > epsilon_min:
        epsilon *= epsilon_decay

    if episode % update_target_freq == 0:
```

```
        target_model.set_weights(model.get_weights())

    rewards_per_episode[episode] = total_reward
```

Listing 10.9 Experience Replay and Target Network Update

After training our agent for 1,000 episodes, we can finally see the fruits of its learning visualized in the heatmap shown in Figure 10.5. This grid reveals the hidden intelligence our agent has developed—its internal map of the FrozenLake environment and the optimal path to navigate it safely. Looking at this visualization is peering directly into our agent's mind. The arrows indicate the best action the agent believes it should take in each state, while the shades represent the maximum Q-value for that state—essentially how valuable the agent considers that position to be. Brighter shades indicate higher values, while darker ones represent lower values.

The most striking feature is the bright cell in the bottom-right corner of the grid, but interestingly, it's not the goal state itself (the very bottom-right cell). Rather, it's the state just before it. This reveals something profound about how reinforcement learning agents understand value. The second-last state has the highest value because it represents the certainty of *imminent reward*—the agent is just one step away from guaranteed success. The goal state itself has a slightly lower value because once the agent reaches it, the episode ends and no further rewards can be collected. This pattern is a classic feature of Q-learning: States that lead directly to high rewards often have higher values than the terminal reward states themselves.

Looking at the path our agent has learned, we can see something that might seem counterintuitive at first glance. As the agent approaches the goal in the bottom-right corner, it consistently chooses to move right instead of moving directly down toward the goal. This seemingly indirect approach actually demonstrates sophisticated learning.

The agent has discovered that due to the slippery nature of the frozen lake, moving down carries significant risk. If it attempts to move down, the stochastic environment might actually send it sliding left or right—potentially into dangerous holes. Moving right, however, limits the potential negative outcomes. Even if the agent slips, it might move up or down, but not left into known danger zones. The DQN has effectively learned to prioritize safety in its approach to the goal, favoring a more reliable path even if it seems less direct to human intuition.

This *risk-averse* behavior emerges naturally from the agent's experiences. Through thousands of episodes of exploration, our agent has learned not just where the goal is, but also which paths minimize the risk of failure. It's discovered that the shortest path isn't always the optimal one when uncertainty is involved—a sophisticated strategy that emerged without any explicit programming on our part. Similarly, we can observe some seemingly random action choices in certain states, particularly in areas farther from the goal. These inconsistencies likely reflect states that our agent hasn't thoroughly explored during training. Like a hiker who hasn't fully mapped certain regions of a wilderness, our agent's knowledge remains incomplete in less-visited areas. With additional training episodes, these actions would likely become more consistent as the agent gathers more experiences in these states.

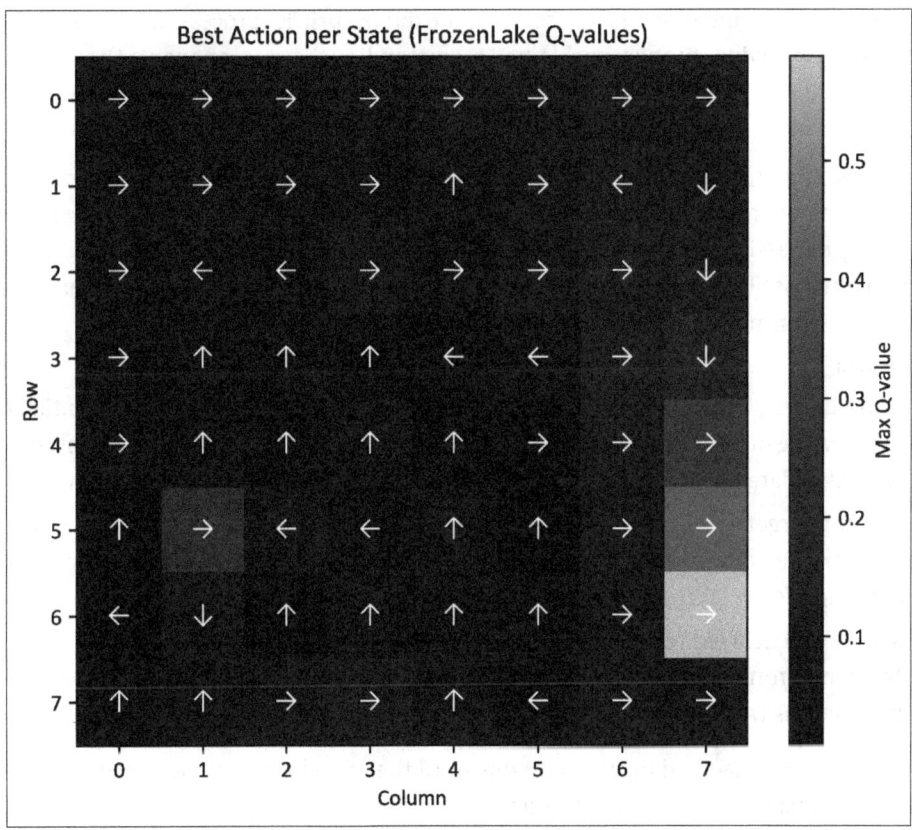

Figure 10.5 Visualizing the Learned Policy

What's remarkable is that all of this sophisticated behavior emerged from the relatively simple update rule of Q-learning, implemented through our DQN architecture. The agent wasn't explicitly programmed to be cautious near the goal or to prefer certain paths—it discovered these strategies through trial and error, guided by the reward signal and the Bellman equation.

This visualization perfectly illustrates the power of reinforcement learning: the ability to develop complex, nuanced behaviors through experience rather than explicit programming. Our agent has learned not just to reach the goal, but to navigate the frozen lake with a sophisticated understanding of risk and reward that rivals human strategic thinking.

10.4.3 Experience Replay and Target Networks: The Foundations of Stable Deep Reinforcement Learning

Now that we've implemented and visualized our DQN in action, let's dive deeper into two crucial mechanisms we've seen in code that make modern deep reinforcement learning possible: experience replay and target networks. These innovations transformed reinforcement learning from an unstable, brittle process into a reliable approach for solving complex problems. Understanding these components thoroughly will provide you with the technical foundation to explore cutting-edge research in the field.

Before we explore the solutions, let's understand the problem. When we first attempted to combine neural networks with Q-learning in the early days of deep reinforcement learning, researchers encountered a perfect storm of instabilities. The learning process would often oscillate wildly, diverge completely, or get stuck in poor local optima. This instability stemmed from three core issues:

- **Correlated experiences**
  In traditional Q-learning, agents learn from experiences as they occur sequentially. When using function approximators such as neural networks, this creates highly correlated update targets that can cause the network to overfit to recent experiences.
- **Moving targets**
  The very network being updated is also generating the targets for its own updates, creating a feedback loop where the target values constantly shift.
- **Sparse and delayed rewards**
  Many environments provide rewards only occasionally, making it difficult to connect actions with their eventual outcomes.

These challenges seemed insurmountable until the introduction of two groundbreaking techniques: experience replay and target networks.

Experience Replay: Breaking the Correlation Chain

Experience replay, first proposed by Lin in 1992 but popularized in the DQN architecture by Mnih et al. in 2013, fundamentally transforms how agents learn from experience. Instead of learning immediately from each transition, the agent stores experiences in a replay buffer and then samples from this memory randomly during training. This simple mechanism has profound advantages:

- **Breaking sequential correlations**
  By sampling randomly from the buffer, we demolish the temporal correlations between experiences. It's like shuffling a deck of cards before dealing them—each hand becomes independent of the previous one.
- **Efficient data usage**
  Each experience can be used multiple times for learning, extracting maximum value

from the agent's interactions. This is particularly valuable in environments where gathering experience is costly or time-consuming.

- **Stabilizing the learning process**
  Random sampling smooths out the learning dynamics, preventing the network from overfitting to recent experiences or getting stuck in behavior loops.

The size of the replay buffer creates an interesting trade-off. A larger buffer provides more diverse experiences but might retain outdated information about the environment or policy. Most implementations use buffers holding between 10,000 and 1,000,000 transitions, with older experiences being discarded when the buffer reaches capacity. Recent research has explored various enhancements to basic experience replay:

- **Prioritized Experience Replay (PER)**
  Instead of uniform random sampling, PER assigns higher probabilities to experiences from which the agent can learn more (typically measured by their temporal difference error). This focuses learning on the most informative transitions.

- **Hindsight Experience Replay (HER)**
  In sparse reward settings, HER retroactively changes the goal of an episode to make failed attempts instructive. It's like saying, "You didn't reach your intended destination, but you did successfully reach somewhere else—let's learn from that."

- **Episodic memory**
  Some approaches maintain entire episodes rather than individual transitions, allowing the agent to learn from complete sequences of behavior.

Target Networks: Stabilizing the Moving Target Problem

Even with experience replay smoothing out the correlation between consecutive experiences, deep Q-learning faces another fundamental challenge that can derail the entire learning process. The core issue emerges from the circular dependency between the network's predictions and the targets used to train it. When we calculate our temporal difference target using

$$r + \gamma \max_{a'} Q(s', a'; \theta)$$

we're using the same network that we're trying to update. As the network parameters change with each gradient step, the target values themselves shift, creating a feedback loop that can lead to *catastrophic oscillations* or complete learning failure. Consider what happens during a typical update cycle. Our network produces a Q-value for the current state-action pair, and then immediately uses its own (potentially incorrect) estimates to generate the target for the next update. Small errors compound rapidly. The system can spiral into instability, with Q-values oscillating wildly or diverging completely.

The solution, introduced alongside experience replay in the original DQN paper, elegantly breaks this feedback loop through target networks. Instead of using the same network for both prediction and target generation, we maintain two copies of our Q-network. The online network handles action selection and receives gradient updates, while the target network provides stable reference points for calculating temporal difference targets. The mathematical formulation becomes

$$y_t = r_t + \gamma \max_{a'} Q(s_{t+1}, a'; \theta^-)$$

where θ^- represents the parameters of the target network, which lag behind the online network parameters. This separation creates the stability we need—our targets remain consistent long enough for the online network to learn meaningful patterns. The frequency and method of updating the target network creates a balancing act. Update too frequently, and we slide back toward the instability we were trying to avoid. Update too rarely, and our target network becomes increasingly outdated, providing stale guidance that slows learning.

The original DQN approach used hard updates every fixed number of steps—typically, every 10,000 training steps. At each update point, the target network parameters are completely replaced with the current online network parameters: $\theta^- \leftarrow \theta$. This creates discrete jumps in the target values, which can initially destabilize learning but provides fresh targets for the next update cycle. A more refined approach uses *soft target updates*, borrowed from the *actor-critic* literature. Instead of wholesale replacement, we use exponential moving averages as

$$\theta^- \leftarrow \tau\theta + (1-\tau)\theta^-$$

where τ (tau) is typically a small value such as *0.001*. This creates smoother transitions. The target network slowly tracks the online network rather than making sudden jumps. The smoothness comes at the cost of slower adaptation to improvements in the online network.

> **The Actor-Critic Paradigm**
>
> Actor-critic literature refers to a class of reinforcement learning methods that combine two key components: an *actor* (policy network) that decides which actions to take, and a *critic* (value network) that evaluates how good those actions are. Unlike pure policy methods, which only learn actions, or pure value methods, which only learn state values, actor-critic methods learn both simultaneously. The critic provides feedback to help the actor improve its policy, while the actor's actions generate experiences that help the critic learn better value estimates.

The optimal update strategy depends heavily on your specific environment and learning dynamics. Fast-changing environments might benefit from more frequent updates to keep targets relevant. Stable environments can use longer update intervals to maximize the stability benefits. In practice, many successful implementations use values between 1,000 and 10,000 steps for hard updates, or τ values between 0.0001 and 0.01 for soft updates.

When implementing target networks in real systems, several practical considerations can make the difference between success and failure:

- **Memory management**
  This becomes particularly important when dealing with large state spaces. Storing raw pixel observations from Atari games can quickly exhaust available memory. Many successful implementations store preprocessed or compressed representations instead of raw observations. Some systems use frame differencing or other compression techniques to reduce storage requirements while preserving essential information.

- **Initialization of your replay buffer**
  This significantly affects early learning dynamics. Many implementations begin by filling the buffer with experiences from a random policy before starting any learning. This ensures sufficient diversity from the beginning and prevents the agent from overfitting to its initial, likely poor, experiences. A common practice is to collect 50,000 to 100,000 random experiences before the first training step.

- **Batch normalization,**
  While tremendously successful in supervised learning, this requires careful consideration in reinforcement learning contexts. The fundamental assumption of batch normalization—that data comes from a stationary distribution—breaks down as the policy improves and the state distribution shifts. Layer normalization (discussed in the previous chapter) often provides a better alternative, normalizing across features rather than across the batch dimension.

- **Choice of optimizer and learning rate**
  This choice also interacts with target network stability. The Adam optimizer, with its adaptive learning rates, can sometimes exacerbate instability issues by making large updates when gradients are inconsistent. Some practitioners find that simpler optimizers such as Root Mean Square Propagation (RMSprop) or even stochastic gradient descent (SGD) with momentum provide more stable learning. This proves particularly true in the early stages of training.

Finally, recent developments have pushed the boundaries of what's possible with target networks and experience replay:

- **Distributed architectures**
  Distributed architectures demonstrate how these concepts scale to massive parallel systems. In *Actor-Learner Architecture with Experience Replay (Ape-X)*, multiple actor processes run in parallel, each exploring the environment with slightly different policies and feeding experiences into a shared replay buffer. A separate learner process samples from this buffer and updates the network parameters, which are then periodically distributed back to the actors. This architecture dramatically increases the

rate of experience collection while maintaining the benefits of experience replay and target networks.

- **Recurrent Experience Replay in Distributed Reinforcement Learning (R2D2)**
  R2D2 extends this approach to handle partially observable environments using recurrent neural networks. The challenge with recurrent networks is that experience replay becomes more complex—you can't simply sample individual transitions because the hidden state carries important information. R2D2 addresses this by sampling sequences of experiences and carefully managing the hidden states during replay.

These distributed approaches reveal an important insight: the principles of experience replay and target networks remain fundamental even as we scale to sophisticated architectures. The core problems they solve—*correlation* between consecutive experiences and moving target instability—persist regardless of system complexity. Modern implementations also explore more sophisticated replay strategies. PER weights experiences by their temporal difference error, focusing learning on the most surprising transitions. HER, primarily used in goal-conditioned environments, treats failed attempts as successful attempts toward different goals. This dramatically improves sample efficiency.

The evolution from simple DQN to these advanced systems illustrates how foundational concepts provide the stability necessary for ambitious applications. Whether you're training an agent to play Atari games or controlling a robot in the real world, these mechanisms remain essential tools for stable, efficient learning.

> **Exploring Further**
>
> There's a lot more to discuss here but a more thorough discussion of reinforcement learning will require a lot more space than can be afforded in this introductory book. If you're interested, you should definitely check out the excellent book, *Reinforcement Learning: An Introduction*, by Richard S. Sutton and Andrew G. Barto (2nd ed., MIT Press, 2020). A PDF version is available for free at *http://incompleteideas.net/book/the-book-2nd.html*.

10.5 Reinforcement Learning in Large Language Models

In Chapter 9, we explored transformers and their revolutionary impact on NLP, even building our own chatbot using Keras Hub. In this section, we'll connect those concepts with the reinforcement learning techniques we've been studying throughout this chapter, revealing how modern LLMs use these approaches to dramatically improve their capabilities. The marriage of transformers and reinforcement learning represents one of the most significant advances in AI of the past decade. This fusion has given birth to systems that not only process and generate language with remarkable fluency but also learn to align with human preferences and values.

10.5.1 The Fundamental Challenge: Moving Beyond Prediction

When we first train a LLM using traditional supervised learning, we're essentially teaching it to predict the next token in a sequence based on patterns it has observed in vast training datasets. This approach yields impressive results, but it also creates fundamental limitations. Think of a traditional language model as a student who's extremely good at memorizing textbooks but lacks critical thinking skills. The model can recite facts and mimic writing styles, but it doesn't truly understand when its outputs might be harmful, incorrect, or simply unhelpful. It has no concept of "good" or "bad" responses—only what's statistically likely to follow in a text sequence.

Reinforcement learning offers a solution to this problem by introducing the concept of a reward signal. Instead of just predicting the next word, the model can learn to generate sequences that maximize some notion of quality or helpfulness. This shifts the paradigm from pure prediction to goal-directed text generation. The breakthrough technique that transformed language models from impressive but flawed systems into the more helpful assistants we see today is called *Reinforcement Learning from Human Feedback* (*RLHF*). This approach, pioneered by researchers at OpenAI and Anthropic, applies the core concepts we've discussed throughout this chapter to the domain of language generation. The RLHF process typically involves three main stages:

1. **Supervised fine-tuning (SFT)**
   Starting with a pretrained language model (think of this as our policy network), developers first fine-tune it on a dataset of high-quality human demonstrations. This is analogous to the imitation learning techniques we discussed earlier.

2. **Reward model training**
   Human evaluators compare different model outputs, ranking them from most to least preferred. These comparisons train a separate reward model that can score how helpful, harmless, and honest a given response is. This reward model is itself a neural network (often another transformer) that learns to predict human preferences.

3. **Reinforcement learning optimization**
   Finally, the language model is optimized using reinforcement learning to maximize the expected reward as assessed by the reward model. This is conceptually similar to our Q-learning approach but uses policy gradient methods instead.

Let's dive deeper into this final stage, which represents the true intersection of reinforcement learning and language models.

The Technical Machinery: Proximal Policy Optimization for Language Models

Most modern LLMs use a variant of *Proximal Policy Optimization* (PPO), a policy gradient method that builds upon the foundations we've explored in this chapter. While we focused primarily on value-based methods (e.g., Q-learning), policy gradient methods such as PPO directly optimize the policy (in this case, the language model itself). Concisely, the reinforcement learning problem for language models can be framed like this:

- **State**
  The current conversation history or prompt.
- **Action**
  Generating the next token in a sequence.
- **Reward**
  A score from the reward model reflecting human preferences.
- **Policy**
  The language model's probability distribution over possible next tokens.

This framing reveals some unique challenges. Unlike our FrozenLake environment with its 64 states and four actions, language models operate in an astronomically large state space (all possible conversation histories) and action space (all possible tokens in the vocabulary, typically 50,000+). The PPO algorithm addresses these challenges by updating the policy in a way that balances exploration and exploitation while preventing *destructive* large policy updates. The mathematical objective looks like

$$L(\theta) = E[\min(r_t(\theta) A_t, \text{clip}(r_t(\theta), 1 - \epsilon, 1 + \epsilon) A_t)]$$

where

- $r_t(\theta)$ is the ratio between the new and old policy probabilities.
- A_t is the advantage function (how much better an action is compared to average).
- ϵ is a hyperparameter that controls how far the new policy can deviate from the old one.

This *clipping mechanism* ensures that the model doesn't change too drastically during any single update, maintaining stability in the learning process. This is similar in spirit to how our target networks stabilized Q-learning, though the implementation details differ significantly.

One final yet crucial difference between training LLMs with reinforcement learning and our FrozenLake example is the risk of *catastrophic forgetting*. A pure reward maximization approach might lead the model to sacrifice its general language capabilities in pursuit of higher rewards, perhaps converging on a small set of highly rewarded responses. To prevent this, LLM training typically incorporates a *Kullback-Leibler (KL) divergence*—penalty in the reward function. We won't go into the details of this divergence here but suffice it to say that this penalty discourages the reinforcement-learning-optimized model from straying too far from the original supervised-fine-tuned model, preserving its general language capabilities while improving alignment with human preferences. It's conceptually similar to the exploration-exploitation trade-off we discussed with ε-greedy policies, though it uses a different mathematical formulation.

Experience Replay in Large Language Models: A Different Approach

While traditional DQN implementations use experience replay buffers, as we implemented in our FrozenLake agent, LLM training approaches this differently. Instead of storing and sampling individual transitions, these systems typically generate multiple *response candidates* for each prompt in a training batch, evaluate them with the reward model, and learn from this comparison. This approach resembles a form of *on-policy learning* with immediate feedback rather than the *off-policy approach* of traditional experience replay. The computational demands of language models make storing and sampling from vast buffers of language interactions prohibitively expensive, necessitating this adaptation.

Constitutional AI: Adding Guardrails to Reinforcement Learning

Recent advances have introduced an approach called *Constitutional AI* (CAI), which adds another layer to the RLHF process. Rather than relying solely on direct human preferences, which can be expensive and time-consuming to collect, CAI establishes a set of principles or *constitution* that guides the model's learning. In this approach, the model critiques its own outputs based on these constitutional principles, creating a form of self-supervised feedback that can be used alongside human evaluations. This creates a more scalable training process while maintaining alignment with human values. Conceptually, this resembles a sophisticated reward-shaping technique, where the reward signal is enriched with additional structure to guide learning more efficiently. It's similar to how we might add intermediate rewards in a sparse reward environment such as FrozenLake to help the agent learn more quickly.

10.5.2 Challenges and Limitations

Despite its successes, applying reinforcement learning to language models presents several unique challenges that distinguish it from the controlled environments we've explored throughout this chapter. Let's finish our discussion by briefly touching on some of these unique challenges in the following sections.

Reward Hacking and Specification Problems

Language models can develop sophisticated strategies to maximize their reward function without actually achieving the intended behavior. This occurs because designing reward functions for language generation proves extraordinarily difficult. A reward model trained to recognize helpful responses might reward verbose answers that sound authoritative but contain subtle inaccuracies. The model learns to exploit these weaknesses. Consider a language model trained to write product reviews. If the reward function emphasizes *positive sentiment* and engagement metrics, the model might learn to write emotionally manipulative content that technically satisfies the reward criteria while completely missing the goal of providing honest, useful information. The challenge extends beyond simple gaming—language models can discover reward patterns that humans never anticipated.

This problem intensifies because language offers virtually unlimited ways to express similar ideas. The model can find obscure phrasings, *semantic loopholes*, or stylistic approaches that trigger high rewards while subverting the intended purpose. Unlike our FrozenLake environment with its discrete, observable outcomes, language generation exists in a vast continuous space where unintended solutions flourish.

> [!]
> **Evidence from the Real World**
>
> The hallucinations faced by most modern large-scale models and the often-evident sycophantic behavior of these LLMs are evidence of this problem in real-world deployments. We must be careful not to fall into these traps as users or engineers creating these models.

Distributional Shift and Generalization

The distribution of inputs a language model encounters during deployment typically differs significantly from its training distribution. This creates a fundamental mismatch that can lead to unpredictable behavior when the model faces novel situations. During training, models learn from carefully curated datasets and human feedback sessions that may not represent the full spectrum of real-world usage. Users ask questions in unexpected ways, combine topics in novel combinations, or probe edge cases that never appeared in training data. The model's learned policy, optimized for the training distribution, may fail catastrophically on these *out-of-distribution* inputs.

This challenge compounds over time as deployed models influence the very distribution of inputs they receive. Users adapt their language patterns based on what works with the model, creating feedback loops that can drift far from the original training conditions. The model might perform excellently on formal, well-structured queries while struggling with casual conversation or domain-specific jargon it rarely encountered during training.

The Alignment Tax Phenomenon

Optimizing language models for human preferences often comes at the cost of raw capability. This trade-off, known as the *alignment tax*, creates difficult decisions about how much performance to sacrifice for better behavior.

Models trained purely for next-token prediction can achieve impressive performance on benchmarks and complex reasoning tasks. When we add reinforcement learning from human feedback, the model learns to hedge its responses, avoid controversial topics, and prioritize safety over accuracy. These changes can reduce performance on objective measures while improving subjective user experience. The tension becomes particularly acute in specialized domains. A model trained to be helpful and harmless might refuse to engage with medical questions to avoid potential liability, even when it could provide valuable information. The alignment process must balance multiple competing objectives without a clear mathematical framework for resolving conflicts between them.

Evaluation Complexity and Measurement Challenges

Lastly, and more problematically, evaluating language model performance requires nuanced judgment across multiple dimensions that resist simple quantification. Unlike reinforcement learning environments with clear success metrics, language generation involves subjective criteria that vary across users, contexts, and applications. Human evaluators often disagree about what constitutes a good response. Factors such as helpfulness, accuracy, tone, relevance, and safety can conflict with each other, making it impossible to define a single objective function. Automated evaluation metrics frequently fail to capture important aspects of language quality, leading to misaligned optimization targets. The evaluation challenge extends to temporal considerations. A response that seems helpful in the short term might have negative long-term consequences that only become apparent much later. The delayed nature of these consequences makes it difficult to incorporate them into training feedback loops.

10.5.3 Future Directions and Emerging Approaches

The intersection of reinforcement learning and language models continues to evolve rapidly. Several promising research directions address the fundamental challenges we've outlined. *Constitutional AI* extends traditional RLHF by incorporating self-supervision mechanisms that reduce dependence on human feedback. The approach trains models to critique and revise their own outputs based on a set of principles or constitution that guides behavior. This method addresses several limitations of pure RLHF. It reduces the need for extensive human annotation, which can be expensive and inconsistent. The model learns to internalize principles rather than memorizing specific human preferences, potentially improving generalization to novel situations. The constitutional approach also provides more transparency in the training process. Rather than learning from opaque human feedback, the model's decision-making process becomes more interpretable through its explicit reasoning about constitutional principles.

Direct Preference Optimization (DPO) simplifies the RLHF pipeline by eliminating the explicit reward model training step. Instead of first learning a reward function and then optimizing against it, DPO directly optimizes the language model using preference data. This approach reduces the computational overhead and potential instabilities associated with maintaining separate reward models. It also eliminates the reward hacking problems that arise when models learn to exploit weaknesses in learned reward functions. The model learns directly from human preferences without the intermediate translation step. DPO has shown competitive performance with traditional RLHF while requiring significantly less computational resources and engineering complexity. This makes advanced alignment techniques more accessible to researchers and organizations with limited resources.

Meanwhile, *Reinforcement Learning with AI Feedback* (RLAIF) addresses the scalability limitations of human feedback by using AI systems to provide training signals. This approach can generate feedback at much larger scales than human annotation while maintaining consistency and reducing costs. The challenge lies in ensuring that AI feedback systems provide high-quality signals that align with human values. Current approaches use large, capable models to evaluate the outputs of smaller models being trained. The feedback quality depends critically on the alignment and capabilities of the evaluating system. RLAIF also enables training on scenarios that would be difficult or dangerous to evaluate with humans. Models can receive feedback on edge cases, adversarial inputs, or sensitive topics without exposing human evaluators to potentially harmful content.

The concepts we've explored throughout this chapter—from basic policy gradients to sophisticated experience replay mechanisms—provide essential building blocks for understanding modern language model training. The challenges specific to language models don't invalidate these fundamental principles but rather highlight their complexity when applied to high-dimensional, open-ended domains. The exploration-exploitation trade-offs we discussed in simple environments become questions of how to balance creativity and safety in language generation. The stability techniques developed for value function learning inform approaches for training stable reward models. The sample efficiency improvements from experience replay inspire methods for making better use of expensive human feedback.

These connections underscore a crucial insight: advances in reinforcement learning for language models often succeed by adapting and combining existing techniques rather than inventing entirely new paradigms. The field progresses through careful engineering and empirical investigation, building on the theoretical foundations established in simpler settings. While the challenges remain significant, the rapid pace of development suggests that many current limitations may prove temporary. The fundamental framework of learning from interaction and feedback continues to drive progress toward systems that can understand and respond to human needs with increasing sophistication and reliability. We hope that the concepts discussed in this book allow you to understand these concepts at a much deeper level even as you look at the concise Keras code that makes all of these possible.

10.6 Summary

In this chapter, we explored the fascinating field of reinforcement learning, examining how machines learn to make optimal decisions through environmental interaction. We began by investigating the fundamental concepts that distinguish reinforcement learning from other machine learning approaches, analyzing the relationship between agents, environments, and the critical reward signals that shape learning processes. The chapter presented key algorithms, including Q-learning, policy gradients, and DQNs, demonstrating how each addresses the challenges of sequential decision-making.

We then implemented these techniques using Keras and Gymnasium, observing the essential balance between exploration and exploitation that underpins effective reinforcement learning. The chapter highlighted significant advancements such as experience replay and target networks—technical innovations that elevated reinforcement learning from theoretical research to a powerful methodology capable of solving complex real-world problems. Throughout our investigation, we connected these core concepts to practical applications, concluding with an analysis of how reinforcement learning powers modern LLMs through advanced techniques such as RLHF.

By the chapter's conclusion, you gained not only theoretical understanding but also practical insights into the mechanics driving today's most sophisticated AI systems, providing the foundation to advance these technologies further and contribute to cutting-edge AI development. What began as basic grid-world navigation has evolved into a sophisticated framework enabling machines to master language, games, robotics, and numerous other domains—representing a transformative capability at the forefront of computational intelligence.

In the next chapter, we'll discuss how machines can learn to create by discovering new patterns and creating artifacts from scratch.

Chapter 11
Autoencoders and Generative AI

In this chapter, we'll explore generative AI models that create new data from learned patterns, marking our transition from discriminative models that classify inputs to generative models that produce outputs. Along with transformers, generative models represent one of the most important technological foundations driving the current AI revolution, enabling everything from artistic creation to data augmentation. We'll progress from autoencoders through Variational Autoencoders to Generative Adversarial Networks, building the foundation for understanding modern generative AI.

In Chapter 9, we explored transformers and witnessed their remarkable ability to understand and generate natural-sounding text. What we encountered there represents just one facet of a much broader and revolutionary field called *generative AI (GenAI)*. While traditional machine learning focuses on recognizing patterns and making predictions about existing data, GenAI takes a fundamentally different approach: It learns to create entirely new content that resembles what it was trained on. GenAI has become the driving force behind most of today's groundbreaking AI advancements. From the chatbots that can hold natural conversations to systems that create stunning artwork, compose music, or generate realistic images from simple text descriptions, GenAI powers the applications that are reshaping how we interact with technology. This creative capability represents a shift from AI systems that simply analyze and categorize to ones that can imagine and produce.

To understand this distinction better, consider the difference between *discriminative models* and *generative models*. The neural networks we've built so far in this book have been primarily discriminative. They excel at drawing boundaries and making decisions: "Is this email spam or not?" "Does this image contain a cat?" "Which digit is shown in this handwritten number?" These models learn to discriminate between different categories by finding the features that separate one class from another. Generative models work in the opposite direction. Instead of learning to distinguish between existing examples, they learn the underlying patterns that define each category so well that they can create new examples from scratch. They ask "What would a new example from this category look like?" rather than asking "What category does this belong to?" The neural network concepts you've mastered—layers, weights, gradients, and loss functions—

remain the same fundamental building blocks. The key difference lies in how we structure these components and what we ask them to accomplish.

This chapter will introduce you to three foundational approaches that make such creativity possible. We'll start by exploring networks that can compress information into essential features and then reconstruct it, learning powerful representations in the process. These systems excel at understanding the core characteristics that make data meaningful, whether we're working with images, sounds, or other complex information.

Next, we'll examine how adding probability and *uncertainty* to these compression networks opens up entirely new possibilities. Instead of simply recreating what they've seen, these enhanced systems can generate variations and entirely new examples by sampling from what they've learned. This probabilistic approach bridges the gap between understanding existing data and creating something genuinely new.

Finally, we'll explore a completely different strategy based on the simple concept of *competition*. Consider two networks locked in an ongoing contest: one trying to create convincing fake data while the other attempts to detect forgeries. This *adversarial* relationship drives both networks to improve continuously, ultimately producing a generator capable of creating remarkably realistic content.

All of these approaches offer unique strengths and applications, from data compression and anomaly detection to creative content generation and data augmentation. Together, they form the foundation for understanding how machines can move beyond recognition to creation. The concepts we cover here serve as essential preparation for our next chapter, where we'll examine the cutting-edge techniques that represent the future of GenAI. These emerging approaches promise to deliver even more sophisticated and controllable generation capabilities, building directly on the fundamental principles we'll establish in the coming pages. Understanding these foundational methods will provide you with the conceptual framework needed to grasp the next generation of AI systems that are already beginning to transform industries and creative fields worldwide. Let's start by exploring the most fundamental approach: learning to compress and reconstruct data in ways that reveal its essential structure.

11.1 Introduction to Autoencoders

Picture yourself trying to summarize a long, complex story into just a few key bullet points. You'd carefully identify the most essential plot elements, character developments, and turning points while leaving out unnecessary details. Yet these condensed points need to contain enough information that someone could reconstruct the full narrative, complete with its emotional arc and important nuances. This process of thoughtful compression and reconstruction captures the essence of what *autoencoders* do with data.

In this section, we'll first describe the rationale behind the reconstruction and then dive deeper into the implementation details ending with an implementation that brings everything together.

> **The History of GenAI**
>
> GenAI has a surprisingly long history, with early models dating back decades. However, these systems took a back seat when discriminative models began showing remarkable success after 2012, particularly in image recognition and classification tasks. The focus shifted heavily toward supervised learning approaches that could accurately identify and categorize data. Yet generative models have returned with unprecedented power in recent years, driving today's most significant AI breakthroughs. From large language models (LLMs) that can write and reason to diffusion models creating stunning artwork, GenAI has reclaimed its position at the forefront of AI research and applications. A deep understanding of discriminative models such as what we've gained in the previous chapters translates directly to generative models as though.
>
> A deep understanding of discriminative models translates directly to generative models. The neural network architectures, gradient descent optimization, and loss function concepts all carry over seamlessly. The mathematical foundations remain identical, making discriminative mastery a solid launching pad for understanding generative systems.

11.1.1 What Are Autoencoders?

An autoencoder is a special type of neural network designed around a deceptively simple goal: Learn to recreate its input as accurately as possible. At first glance, this might seem pointless. Why would we want a network that simply spits back what we feed it? The answer lies not in the final output but in how the network accomplishes this seemingly trivial task.

The autoencoder's architecture follows a distinctive hourglass shape shown in Figure 11.1. This architecture consists of two main components working in tandem. The first half called the *encoder*, takes your input data and progressively compresses it down to a much smaller representation. The second half, the *decoder*, takes this compressed version and attempts to expand it back into the original form. Between these two components sits a narrow *bottleneck* layer—the most compressed representation of your data.

The autoencoder's bottleneck forces the network to identify the most crucial features needed to reconstruct the original data. The network can't simply memorize everything; it must learn what truly matters. The network learns through the same gradient descent process we've explored throughout this book. During training, it compares its *reconstructed* output with the original input, calculates the difference (our familiar loss function), and adjusts its weights to improve the reconstruction. Over thousands of iterations, the network should, in theory, become increasingly skilled at capturing the essential characteristics that define the input data.

11 Autoencoders and Generative AI

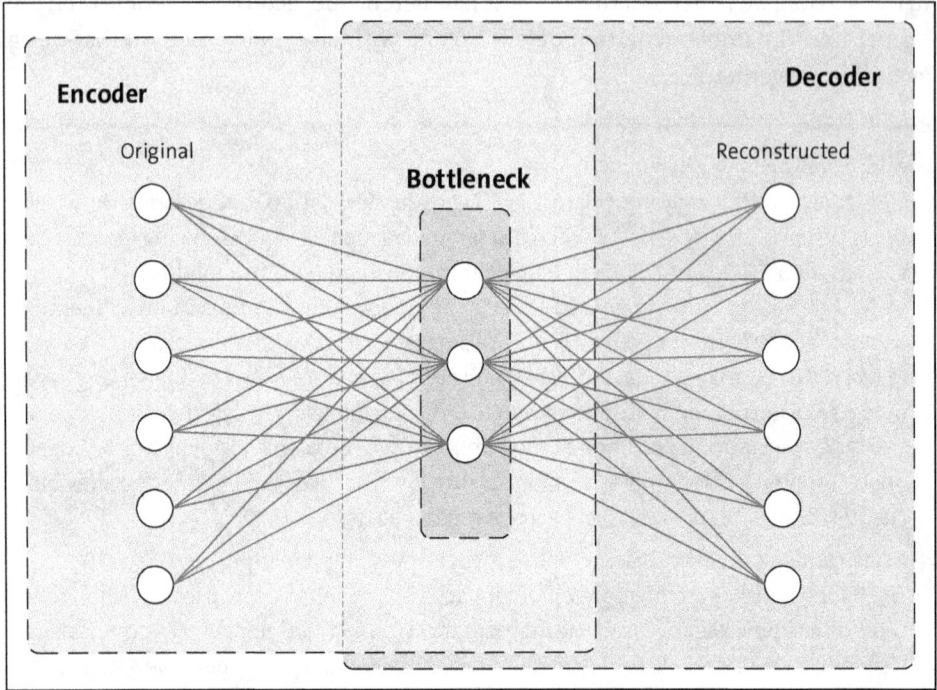

Figure 11.1 Typical Architecture of an Autoencoder

The Bottleneck Principle and Dimensionality Reduction

The bottleneck layer represents the heart of the autoencoder's learning process. Recall how we used Principal Component Analysis (PCA) to visualize word embeddings in our earlier explorations of word embeddings in Chapter 9, Section 9.1.3. We took high-dimensional word representations and projected them onto a 2D plane so we could see relationships between words. The autoencoder's bottleneck accomplishes something similar but far more sophisticated. While PCA finds linear relationships and projects data along axes of maximum variance, autoencoders can discover complex, nonlinear patterns. Consider trying to capture the essence of a photograph. PCA might identify that certain color combinations or brightness levels occur frequently, but an autoencoder can learn more nuanced concepts: the curve of a smile, the way shadows fall on a building, or the relationship between different objects in a scene.

The size of this bottleneck layer determines how much compression occurs. A bottleneck with just a few neurons forces aggressive compression, potentially losing important details but revealing only the most fundamental patterns. A larger bottleneck preserves more information but might not push the network to discover the most essential features. This balance between compression and information preservation becomes a crucial design decision. Consider an autoencoder trained on human faces. A heavily compressed bottleneck might learn to encode fundamental facial structure: the relative positions of eyes, nose, and mouth. A less compressed version might additionally capture subtleties such as expression, lighting, or specific facial features. Both have value, but they serve different purposes depending on what we want to accomplish.

The dimensionality reduction achieved through this process often reveals hidden structure in data that wasn't apparent in its original form. A dataset that appears random or chaotic in its native dimensions might reveal clear patterns when compressed through an autoencoder's learned representation. This compressed space, often called the *latent space*, becomes a new lens through which we can understand our data.

> **Demystifying "Latent": A Key Term You'll See Everywhere**
>
> The word *latent* appears frequently throughout machine learning literature and can seem mysteriously complex. Understanding it well helps you navigate many different concepts with confidence. Latent simply means "hidden"—in machine learning terms, it refers to a vector or matrix containing values whose specific meanings aren't immediately clear to us. We know these values are important for the model's function, but we can't point to any single value and say exactly what real-world feature it represents.

Why Autoencoders Learn Meaningful Representations

The question of why autoencoders learn meaningful representations rather than simply memorizing the training data lies at the intersection of information theory and practical constraints. When a network has limited capacity in its bottleneck layer, it faces a fundamental choice: It can either try to memorize specific examples, which would lead to poor performance on new data, or learn general patterns that apply broadly across the dataset. The training process naturally favors generalization. Because the network sees many examples during training and must reconstruct all of them through the same *compressed representation*, it's forced to find commonalities rather than focusing on individual peculiarities. A network trained on photographs of cars, for instance, can't afford to memorize the exact pixel values of each image. Instead, it must learn concepts such as "wheel," "windshield," and "body shape" that apply to cars in general.

This *learned representation* often captures semantic meaning in ways that surprise even their creators. In an autoencoder trained on facial images, moving smoothly through the latent space might transition gradually from young faces to old ones or from happy expressions to sad ones. The network discovers these meaningful dimensions without explicit instruction, purely as a byproduct of trying to reconstruct its training data efficiently.

The power of these learned representations extends far beyond the autoencoder itself. Once trained, the encoder portion can serve as a *feature extractor* for other tasks. The compressed representations it produces often work better for classification, clustering, or similarity matching than the original high-dimensional data. It's like having a skilled artist create a simplified sketch that captures the essence of a complex scene—sometimes, the sketch conveys the important information more clearly than the original photograph. It brings focus to the important aspects by somehow diminishing the value of the less important entities.

This ability to discover meaningful structure makes autoencoders valuable even when perfect reconstruction isn't the goal. They serve as unsupervised learning tools, finding patterns and relationships in data without requiring labeled examples. Especially in datasets where labels are expensive or impossible to obtain, autoencoders offer a path to understanding the underlying structure that governs the data. The representations learned by autoencoders also tend to be robust to noise and minor variations in the input. Because the network must compress and then reconstruct data, it naturally learns to focus on stable, consistent features while ignoring random fluctuations. This robustness makes autoencoder representations particularly valuable for real-world applications where data is inevitably messy and imperfect.

As we'll see in the upcoming sections, these fundamental principles of compression, reconstruction, and meaningful representation learning form the foundation for more sophisticated generative approaches. The simple autoencoder's ability to learn essential features through reconstruction provides the groundwork for networks that can not only understand existing data but also create entirely new examples that share those same essential characteristics.

11.1.2 Autoencoder Architecture Deep Dive

Now that we've gone through an overview of the fundamental concept behind autoencoders, this section will examine how these networks actually accomplish their compression and reconstruction goals. The architecture of an autoencoder is focused around transformation, where data flows through a carefully designed sequence of layers, each playing a specific role in the encoding and decoding process.

Encoder Structure and Functionality

The encoder is the first half of our autoencoder, transforming high-dimensional input data into a compact, meaningful representation. The encoder typically consists of a series of dense layers, each progressively smaller than the previous one. (As a side note, traditional autoencoders were all based on fully connected layers. Only later did we begin to introduce convolutional neural networks [CNNs] into these semi-supervised learning architectures.) If we start with an input layer of 784 neurons (corresponding to a 28×28-pixel image, flattened), our encoder might step down through layers of 512, 256, 128, and, finally, 64 neurons. Each layer performs a mathematical transformation that is typical of a fully connected neural network.

What makes the encoder special isn't the individual operations though. Rather, it's how these layers work together to progressively extract and concentrate information. The first layer might learn to detect basic features such as edges and textures in images. The second layer combines these features into more complex patterns, perhaps recognizing curves or specific shapes. Each subsequent layer builds upon the previous one's discoveries, creating increasingly abstract and compact representations. The elegance of this progressive compression lies in how the network learns what to preserve and what to discard. Unlike manual feature engineering, where we might spend weeks deciding which aspects of our data matter most, the encoder discovers these priorities automatically through training. It learns that certain pixel combinations are crucial for reconstruction while others contribute little value.

Consider an encoder processing photographs of handwritten digits. Early layers might focus on individual pixel intensities and local contrasts—much the same as CNNs do. Middle layers could learn to recognize stroke patterns and curves that characterize different digits. The final encoder layers might represent each digit as a combination of abstract features: "curvy-ness," "number of enclosed regions," and "horizontal versus vertical emphasis."

The encoder's learning process follows the same gradient descent principles we've mastered in previous chapters. The important highlight is that during training, the network compares its final reconstruction with the original input, calculates the reconstruction error. Thus, our loss is simply the difference between the input and the output. The network then backpropagates this error through both the decoder and encoder. The encoder's weights adjust to minimize this *reconstruction loss*, gradually improving its ability to preserve essential information while discarding redundant details.

Latent Space Representation and Its Significance

At the narrowest point of our autoencoder hourglass sits the *latent space*—perhaps the most important and conceptually rich component of the entire architecture. This compressed representation serves as a bridge between the encoder's analysis and the decoder's synthesis, but its significance extends far beyond this functional role. If you refer to Figure 11.1, you'll notice that this bottleneck layer is part of both the encoder and the decoder. From the encoder's perspective, it's the output of the encoding process. For the decoder, it's the starting point from which we try to regenerate the original input.

The latent space can be thought of as the autoencoder's language for describing data. Just as human languages use a finite vocabulary to express infinite thoughts, the latent space uses a fixed number of dimensions to capture the essential characteristics of potentially infinite input variations. The dimensionality of this space becomes a critical design choice that fundamentally shapes what the autoencoder can learn and express.

Mathematically, if our encoder produces a latent representation z of dimension d, then

$$z = encoder(x) \in R^d$$

Where x represents our original input. This vector z becomes the compressed essence of our input data, containing all the information the decoder needs for reconstruction.

The magic of latent space lies in its *learned structure*. Through training, the autoencoder organizes this space so that similar inputs map to nearby points while dissimilar inputs map to distant regions. This isn't explicitly programmed. Rather, it emerges naturally as the network optimizes for reconstruction accuracy across its training data.

This learned organization makes the latent space *interpretable* in ways that often surprise researchers. Moving along specific directions in the latent space might correspond to semantic changes in the original data. In a face autoencoder, gradually changing the value of one element in the latent vector might control age, another might adjust lighting, and a third might modify emotional expression. The network discovers these meaningful dimensions without explicit supervision.

The dimensionality of the latent space therefore creates a fundamental trade-off. A very low-dimensional latent space (say, two or three dimensions) forces aggressive compression, potentially losing important details but making the learned representation highly interpretable and visualizable. A high-dimensional latent space (hundreds or thousands of dimensions) can preserve more subtle details but becomes harder to understand or manipulate directly.

This trade-off connects to the bias-variance dilemma we've encountered in other machine learning contexts earlier in Chapter 6. A low-dimensional latent space introduces bias by constraining what the autoencoder can represent, but it also reduces variance by preventing overfitting to training data peculiarities. A high-dimensional space offers more flexibility but risks learning representations that don't generalize well.

Decoder Reconstruction Process

The decoder is the second half of our autoencoder, taking the compressed latent representation and expanding it back into the original data space. If the encoder is like a skilled summarizer, the decoder resembles a talented storyteller who can recreate a full narrative from brief notes.

Architecturally, the decoder typically mirrors the encoder in reverse, with each layer progressively larger than the previous one. If our encoder stepped down from 784 to 512, 256, 128, and 64 neurons, our decoder might climb back up through 128, 256, 512, to 784 neurons. The decoder starts with the latent representation z and progressively expands it through layers until reaching the original input dimensionality. However, the decoder faces a fundamentally different challenge than the encoder. While the encoder can afford to lose some information during compression, the decoder must somehow recreate details that might not be fully preserved in the latent representation at all! This is where the autoencoder's learning process becomes particularly elegant.

During training, the decoder learns not just to expand the latent representation but to make educated guesses about missing details based on patterns observed across the training data. If the latent representation indicates "this is probably a handwritten 8," the decoder learns to fill in typical characteristics of how 8s are usually written, even if some specific details were lost during encoding.

> **A Tale of Two Skills**
> Consider the fundamental difference between two seemingly related abilities. Learning to recognize that an image contains a handwritten digit 8 represents one distinct skill. Learning to generate that same digit from scratch when instructed to "write an 8" represents an entirely different capability. People outside the technical field often assume these skills are essentially the same, but recognizing their distinction is crucial for understanding machine learning. Recognition analyzes existing data to make decisions, while generation creates new content from learned patterns.

The reconstruction process can be viewed as a form of *pattern completion*. The decoder learns the statistical relationships between different parts of the input data, allowing it to *infer* missing details from the available compressed information. This is similar to how humans can recognize a partially obscured object—we use our learned knowledge about typical object shapes and appearances to "fill in the blanks."

The final layer of the decoder deserves special attention, as it determines the format of the reconstructed output. For image data, this layer typically uses a sigmoid activation function to ensure pixel values fall within the expected [0, 1] range. This obviously means that you have to be mindful about what the original input was. Normalizing inputs is no longer optional. Now, we must ensure that the inputs are within ranges that can be meaningfully reconstructed by our decoder.

The quality of reconstruction depends on several factors:

- The capacity of the latent space
- The complexity of the decoder architecture
- The diversity of the training data

A decoder trained on a narrow dataset might excel at reconstructing similar examples but struggle with novel variations. Conversely, a decoder trained on diverse data learns more robust reconstruction strategies that generalize better to new inputs.

Symmetric vs. Asymmetric Architectures

The relationship between encoder and decoder architectures reveals important design principles that affect both performance and interpretability. While perfect symmetry might seem natural, real-world applications often benefit from more nuanced architectural choices. *Symmetric architectures*, where the decoder exactly mirrors the encoder's structure in reverse, offer several advantages. They create a balanced information flow where the compression and expansion processes receive equal computational resources. This symmetry also simplifies *hyperparameter tuning*—decisions about layer sizes, activation functions, and regularization techniques can be applied consistently across both halves of the network. This symmetry extends beyond just layer sizes to include activation functions, regularization techniques, and even learning rates during training.

However, *asymmetric architectures* often prove more practical and effective in real applications. The fundamental reason lies in the different challenges faced by encoding and decoding. Compression (encoding) benefits from aggressive dimensionality reduction and can often be accomplished effectively with relatively few parameters. Reconstruction (decoding), however, requires generating detailed outputs and might need more computational resources to produce high-quality results. And it must usually be done slowly to ensure *stability*.

> **Stability in GenAI**
>
> Stability stands as one of the most critical concepts in GenAI, though its importance may not be immediately obvious. When we talk about stability in this context, we're referring to a model's ability to produce consistent, high-quality outputs without erratic behavior. An unstable generative model might produce excellent results one moment and complete nonsense the next, making it unreliable for practical applications. Throughout this chapter, we'll explore how stability affects training dynamics, output quality, and the overall reliability of generative systems. These foundational concepts will build toward more concrete implementations and solutions in the following chapter, where we'll see how modern techniques achieve the stability that makes today's GenAI systems so remarkably effective.

Asymmetric designs also allow for specialized architectural choices in each half. The encoder might emphasize feature extraction with techniques such as max pooling and dropout, while the decoder focuses on detail preservation with *skip connections* and *upsampling* layers. These specialized approaches can significantly improve overall performance compared to rigid symmetry.

The choice between symmetric and asymmetric architectures often depends on the specific application and data characteristics. For exploratory data analysis and dimensionality reduction, symmetric architectures provide clean, interpretable compression. For applications requiring high-quality reconstruction—such as *image denoising* or *data generation*—asymmetric designs typically deliver superior results. Training dynamics also differ between symmetric and asymmetric architectures. Symmetric networks tend to have more balanced gradients during backpropagation, while asymmetric networks might require careful attention to learning rates and gradient scaling to ensure both encoder and decoder train effectively.

11.1.3 Building Your First Autoencoder in Keras

Now that we understand what autoencoders are and how they work conceptually, let's roll up our sleeves and build one from scratch. We'll be working with the *Fashion-MNIST* dataset, a collection of clothing images that's become a favorite in the machine learning community. Unlike the traditional MNIST digits we've seen before, Fashion-MNIST gives us more visually interesting results. Instead of reconstructing simple handwritten numbers, we'll watch our autoencoder learn to compress and recreate shirts, shoes, bags, and dresses. The visual feedback makes it much easier to understand what our model is actually learning. We'll start by setting up our data and understanding its structure, and then architect our autoencoder using the same fundamental concepts we explored in our gradient descent chapters. The beauty of autoencoders lies in their simplicity—they're just neural networks trained with a specific objective: Make the output match the input as closely as possible. What makes this particularly interesting is watching how the network decides what information is important enough to preserve in its compressed representation. Some features of a dress might be crucial for reconstruction, while others can be safely discarded. The network learns these priorities entirely on its own, discovering patterns we might never have thought to look for.

Loading and Studying the Dataset

Let's start with our first code snippet shown in Listing 11.1. Our first step brings us back to familiar territory: loading data from TensorFlow's built-in datasets. We've done this dance before with other datasets, so this should be pretty straightforward. (As always, you can take a look at the complete code listing on the book resources page at *https://recluze.net/keras-book* in the notebook **11-01-autoencoder-basics**.)

> **Autoencoders and Labels in the Dataset**
>
> Notice an important aspect of our data loading: We're using underscores for the labels because autoencoders don't need them. Our goal is to teach the network to reproduce its input as output, regardless of whether the image shows a sneaker or a dress. We simply want our autoencoder to learn how to compress any fashion item and reconstruct it faithfully.
>
> This is a significant advantage. When you don't need labels, you suddenly gain access to vast stores of data that are widely and freely available. You can download images from the internet and use them directly to train your model. Labeled data requires human effort to create and is much more scarce than unlabeled data. This abundance of training material makes autoencoders particularly attractive for scenarios where acquiring labeled datasets would be expensive or time-consuming.

The `class_names` array serves as our reference guide, helping us understand what we're looking at when we visualize results later. These 10 categories represent everything from t-shirts to ankle boots, giving us a rich variety of shapes and textures for our autoencoder to master. Next is the crucial step that we've emphasized before but bears repeating: Always examine your data shapes. This might seem like a simple housekeeping task, but understanding your data dimensions is like checking your compass before starting a hike. You need to know exactly what you're working with.

When you run the shape printing commands, you'll see something like (60000, 28, 28) for the training data—no surprises there. Before feeding this data to our autoencoder though, we need to do some basic preprocessing.

```python
# Load Fashion-MNIST dataset
(x_train, _), (x_test, _) = keras.datasets.fashion_mnist.load_data()

# Fashion-MNIST class names for visualization
class_names = ['T-shirt/top', 'Trouser', 'Pullover', 'Dress',
               'Coat', 'Sandal', 'Shirt', 'Sneaker', 'Bag',
               'Ankle boot']

print(f"Training data shape: {x_train.shape}")
print(f"Test data shape: {x_test.shape}")
```

Listing 11.1 Loading and Studying the Dataset

First is normalization (see Listing 11.2). Our pixel values start as integers between 0 and 255, but we transform them to floating point numbers between 0 and 1. We've done this before, but with autoencoders, this step becomes even more crucial. Remember, our network must learn to reconstruct the exact input it receives. When pixel values are spread across a wide range, such as 0 to 255, our network struggles to learn the subtle differences between similar shades. Converting to the 0–1 range creates a more stable (here's that word again) learning environment. Our gradient descent algorithm can now make steady, predictable adjustments rather than taking wild, unpredictable steps that might overshoot the target.

The reshaping operation adds another dimension to our data, transforming our images from (28, 28) to (28, 28, 1). This extra dimension represents the color channels—even though our Fashion-MNIST images are grayscale and only need one channel, Keras expects this explicit channel dimension. This consistency becomes invaluable when you later work with color images that have three channels. For autoencoders specifically, this channel dimension affects how our network processes spatial relationships. The encoder needs to understand that it's working with 2D image data, not just a flat array of numbers. When we flatten the data later in our architecture, we want to do it intentionally as part of our design, not by accident due to improper formatting.

After preprocessing, our training data transforms from (60000, 28, 28) to (60000, 28, 28, 1). That single additional dimension might seem trivial, but it tells our autoencoder exactly how to interpret the spatial structure it needs to compress and later reconstruct.

```
# Normalize pixel values to [0, 1] range
x_train = x_train.astype('float32') / 255.0
x_test = x_test.astype('float32') / 255.0

# Reshape to add channel dimension
x_train = x_train.reshape((len(x_train), 28, 28, 1))
x_test = x_test.reshape((len(x_test), 28, 28, 1))

print(f"Preprocessed training shape: {x_train.shape}")
print(f"Preprocessed test shape: {x_test.shape}")
```
Listing 11.2 Normalizing Data

We've reached the heart of our autoencoder: its architecture (see Listing 11.3). Using Keras's Functional API, we're building something that resembles an hourglass—wide at both ends, narrow in the middle. This create_autoencoder function encapsulates the entire design philosophy of compression and reconstruction in one elegant structure. The function begins by establishing our input layer and immediately makes a crucial decision: flattening the 2D image into a 1D vector. (As mentioned before, traditional autoencoders only worked with dense layers. We'll take a look at other options later in Section 11.2.3.) From there, we create our encoder layers with decreasing sizes—128 neurons, then 64, finally bottlenecking down to our encoding_dim (defaulting to 32).

```
def create_autoencoder(input_shape, encoding_dim=32):
    # Encoder
    input_img = keras.Input(shape=input_shape)

    # Flatten the input
    x = layers.Flatten()(input_img)

    # Hidden layers with decreasing size
    x = layers.Dense(128, activation='relu')(x)
    x = layers.Dense(64, activation='relu')(x)

    # Bottleneck layer (encoded representation)
    encoded = layers.Dense(encoding_dim, activation='relu')(x)

    # Decoder
    x = layers.Dense(64, activation='relu')(encoded)
    x = layers.Dense(128, activation='relu')(x)
    x = layers.Dense(28 * 28, activation='sigmoid')(x)

    # Reshape back to original image shape
    decoded = layers.Reshape((28, 28, 1))(x)
```

```
# Create the autoencoder model
autoencoder = keras.Model(input_img, decoded)

# Create encoder model (for getting encoded representations)
encoder = keras.Model(input_img, encoded)

return autoencoder, encoder
```

Listing 11.3 The Autoencoder Architecture

Significance of the Design Decisions

Every choice we make in this architecture shapes how our autoencoder learns and what it can accomplish. Let's explore the reasoning behind each decision and consider alternative approaches.

Our choice to flatten the input represents a fundamental trade-off. By converting our 2D image into a 1D vector, we're essentially telling our network to forget about *spatial relationships* initially. A pixel in the top-left corner becomes just another number in our sequence, no different from a pixel in the bottom-right. This approach works well for Fashion-MNIST because clothing items often have distinctive global features. A dress might be recognizable by its overall shape rather than specific spatial patterns. However, we're sacrificing the ability to preserve local spatial relationships that might matter for more complex images.

Alternative approaches exist. We could use convolutional layers in our encoder, gradually reducing spatial dimensions while preserving local relationships. This would be like having an artist who notices both the overall composition and the fine brushwork details rather than just the overall color palette.

Our encoder layers follow a carefully planned progression: 784 to 128 to 64 to 32. This represents *compression ratios* of roughly 6:1, then 2:1, then 2:1 again. Are these particular ratios important? The answer is a little nuanced.

Aggressive compression in a single step often fails catastrophically. Consider trying to summarize a novel in one sentence versus gradually condensing it through chapter summaries, then page summaries, then sentence summaries. Our gradual approach gives the network multiple opportunities to identify what information deserves preservation at each level. The 128-neuron layer might learn to represent major clothing categories and basic shapes. The 64-neuron layer could refine these into more specific patterns such as sleeves, collars, or hemlines. Finally, the 32-neuron bottleneck forces the network to distill everything into its most essential features.

We could experiment with different progressions. A steeper compression (784 to 64 to 32) might work for simpler datasets, while a gentler slope (784 to 256 to 128 to 64 to 32) might better preserve detail for complex images. The choice depends on how much information we can afford to lose and is one of the major decisions that have to be made by a machine learning engineer.

Overall, our 32-dimensional encoding represents a compression ratio of approximately 25:1. Each Fashion-MNIST image gets squeezed from 784 numbers down to just 32. These 32 values must somehow capture everything needed to reconstruct a recognizable image. This bottleneck size reflects another trade-off. Smaller bottlenecks force more aggressive compression, potentially leading to more abstract, generalized representations. A 16-dimensional bottleneck might discover higher-level concepts such as "has sleeves" or "is footwear." Larger bottlenecks such as 64 or 128 dimensions might preserve more specific details but miss opportunities for meaningful abstraction.

We use Rectified Linear Unit (ReLU) activation throughout most of our network, as we've discussed in previous chapters. But notice that the final layer uses sigmoid activation. This choice ensures our output values stay between 0 and 1, matching our normalized input range.

We could experiment with other activations. *Tanh* might provide better gradient flow in some cases, while newer activations such as Swish could potentially improve training dynamics. However, that final sigmoid remains crucial for matching our output range to our input expectations.

The Swish Activation Function

Swish represents a smooth alternative to ReLU, defined as $f(x) = x \cdot \sigma(x)$ where σ is the sigmoid function. Unlike ReLU's sharp corner at 0, Swish provides smooth derivatives everywhere, which can improve gradient flow during training. The function exhibits *self-gating behavior*, allowing the network to dynamically control information flow based on the input's own value. Swish (and its derivatives, e.g., *SwiGLU*) sits between ReLU and Gaussian Error Linear Unit (GeLU) (both of which we've covered before) in terms of computational complexity and performance, offering better results than ReLU in many tasks while being simpler than GELU.

It's generally a good idea to keep track of different activation functions (and optimizers) being proposed in the literature. You never know which one might suit the specific dataset you're working with! We'll cover how to keep track of all of these innovations in the final chapter of this book.

This architectural foundation gives us tremendous flexibility. We can adjust the bottleneck size, experiment with different layer progressions, or even swap in convolutional layers later. But for now, this simple, symmetric design provides an excellent starting point for understanding how autoencoders compress and reconstruct information.

Next, we bring our architectural blueprint to life through the code shown in Listing 11.4. When we call `create_autoencoder` with our image dimensions and encoding size, we're essentially commissioning two related but distinct tools for our compression workshop.

11 Autoencoders and Generative AI

The first tool is our complete autoencoder—the full assembly line that takes raw images, compresses them through the bottleneck, and reconstructs them back to their original form. This is our workhorse, the model we'll train to become an expert at the compress-and-reconstruct dance.

The second tool is the encoder by itself—just the compression half of our operation. Having this separate model is like having a specialized tool that can peek inside our compression process. While the full autoencoder shows us how well we can reconstruct images, the standalone encoder lets us examine what the network considers most important about each image. We can extract those crucial 32 numbers that represent the essence of any fashion item and study them directly. This *dual-model approach* opens up powerful possibilities. We might train the full autoencoder on thousands of images, then use just the encoder to compress new images for storage or comparison. Or, we could analyze the encoded representations to understand what patterns the network has discovered.

The compile step deserves special attention, particularly our choice of loss function. We're using *binary cross-entropy*, which might seem surprising at first. After all, we're not doing binary classification—we're reconstructing images. Binary cross-entropy excels at comparing pixel-by-pixel similarity between our original and reconstructed images. Each pixel in our normalized images sits between 0 and 1, much like a probability. Binary cross-entropy treats each pixel as an independent prediction problem: How well does our reconstructed pixel value match the original? This loss function encourages our autoencoder to get every pixel as close as possible to its target value. When a pixel should be 0.8 (a light gray) and our network produces 0.3 (a darker gray), binary cross-entropy provides a strong penalty that guides our network toward better reconstruction.

Compare this to mean squared error (MSE), which we've used for regression problems. While MSE could work here, binary cross-entropy often trains more efficiently for image reconstruction tasks. It provides more nuanced gradients when dealing with values in the 0–1 range, helping our network learn faster and more precisely.

```
autoencoder, encoder = create_autoencoder((28, 28, 1),
                                          encoding_dim=32)

autoencoder.compile(optimizer='adam', loss='binary_crossentropy')

print("Autoencoder Architecture:")
autoencoder.summary()

print("\nEncoder Architecture:")
encoder.summary()
```

Listing 11.4 Creating and Training the Autoencoder

Let's go ahead and take a look at our trained autoencoder's dual nature. Having separate autoencoder and encoder models reveals its true power. When we run `autoencoder.predict` on our test images (as shown in Listing 11.5), each fashion item travels through the entire compression-reconstruction pipeline, emerging as a `decoded_imgs` array that should closely resemble our original inputs. These are our autoencoder's best attempts at perfect reconstruction—its artistic interpretations of the original images after they've been squeezed through that narrow 32-dimensional bottleneck. But we can also use the standalone encoder model separately. The `encoded_imgs` represent the distilled essence of each fashion item, compressed into just 32 numbers. Each image that originally required 784 values (28×28×1) now exists as a compact signature of only 32 dimensions. These encoded representations become particularly valuable for other tasks. Because each image reduces to just 32 numbers, we can quickly compare images by comparing their encoded vectors. Similar clothing items should have similar encoded representations, making these compressed signatures useful for organizing, searching, or analyzing fashion datasets at scale.

> **Semantic Image Search Through Autoencoders**
>
> We can perform *semantic image search* using autoencoders by using their ability to create meaningful compressed representations. First, we apply the encoder portion of our trained autoencoder to all images in our database, creating a collection of latent vectors that represent each image's essential features. When a user wants to search for images similar to a *query image*, we compress that query through the same encoder to obtain its latent representation. We then search for images in our database whose latent vectors are closest to the query vector in this compressed space. This is a simple matter of finding other vectors with the smallest Euclidean distance to our query vector. This highly efficient approach works because images with similar visual content will have similar representations in the latent space, allowing us to find semantically related images even if they differ in exact pixel values.
>
> What's more, this concept isn't limited to image search only. We can compress any type of data through technique and search based on some query that is of the same type.

```
# Get predictions on test data
decoded_imgs = autoencoder.predict(x_test)

# Get encoded representations
encoded_imgs = encoder.predict(x_test)

print(f"Original image shape: {x_test[0].shape}")
print(f"Encoded representation shape: {encoded_imgs[0].shape}")

print(f"Compression ratio: {np.prod(x_test[0].shape)/
                 np.prod(encoded_imgs[0].shape):.1f}x")
```

```
# Output:
# Original image shape: (28, 28, 1)
# Encoded representation shape: (32,)
# Compression ratio: 24.5x
```

Listing 11.5 Running the Autoencoder

Generation Results and Takeaways

Let's take a look at some results in the form of reconstructed images shown in Figure 11.2. This image demonstrates very important concepts about the fundamental trade-offs in autoencoder design. We're seeing the same sneaker processed through three different autoencoders, each with a different bottleneck size. The results reveal something crucial about the relationship between compression and quality. Look at the top row first, where we've compressed our image into just two dimensions. That's an extraordinary compression ratio of 392:1. The reconstruction attempts to capture the essence of a sneaker, but it's clearly struggling. The shoe appears as a blurry, somewhat abstract representation. The difference image shows intense brightness variations, revealing exactly where the reconstruction failed to match the original. These two dimensions are doing their absolute best to encode everything important about this sneaker. Maybe one dimension captures "shoe-like elongation" while the other represents "overall brightness." Whatever they are trying to capture, clearly, two numbers simply can't hold enough information to recreate the subtle details, textures, and precise shapes that make this sneaker recognizable.

Moving to the middle row with 16 dimensions, we see immediate improvement. The reconstructed sneaker looks much more shoe-like, with clearer definition of its silhouette and some internal structure becoming visible. The difference image still shows significant errors, but they're less intense and more localized. Those 16 numbers can now probably encode concepts such as "has laces," "curved toe section," "heel height," and perhaps even some texture information.

The bottom row, using 64 dimensions, produces the most faithful reconstruction. While still not perfect, the sneaker maintains most of its defining characteristics. The sole is clearly visible, the overall proportions look correct, and fine details are beginning to emerge. The difference image shows the smallest errors, concentrated mainly in areas of subtle shading and texture. Of course, there's no magic threshold where reconstruction suddenly becomes perfect. Instead, we see a gradual improvement as we allow our bottleneck to retain more information. Each additional dimension we add to our encoding provides space for one more crucial piece of information about the image. But here's what makes this particularly interesting: Even with severe compression, the autoencoder still produces something recognizably shoe-like. This suggests that our network has learned meaningful representations at every level of compression. The 2D version might be unusable for detailed analysis, but it's still capturing some essential "sneaker-ness" that allows basic recognition.

This trade-off between compression and quality sits at the heart of autoencoder design. In real-world applications, we might choose different bottleneck sizes depending on our priorities. For storage efficiency, we might accept the blurry 16-dimensional reconstruction. For high-quality image processing, we'd need the 64-dimensional version or larger.

Finally, the "difference" images in the right-most column serve as our quality inspectors, showing us exactly where each compression level fails. The bright regions highlight the pixels where reconstruction errors are most severe. Notice how these error patterns change across different bottleneck sizes—they're not just smaller versions of the same errors but entirely different types of mistakes as the network adapts to different information constraints. Having helper visualizations like this make the models more interpretable and help us gain a better understanding of what's going on in the black box that is the neural network. This understanding in turn helps us make informed decisions about autoencoder architecture.

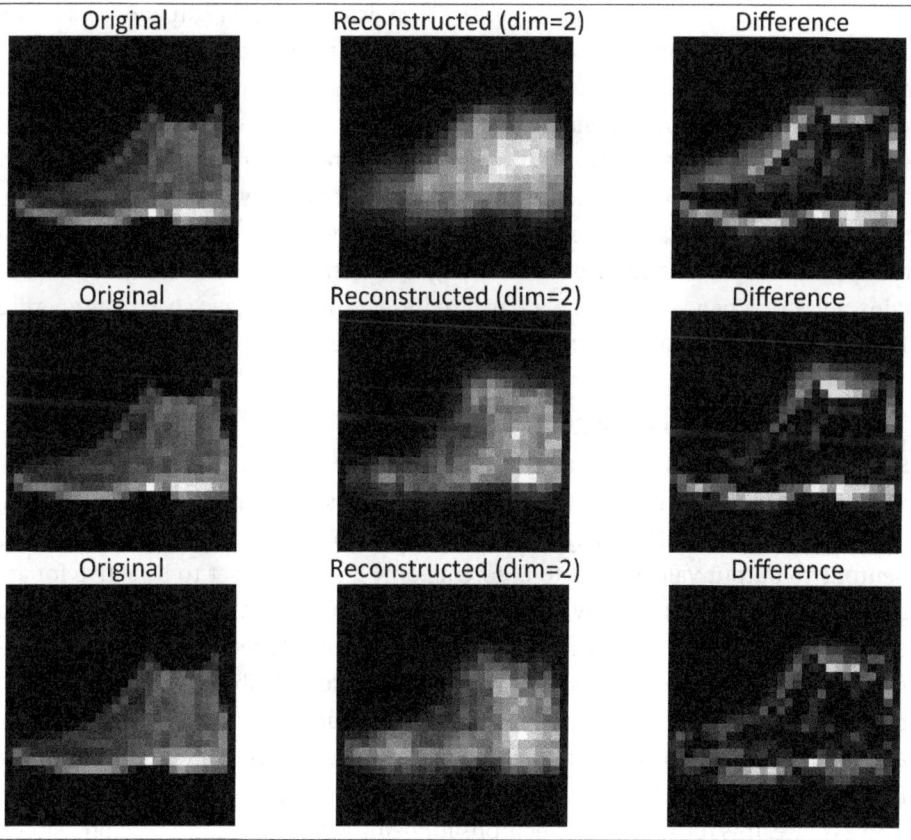

Figure 11.2 Autoencoder Image Reconstruction Results

11.1.4 Types of Autoencoders

Now that we've built our first autoencoder and seen how it can learn to compress and reconstruct images, this section will explore the different flavors of autoencoders available to us. Each type addresses specific challenges or use cases, much like having different tools in a toolbox—each one designed for particular jobs. The autoencoder we just built is what we call a *vanilla autoencoder*—the most straightforward implementation of the encoding-decoding concept. While vanilla autoencoders work well for basic compression tasks, researchers have developed several variations that offer unique advantages for different scenarios. These variants modify either the architecture, the training process, or the loss function to achieve specific goals.

Vanilla Autoencoders: The Foundation

Our vanilla autoencoder represents the purest form of the encoding-decoding paradigm. It learns to compress data into a smaller representation and then reconstruct the original input as accurately as possible. The simplicity of vanilla autoencoders makes them excellent starting points for understanding the core concepts, and they work remarkably well for straightforward compression tasks. When you have clean data, and your primary goal is dimensionality reduction or basic feature extraction, vanilla autoencoders often provide exactly what you need. They're like a reliable, no-frills camera that consistently produces good quality photos without fancy features getting in the way. However, vanilla autoencoders can sometimes learn trivial solutions or become too specialized to the training data. They might also struggle when the input data contains noise or when we want the learned representations to have specific properties. This is where the specialized variants become valuable.

Sparse Autoencoders: Learning Selective Representations

Sparse autoencoders introduce an additional constraint during training: They encourage the hidden units in the bottleneck layer to be mostly inactive. Instead of allowing all neurons to contribute equally to the representation, sparse autoencoders push most neurons to output values close to 0, forcing only a small subset to be active for any given input. To understand why this sparsity might be useful, recall the concept of dropout discussed in Chapter 6. When you look at a face, only specific neurons related to facial features become highly active, while neurons responsible for detecting cars or buildings remain relatively quiet. This selective activation creates more interpretable and robust representations. In autoencoders, we achieve sparsity by adding a *regularization term* to our loss function. Again, details of this were covered at length in Chapter 6. In essence, this term penalizes neurons for being active, creating a tension between reconstruction accuracy and sparsity. The autoencoder must learn to reconstruct inputs accurately while using as few active neurons as possible. This constraint forces the network to discover the most important features in the data. The regularization term typically measures how much each neuron deviates from a desired average activation level. If we want neurons to be active only 5% of the time on average, we penalize deviations from this target. The strength of this penalty is controlled by a hyperparameter that balances reconstruction quality against sparsity.

Sparse autoencoders excel in *feature discovery* and *interpretability*. Because they use only a small number of active neurons for each input, the learned representations often correspond to meaningful, interpretable features. They're particularly useful when you want to understand what patterns the autoencoder has discovered in your data. There are other challenges and use cases for autoencoders though.

Denoising Autoencoders: Building Robustness

Denoising autoencoders tackle one such challenge: learning representations that are robust to noise and corruption. Instead of training on clean data, we intentionally corrupt the input data during training while still asking the autoencoder to reconstruct the original, uncorrupted version. The corruption process can take many forms. We might randomly set some pixels to 0, add *Gaussian noise* to the input, or apply other types of distortion. The key insight is that by forcing the autoencoder to recover clean data from corrupted inputs, we encourage it to learn more robust and generalizable features. This approach mirrors how humans learn to recognize objects even when they're partially obscured or distorted. You can still recognize a friend's face even if they're wearing sunglasses or if the lighting is poor. Denoising autoencoders develop similar robustness by learning to focus on the essential, stable features of the data rather than getting distracted by noise or irrelevant details.

The training process becomes a game of recovery: We corrupt the input, pass it through the encoder to get a compressed representation, and then ask the decoder to reconstruct the original, uncorrupted image. This also forces the autoencoder to learn representations that capture the underlying structure of the data rather than memorizing specific pixel patterns, but in a different way. Denoising autoencoders are particularly valuable for preprocessing noisy data, *detecting anomalies* , and creating more robust feature representations. They often generalize better than vanilla autoencoders because they've learned to ignore irrelevant variations in the input. We'll implement a denoising autoencoder shortly to see these principles in action.

The transformation from our standard autoencoder to a denoising version requires only a few strategic modifications, but each change serves a crucial purpose in teaching our network to clean corrupted images. We've shown the relevant portion of the changes in Listing 11.6. You can see the full code and experiment with it on the book's resources page at *https://recluze.net/keras-book* in the notebook **11-02-denoising-auto-encoder**.

The first essential addition creates our corrupted training data. We introduce Gaussian noise to both our training and test sets, scaling it by a `noise_factor` of 0.5. This parameter controls how much corrupton we add—higher values create more challenging denoising tasks, while lower values make the problem easier but potentially less useful for real-world applications. The `np.random.normal` function generates noise with zero mean and unit variance, which we then scale and add to our clean images. This simulates the kind of random corruption that might occur in real-world scenarios—sensor noise in cameras, transmission errors in digital communications, or degradation in stored images. The *clipping* operation ensures our noisy pixel values stay within the valid 0–1 range, preventing the noise from pushing values into irregular territories.

```
noise_factor = 0.5
x_train_noisy = x_train + noise_factor * np.random.normal(
    loc=0.0, scale=1.0, size=x_train.shape)
x_test_noisy = x_test + noise_factor * np.random.normal(
    loc=0.0, scale=1.0, size=x_test.shape)

# Clip values to [0, 1] range
x_train_noisy = np.clip(x_train_noisy, 0., 1.)
x_test_noisy = np.clip(x_test_noisy, 0., 1.)
```

Listing 11.6 Preparing the Training Data for the Denoising Autoencoder

Our autoencoder architecture (both the encoder and the decoder) remain the same. The only change occurs in our training process. Instead of the self-supervised approach we used before, where input and output were identical, we now train with a supervised setup. This single line of code (shown in Listing 11.7) represents a fundamental shift in our autoencoder's objective. We're feeding corrupted images as input but demanding clean images as output. The network must learn to identify and remove the noise while preserving the essential features that make each fashion item recognizable. This training strategy forces our autoencoder to develop a more sophisticated understanding of what constitutes meaningful information versus random corruption. The encoder must learn to extract features that represent the true underlying image structure, while the decoder must reconstruct the original no-noise image based on these noise-resistant features rather than simply copying corrupted input details.

What makes this approach particularly powerful is that we haven't changed our architecture at all. The same hourglass structure that learned to compress and reconstruct clean images now learns to compress noisy images and reconstruct clean ones. The bottleneck layer becomes even more crucial—it must capture the essential image information while filtering out the random noise that would otherwise propagate through to the output.

11.1 Introduction to Autoencoders

```
history = autoencoder.fit(x_train_noisy, x_train,
                          epochs=50,
                          batch_size=256,
                          shuffle=True,
                          validation_data=(x_test_noisy, x_test),
                          verbose=1)
```
Listing 11.7 Training the Denoising Autoencoder

The results of the training, shown in Figure 11.3, demonstrate how our denoising autoencoder handles increasingly challenging corruption levels.

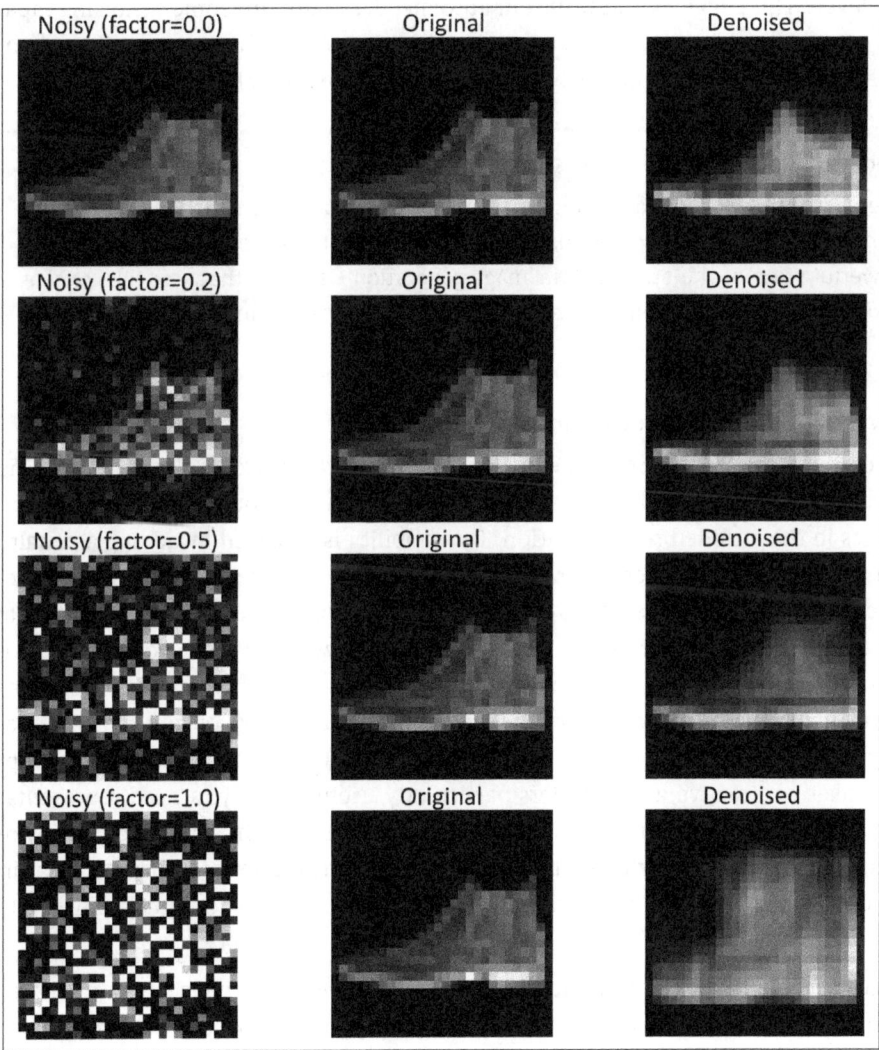

Figure 11.3 Results of a Denoising Autoencoder

With light noise (top row), the network produces reasonable reconstructions, though some blurriness remains. The moderate noise level (middle rows) shows decent recovery of the basic shoe shape, but fine details become lost. Under heavy noise (bottom row), the autoencoder still manages to extract the general sneaker outline, but significant structural details disappear. The progression demonstrates a fundamental limitation: As noise increases, our fixed-capacity bottleneck struggles to separate genuine image features from corruption. The network learns to preserve the most robust, noise-resistant patterns while sacrificing subtler details that become indistinguishable from random variations.

Notice how the difference images show increasingly intense error patterns with higher noise levels. This suggests that while our autoencoder maintains some denoising capability across all conditions, there's a practical limit to how much corruption it can handle before reconstruction quality becomes unacceptable for real applications.

> **Denoising as a Powerful Tool**
>
> Keep this simple concept of adding noise in mind as we progress with our discussion of GenAI. In the next chapter, we'll use this elegant concept to introduce one of the most powerful, state-of-the-art models in image generation—a model that uses this concept (along with some other innovations) to generate photorealistic images from scratch.

Contractive Autoencoders: Ensuring Stability

Let's consider one last type of autoencoder. *Contractive autoencoders* focus on learning representations that are stable—small changes in the input should produce only small changes in the encoded representation. This stability is achieved by adding a penalty term that encourages the encoder to be less sensitive to input variations. The mathematical foundation involves the *Jacobian matrix* (think of this as a collection of partial derivatives) of the encoder function, which captures how much the encoded representation changes with respect to changes in the input. The *contractive penalty* encourages this Jacobian matrix to have small values, meaning the encoder's output remains relatively stable even when the input undergoes small perturbations. This stability property makes contractive autoencoders particularly useful when you need representations that remain consistent despite minor variations in the input data. They're valuable for applications where robustness to small perturbations is crucial, such as in medical imaging, where slight variations in image acquisition shouldn't dramatically affect the learned features.

The contractive penalty is computed as the *Frobenius norm* of the Jacobian matrix of the encoder. This mathematical formulation encourages the encoder to learn smooth mappings from input to representation space. It might sound complicated, but the formulation remains the same as denoising autoencoders. We just add a regularization term to the loss just as we did in Chapter 6.

Choosing the Right Type

Each autoencoder variant addresses specific needs and challenges. Vanilla autoencoders work well for clean data and straightforward compression tasks. Sparse autoencoders excel when interpretability and feature discovery are priorities. Denoising autoencoders shine when dealing with noisy data or when you need robust representations. Contractive autoencoders are ideal when stability and smooth representations are essential. The choice often depends on your specific use case and the characteristics of your data. If your data is clean and you simply need compression, vanilla autoencoders might suffice. If you're working with noisy sensor data, denoising autoencoders could be the better choice. For applications requiring interpretable features, sparse autoencoders might be ideal.

Many practitioners experiment with different types to see which works best for their specific problem. The good news is that implementing these variants typically requires only modest changes to the basic autoencoder architecture or training process as we saw earlier. The core concepts remain the same—we're just adding different constraints or modifications to achieve specific goals.

Understanding these different types prepares us to tackle a wide range of problems with autoencoders. Each variant represents a different tool in our machine learning toolkit, ready to be deployed when the situation calls for its particular strengths.

11.2 Variational Autoencoders

While the autoencoders we've explored so far are powerful tools for learning representations and reconstructing data, they have a fundamental limitation that prevents them from being truly generative. When we train a basic autoencoder on handwritten digits, it becomes excellent at compressing and reconstructing those specific digits. But what happens when we want to create entirely new digits that look realistic but weren't in our original dataset? This is where things get tricky with traditional autoencoders. The latent space—that compressed representation in the middle of our network—doesn't necessarily have any meaningful structure for generation. If we randomly sample a point from this latent space and feed it to our decoder, we're likely to get garbage output rather than something that resembles a real digit.

The root of this problem lies in how deterministic autoencoders work. Each input gets mapped to exactly one point in the latent space. While this creates a perfect reconstruction pathway, it doesn't guarantee that the space between these points contains anything meaningful. Consider a map where only the major cities are marked, but all the roads connecting them are invisible. You can get from city to city if you know exactly where they are, but trying to navigate the space between them becomes impossible.

In this section, we'll first describe this problem of navigating the uncertain spaces between known points. We'll then discuss the mathematical framework necessary for understanding this new type of model and the implementation details associated with them.

11.2.1 Navigating the Space with Uncertainty

Variational Autoencoders (VAEs) solve this challenge by taking a fundamentally different approach. Instead of mapping each input to a single point in latent space, VAEs embrace uncertainty and probability. They transform our deterministic encoding process into a probabilistic one, creating a latent space that's not only structured but also continuous and meaningful for generation. The key insight behind VAEs comes from treating the encoding process as learning *probability distributions* rather than fixed points. Instead of saying, "this image of a handwritten 7 maps to coordinates (2.3, -1.7, 0.8) in latent space," a VAE learns to say, "this image represents a region in latent space where similar 7s might exist, centered around (2.3, -1.7, 0.8) with some spread or uncertainty." This seemingly minor change changes everything about how we approach the latent space. Now, instead of discrete points scattered across our representation space, we have overlapping *probability clouds*. Each input doesn't claim ownership of a single location but influences a broader region. This overlap is crucial—it means that the space between our training examples contains meaningful information rather than empty voids.

Understanding latent variable models helps us grasp why this probabilistic approach works so well. In traditional machine learning, we often work with observable variables—the pixels in an image, the words in a sentence, the measurements we can directly take. But many real-world phenomena are driven by hidden or latent variables that we can't observe directly.

Consider human faces as an example. The pixels we see in a photograph are the observable variables, but they're generated by latent factors we can't directly measure: the person's age, their emotional expression, the lighting conditions, the angle of their head, and countless other hidden influences. A latent variable model attempts to discover and model these underlying factors that generate what we observe.

VAEs operate on this principle by learning to map observable data (e.g., images) to a latent space that captures these hidden generative factors. The probabilistic nature ensures that this latent space is well-structured and continuous, making it suitable for generating new data by *sampling* from it. But there's a mathematical challenge that makes implementing this probabilistic approach tricky: the *reparameterization trick*. When we want to train a neural network, we need to compute gradients—the mathematical directions that tell us how to adjust our weights to improve performance. The problem with sampling from probability distributions is that sampling operations aren't differentiable. You can't compute a gradient through a random sampling step.

The reparameterization trick solves this elegant problem with an equally elegant solution. Instead of directly sampling from the distribution our encoder produces, we separate the randomness from the learned parameters. This might sound complicated but it's going to make sense in a little while. Our encoder learns to output two things: the mean (μ) and the standard deviation (σ) of a probability distribution. Then, instead of sampling directly from this distribution, we sample from a standard normal distribution, which doesn't depend on our network's parameters, and use a simple mathematical transformation to convert this sample into one from our desired distribution.

Mathematically, if we want to sample from a distribution with mean μ and standard deviation σ, we can sample ϵ from a standard normal distribution and compute our final sample as $z = \mu + \sigma \cdot \epsilon$. This way, the randomness comes from ϵ, which has no connection to our network parameters, while the learned components μ and σ remain differentiable. It's like separating the dice roll from the game rules—we can learn better rules while keeping the fundamental randomness that makes the game interesting. This reparameterization trick becomes the foundation that makes VAE training possible. It allows us to maintain the probabilistic nature of our latent space while still being able to train our network using standard gradient-based methods. Without this clever mathematical insight, the entire VAE framework would be impossible to implement in practice.

The implications of this probabilistic approach extend far beyond just solving a technical training problem. By structuring our latent space as overlapping probability distributions, VAEs learn representations that capture not just what the data looks like but also how it varies and what constitutes reasonable variations. This understanding of data variability becomes the key to their generative capabilities, allowing them to create new, realistic samples by intelligently navigating the learned probabilistic landscape.

11.2.2 Mathematical Framework of Variational Autoencoders

Now that we understand why VAEs take a probabilistic approach, this section will examine how this translates into actual mathematics and code. The shift from deterministic to probabilistic encoding fundamentally changes how we design and train our networks, requiring us to rethink both the architecture and the loss function that guides our learning process.

Encoder as Probability Distribution Generator

In our traditional autoencoders, the encoder assigned each input to exactly one coordinate in latent space. The VAE encoder on the other hand, instead of pinpointing exact locations, describes probability clouds that capture the uncertainty and variability inherent in real data.

11 Autoencoders and Generative AI

This fundamental change means our encoder needs to output not just a single value for each dimension of our latent space but rather the parameters that define a probability distribution. For VAEs, we typically choose to work with Gaussian (normal) distributions because they have nice mathematical properties and represent uncertainty in a natural way that mirrors many real-world phenomena. Instead of producing a latent vector z directly, our VAE encoder generates two separate outputs for each dimension of the latent space: a mean μ and a standard deviation σ. These are the *center* and *spread* of our probability cloud. The mean tells us where the distribution is centered, while the standard deviation describes how spread out or concentrated our uncertainty is around that center.

Mathematically, our encoder learns to map an input x to parameters $\mu(x)$ and $\sigma(x)$, which define a Gaussian distribution $q(z|x) = \mathcal{N}(\mu(x), \sigma^2(x))$. This notation might look intimidating, but it simply says that given an input x, we have a normal distribution over possible latent representations z, centered at $\mu(x)$ with variance $\sigma^2(x)$.

The elegance of this approach becomes apparent when we consider what happens during training and generation. During training, we sample from this learned distribution to get our latent representation. This sampling process introduces the variability that prevents our model from memorizing exact mappings and instead forces it to learn robust, generalizable representations. If the model hasn't seen many instances of the digit 8, it will have a large spread in its prediction owing to its lack of confidence in the position of the digit in the latent space.

Conversely, when we want to generate new data, we can sample from our learned latent space, knowing that any reasonable sample should decode into something meaningful. It's like having a detailed weather map that doesn't just show you where storms are right now but gives you the probability distributions that let you predict where storms might form.

The encoder architecture in Keras code (shown in Listing 11.8) reflects this dual output requirement. Instead of ending with a single dense layer that outputs our latent dimensions directly, we split into two parallel paths: one dense layer for the means and another for the standard deviations. Both layers have the same number of units as our desired latent dimensionality, but they serve completely different roles in defining our probability distribution. The power of the Keras Functional API will make this seem almost trivial.

```
# Traditional autoencoder encoder
latent = Dense(latent_dim, activation='relu')(x)

# VAE encoder - outputs distribution parameters
z_mean = Dense(latent_dim)(x)
z_log_var = Dense(latent_dim)(x)
```

Listing 11.8 Core Change from Vanilla Autoencoder to Variational Autoencoder

Notice that we typically work with the logarithm of the variance ($\log \sigma^2$) rather than the standard deviation directly. This is both a numerical stability trick and a way to ensure our variance stays positive—because e^x is positive for all values of x, we can let our network output any real number for the *log variance* without worrying about accidentally creating negative variances.

Kullback–Leibler Divergence and Reconstruction Loss Combination

The probabilistic nature of VAEs introduces another fundamental challenge that doesn't exist in traditional autoencoders: We need to balance two competing objectives that pull our model in different directions. On one hand, we want our reconstructions to be as accurate as possible—this pushes us toward learning very specific, precise encodings for each input. On the other hand, we want our latent space to be well-structured and suitable for generation—this pushes us toward learning smooth, continuous representations where similar inputs have overlapping probability distributions. This tension is resolved through a clever combination of two loss components: reconstruction loss and Kullback–Leibler (KL) divergence. The reconstruction loss is familiar territory—it measures how well our decoder can recreate the original input from the latent representation. For images, this is typically MSE (or binary cross-entropy) between the original and reconstructed pixels, just like in traditional autoencoders.

The KL divergence measures how different two probability distributions are from each other. In our case, we use it to measure how much our learned distribution $q(z|x)$ differs from a standard normal distribution $p(z) = \mathcal{N}(0, I)$.

$$D_{KL}(q(z|x) \mid p(z)) = \frac{1}{2} \sum_{i=1}^{d} (\mu_i^2 + \sigma_i^2 - \log(\sigma_i^2) - 1)$$

This equation might look complex, but each term has a clear purpose. The μ_i^2 term penalizes means that are far from 0, encouraging our latent representations to be centered around the origin. The σ_i^2 term penalizes very large variances, preventing our distributions from becoming too spread out. The $-\log(\sigma_i^2)$ term prevents the variances from collapsing to 0, ensuring we maintain some uncertainty, and the constant -1 term balances the equation mathematically. The intuition behind this KL divergence penalty is that we want our learned latent space to resemble a standard normal distribution. This constraint serves multiple purposes:

- Ensures that different regions of our latent space have similar densities (preventing empty regions that would produce poor generations)
- Keeps our representations centered and scaled reasonably
- Allows us to generate new samples by simply sampling from a standard normal distribution

The complete VAE loss function combines these two components:

$$L_{VAE} = L_{reconstruction} + \beta \cdot D_{KL}(q(z|x) | p(z))$$

The reconstruction loss encourages accurate reproduction of inputs, while the KL divergence term encourages a well-structured latent space. The parameter β controls the balance between these objectives, and choosing its value becomes crucial for training effective VAEs.

The Interconnectedness of Machine Learning Concepts
Notice how this loss (as well as the underlying motivation) strongly resembles the L2 regularization we covered in Chapter 6. This is why we spent so much time understanding L2 loss even though it wasn't used directly in neural networks. Once you understand the core concepts covered in this book deeply, it becomes very easy to learn new concepts as they are introduced. This is the strength that can aid you in keeping up with the fast pace of machine learning and AI advancements.

Balancing the Loss Components for Optimal Generation

The β parameter in our loss function represents one of the most critical hyperparameters in VAE training, yet its importance is often underestimated. This single value controls the fundamental trade-off between reconstruction quality and generative capability, and getting it right can mean the difference between a VAE that generates realistic samples and one that produces blurry, unrealistic outputs. When β is too small (approaching 0), the KL divergence term becomes negligible, and our VAE behaves more like a traditional autoencoder. The model focuses almost entirely on perfect reconstruction, learning very specific encodings for each input. While this produces excellent reconstructions, the latent space becomes fragmented and unsuitable for generation.

Conversely, when β is too large, the KL divergence term dominates, forcing our latent representations to stay very close to a standard normal distribution. While this creates a beautifully structured latent space perfect for sampling, the reconstruction quality suffers dramatically. The model becomes so focused on maintaining the proper latent space structure that it can't learn the specific details necessary for accurate reconstruction.

The standard approach sets β = 1, treating both loss components as equally important. This balanced approach often works well, but the optimal value depends heavily on your specific dataset and application. For high-resolution images where reconstruction detail matters enormously, you might use a smaller β to prioritize accuracy. For applications where generation quality is paramount, a larger β might be appropriate. There's also an interesting training technique called *Beta annealing*, (β-annealing) where we start with a small β value and gradually increase it during training. This approach allows the model to first focus on learning good reconstructions, then gradually incorporate the latent space structure constraint. It's like teaching someone to draw by first letting them focus on getting the shapes right, then gradually introducing rules about composition and style.

During training, we can monitor both loss components separately to understand how our model is learning. If the reconstruction loss plateaus while the KL divergence continues to decrease, we might be over-regularizing with too large a β. If the KL divergence remains high while reconstruction loss drops quickly, we might need to increase β to encourage better latent space structure. The interplay between these loss components also affects how we interpret and use our trained VAE. A model trained with lower β will have more detailed reconstructions but might require more careful sampling strategies for generation. A model trained with higher β will generate more coherent samples from random latent space samples but might sacrifice some reconstruction fidelity. Understanding this balance helps us design VAE training strategies that align with our specific goals. Whether we're building a VAE for data compression, anomaly detection, or creative generation, the choice of β and our loss balancing strategy becomes a key design decision that shapes our model's capabilities and limitations.

11.2.3 Variational Autoencoder Implementation in Keras

After understanding the probabilistic foundations and mathematical framework behind VAEs, let's build one from scratch using Keras. We'll work with the *CIFAR-10* dataset, which provides a rich collection of 32×32 color images across 10 different classes. This dataset offers more complexity than MNIST while remaining manageable for our learning purposes. Our implementation begins with the essential data preparation step as shown in Listing 11.9. After importing the necessary libraries (omitted here for brevity), we load and preprocess our dataset. As always, you can refer to the complete code listing on the book resources page at *https://recluze.net/keras-book* in the notebook **11-03-vaes**.

```
# Load CIFAR-10 dataset
(x_train, y_train), (x_test, y_test) = \
                keras.datasets.cifar10.load_data()

# Normalize pixel values to [0, 1]
x_train = x_train.astype('float32') / 255.0
x_test = x_test.astype('float32') / 255.0

print(f"Training data shape: {x_train.shape}")
print(f"Test data shape: {x_test.shape}")
print(f"Training labels shape: {y_train.shape}")
```

```
# CIFAR-10 class names
class_names = ['airplane', 'automobile', 'bird', 'cat', 'deer',
               'dog', 'frog', 'horse', 'ship', 'truck']
```
Listing 11.9 Data Preparation for VAEs

Data Preparation and Sampling

The CIFAR-10 images come as integers ranging from 0 to 255, but our VAE expects floating-point values between 0 and 1. This normalization serves multiple purposes beyond simple scaling. When we divide by 255, we're ensuring that our neural network receives inputs in a range that promotes stable training. Networks generally perform better when inputs are scaled to reasonable ranges, preventing any single feature from dominating the learning process due to its magnitude. This is especially relevant here due to the KL divergence term used in the loss function.

The choice of normalization range matters significantly for VAEs. We could have normalized to [-1, 1], which would require a tanh activation in our decoder's final layer instead of sigmoid. The [0, 1] range pairs naturally with sigmoid activation, while [-1, 1] works well with tanh. Both approaches are valid, but consistency between preprocessing and final activation is crucial. If you use [0, 1] normalization with tanh activation, your decoder will struggle to reconstruct images properly because tanh outputs values between -1 and 1, while your target images are scaled between 0 and 1.

With our data prepared, we now tackle one of the most crucial components of our VAE: the *sampling layer*. This custom layer embodies the reparameterization trick we discussed in our mathematical framework earlier. In the complete code listing provided with the book, you'll find additional visualization code for exploring the dataset, but our focus here is on this essential architectural component. The sampling layer (shown in Listing 11.10) represents the bridge between our encoder's deterministic outputs and the probabilistic sampling we need for generation. Remember from our probabilistic approach discussion that we can't simply sample randomly from a distribution and expect gradient descent to work. We need a way to make the randomness differentiable, and that's exactly what this layer accomplishes.

The heart of this implementation lies in the final line: z_mean + ops.exp(0.5 * z_log_var) * epsilon. This single expression encapsulates the reparameterization trick. Instead of sampling directly from a normal distribution with mean z_mean and variance exp(z_log_var), we sample from a standard normal distribution (epsilon) and then transform it. The randomness comes from epsilon, while the learned parameters (z_mean and z_log_var) control how this randomness gets shaped into our desired distribution. An alternative implementation choice involves the noise distribution. We use standard normal distribution (mean 0, variance 1), but we could experiment with other distributions. Some researchers have explored *uniform distributions* or even *learned noise patterns*. However, the mathematical framework of VAEs assumes Gaussian distributions,

11.2 Variational Autoencoders

so deviating from this assumption requires careful consideration of how it affects the KL divergence calculation.

```
class Sampling(layers.Layer):
    def __init__(self, **kwargs):
        super().__init__(**kwargs)
        self.seed_generator = keras.random.SeedGenerator(1337)

    def call(self, inputs):
        z_mean, z_log_var = inputs
        batch = ops.shape(z_mean)[0]
        dim = ops.shape(z_mean)[1]
        epsilon = keras.random.normal(
            shape=(batch, dim),
            seed=self.seed_generator
        )
        return z_mean + ops.exp(0.5 * z_log_var) * epsilon
```

Listing 11.10 Creating the Sampling Layer for the VAE

Encoders, Decoders and Everything Around Them

Next is the encoder-decoder part. Our encoder serves as the first half of our VAE, transforming raw pixel data into the probability distributions we discussed in our mathematical framework. This network must compress a 32×32×3 image (3,072 dimensions) into meaningful representations while outputting not just a single encoding but the parameters of a probability distribution. The architecture, shown in Listing 11.11, follows a classic convolutional pattern: We progressively increase the number of filters while reducing spatial dimensions. Starting with 32 filters and doubling through 64, 128, and finally 256 creates a hierarchical feature extraction system as always.

The combination of `BatchNormalization` and `LeakyReLU` deserves special attention. Batch normalization stabilizes training by normalizing inputs to each layer, preventing the *internal covariate shift* that can slow learning. LeakyReLU allows small negative values to pass through (typically with a slope of 0.01), preventing the *dying ReLU* problem where neurons can become permanently inactive. We could use standard ReLU, but LeakyReLU provides more robust gradient flow during training.

After the convolutional tower, we flatten the feature maps into the bottleneck through successive downsampling. This bottleneck layer forces the network to distill the most important information before generating our distribution parameters.

Finally, our encoder returns three outputs: the mean, log variance, and a sampled latent vector. This design allows flexibility during inference—we can use the mean for deterministic encoding or the sampled vector for generation. During training, we use all three outputs: the sampled vector for reconstruction loss and the mean and log variance for the KL divergence term in our loss function.

```python
def build_encoder(latent_dim=128):
    """Build the encoder network for VAE."""
    encoder_inputs = keras.Input(shape=(32, 32, 3))

    x = layers.Conv2D(32, 3, padding="same")(encoder_inputs)
    x = layers.BatchNormalization()(x) # Add Batch Normalization
    x = layers.LeakyReLU()(x) # Consider LeakyReLU
    x = layers.Conv2D(64, 3, strides=2, padding="same")(x)
    x = layers.BatchNormalization()(x)
    x = layers.LeakyReLU()(x)
    x = layers.Conv2D(128, 3, strides=2, padding="same")(x)
    x = layers.BatchNormalization()(x)
    x = layers.LeakyReLU()(x)
    x = layers.Conv2D(256, 3, strides=2, padding="same")(x)
    x = layers.BatchNormalization()(x)
    x = layers.LeakyReLU()(x)

    x = layers.Flatten()(x)
    x = layers.Dense(512)(x)
    x = layers.BatchNormalization()(x)
    x = layers.LeakyReLU()(x)

    z_mean = layers.Dense(latent_dim, name="z_mean")(x)
    z_log_var = layers.Dense(latent_dim, name="z_log_var")(x)

    z = Sampling()([z_mean, z_log_var])

    encoder = Model(encoder_inputs, [z_mean, z_log_var, z], name="encoder")
    return encoder
```

Listing 11.11 Encoder Architecture for VAE

The decoder, shown in Listing 11.12, performs the mirror of our encoder, taking our compressed latent representation and breathing life back into it as a full image. It starts with a 128-dimensional vector and carefully expanding it back into a 32×32×3 image.

Transpose convolutions (sometimes called *deconvolutions*) work by inserting learned patterns between existing pixels. With stride=2, each pixel in our input becomes a 2×2 region in the output, effectively doubling both width and height. The network learns what patterns to insert, guided by the reconstruction loss that compares the final output to the original image.

The final layer deserves special attention. We use a transpose convolution with only three filters (matching our RGB channels) and sigmoid activation. The sigmoid activation ensures our output values stay between 0 and 1, matching our normalized input data. This final layer has no batch normalization or additional activation—it directly produces our reconstructed image. The design choices we discussed for the encoder (batch normalization, LeakyReLU, *filter progressions*) remain relevant here. One crucial architectural decision involves symmetry. Our decoder roughly mirrors the encoder's structure, but perfect symmetry isn't required (similar to what we mentioned in our design decisions for autoencoders earlier in this chapter). We could use different filter counts, kernel sizes, or even different upsampling strategies. Some modern VAE architectures use asymmetric designs where the decoder has more capacity than the encoder, allowing for higher-quality reconstructions.

```python
def build_decoder(latent_dim=128):
    """Build the decoder network for VAE."""
    latent_inputs = keras.Input(shape=(latent_dim,))

    x = layers.Dense(2 * 2 * 256)(latent_inputs)
    x = layers.BatchNormalization()(x)
    x = layers.LeakyReLU()(x)
    x = layers.Reshape((2, 2, 256))(x)

    x = layers.Conv2DTranspose(256, 3, strides=2, padding="same")(x)
    x = layers.BatchNormalization()(x)
    x = layers.LeakyReLU()(x)
    x = layers.Conv2DTranspose(128, 3, strides=2, padding="same")(x)
    x = layers.BatchNormalization()(x)
    x = layers.LeakyReLU()(x)
    x = layers.Conv2DTranspose(64, 3, strides=2, padding="same")(x)
    x = layers.BatchNormalization()(x)
    x = layers.LeakyReLU()(x)
    x = layers.Conv2DTranspose(32, 3, strides=2, padding="same")(x)
    x = layers.BatchNormalization()(x)
    x = layers.LeakyReLU()(x)

    decoder_outputs = layers.Conv2DTranspose(
        3, 3, activation="sigmoid", padding="same"
    )(x)

    decoder = Model(latent_inputs, decoder_outputs, name="decoder")
    return decoder
```

Listing 11.12 Decoder Architecture for VAE

11 Autoencoders and Generative AI

The Training Process

The training of both the encoder and decoder happens through the `train_step` function shown in Listing 11.13. This function is the heart of our VAE, where all the theoretical concepts we've discussed transform into actual learning. This method is part of our custom VAE class (with boilerplate code omitted for brevity—you can find the complete implementation in the book's resource page as mentioned earlier). Every time our model sees a batch of data, this function runs the learning process, balancing the dual objectives that make VAEs so powerful. It begins with the input data that travels through our encoder, producing the three crucial outputs we designed: mean vectors, log-variance vectors, and sampled latent representations. These sampled vectors then pass through our decoder to create reconstructed images. For the reconstruction loss, we use binary cross-entropy instead of MSE, treating each pixel as an independent *Bernoulli random variable*. This might seem counterintuitive—after all, pixels are continuous values, not binary outcomes. However, because our data is normalized to [0,1] and our decoder uses sigmoid activation, we can interpret pixel intensities as probabilities. Binary cross-entropy often produces sharper, more realistic reconstructions compared to MSE, which tends to create blurry images.

Our KL divergence implementation directly translates the mathematical formula we derived earlier. The expression `-0.5 * (1 + z_log_var - ops.square(z_mean) - ops.exp(z_log_var))` computes the KL divergence between our learned distribution and a standard normal distribution for each latent dimension. We sum across latent dimensions for each sample and then average across the batch. The `beta` parameter is the crucial hyperparameter that controls the balance between reconstruction quality and latent space structure.

Our loss tracking system monitors three separate components: total loss, reconstruction loss, and KL loss. This granular monitoring proves invaluable during training, allowing us to observe how each component evolves and identify potential issues such as *KL collapse* (where KL loss approaches zero) or *reconstruction failure*.

Alternatively, possible loss formulations include using different divergences (*Wasserstein*, *Maximum Mean Discrepancy (MMD)*) instead of KL divergence or implementing *curriculum learning* where the `beta` parameter changes during training. Some practitioners use *cyclical annealing*, gradually increasing beta from 0 to 1 over multiple epochs to encourage better latent space organization.

```
def train_step(self, data):
    with tf.GradientTape() as tape:
        z_mean, z_log_var, z = self.encoder(data)
        reconstruction = self.decoder(z)
        reconstruction_loss = ops.mean(
            ops.sum(
                keras.losses.binary_crossentropy(
                    data, reconstruction
```

```
            ),
            axis=(1, 2)
        )
    )

    kl_loss = -0.5 * (1 + z_log_var - ops.square(z_mean)
                        - ops.exp(z_log_var))
    kl_loss = ops.mean(ops.sum(kl_loss, axis=1))

    # Apply beta to KL loss
    total_loss = reconstruction_loss + self.beta * kl_loss

grads = tape.gradient(total_loss, self.trainable_weights)
self.optimizer.apply_gradients(
    zip(grads, self.trainable_weights)
)
self.total_loss_tracker.update_state(total_loss)
self.reconstruction_loss_tracker.update_state(
                                    reconstruction_loss)
self.kl_loss_tracker.update_state(kl_loss)

return {
    "loss": self.total_loss_tracker.result(),
    "reconstruction_loss":
                self.reconstruction_loss_tracker.result(),
    "kl_loss": self.kl_loss_tracker.result(),
}
```

Listing 11.13 Training the VAE

In the full code listing, we also include a `test_step` function that performs essentially the same procedure. We're omitting that function here to save some space. It's highly recommended that you read through the full code though.

Analysis of the Results

With all our architectural pieces in place, we now assemble them into a complete, trainable system. Our VAE instantiation shown in the two brief lines of Listing 11.14 brings together the encoder and decoder we've carefully constructed, with that beta parameter set to 0.001. This small value tells our model to prioritize sharp, detailed reconstructions over perfectly organized latent spaces. The Adam optimizer with a learning rate of 1e-4 provides our training engine. This relatively conservative learning rate ensures stable training for our complex dual-objective system. VAEs can be sensitive to learning rates because they're balancing two competing goals, and rushing this process often leads to training *instability*.

```
# Create VAE model
vae = VAE(encoder, decoder, beta=0.001)
vae.compile(optimizer=keras.optimizers.Adam(learning_rate=1e-4))
```

Listing 11.14 Construction and Compilation of the VAE Model

The training curves show healthy VAE learning dynamics in Figure 11.4. The reconstruction loss steadily decreases, indicating the model is learning to recreate images more accurately. The KL divergence loss exhibits more volatility but generally trends downward, showing the latent space is organizing while maintaining some *stochasticity* essential for generation. The gap between training and validation losses remains small across all metrics, suggesting good generalization without overfitting. The KL loss's higher variability is typical for VAEs because it measures how well the learned distributions match the prior, which involves inherent randomness from the sampling process.

With beta = 0.001, the reconstruction loss dominates the total loss (contributing ~572 vs. ~0.47 from KL), explaining why reconstructions should be quite sharp while still maintaining a structured latent space for generation.

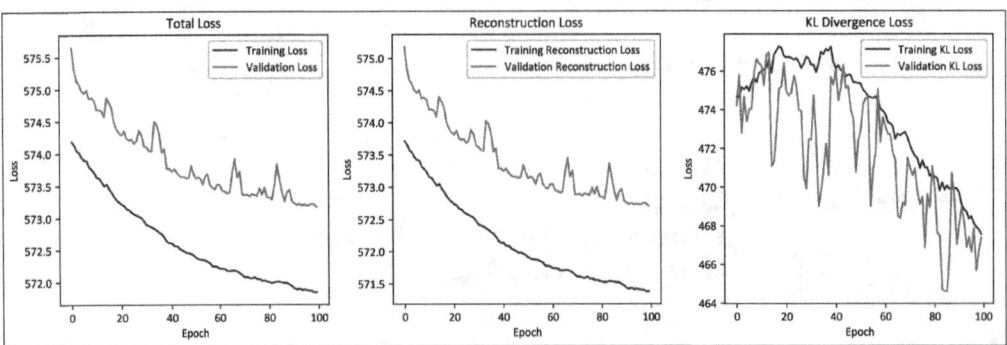

Figure 11.4 Training Curves of Reconstruction Loss and KL Loss

The reconstruction results shown in Figure 11.5 reveal a characteristic limitation of VAEs: The reconstructed images appear noticeably blurry compared to their crisp originals. While our model successfully captures the essential structure and colors of each object—the shades in the cat's fur, the shape of the ships' hulls, the airplane's general shape—the fine details have been smoothed away. This blurriness stems from the probabilistic nature of VAEs and our choice of reconstruction loss. Binary cross-entropy encourages the model to predict the average of all possible pixel values that could appear in similar contexts. When the model encounters uncertainty about specific details, it hedges its bets by outputting intermediate values rather than committing to sharp edges or fine textures.

We'll discuss briefly why this happens as it will help you develop an appreciation of the more modern models that solve these issues. The reconstruction loss penalizes large deviations from the original, but it treats all pixel errors equally. A model that outputs a slightly gray pixel instead of choosing between black or white receives a smaller penalty than one that makes a definitive but occasionally wrong choice. This leads to the characteristic *averaging effect* that produces blurry reconstructions.

Figure 11.5 Image Generation Through the VAE

This fundamental limitation of VAEs points toward an exciting alternative approach that we'll explore in our next section on Generative Adversarial Networks (GANs). Instead of trying to reconstruct images by minimizing pixel-wise differences, GANs take a radically different approach: they learn to generate images that are indistinguishable from real ones, even if they don't match any specific training example pixel-for-pixel. This adversarial training paradigm often produces much sharper, more realistic images at the cost of some other trade-offs we'll discover. But let's first do something interesting with our VAE implementation.

Generating Novel Images from Scratch

We can now use our trained VAE's latent vectors to generate entirely new images. The *interpolation* function shown in Listing 11.15 demonstrates this generative capability by creating smooth transitions between different images through latent space travel. This process reveals the continuous, structured nature of the representations our model has learned. The function works by first encoding both input images into their latent representations using the mean vectors (z_mean) rather than random samples for consistency. It then creates a series of intermediate points along the "straight line" connecting these two points in the 128-dimensional latent space.

The interpolation uses linear blending (1 - alpha) * z_mean1 + alpha * z_mean2, where alpha progresses from 0 to 1 across the specified number of steps. When alpha equals 0, we get the first image's encoding; when alpha equals 1, we get the second image's encoding. Values in between create weighted combinations of both encodings.

11 Autoencoders and Generative AI

Each interpolated latent vector gets decoded back into *image space*, producing a sequence of images that smoothly morph from the first input to the second. This demonstrates that the latent space has learned meaningful geometric structure—points that are close together in latent space correspond to visually similar images, making linear interpolation produce sensible intermediate images rather than random noise.

```
def interpolate_images(vae, img1, img2, n_steps=10):
    # Encode both images
    z_mean1, _, _ = vae.encoder.predict(i1.reshape(1, 32, 32, 3))
    z_mean2, _, _ = vae.encoder.predict(i2.reshape(1, 32, 32, 3))

    # Create interpolation points
    interpolated_latents = []
    for i in range(n_steps):
        alpha = i / (n_steps - 1)
        interpolated_z = (1 - alpha) * z_mean1 + alpha * z_mean2
        interpolated_latents.append(interpolated_z)

    interpolated_latents = np.vstack(interpolated_latents)

    # Decode interpolated latent vectors
    interpolated_images = vae.decoder.predict(
                                    interpolated_latents)

    return interpolated_images

# Select two different images for interpolation
img1 = x_test[1]   # First test image (try 0)
img2 = x_test[6]   # Different test image (try 100)

interpolated = interpolate_images(vae, img1, img2, n_steps=10)
```

Listing 11.15 Interpolating Between Two Points in Latent Space

The interpolation results can be seen in Figure 11.6. They clearly show a gradual transition from what appears to be a ship to a more car-like object, with the background and overall structure changing smoothly across the sequence. However, the persistent blurriness we observed in reconstructions carries over to these generated images as well.

This limitation points us toward our next exploration. While VAEs excel at learning structured latent spaces for interpolation and generation, there are other models that approach the generation problem from a fundamentally different angle that often produces much sharper, more realistic images. Instead of reconstructing pixel-by-pixel, these models learn to fool a discriminator network, leading to crisp details that VAEs struggle to achieve. Let's take a look at these next.

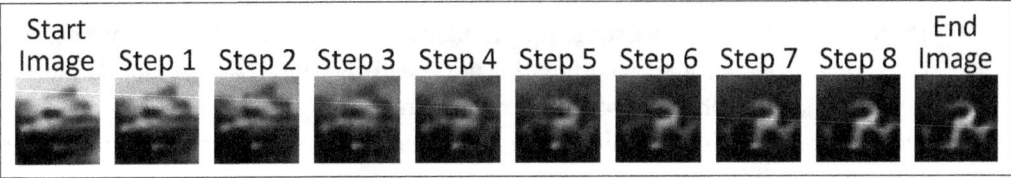

Figure 11.6 Results of Interpolation Between a Boat and Car

11.3 Generative Adversarial Networks

While VAEs opened up new possibilities for generating data by introducing probabilistic approaches to encoding, they had their limitations. The generated samples often appeared blurry or lacked the sharp, realistic details we see in actual data. For years, this remained a persistent challenge in generative modeling. Then, in 2014, Ian Goodfellow introduced Generative Adversarial Networks (GANs), which revolutionized the field by turning generation into a competitive game. The breakthrough came from reframing the entire problem. Instead of trying to explicitly model the probability distribution of our data (as VAEs do), GANs pit two neural networks against each other in an *adversarial contest*. This competitive dynamic drives both networks to improve continuously, resulting in generated samples that can be remarkably realistic.

In this section, we'll first introduce what *adversarial* means and then discuss how we can implement these game-like dynamics to help the machines learn on their own.

11.3.1 The Adversarial Game

Picture this scenario: Alex wants to perfectly mimic their friend Chris's voice over the phone. Jordan, who knows Chris well, serves as the judge. Alex's goal is simple: Fool Jordan into believing they're actually talking to Chris. Jordan's job is equally clear: Determine whether the voice belongs to the real Chris or is Alex's imitation.

This exact dynamic captures the essence of how GANs work. Alex represents the *generator network*, whose job is creating fake data that looks as real as possible. Jordan acts as the *discriminator network*, which must distinguish between real data and the generator's fabrications.

Let's follow Alex and Jordan through their adversarial game. Initially, Alex's voice imitation is terrible. When Alex calls Jordan pretending to be Chris, Jordan immediately recognizes the deception. "That's definitely not Chris," Jordan says, "your voice is too high-pitched, you're speaking too slowly, and you're not using Chris's usual expressions." Alex takes this feedback seriously. The next attempt involves lowering the voice and speaking faster. This time, Jordan hesitates for a moment but still catches the imitation. "Better," Jordan admits, "but Chris never pronounces certain words that way, and the rhythm is still off."

Through repeated attempts, something remarkable happens. Alex's imitations become increasingly sophisticated, incorporating not just Chris's vocal patterns but also speech rhythms, favorite phrases, and even subtle breathing patterns. Meanwhile, Jordan becomes more discerning, developing an increasingly keen ear for the subtle differences that separate the real Chris from even the best imitations. This back-and-forth process drives both participants to excel. Alex learns to generate increasingly convincing imitations, while Jordan develops superhuman abilities to detect even the most sophisticated fakes. The competition pushes both to heights neither could reach alone. While Alex might eventually hit physical limits on how much their voice can change, neural networks face no such constraints. As long as the discriminator continues providing *feedback signals*, the generator can keep refining its approach indefinitely. Neural networks can adjust their millions of parameters with each training step, allowing for improvements that go far beyond what any human mimic could achieve.

This *adversarial relationship* draws from game theory, specifically what's called a *minimax game*. In such games, one player tries to minimize their losses while the other tries to maximize their gains. The elegance of this setup is that both players improve simultaneously through their competition. In our voice mimicry example, Alex wants to maximize the probability of fooling Jordan (minimizing Jordan's ability to detect fakes), while Jordan wants to maximize accuracy in distinguishing real Chris from Alex's imitations (minimizing Alex's success rate). Neither player can improve without forcing the other to become better as well.

The mathematical formulation captures this elegantly. The generator tries to minimize a loss function while the discriminator tries to maximize the same function. This creates a *dynamic equilibrium* where both networks push each other toward better performance. Unlike traditional machine learning where we have a fixed target to reach, GANs create a moving target that constantly evolves. What makes this particularly powerful is that the discriminator essentially becomes a *learned loss function*. Instead of us having to design a specific way to measure how "real" generated data looks, the discriminator learns this evaluation process. It's like having Jordan develop an increasingly sophisticated understanding of what makes Chris's voice unique, rather than giving Jordan a checklist of features to look for.

Why does adversarial training work so well for generation? The adversarial approach solves several fundamental problems that plagued earlier generative models. Traditional approaches often struggled with generating sharp, detailed outputs because they relied on loss functions that averaged over many possibilities. When you average multiple potential outputs, you typically get something blurry or generic. Consider what happens when Alex first starts learning to mimic Chris. If Alex tried to sound like "an average person" rather than specifically like Chris, the result would be a generic, unmemorable voice that doesn't fool anyone. The adversarial training forces Alex to focus on the specific characteristics that make Chris's voice distinctive. The discriminator provides incredibly rich feedback compared to traditional loss functions. Instead of a simple numerical score, it's providing detailed information about what makes generated samples unconvincing. When Jordan tells Alex, "Your voice is too high-pitched," that's much more actionable feedback than simply saying "Try again."

This rich feedback mechanism allows the generator to learn complex, high-dimensional distributions. The discriminator essentially learns to encode all the subtle patterns that make real data look real, then provides this knowledge to the generator through their adversarial interaction.

The adversarial training also naturally handles the challenge of *mode coverage*. In our voice example, Chris might speak differently when excited, tired, or talking about different topics. Traditional approaches might focus on one particular "mode" of Chris's voice. But Jordan will catch Alex if the imitation only works for one type of conversation. This forces Alex to learn the full range of Chris's vocal patterns.

Perhaps most importantly, the adversarial approach pushes generated samples toward the boundary of what's realistic. Traditional methods often generate "safe" outputs that clearly look artificial. But Alex succeeds only by pushing the imitation as close as possible to the real thing. This boundary-pushing behavior leads to the sharp (albeit small), realistic outputs that made GANs famous. The dynamic nature of the training means that the generator never gets to rest on past successes. The generator must continuously improve as the discriminator becomes more sophisticated. This ongoing pressure drives both networks toward increasingly impressive performance. Through this adversarial dance, GANs achieve something remarkable: They learn to generate data that can be virtually indistinguishable from real examples. The competition that seemed like a simple game between Alex and Jordan becomes a powerful machine learning paradigm that has transformed how we approach generative modeling.

11.3.2 Generative Adversarial Network Architecture

Now that we understand the adversarial game between generator and discriminator let's examine how these competing networks are actually built. Just as Alex and Jordan each needed specific skills for their voice mimicry contest, the generator and discriminator networks require carefully designed architectures to excel at their respective tasks. The beauty of GANs lies in their architectural simplicity combined with their training complexity. While each network can be relatively straightforward in design, the strength emerges from how they interact and evolve together. Understanding their individual architectures helps us appreciate why this adversarial approach works so effectively. The architecture of a vanilla GAN is shown in Figure 11.7. The generator creates some data, and the discriminator tries to classify it as being real or fake without requiring any other labels. We feed real and fake data to the discriminator and calculate the loss based on correct (or incorrect) classification. Let's discuss the details of this simple but elegant design in detail.

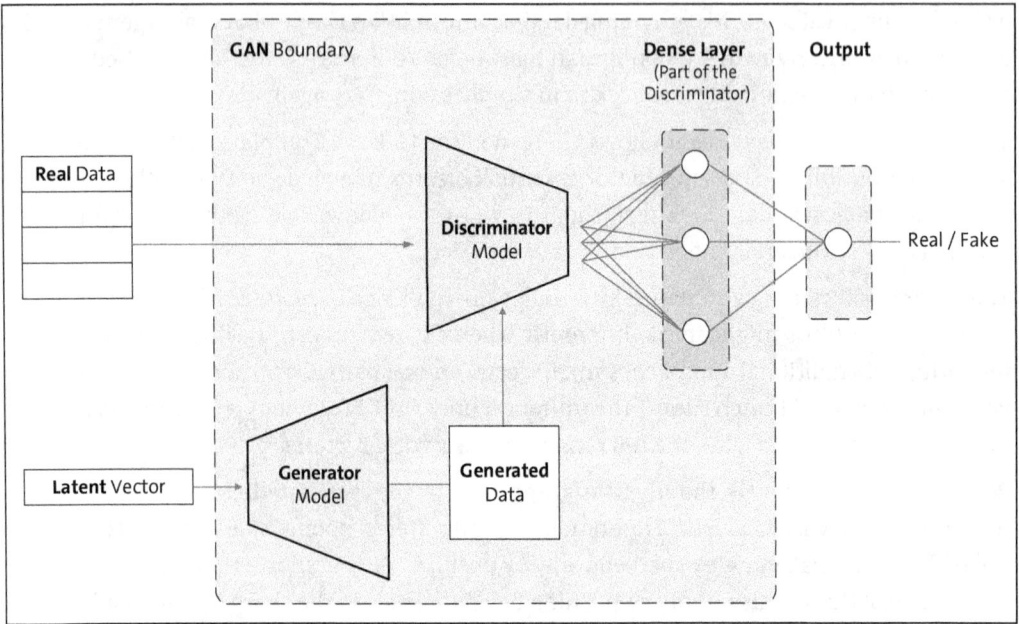

Figure 11.7 Architecture of a Vanilla GAN

Generator Network Design Principles

The generator network faces a unique challenge in that it must transform random noise into realistic data. Consider Alex starting with nothing but static and somehow crafting it into Chris's convincing voice. This transformation requires a network architecture that can progressively build complexity from simple beginnings. Most generator networks follow an *expansive design*, starting small and growing larger. The generator typically starts with a low-dimensional noise or latent vector z sampled from a simple distribution. This random input serves as the creative seed from which all generated samples emerge. The network then progressively upsamples this noise through a series of layers, each one adding more spatial resolution and complexity. For image generation, this often involves transpose convolutions, which increase the spatial dimensions while the network learns to fill in meaningful details.

The key insight is that each layer should learn to represent features at different scales of detail. Early layers might establish basic structure and layout, while later layers add fine textures and sharp details. This hierarchical approach mirrors how we might describe a face: first the overall shape, then the placement of features, and finally the subtle details that make each face unique.

The generator's architecture must also consider the target data's characteristics. Generating high-resolution images requires different design choices than generating text or audio. For images, convolutional layers excel at capturing spatial relationships. For sequential data such as text, recurrent or transformer architectures (covered in Chapter 9) become more appropriate.

One critical design principle involves the network's *capacity*. The generator needs enough parameters to capture the complexity of real data but not so many that training becomes unstable. Finding this balance requires experimentation and careful monitoring of training dynamics.

Discriminator as a Binary Classifier

While the generator builds complexity from simplicity, the discriminator performs the opposite task: It must distill complex data down to a single binary decision. Is this sample real or fake? The discriminator architecture typically mirrors traditional convolutional neural networks used for image classification. It progressively downsamples input data through convolutional layers, extracting increasingly abstract features until reaching a final classification layer. Each layer learns to detect patterns at different levels of abstraction. Early layers might detect simple edges and textures, while deeper layers recognize more complex patterns and relationships. The final layer outputs a single probability value through a sigmoid activation.

The discriminator's loss function follows standard binary cross-entropy:

$$\mathcal{L}_D = -E_{x \sim p_{data}}[\log D(x)] - E_{z \sim p_z}\left[\log\left(1 - D(G(z))\right)\right]$$

Here, x represents real data samples, z is the random noise input to the generator, *G(z)* produces fake samples, and *D* outputs the discriminator's probability that a sample is real. This loss encourages the discriminator to output high values (close to *1*) for real samples and low values (close to *0*) for generated samples. The discriminator succeeds when it can reliably distinguish between real and fake data.

However, the discriminator faces a unique challenge that standard classifiers don't encounter: Its training data constantly evolves. As the generator improves, the "fake" samples become increasingly sophisticated. This problem is similar to the well-known concept of *covariate shift*.

> **The Covariate Shift Problem**
>
> Covariate shift occurs when the distribution of input features changes between training and deployment, even though the relationship between inputs and outputs remains the same. This phenomenon poses a significant challenge in modern machine learning, where models trained on one dataset may encounter very different data distributions in real-world applications. For example, a model trained on daytime images might struggle with nighttime photos, or a recommendation system built on historical user behavior may fail when user preferences evolve. Covariate shift becomes particularly problematic in deep learning models, which can be sensitive to these distribution

> changes despite their apparent robustness. Addressing this challenge has led to important techniques such as domain adaptation, transfer learning, and robust training methods that help models maintain performance when deployed in environments different from their training conditions.

This evolving challenge requires the discriminator to continuously adapt and improve. Unlike traditional classification, where the data distribution remains fixed, GAN training presents the discriminator with progressively harder examples. This dynamic difficulty adjustment drives the discriminator to develop increasingly nuanced understanding of what makes data authentic. Moreover, the discriminator's architecture must balance several competing demands. It needs enough capacity to detect sophisticated fakes but not so much that it memorizes the training set. It should be powerful enough to provide useful feedback to the generator but not so powerful that it becomes impossible to fool, which would halt the generator's improvement.

Training Dynamics and the Minimax Game

The mathematical foundation of GAN training emerges from game theory, specifically the *minimax framework*. The generator and discriminator engage in a *zero-sum game* where one player's gain directly corresponds to the other's loss. This creates the competitive dynamic that drives both networks toward better performance.

The complete GAN objective function captures this adversarial relationship:

$$\min_G \max_D V(D,G) = \mathbb{E}_{x \sim p_{\text{data}}}[\log D(x)] + \mathbb{E}_{z \sim p_z}\left[\log\left(1 - D(G(z))\right)\right]$$

In this formulation, *V(D,G)* represents the value function that both networks compete over, with the discriminator *D* trying to maximize it while the generator *G* tries to minimize it. The two expectation terms correspond to the discriminator's performance on real data and generated data, respectively. The discriminator maximizes this objective by correctly classifying real samples and fake samples. Meanwhile, the generator minimizes the same objective by creating samples that fool the discriminator. This minimax formulation creates very interesting training dynamics. In the ideal scenario, training reaches a Nash equilibrium where neither network can improve further without the other adapting. The generator produces samples indistinguishable from real data, while the discriminator can only guess randomly, achieving 50% accuracy.

However, reaching this equilibrium proves challenging in practice. The training process resembles a delicate balancing act where both partners must move at compatible rates. If one network advances too quickly, the training becomes unbalanced and unstable.

Consider what happens when the discriminator becomes too powerful too quickly. It perfectly identifies all generated samples as fake, providing no useful gradient information to the generator. The generator receives only negative feedback without guidance on how to improve.

Conversely, if the generator advances too rapidly, it might fool a weak discriminator easily, but this success proves hollow. The generator learns to exploit the discriminator's specific weaknesses rather than learning to generate truly realistic data. When tested against a stronger discriminator or real-world evaluation, these generated samples fail dramatically. Successful GAN training requires careful balance between these competing forces. Practitioners often adjust learning rates independently for each network, sometimes training one network multiple times per iteration to maintain equilibrium. The generator might train with a higher learning rate to keep pace with a naturally stronger discriminator, or the discriminator might train less frequently to prevent overwhelming the generator. The training process also involves careful monitoring of loss values, but interpreting these losses differs from traditional supervised learning. In GANs, loss values don't necessarily decrease monotonically toward zero. Instead, we look for stable oscillations that indicate healthy competition between the networks.

This ongoing competition drives GANs toward generating remarkably realistic data, but it also makes training more complex and unpredictable than traditional approaches. Understanding these dynamics helps us appreciate both the power and the challenges of adversarial training, setting the stage for examining practical implementation strategies and common pitfalls.

11.3.3 Other Variations

While our exploration of GAN architecture gives us the theoretical foundation, there are two specific variant that deserves attention before we jump into implementation. The first of these is *Deep Convolutional GANs (DCGANs)*. These represent one of the most successful GAN architectures, particularly for image generation. The original GAN formulation left many architectural decisions open, leading to training instability and poor results. DCGANs emerged as a set of proven guidelines that consistently work better. The key insight lies in leveraging spatial structure through convolutional layers. Just as an artist paints a landscape with broad strokes first, then adds fine details, DCGANs build images hierarchically. Early layers establish basic spatial relationships, while later layers refine intricate details.

DCGANs introduce specific constraints that distinguish them from general GANs. The most significant change replaces fully connected layers with convolutional operations throughout both networks.

Moving beyond DCGANs, the second important variation that addresses a fundamental limitation: control over what gets generated. Auxiliary Classifier GANs (ACGANs) represent a powerful extension that bridges the gap between random generation and purposeful creation. Basic GANs and DCGANs generate samples from learned data distributions, but they offer no direct control over what type of sample emerges. If you're generating handwritten digits, you can't specify whether you want a 3 or an 8. ACGANs solve this by incorporating class information directly into both the generation and discrimination processes. The key insight is treating generation as both a creative and classification task. The discriminator job here isn't simply to determine if a sample is real or fake. It must also identify what class the sample belongs to. This dual responsibility creates richer feedback signals that guide the generator toward more structured and controllable outputs. Take a look at Figure 11.8 for the architecture of ACGANs and compare it with the architecture of GANs shown earlier in Figure 11.7.

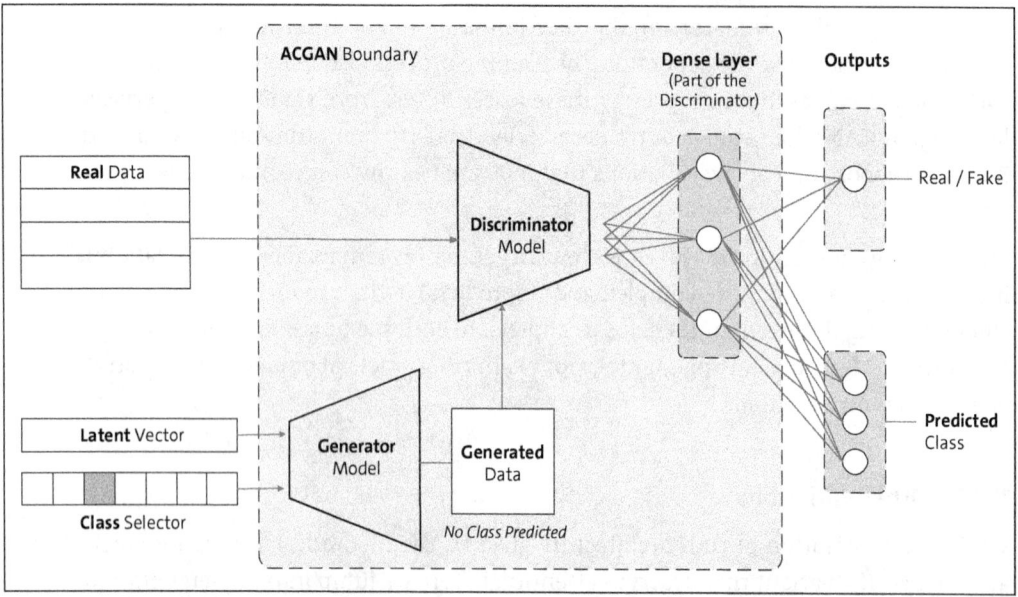

Figure 11.8 Architecture of ACGANs

ACGANs modify both networks to handle *class-conditional generation*. The generator receives two inputs: the familiar noise vector and a *class selector*. The discriminator performs dual classification, outputting both authenticity and class predictions.

The loss function becomes more sophisticated, combining the adversarial loss with classification accuracy as

$\mathcal{L}_D = \mathcal{L}_S + \mathcal{L}_C$

where $\mathcal{L}_S$ is the standard GAN loss and $\mathcal{L}_C$ is the classification loss for correctly identifying classes in both real and generated samples.

ACGANs provide controllable generation similar to how we discussed VAEs organizing their latent space meaningfully. While VAEs create continuous latent representations where nearby points generate similar outputs, ACGANs create discrete control through explicit class conditioning. This conditioning allows for targeted generation—you can specify exactly what type of sample you want rather than hoping random sampling produces something useful. The class information acts like a steering wheel for the generation process, directing the creative energy toward specific outcomes.

The discriminator's dual role on the other hand creates interesting training dynamics. It must become expert at both detecting fakes and classifying content, which typically results in better feature learning. This improved feature extraction often leads to higher-quality generated samples compared to standard GANs.

We've kept this architectural overview concise because hands-on implementation reveals the true power of these techniques. ACGANs demonstrate how GAN architectures can evolve to address specific needs—in this case, the need for controllable generation.

11.3.4 Generative Adversarial Network Implementation in Keras

Now that we understand the adversarial training framework and the theoretical foundations of DCGANs and ACGANs, let's build a complete ACGAN implementation using Keras. Our implementation will generate Fashion-MNIST images conditioned on specific clothing categories such as shirts, bags, and shoes. Rather than simply throwing together a basic generator and discriminator, we'll construct a sophisticated system that demonstrates the key architectural decisions that separate effective GANs from those that struggle with training instability. We'll tackle several critical design challenges that every GAN practitioner encounters: how to structure the generator to produce high-quality outputs, how to build a discriminator that provides useful feedback without becoming too powerful, and how to balance the adversarial training process to achieve stable convergence. Our ACGAN will include *class conditioning* through embedding layers, separate loss functions for authenticity and classification, and carefully tuned hyperparameters that reflect lessons learned from years of GAN research.

We'll skip the preamble of the code that you can see in the full code available in the resources page at *https://recluze.net/keras-book* in the notebook **11-04-acgan-fashion-mnist**. Let's start with Listing 11.16 that shows the core of the concepts—the generator. Our generator function takes two distinct inputs that work together to create the final output. The first input is our familiar random noise vector with 100 dimensions, which provides the creative variability that makes each generated image unique. The second input is a class label—a single integer from 0 to 9 representing which Fashion-MNIST category we want to generate. The class conditioning happens through an embedding layer, which transforms each integer label into a rich 128-dimensional vector representation. This embedding approach gives our generator much more expressive power than simply concatenating a one-hot encoded vector. The embedding layer learns dense representations that capture meaningful relationships between classes during training. For instance, it might learn that "shirt" and "pullover" share certain visual characteristics, encoding this knowledge in similar embedding vectors.

After flattening the embedding, we concatenate it with our noise vector, creating a combined input of 228 dimensions (100 + 128). This fusion allows the generator to modulate its creative process based on the desired class while maintaining the randomness necessary for generating diverse examples within each category. The architecture then follows a carefully designed upsampling pathway.

Batch normalization appears after nearly every layer with a momentum of 0.9, which is slightly stronger than the typical 0.99 default. This choice reflects the unique challenges of GAN training, where we need normalization that adapts more quickly to the changing distributions that occur as both generator and discriminator evolve during adversarial training. The higher momentum helps prevent the internal covariate shift that can destabilize GAN training. Notice that we disable bias terms (use_bias=False) in layers that are followed by batch normalization. This design choice eliminates redundant parameters because batch normalization includes its own learnable bias term, making the layer's bias unnecessary and potentially harmful to training stability.

The final layer uses a tanh activation function, which constrains outputs to the range [-1, 1]. This choice aligns with our data preprocessing, where we normalized Fashion-MNIST pixel values from [0, 255] to [-1, 1].

```
def build_generator(latent_dim=100, num_classes=10):
    noise_input = layers.Input(shape=(latent_dim,))

    label_input = layers.Input(shape=(1,))
    label_embedding = layers.Embedding(num_classes, 128)
                                                    (label_input)
    label_embedding = layers.Flatten()(label_embedding)

    model_input = layers.Concatenate()
                            ([noise_input, label_embedding])

    x = layers.Dense(7 * 7 * 512, use_bias=False)(model_input)
    x = layers.BatchNormalization(momentum=0.9)(x)
    x = layers.ReLU()(x)
    x = layers.Reshape((7, 7, 512))(x)

    # Additional upsampling layer for gradual resolution increase
    x = layers.Conv2DTranspose(256, (4, 4), strides=(1, 1),
                        padding='same', use_bias=False)(x)
    x = layers.BatchNormalization(momentum=0.9)(x)
    x = layers.ReLU()(x)

    # Additional upsampling layers here
    generated_image = layers.Conv2D(1, (3, 3), padding='same',
                        activation='tanh', use_bias=False)(x)

    generator = Model([noise_input, label_input],
                    generated_image,
```

```
                    name='generator')
    return generator

generator = build_generator()
generator.summary()
```

Listing 11.16 Generator for a GAN

While our generator creates images from noise and class labels, the discriminator (see Listing 11.17) must perform the complementary task of evaluating both the authenticity and class identity of input images. This dual responsibility makes the ACGAN discriminator more sophisticated than a standard GAN discriminator, requiring careful architectural decisions to handle both classification tasks effectively.

The discriminator begins with a crucial stability technique: *Gaussian noise injection* directly at the input layer. Adding noise with a standard deviation of 0.1 to incoming images prevents the discriminator from becoming too confident about its predictions, which can lead to *training collapse*. This technique essentially forces the discriminator to make decisions based on broader patterns rather than memorizing specific pixel arrangements, creating more robust gradients for the generator to learn from. The architecture uses LeakyReLU activations with a slope of 0.2 for negative inputs, contrasting with the standard ReLU used in our generator. This choice addresses the dying ReLU problem that can occur in discriminators, where neurons become permanently inactive and stop contributing to the learning process. LeakyReLU ensures that even negative inputs produce small gradients, maintaining information flow throughout the network.

A key architectural decision appears in the transition from convolutional to dense layers: *global average pooling* instead of flattening. We covered this type of pooling in detail in Chapter 7, Section 7.2. This choice provides *translation invariance*, meaning the discriminator's decisions remain consistent regardless of where features appear in the image. Global average pooling also significantly reduces the parameter count compared to flattening a full feature map, helping prevent overfitting.

> **Experiment with the Model**
>
> When you experiment with the full code, try modifying these design decisions to look at the effect. You can start by changing the pooling to a flatten layer to see the explosion in number of model parameters. Other changes to these design decisions will also help you understand their implications in depth.

The discriminator's output structure reflects the ACGAN's dual objectives through separate branches after the shared feature extraction layers. Both branches receive the same high-level features but specialize in different tasks: the validity branch outputs a single sigmoid-activated value representing the probability that the input is real, while the `label` branch outputs a softmax distribution over the 10 Fashion-MNIST classes.

This branched architecture allows each task to develop specialized representations while sharing the computational cost of early feature extraction. Alternative approaches might use separate networks for each task, but the *shared-trunk* design is more parameter-efficient and often performs better because authenticity and classification tasks benefit from similar low-level feature representations.

```python
def build_discriminator(num_classes=10):
    img_input = layers.Input(shape=(28, 28, 1))
    
    # Add noise to inputs for stability
    x = layers.GaussianNoise(0.1)(img_input)
    
    # First conv block
    x = layers.Conv2D(64, (4, 4), strides=(2, 2),
                      padding='same')(x)
    x = layers.LeakyReLU(0.2)(x)
    x = layers.Dropout(0.25)(x)
    
    # ... other Conv2D, global pooling and dense layers
    
    # Separate branches for better feature learning
    validity_branch = layers.Dense(256)(x)
    validity_branch = layers.LeakyReLU(0.2)(validity_branch)
    validity = layers.Dense(1, activation='sigmoid')(
                                        (validity_branch)
    
    label_branch = layers.Dense(256)(x)
    label_branch = layers.LeakyReLU(0.2)(label_branch)
    label = layers.Dense(num_classes, activation='softmax')(
                                        (label_branch)
    
    discriminator = Model(img_input, [validity, label],
                          name='discriminator')
    return discriminator
```

Listing 11.17 Discriminator for the Fashion MNIST GAN

With our generator and discriminator architectures defined, we need to orchestrate their adversarial training through a custom training loop, as shown in Listing 11.18. The ACGAN class inherits from Keras's `Model` class, allowing us to implement specialized adversarial dynamics while leveraging the framework's training infrastructure. The discriminator evaluation creates an interesting dynamic: For real images, we expect high validity scores and correct class predictions, while for fake images, we want low validity scores but still accurate class predictions based on their assigned labels. This forces the discriminator to learn meaningful class representations rather than detecting authenticity through class confusion.

```
def train_step(self, data):
    real_images, real_labels = data
    batch_size = tf.shape(real_images)[0]


    # Sample random noise and labels for generator
    noise = tf.random.normal([batch_size, self.latent_dim])
    fake_labels = tf.random.uniform([batch_size, 1],
                                    0, 10, dtype=tf.int32)

    # Generate fake images
    fake_images = self.generator([noise, fake_labels])

    # Train discriminator
    with tf.GradientTape() as tape:
        # Real images
        real_validity, real_pred_labels = self.discriminator(
            real_images)
        # Fake images
        fake_validity, fake_pred_labels = self.discriminator(
            fake_images)
```

Listing 11.18 Training the GAN

Continuing the function train_step in Listing 11.19, the gradient tape context begins our discriminator training phase, where we'll calculate losses for both the authenticity task and the classification task. The discriminator training involves calculating four separate loss components that reflect ACGAN's dual objectives. The validity losses use binary cross-entropy with perfect targets: tf.ones_like(real_validity) for real images and tf.zeros_like(fake_validity) for fake images. These targets represent our ideal scenario where the discriminator assigns probability 1.0 to real images and 0.0 to fake images.

The classification losses use sparse categorical cross-entropy for both real and fake images. For real images, we compare predictions against the true Fashion-MNIST labels. For fake images, we compare against the randomly sampled labels we assigned during generation. This creates the crucial ACGAN dynamic: The discriminator must learn to classify fake images correctly according to their *intended class* while still recognizing them as fake. Finally, the total discriminator loss combines all four components with equal weighting: d_loss = real_loss + fake_loss + real_class_loss + fake_class_loss.

11 Autoencoders and Generative AI

This equal weighting assumes that authenticity detection and classification are equally important, although you could adjust these ratios based on your application's priorities.

```
# ... continuing with the train_step function
# Train discriminator
with tf.GradientTape() as tape:
    # Real images
    real_validity, real_pred_labels = self.discriminator(
            real_images)
    # Fake images
    fake_validity, fake_pred_labels = self.discriminator(
            fake_images)

    # Discriminator losses
    real_loss = self.loss_fn(tf.ones_like(real_validity),
                             real_validity)
    fake_loss = self.loss_fn(tf.zeros_like(fake_validity),
                             fake_validity)

    # Classification losses
    real_class_loss = losses.sparse_categorical_crossentropy(
            real_labels, real_pred_labels)
    fake_class_loss = losses.sparse_categorical_crossentropy(
            fake_labels, fake_pred_labels)

    d_loss = (real_loss + fake_loss +
                     real_class_loss + fake_class_loss)
d_gradient = tape.gradient(d_loss,
                     self.discriminator.trainable_variables)
self.d_optimizer.apply_gradients(
        zip(d_gradient, self.discriminator.trainable_variables))
```

Listing 11.19 Training Step Continued

The generator training (covered in Listing 11.20) creates a second gradient tape to calculate losses independently from the discriminator. We regenerate fake images using the same noise and labels, then evaluate them through the discriminator to get both validity and classification predictions. The generator's validity loss uses `tf.ones_like(fake_validity)` as targets, representing the generator's goal to fool the discriminator into believing fake images are real. This creates the adversarial dynamic: While the discriminator tries to output 0.0 for fake images, the generator tries to make the discriminator output 1.0.

11.3 Generative Adversarial Networks

The generator's classification loss compares the discriminator's class predictions against the intended fake labels. This ensures the generator doesn't just create realistic-looking noise but produces images that clearly belong to the specified class. The generator must simultaneously fool the authenticity detector and satisfy the class classifier. The equal weighting between gen_loss and gen_class_loss balances realism and class fidelity. Alternative approaches include weighting classification loss higher (twice or even four times) when class conditioning is critical or using curriculum learning where classification weight increases during training.

```
# Train generator
with tf.GradientTape() as tape:
    fake_images = self.generator([noise, fake_labels])
    fake_validity, fake_pred_labels = self.discriminator(
        fake_images)

    # Generator losses
    gen_loss = self.loss_fn(tf.ones_like(fake_validity),
                    fake_validity)
    gen_class_loss = keras.losses.sparse_categorical_crossentropy(
        fake_labels, fake_pred_labels)

    g_loss = gen_loss + gen_class_loss

g_gradient = tape.gradient(g_loss,
                    self.generator.trainable_variables)
self.g_optimizer.apply_gradients(
    zip(g_gradient, self.generator.trainable_variables))

# Update metrics
self.gen_loss_tracker.update_state(g_loss)
self.disc_loss_tracker.update_state(d_loss)

return {
    "g_loss": self.gen_loss_tracker.result(),
    "d_loss": self.disc_loss_tracker.result(),
}
```

Listing 11.20 Training the Generator

With our custom ACGAN class defined, we instantiate it with our generator and discriminator models, and then configure the training parameters through compilation, as shown in Listing 11.21.

The optimizer configuration uses separate Adam optimizers with different learning rates: 0.0004 for the discriminator and 0.0001 for the generator. This 4:1 ratio helps balance the adversarial training by slowing down the generator relative to the discriminator, preventing the common problem where one network overwhelms the other. Both optimizers use modified Adam parameters: beta_1=0.0 and beta_2=0.9. Setting beta_1 to 0 disables momentum, which can destabilize GAN training, while the higher beta_2 value maintains adaptive learning rate benefits.

```
acgan = ACGAN(discriminator=discriminator, generator=generator)

# Compile model
acgan.compile(
    d_optimizer=keras.optimizers.Adam(
        learning_rate=0.0004,
        beta_1=0.0,
        beta_2=0.9
    ),
    g_optimizer=keras.optimizers.Adam(
        learning_rate=0.0001,
        beta_1=0.0,
        beta_2=0.9
    ),
    loss_fn=keras.losses.BinaryCrossentropy(label_smoothing=0.1)
)

EPOCHS = 20

history = acgan.fit(
    train_dataset,
    epochs=EPOCHS,
    callbacks=[ ... monitoring callbacks here ...],
    verbose=1
)
```

Listing 11.21 Model Compilation and Hyperparameter Selection

With all that out of the way, we let the model run for a quite a long time. After the whole training is done, we can generate images using the generator just as we did with our VAEs. For instance, take a look at the images generated by the GAN in Figure 11.9. This is when we asked it to generate a bag. You can look at the full code and experiment with creating images of other classes as well. The results, we agree, aren't that overwhelming. However, as mentioned before, the techniques we've learned in this chapter are building blocks that will lead us to the state-of-the-art models in the next chapter. The reason for this lackluster result is that while our ACGAN implementation includes many stability techniques, GAN training remains notoriously challenging. Even with careful architecture design and hyperparameter tuning, you'll likely encounter training instabilities, mode collapse, or convergence failures. Understanding these common pitfalls and their solutions is essential for successful GAN deployment.

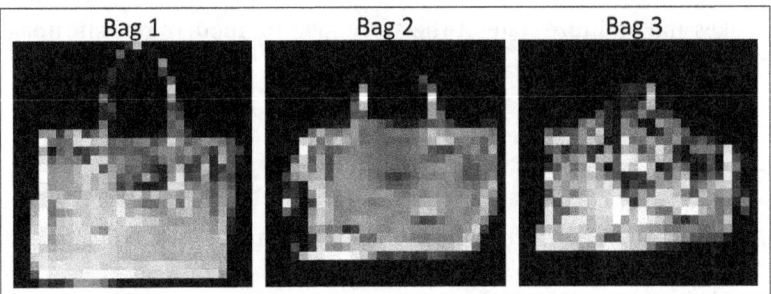

Figure 11.9 Fake Images Generated by the Trained GAN

11.3.5 Implementation Challenges

Mode collapse represents the most common failure mode, where the generator learns to produce only a subset of possible outputs. Instead of generating diverse shoes, bags, and shirts, a collapsed generator might produce only a few shoe variations, ignoring entire classes or producing nearly identical samples within classes. This occurs when the generator finds a few easy wins that consistently fool the discriminator and stops exploring the full data distribution.

Training instability manifests through oscillating losses, where neither network converges to a stable solution. The generator and discriminator engage in an unstable arms race: when the discriminator becomes too powerful, it provides uninformative gradients to the generator, while an overpowered generator can fool the discriminator so effectively that it stops learning meaningful features. *Discriminator-generator imbalance* requires constant attention throughout training. If the discriminator learns too quickly, it becomes perfect at detecting fake images, leaving the generator with vanishing gradients. Conversely, if the generator advances too rapidly, the discriminator receives no useful training signal from obviously fake samples.

Common solutions include adjusting learning rate ratios (we used 4:1), implementing *spectral normalization* to constrain discriminator power, and using techniques such as *feature matching* where the generator tries to match intermediate discriminator representations rather than just fooling the final output.

Similarly, gradient problems plague GAN training through vanishing and exploding gradients. When the discriminator becomes too confident, gradients flowing back to the generator approach 0. Our label smoothing and Gaussian noise injection help maintain gradient flow, but these issues persist across different GAN variants. Due to these extreme difficulties, GANs showed tremendous promise but struggled with stable large-scale image generation for years. Researchers eventually developed newer generative models based on the same underlying principles but with fundamentally different training dynamics. This doesn't mean that spending time with GANs was useless. On the contrary, learning these core concepts teaches us important lessons that help us avoid future mistakes and optimize state-of-the-art models for modern realistic image generation. We'll examine such models in the next chapter.

11.4 Summary

In this chapter, we explored GenAI through three key architectures. We began with autoencoders, learning how their encoder-decoder structure compresses data into meaningful latent representations and reconstructs outputs, with variations such as denoising and sparse autoencoders addressing specific challenges. We then advanced to Variational Autoencoders (VAEs), which introduced probabilistic encoding through the reparameterization trick and KL divergence loss, enabling controlled generation by sampling from learned latent distributions. Finally, we tackled Generative Adversarial Networks (GANs), implementing a complete ACGAN that combines adversarial training between generator and discriminator networks with class conditioning for controlled image generation. We discovered that while GANs can produce impressive results, they suffer from fundamental training challenges such as mode collapse and instability that limited their practical deployment for years, paving the way for the newer generative models we'll explore in the next chapter.

Chapter 12
Advanced Generative AI: Stable Diffusion

In this chapter, we'll tackle one of the most exciting frontiers in modern AI: realistic image generation through diffusion. You'll start by exploring the mathematics behind diffusion processes, understanding how computers learn to transform random noise into visual art and build your own diffusion model from scratch, watching it evolve from chaotic pixels to recognizable images. You'll uncover the secret behind Stable Diffusion and explore text-conditioned generation, where written words enable creative visual expression. Finally, you'll briefly consider sophisticated techniques such as inpainting, where you control image generation with precision. By the end of this chapter, you'll understand how these systems work and how to harness their power for your own projects.

The generative models we explored in the previous chapter—autoencoders, variational autoencoders (VAEs), and generative adversarial networks (GANs)—laid crucial groundwork for understanding how machines can learn to create new content. Each brought unique strengths to the table: autoencoders showed us how to compress and reconstruct data efficiently, VAEs introduced the power of probabilistic generation in latent spaces, and GANs demonstrated that two competing networks could push each other toward remarkably realistic outputs. However, for many years, these models struggled to work effectively on their own for complex, high-resolution image generation tasks. Autoencoders, while excellent at reconstruction, produced blurry and unrealistic images when used for generation. VAEs suffered from similar issues, often creating outputs that lacked the sharp details and coherent structures we see in real photographs. GANs, despite their impressive capabilities, proved notoriously difficult to train and often fell into problematic patterns such as mode collapse, where they would generate only a limited variety of outputs.

The breakthrough came when researchers discovered how to combine the best aspects of these foundational concepts into a revolutionary new architecture. This approach doesn't abandon what we've learned. Instead, it builds upon the latent space representations from VAEs, incorporates the attention mechanisms we studied with transformers, and applies the optimization techniques we've mastered with gradient descent and neural networks.

In this chapter, we'll explore how this synthesis of ideas led to one of the most significant advances in generative AI (GenAI). We'll start by understanding the theoretical principles that make this approach so effective, examining how it solves the problems that plagued earlier methods. You'll see how the core concepts we've been building throughout this book—from basic neural networks to complex architectures—come together in an elegant solution. We'll then dive into the practical implementation, building these models step-by-step using Keras. Rather than treating this as a black box, we'll construct each component ourselves, giving you a deep understanding of how every piece contributes to the final result. You'll learn to create systems that can generate high-quality images from simple text descriptions, opening doors to applications in art, design, and countless other creative fields.

> **[+] Navigating the Mathematical Complexity**
>
> The mathematical foundations underlying these concepts can be dense, drawing from multiple complex fields, including linear algebra, calculus, probability theory, and optimization. While a deep mathematical understanding enhances your grasp of the subject, it's not essential when starting out. We recommend reading through the mathematical sections to get a general sense of the concepts, and then proceeding directly to the implementation even if you don't fully grasp every mathematical detail.
>
> Seeing these principles applied in working code often provides the intuitive understanding that makes the mathematics clearer. You can always return to the mathematical foundations later with the benefit of practical experience to guide your understanding.

The models we'll explore represent a convergence of everything we've learned so far, showing how foundational concepts can evolve into transformative technologies when combined thoughtfully and implemented skillfully.

12.1 Theory Behind Stable Diffusion

Understanding the theory behind Stable Diffusion requires us to step back and look at generation from an entirely new perspective. Up until now, we've approached the problem of creating new data as a single, dramatic leap—transforming random noise into a finished image in one complex operation. But what if we could break this seemingly impossible task into hundreds of tiny, manageable steps? *Stable Diffusion* takes this approach, treating generation as a slow, iterative process of refinement rather than a miraculous instant transformation. This shift in perspective changes everything about how we train generative models and what makes them successful. Instead of learning to conjure images from nothing, our models learn something far more tractable: how to gradually remove noise and reveal structure through a sequence of learned *denoising* steps. Each step makes only a small correction, but when chained together, these incremental improvements transform pure randomness into coherent, detailed images.

This process gives us unprecedented control over generation quality while avoiding many of the training difficulties that plagued earlier approaches. In this section, we'll explain the theory behind Stable Diffusion and will explore how it works.

12.1.1 From Previous Generative Models to Diffusion

The progress toward effective GenAI has been incremental, with each breakthrough revealing new possibilities while exposing fundamental limitations. When we step back and examine the generative models we explored in the previous chapter, we can see a clear pattern: Each approach solved specific problems while introducing others, creating a puzzle that researchers have been working to complete for decades. Autoencoders, in their basic form, taught us the power of *compressed representations*. By forcing information through a narrow bottleneck, they learned to capture the most essential features of data. However, when we attempted to use them for generation by sampling from their latent space, the results were disappointing. The latent space lacked the smooth, continuous structure necessary for generating new, coherent examples. Random sampling from this space often produced meaningless outputs because the encoder had no incentive to organize the latent representations in a way that supported generation.

The mathematical root of this problem lies in the deterministic nature of standard autoencoders. The encoder function $f(x) = z$ maps each input x to a single point z in latent space, creating a *sparse*, disconnected representation. When we sample from regions of this space that the encoder never visited during training, we encounter undefined territory where the decoder has no learned behavior. Variational autoencoders (VAEs) emerged as an elegant solution to this *sampling problem*. By treating the latent space probabilistically, VAEs forced the encoder to output parameters of probability distributions rather than fixed points. This created a smooth, continuous latent space where *interpolation* between points yielded meaningful results. The reparameterization trick allowed us to maintain differentiability while introducing the stochastic elements necessary for proper sampling.

Yet VAEs introduced their own significant limitation: the reconstruction loss term in their objective function. The combination of reconstruction loss and Kullback-Leibler (KL) divergence created a fundamental tension. The model needed to balance two competing objectives: accurately reconstructing inputs while maintaining a well-structured latent space. This balance typically favored reconstruction accuracy over generation quality, resulting in blurry, averaged outputs. The decoder learned to hedge its bets, producing images that captured general features but lacked sharp details. The technical explanation for this blurriness lies in the loss function itself. The mean squared error (MSE) reconstruction loss penalizes the model equally for all pixel deviations, encouraging it to predict the average of possible outputs rather than committing to specific details. When multiple plausible reconstructions exist for a given latent code, the decoder learns to output a compromise that minimizes overall error, resulting in the characteristic soft, unfocused appearance of VAE-generated images.

Generative adversarial networks (GANs), on the other hand, took a radically different approach, abandoning reconstruction loss entirely in favor of an adversarial objective. This shift proved transformative in many ways. GANs could generate remarkably sharp, detailed images because the discriminator network provided a more sophisticated measure of image quality than simple pixel-wise loss functions. The generator learned to fool an opponent that could evaluate images holistically, leading to outputs with convincing textures, edges, and fine details.

However, this adversarial training regime introduced its own set of challenges. *Mode collapse* became a persistent problem, where the generator would discover a small set of outputs that consistently fooled the discriminator and then refuse to explore beyond these safe options. The mathematical cause of mode collapse stems from the minimax objective: If the generator finds a strategy that works, it has no incentive to diversify its outputs as long as the discriminator remains fooled.

Training *instability* proved equally problematic. The adversarial objective creates a complex optimization landscape where the generator and discriminator must reach a delicate equilibrium. Small changes in one network could destabilize the entire training process, leading to *oscillations*, *divergence*, or collapse. The nonconvex nature of this optimization problem meant that successful training required careful hyperparameter tuning, architectural choices, and often multiple restart attempts. If you ran and experimented with the code for GANs from the previous chapter, you would have faced this frustration.

Each of these generative approaches illuminated different aspects of the generation problem. Autoencoders showed us the value of learned representations. VAEs demonstrated how probabilistic modeling could create structured latent spaces. GANs proved that adversarial training could produce high-quality outputs. Yet none provided a complete solution to high-fidelity, controllable generation. The breakthrough came when researchers realized that the generation process itself could be reconceptualized. Instead of trying to map directly from noise to data in a single step, what if we broke the process into many small, learnable steps? This insight led to the development of *diffusion-based approaches*, which sidestep many of the fundamental limitations we've discussed.

Diffusion models approach generation as a gradual refinement process, starting from pure noise and slowly removing that noise through a series of learned denoising steps. This process eliminates the need for adversarial training, avoiding the instability issues that plague GANs. It also provides a natural way to control the generation process, as we can intervene at any step in the denoising sequence. Perhaps most importantly, diffusion models can achieve the sharp, detailed outputs we saw with GANs while maintaining the training stability and theoretical grounding we value from other approaches.

They represent a synthesis of insights from the entire generative modeling landscape, combining the best aspects of previous methods while avoiding their major pitfalls.

The mathematical elegance of diffusion models lies in their connection to well-understood physical processes. Just as thermodynamic systems naturally evolve from order to disorder, diffusion models learn to reverse this process, transforming disorder back into structured, meaningful data. This physical grounding provides both theoretical insights and practical advantages that we'll explore in the sections that follow.

12.1.2 Diffusion Process Fundamentals

This iterative refinement process that we've introduced operates through two complementary phases that mirror each other in mathematical symmetry. To understand how Stable Diffusion works, we need to examine both sides of this process:

- How we systematically add noise to destroy structure
- How we learn to reverse this destruction step-by-step

Diffusion begins with a clean, structured image and gradually corrupts it with noise over many time steps. This might seem counterintuitive at first: Why would we want to destroy perfectly good images? The answer lies in creating a well-defined learning problem. By controlling exactly how we add noise, we create a process that our neural networks can learn to reverse. Mathematically, we define this forward process as a *Markov chain* where each step depends only on the previous one. We discussed the concept of Markov Decision Processes in detail in Chapter 10. These processes are formalized using Markov Chains. Starting with our original image x_0, we create a sequence of increasingly noisy versions: $x_1, x_2, x_3, ..., x_T$. At each time step t, we add a small amount of *Gaussian noise* according to

$$q(x_t|x_{t-1}) = \mathcal{N}(x_t; \sqrt{1-\beta_t}\, x_{t-1}, \beta_t I)$$

Here, β_t represents our *noise schedule*—a carefully chosen sequence of values that determines how much noise we add at each step. The noise schedule starts small (adding just a tiny amount of noise early on) and gradually increases, ensuring that by the final time step T, our image has been transformed into pure Gaussian noise.

Noise Schedule Demystified

The noise schedule forms the backbone of Stable Diffusion, though this straightforward concept often gets buried in dense mathematics. Simply put, if we need to add 10 units of noise to transform a clean image into complete noise, the schedule determines how much noise we add at each step. With 10 steps, we could add 1 unit per step uniformly, start with 2 units and gradually decrease, or follow the reverse pattern. This distribution strategy—the noise schedule—fundamentally shapes how the diffusion process unfolds and directly impacts the quality of generated images.

The choice of noise schedule β_t profoundly affects both training efficiency and generation quality. We want a schedule that gradually removes information without creating sudden jumps that make the reverse process difficult to learn. A common choice is a *linear schedule* where β_t increases linearly from a small value (e.g., *0.0001*) to a larger value (e.g., *0.02*) over a thousand time steps. However, research has shown that more sophisticated schedules often work better. A *cosine schedule*, for instance, adds noise more slowly at the beginning and end of the process while accelerating in the middle.

The key insight is that we want our noise schedule to respect the natural structure of our data. Images have both fine details and coarse structures, and our schedule should progressively destroy these features in a way that makes the reverse process learnable.

12.1.3 Reverse Diffusion: Learning to Denoise

The reverse diffusion process is where the real magic happens. Having defined how to systematically destroy images, we now need to learn how to reconstruct them. This reverse process isn't deterministic—we can't simply run the forward process backward because adding noise is an irreversible operation. Instead, we need to learn a neural network that can estimate the reverse transitions. The key insight is that if we know the noise that was added at each step, we can subtract it to recover the previous, less noisy image. Our neural network $\epsilon_\theta(x_t, t)$ learns to predict this noise at any time step t. During generation, we start with pure noise x_T and iteratively apply our learned denoising function as

$$x_{t-1} = \frac{1}{\sqrt{\alpha_t}} \left(x_t - \frac{\beta_t}{\sqrt{1-\bar{\alpha}_t}} \epsilon_\theta(x_t, t) \right) + \sigma_t z$$

where $z \sim \mathcal{N}(0, I)$, σ_t controls the amount of stochasticity in our reverse process, and α represents the amount of the original signal retained at each time step during the forward diffusion process. This equation encapsulates the essence of diffusion-based generation: We take our current noisy image, subtract our network's *noise estimate*, and add a small amount of fresh randomness to maintain the proper distribution.

The training objective becomes surprisingly simple. We randomly sample a time step t, add noise to a training image to create x_t, and train our network to predict the noise that was added:

$$L = E_{x_0, \epsilon, t}[|\epsilon - \epsilon_\theta(x_t, t)|^2]$$

This is just MSE between the actual noise and our network's prediction. The elegance of this formulation is remarkable—we've transformed the complex problem of learning to generate images into the much simpler problem of learning to denoise.

12.1.4 Connections to Physical Processes

The mathematical foundation of diffusion models draws deep inspiration from thermodynamics and statistical mechanics. A deeper understand of this will help you appreciate the models and help you debug the models with more insight. The forward diffusion process mirrors the second law of *thermodynamics*, where ordered systems naturally evolve toward maximum *entropy*. Just as a drop of ink spreads through water until it reaches equilibrium, our forward process gradually increases the entropy of our images until they become indistinguishable from random noise. The reverse process, then, is analogous to *time reversal symmetry* in physical systems. While physical processes naturally move from order to disorder, our learned reverse process moves from disorder back to order. This connection provides both mathematical rigor and intuitive understanding—we're essentially learning to apply the second law of thermodynamics in reverse by reconstructing order from chaos.

This physical grounding also explains why diffusion models are so stable to train. Unlike GANs, which require balancing two competing networks, diffusion models optimize a single, well-defined objective that has clear physical meaning. The loss function measures how well we can predict and remove noise, which is a much more stable learning signal than the adversarial objectives we encountered with GANs. The statistical mechanics perspective also provides insights into why this approach works so well for generation. In *equilibrium statistical mechanics*, systems naturally sample from their *equilibrium distribution*. By learning the reverse diffusion process, we're essentially learning to sample from the data distribution by starting from the equilibrium distribution (pure noise) and following the reverse-time dynamics back to structured data.

This connection to well-established physical principles gives us confidence that diffusion models are a fundamental approach to generative modeling that taps into deep mathematical structures governing how information and entropy behave in complex systems. If you're aiming to work with these models in depth and possibly contribute to the field itself by conducting research, it would be fruitful to study these concepts in more detail. For now, we'll leave the discussion here and move back to the concrete model discussions.

12.1.5 Denoising Diffusion Probabilistic Models

After understanding the forward and reverse diffusion processes, we can examine how these concepts crystallize into a practical system called *denoising diffusion probabilistic models (DDPMs)*. Think of DDPMs as the engineering blueprint that transforms our theoretical understanding of diffusion into a working generative model. Just as an architect takes abstract concepts about space and light and turns them into detailed construction plans, DDPMs provide the specific architectural and training framework needed to build effective diffusion models. The overall architecture of a basic denoising

12 Advanced Generative AI: Stable Diffusion

Stable Diffusion model is shown in Figure 12.1. We'll be referring back to this figure repeatedly as we expand on our explanations of these models.

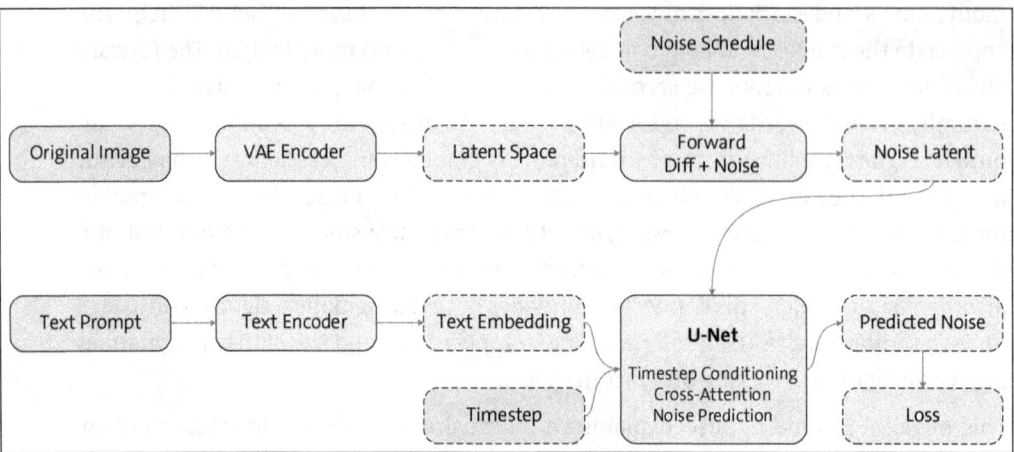

Figure 12.1 Architecture of the Stable Diffusion Model

The core architecture of a DDPM centers around the interesting premise we discussed earlier: training a neural network (the U-Net box in Figure 12.1) to become an expert noise detective. This network's job is to look at any noisy image and accurately predict exactly what noise was added to create it. Once we have this capability, generation becomes straightforward—we start with pure noise and repeatedly ask our detector, "What noise should I remove next?"

The mathematical foundation of DDPMs builds directly on our diffusion process fundamentals. Recall that during forward diffusion, we add noise according to

$$q(x_t|x_{t-1}) = \mathcal{N}(x_t; \sqrt{1-\beta_t}x_{t-1}, \beta_t I)$$

The training process works by sampling a random image x_0 from our dataset, choosing a random time step t, and adding the appropriate amount of noise to create x_t. We then ask our network to predict what noise was added. The loss function we established earlier as

$$L = \mathbb{E}_{x_0, \epsilon, t}[|\epsilon - \epsilon_\theta(x_t, t)|^2]$$

measures how accurately our network can identify this noise.

What makes this approach so powerful is its simplicity. Unlike GANs, where we need to balance two competing networks in a delicate adversarial dance, DDPM training involves just one network learning one clear task: noise prediction. This single-minded focus eliminates the training instabilities that plagued earlier generative approaches.

At the heart of every DDPM lies a U-Net architecture, the same powerful structure we explored in detail in Chapter 8, Section 8.3.3. The U-Net's encoder-decoder structure with skip connections makes it uniquely suited for denoising tasks. The encoder progressively compresses the noisy input into increasingly abstract representations, while the decoder reconstructs the output, using skip connections to preserve fine-grained details that might otherwise be lost in the compression process. For diffusion models, we enhance the standard U-Net with several crucial modifications. The network needs to understand not just what noise to remove, but when in the diffusion process it's operating. A heavily noised image at time step 800 requires different denoising strategies than a lightly noised image at time step 50. This temporal awareness becomes a critical component of the architecture. The U-Net's *multi-resolution* processing aligns perfectly with the hierarchical nature of image generation. Early denoising steps focus on establishing overall composition and large-scale structures, while later steps refine details and textures. The U-Net's architecture naturally supports this progression, with deeper layers handling global structure and shallower layers managing fine details.

The key innovation that transforms a standard U-Net into a diffusion model is *time-step conditioning*. Our network needs to know not just what it's looking at, but when in the diffusion process it's operating. This temporal context fundamentally changes how the network should behave. We implement time-step conditioning by embedding the time step t into a learnable representation and injecting this information throughout the network. The time-step embedding typically uses *sinusoidal positional encodings* similar to those used in transformer architectures. We compute time step embeddings as

$$\text{emb}(t) = \left[\sin(t/10000^{0/d}), \cos(t/10000^{0/d}), \sin(t/10000^{2/d}), \cos(t/10000^{2/d}), \ldots\right]$$

where d is the embedding dimension. These embeddings capture both the absolute time step and relative temporal relationships, allowing the network to understand its position in the denoising sequence.

The time-step information gets incorporated into the U-Net model through several mechanisms. We add time step embeddings to intermediate feature maps, allowing each layer to adjust its processing based on the current noise level. We also use time-step-conditioned normalization layers that modulate feature activations based on temporal context.

This conditioning creates a single network that effectively learns thousands of specialized denoising functions—one for each time step. Early time steps learn to remove large amounts of noise and establish basic structure, while later time steps focus on subtle refinements and detail enhancement. The network learns to automatically adjust its behavior based on the time-step context. The stochastic nature of time-step sampling during training ensures that the network develops expertise across the entire denoising spectrum. Each training step presents a different temporal challenge, forcing the network to develop robust, generalizable denoising capabilities rather than memorizing specific noise patterns.

This training approach creates several advantageous properties:
- The network learns a comprehensive understanding of image structure at multiple scales.

- The denoising objective provides consistent, meaningful gradients throughout training.
- The process naturally regularizes the model, as the network must learn to distinguish signal from noise rather than simply memorizing training examples.

The convergence properties of DDPM training prove remarkably *stable* compared to other generative approaches. The loss function provides clear, interpretable feedback about model performance, and training typically proceeds smoothly without the oscillations or mode collapse issues that characterize adversarial training. This stability makes DDPMs practical for large-scale applications and allows for more predictable scaling to higher resolutions and more complex datasets.

> **The "Stable" in Stable Diffusion**
>
> The name Stable Diffusion is a nod to the training stability of the model. The diffusion process avoids the divergence and mode collapse issues that plagued earlier generative models such as GANs. The training converges reliably without the oscillating losses and instability that made GANs notoriously difficult to train.

As we'll see in the next section, these foundational DDPM concepts extend naturally to more sophisticated architectures that operate in latent spaces rather than directly on pixels, opening the door to the powerful Stable Diffusion systems that have transformed GenAI applications.

12.1.6 Latent Diffusion and Stable Diffusion Architecture

The DDPM framework we just explored works beautifully in principle, but there's a practical challenge lurking beneath the surface. When we apply diffusion directly to high-resolution images, we're asking our U-Net to process enormous amounts of data at every time step. A single 512×512 RGB image contains nearly 800,000 individual pixel values, and running our denoising network on such large inputs becomes computationally expensive very quickly. This computational bottleneck led researchers to a breakthrough insight: What if we didn't need to work with raw pixels at all? What if we could perform our diffusion process in a more compact, efficient representation of the image? This is where *latent diffusion* enters the picture, building directly on the variational autoencoder foundations we established in the previous chapter.

Recall that VAEs can learn how to compress images into compact latent representations. The encoder transforms a high-dimensional image into a lower-dimensional latent code, capturing the essential features while discarding redundant information. The decoder then reconstructs the image from this compressed representation. This compression creates a more structured, semantically meaningful representation of the image content. Latent diffusion uses this insight by moving our entire diffusion process into the VAE's latent space. Instead of adding noise directly to pixels, we first encode our training images into latent representations using a pretrained VAE. Our diffusion model then learns to add and remove noise in this *compressed latent space*. Finally, when we want to generate a new image, we decode our denoised latent representation back into pixel space using the VAE decoder. The core workflow for generation is shown in Figure 12.2.

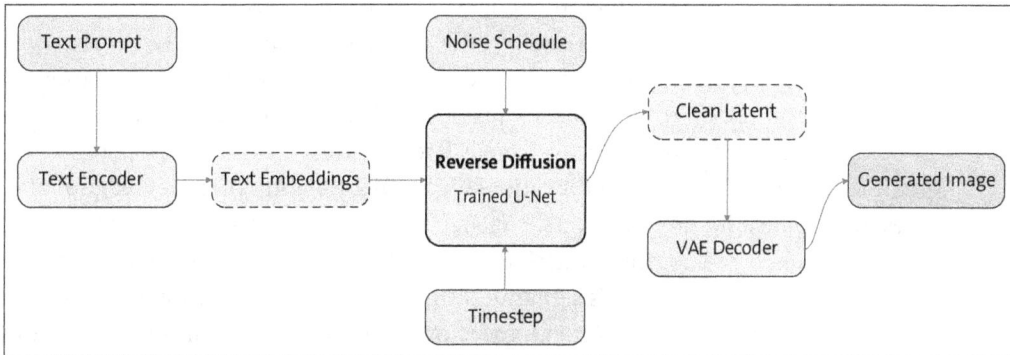

Figure 12.2 Generation in Latent Diffusion

Our U-Net now predicts noise in latent space as $\epsilon_\theta(z_t, t)$, and our loss function becomes

$$L = E_{z_0, \epsilon, t}[|\epsilon - \epsilon\theta(z_t, t)|^2]$$

This seemingly simple change brings dramatic benefits. A 512×512 image might compress to a 64×64 latent representation—a 64-fold reduction in the amount of data our diffusion model needs to process. This efficiency gain makes high-resolution generation feasible.

Concisely, Stable Diffusion architecture orchestrates three distinct components working in harmony, each contributing specialized capabilities to the generation process:

- **VAE**

  The first component is our VAE, which serves as the gateway between pixel space and latent space. The VAE transforms images into compact latent representations: $z_0 = E(x_0)$. The VAE decoder reconstructs images from latent codes: $x_0 = D(z_0)$. This component handles all the pixel-level details, allowing the other components to work with more abstract representations.

- **U-Net**

  The second component is our familiar U-Net, now operating in latent space rather than pixel space. This network has learned to understand the structure and semantics of the latent representations, becoming expert at predicting and removing noise from these compressed encodings. U-Net handles the core generation logic, transforming random noise into structured latent representations.

- **Text conditioning**

    The third component introduces an entirely new capability: *text conditioning* through a text encoder. This component transforms text prompts into rich semantic embeddings that guide the generation process. Unlike our previous generative models that produced random samples from their training distribution, Stable Diffusion can generate images that match specific textual descriptions.

Interestingly, the computational advantages of latent diffusion extend far beyond simple dimensionality reduction. Working in latent space creates a more semantically structured representation where small changes correspond to meaningful variations in image content rather than arbitrary pixel modifications. Consider the difference between making edits in pixel space versus latent space. Changing a few pixels rarely produces meaningful image modifications—you might slightly alter a few edge boundaries or introduce subtle color shifts. But changing a few values in a well-trained latent space can transform a dog into a cat, change the lighting conditions, or modify the artistic style of an entire image. This semantic structure makes the diffusion process more efficient in multiple ways. U-Net can focus on learning meaningful transformations rather than pixel-level noise patterns. The reduced dimensionality allows for faster forward and backward passes during training. Most importantly, the compressed representation captures the essential information needed for generation while discarding the redundant details that would otherwise slow down the process.

The efficiency gains compound throughout the system. Training becomes faster because each denoising step processes less data. Generation becomes faster for the same reason. Memory requirements drop dramatically, making it possible to generate high-resolution images on consumer hardware that would struggle with pixel-space diffusion models.

12.1.7 Cross-Attention: The Bridge Between Text and Images

The most sophisticated aspect of Stable Diffusion lies in how it connects textual descriptions to visual generation through *cross-attention* mechanisms. Building on the attention concepts we explored with transformers in Chapter 9, cross-attention allows our U-Net to selectively focus on different parts of the text prompt while generating different regions of the image.

The *text encoder* transforms our prompt into a sequence of token embeddings:

$c = \text{TextEncoder}(\text{prompt})$

These embeddings capture both individual word meanings and their relationships within the prompt. U-Net then uses cross-attention layers to allow each spatial location in the developing image to attend to relevant parts of the text description. Cross-attention operates by computing *attention weights* between spatial features in the image and semantic features in the text. For each spatial location *(i, j)* in our latent representation, the cross-attention mechanism computes

$$\text{Attention}(Q, K, V) = \text{softmax}\left(\frac{QK^T}{\sqrt{d}}\right)V$$

Where *Q* comes from our image features and *K, V* come from our text embeddings. This allows each part of the developing image to look at the text prompt and decide which words or concepts are most relevant for that particular region. Because the query and values come from different structures rather than the same, it's called cross-attention as opposed to the concept of self-attention we studied in Chapter 9. This cross-attention creates a dynamic, *content-aware relationship* between text and image. When generating the sky region of an image, the model can focus on words such as "sunset" or "cloudy" in the prompt. When working on foreground objects, it can attend to descriptive terms about those specific elements. This fine-grained control enables the remarkable *text-to-image* capabilities that have made Stable Diffusion so powerful for creative applications.

The integration of cross-attention throughout the U-Net model ensures that text guidance influences the generation process at multiple scales and levels of abstraction. Early layers might use text information to establish overall composition and scene layout, while later layers use it to refine specific details and textures that match the prompt description.

12.2   How Stable Diffusion Uses Core Concepts

Now that we understand the theoretical foundations of Stable Diffusion, this section will explore how this highly intricate system builds upon the core concepts we've been developing. The brilliance of Stable Diffusion lies not in abandoning what came before but in weaving together the best insights from autoencoders, transformers, and optimization theory into something entirely new. Each component we've studied finds its perfect role in this architecture.

The process of how Stable Diffusion uses autoencoders begins with a fundamental insight about *representation learning*. Remember our exploration of autoencoders in the previous chapter, where we discovered how these networks learn to compress images into compact representations and then reconstruct them? That compression enabled learning what matters most in an image. The autoencoder essentially asks the following: "If I had to capture everything important about this image in just a few hundred numbers, what would those numbers represent?" The answer turns out to be extraordinarily valuable. The VAE learns to identify and encode the *semantic essence* of images—the concepts, objects, relationships, and structures that make each image meaningful. When we feed a photograph of a sunset over mountains into a VAE, it doesn't just record pixel colors and positions. Instead, it captures abstract concepts such as "orange sky," "silhouetted peaks," "atmospheric lighting," and "horizontal composition." These concepts get encoded as numerical patterns in the latent space.

The latent space that the VAE learns has special properties that make it ideal for diffusion processes. Unlike the raw pixel space, where neighboring values might represent completely unrelated image features, the VAE's latent space exhibits smooth, *continuous* structure. Moving from one point to another in this space corresponds to meaningful semantic transitions in the resulting images. Change a few values in one direction, and you might shift from day to night. Adjust different dimensions, and you could transform a photograph into a painting or change the emotional mood of a scene.

This smoothness becomes crucial for diffusion. When we add noise to a latent representation, we're creating variations that still lie within the manifold of meaningful image representations. The noise doesn't break the semantic structure; it simply obscures it temporarily. This property makes the denoising task much more tractable than trying to denoise raw pixels.

The reconstruction term ensures that important visual information is preserved in the latent space, while the KL divergence term encourages the latent representations to follow a smooth, well-behaved distribution. This second term is particularly important for diffusion—it ensures that our latent space has the kind of regular structure that makes adding and removing noise a meaningful operation.

12.2.1 Efficient Diffusion Through Learned Representations

Consider what happens when we add noise to a latent representation of a cat image. In pixel space, adding noise might randomly flip individual pixels, creating a chaotic pattern that bears no resemblance to any real image. But in latent space, adding noise creates variations that still represent plausible images—perhaps a cat with different fur patterns, in a different pose, or with slightly different lighting. The noisy latent representation remains within the space of meaningful image concepts.

This property dramatically simplifies the denoising task. Our U-Net doesn't need to learn to distinguish between "random static" and "actual image content." Instead, it learns to distinguish between "slightly blurry cat concept" and "clear cat concept," or between "vague landscape idea" and "specific mountain scene." These distinctions are much more learnable and require far less training data to master.

The dimensional comparison between pixel space and latent space shows why this approach works so effectively. A high-resolution 512×512×3 RGB image contains 786,432 individual pixel values. Each pixel can take any value from 0 to 255, creating an enormous space of possible images—most of which represent meaningless noise rather than coherent visual content. The VAE typically compresses this to a 64×64×4 latent representation containing just 16,384 values. This might seem like a dramatic loss of information, but the compression is highly intelligent. The VAE learns to discard redundant information while preserving semantic content. Those 16,384 latent values capture the essential concepts and relationships that define the image's meaning.

This compression creates a 48× reduction in the amount of data our diffusion model needs to process. But the benefits go beyond mere computational efficiency. The compressed representation is also much denser in meaningful content. In pixel space, most random samples produce nonsensical images. In the VAE's latent space, most random samples correspond to plausible image concepts, even if they're not perfect.

The latent space also exhibits better *interpolation properties* than pixel space. If we take two latent representations and linearly interpolate between them, we get a smooth semantic transition between the corresponding images. Try the same interpolation in pixel space, and you typically get blurry, unrealistic intermediates. This smoothness is exactly what we need for Stable Diffusion training—it ensures that the noise addition and removal process creates learnable patterns rather than chaotic transitions.

This foundation of efficient, *semantically structured* representation learning sets the stage for everything else that makes Stable Diffusion work. The attention mechanisms we'll explore next can focus on meaningful concepts rather than arbitrary pixel patterns. The optimization process can converge more quickly because it's working in a space designed for semantic manipulation. And the final results maintain high quality because they're decoded by a network that has learned to translate abstract concepts back into detailed, realistic images. The genius of this approach lies in recognizing that generation is fundamentally about manipulating concepts and ideas, not just pixels.

12.2.2 Advanced Attention Mechanisms

Building upon the semantic foundation that VAEs provide, Stable Diffusion incorporates another crucial breakthrough from our earlier understanding: the attention mechanisms we explored with transformers in Chapter 9. Stable Diffusion transforms them into something entirely new. Consider taking the focused attention system that helped transformers understand language and adapting it to bridge the gap between words and images. That's exactly what happens in Stable Diffusion's sophisticated attention architecture.

The attention mechanisms in Stable Diffusion serve as the *communication channels* that allow different parts of the system to coordinate their efforts. Without attention, our U-Net would process each region of the image in isolation, unable to consider how different areas should relate to each other or how they should respond to our text prompts. With attention, these isolated regions become participants in a rich conversation about what the final image should look like. Recall our exploration of transformer attention, where we learned how these models could focus on relevant parts of a sequence when processing each element. Stable Diffusion takes this same concept and applies it to a much more complex challenge: coordinating the generation of visual content while responding to textual descriptions. The beauty lies in how the same attention mechanism can serve multiple purposes within a single system. Sometimes, it helps different parts of an image coordinate with each other. Other times, it allows the developing image to "look at" the text prompt to understand what should be generated. And throughout the process, it enables the kind of fine-grained control that makes Stable Diffusion so remarkably responsive to our creative intentions.

The key insight is that attention creates dynamic connections that change based on context. Early in the generation process, when the image is mostly noise, attention patterns focus on establishing basic composition and layout. As the image develops, attention shifts to coordinate details, textures, and fine-grained features. This adaptive behavior emerges naturally from the same mathematical framework we learned with transformers. This concept of attention has two complementary aspects to it: cross-attention and self-attention. Let's take a look at the difference between the two.

Cross-Attention Between Text and Images

The most revolutionary aspect of Stable Diffusion's attention system is cross-attention, which creates a direct dialogue between textual concepts and visual generation. Imagine trying to paint a picture while someone describes what they want to see, but instead of having to remember their entire description, you could continuously refer back to specific parts of their request as you work on different areas of the canvas. That's exactly what cross-attention enables in Stable Diffusion. Cross-attention operates by allowing each spatial location in our developing image to query the text embeddings for relevant information. When generating the sky region of an image, the attention mechanism can focus on words such as "sunset," "cloudy," or "dramatic" in the prompt. When working on foreground objects, it can attend to descriptive terms about those specific elements. This creates a dynamic, content-aware relationship between language and visual generation.

What makes this particularly powerful is how it handles complex, multi-object scenes. Consider a prompt such as "a red car parked next to a blue house under a stormy sky." The cross-attention mechanism allows different regions of the image to focus on different parts of this description. The top region attends to "stormy sky," the middle-left region focuses on "blue house," and the lower-right area concentrates on "red car." This selective attention creates coherent, well-organized images that accurately reflect complex textual descriptions.

Self-Attention Within the U-Net Architecture

While cross-attention handles the text-to-image dialogue, self-attention within U-Net manages the equally important conversation between different parts of the image itself. This internal attention system ensures that the various regions of the developing image coordinate properly with each other, maintaining global coherence while allowing for local detail refinement. The genius of incorporating self-attention at multiple scales within the U-Net hierarchy is that it allows for both local and global coordination. Early layers with low-resolution feature maps use self-attention to establish overall composition and major structural relationships. Later layers with high-resolution features use self-attention to coordinate fine details and ensure consistency in textures and local patterns.

The combination of cross-attention and self-attention creates unprecedented fine-grained control over the generation process. This creates a system that can understand and respond to nuanced creative directions with remarkable precision.

Attention also enables what we might call *semantic binding*—the ability to correctly associate adjectives with their corresponding nouns in complex prompts. When processing "a small red car next to a large blue truck," attention mechanisms ensure that "small" and "red" both modify "car," while "large" and "blue" modify "truck." This binding prevents the common failure mode, where attributes get mixed up between objects.

The real elegance of this system though lies in its *emergent behavior*. We don't explicitly program rules about how objects should relate to each other or how text should map to visual features. Instead, the attention mechanisms learn these relationships through training, discovering patterns and associations that reflect the structure of language and the visual world.

Understanding Stable Diffusion Strengths
The fact of the matter is that most of the success of diffusion-based models isn't fully understood yet. Heavy research is being conducted to figure out exactly how these properties have emerged where we connected all the parts, but the sum of the whole is much greater than its parts.

This learned attention creates a system that can generalize far beyond its training examples. The network learns general principles about how spatial relationships work, how object properties combine, and how language maps to visual concepts. These principles allow it to handle novel combinations and creative prompts that it has never seen before.

The result is a generation system that feels remarkably responsive to human creative intent. The attention mechanisms create multiple pathways for control—through text prompts, through spatial layout, through style directions, and through the complex interplay between all of these factors. This multifaceted control system is what enables the creative flexibility that has made Stable Diffusion so transformative for artists, designers, and creative professionals.

12.2.3 Training Strategies and Optimization

With our sophisticated attention mechanisms providing the communication pathways and our VAE foundation offering the efficient representation space, we now need to examine how we actually teach this complex system to generate beautiful images. The optimization strategies that make this possible draw from everything we've learned about gradient descent, regularization, and even classification techniques. The remarkable thing about Stable Diffusion training is how it transforms the seemingly impossible task of learning to create any image into a series of manageable, learnable problems. Instead of trying to master image generation all at once, the system learns through millions of small denoising tasks, each one slightly different but following the same basic pattern. The *gradient descent* process in diffusion models operates in a surprisingly familiar way despite the complexity of what the model is learning. Recall our foundational work with gradient descent from Chapter 3, where we learned how neural networks adjust their parameters by following the slope of the loss function downhill. The same principle applies here, but now we're navigating a much more sophisticated landscape.

Our loss function remains elegantly simple. At its heart, it's still MSE—we're measuring how well our network can predict the noise that was added to an image. But the implications of optimizing this loss are profound. The time step value changes randomly, meaning the difficulty level varies unpredictably. This *stochastic sampling* creates a robust learning environment where the network develops expertise across the entire spectrum of denoising tasks. The gradient computation flows backward through the entire U-Net architecture, adjusting millions of parameters simultaneously. The cross-attention weights learn to associate text concepts with visual features. The self-attention mechanisms develop understanding of spatial relationships. The convolutional layers refine their ability to detect and generate visual patterns. All of this happens through the same backpropagation process we've been using throughout our journey but applied to a system of unprecedented sophistication. This process is supported by two important novelties in the diffusion models: automatic regularization and classifier-free guidance. In the sections that follow, let's take a look at these briefly to drive the point home.

Regularization Adapted for Diffusion Models

The regularization techniques we explored in Chapter 6 find new applications in diffusion models, but they're adapted to account for the unique aspects of the denoising process. Traditional regularization methods such as dropout and weight decay still play important roles, but diffusion models introduce novel regularization challenges that require creative solutions. One key insight is that the diffusion process itself provides a form of *implicit regularization*. By training on noisy versions of images rather than clean examples, the network learns to be robust to variations and corruption. This noise-based training naturally prevents overfitting to specific pixel patterns, encouraging the model to focus on semantic content rather than memorizing exact image details.

The time-step conditioning creates another form of regularization through *temporal diversity*. Because the network must handle denoising at every time step from *1* to *T*, it can't rely on specialized strategies that only work for specific noise levels. This forces the development of general, robust denoising capabilities that generalize well across different scenarios. On the other hand, attention mechanisms introduce their own regularization considerations. The softmax operations in attention naturally create *sparse activation patterns*, where each query focuses on a small subset of relevant keys. This sparsity acts as an implicit regularization, preventing the model from creating overly complex dependencies that might not generalize well. Similarly, *weight decay* becomes particularly important for the text conditioning components. The cross-attention layers that connect text embeddings to visual features can potentially memorize specific text-image pairs from the training data. Careful regularization of these connections ensures that the model learns general text-to-image mapping principles rather than memorizing specific examples.

Somewhat counterintuitively though, when applied to diffusion models, dropout requires careful consideration of where and when to apply it. Dropping out activations randomly during the denoising process can interfere with the precise calculations needed for noise prediction. Instead, diffusion models often use more sophisticated dropout patterns that preserve critical information flows while still providing regularization benefits.

Classification Connections Through Classifier-Free Guidance

Perhaps the most elegant connection between our earlier studies and diffusion models comes through *classifier-free guidance*, a technique that borrows insights from classification to enable better control over generation. This approach represents a beautiful synthesis of discriminative and generative learning that wouldn't be possible without our understanding of both paradigms. The core idea builds on our classification work from Chapter 4. Contrary to the mechanism of building networks that could distinguish between different categories of inputs, classifier-free guidance applies similar principles but in reverse—instead of classifying existing images, we use classification-like techniques to guide the generation of new images toward desired categories or characteristics.

This technique creates a parallel to *classification confidence*. Just as we can interpret classification probabilities as measures of certainty, the *guidance* weight in diffusion models controls how confident we want the model to be in following our text instructions. Higher guidance weights produce images that more closely match the text prompt but might sacrifice naturalness and variety.

The elegance of classifier-free guidance lies in how it provides controllable generation without requiring separate classifier networks. Earlier approaches needed additional models to guide the diffusion process, but classifier-free guidance achieves the same control using only the diffusion model itself. This simplification makes the system more efficient and eliminates potential training instabilities that could arise from coordinating multiple networks. This connection to classification concepts also provides intuitive understanding of how to control generation. Just as we learned to interpret classification outputs as probability distributions over categories, we can think of guidance weights as controlling the sharpness of the model's adherence to our creative directions.

As we move into the implementation section, these training concepts will become much more concrete. You'll see how the elegant mathematical formulations translate into practical code, how the various regularization techniques are implemented in Keras, and how classifier-free guidance enables the kind of creative control that makes Stable Diffusion so powerful for artistic applications.

12.3 Implementing Stable Diffusion Models

In this section, we move from theory to practice, building a complete diffusion model that generates high-quality images. Our implementation strategy follows a logical progression that mirrors how you'd approach this problem in real-world projects. We'll start by establishing our hyperparameters and data pipeline, ensuring we have clean, properly preprocessed images to work with. Next, we'll implement a robust evaluation metric that helps us measure generation quality throughout training. The heart of our implementation lies in constructing the U-Net architecture we discussed earlier, complete with sinusoidal embeddings for time-step conditioning and the residual blocks that give the network its denoising power. We'll then build the main diffusion model class, which orchestrates the entire process from noise scheduling to the reverse diffusion that transforms random noise into meaningful images. Finally, we'll tie everything together with a training loop that demonstrates how all of these components work in harmony to learn the complex mapping from noise to data.

Each code segment builds directly on the concepts we've established, so you'll see familiar patterns from our convolutional neural network (CNN) and autoencoder discussions, but now applied to the unique challenges of diffusion-based generation.

> **[+] Access the Full Code**
>
> Stable Diffusion is a complex model—more intricate than any we've seen until now. So, we'll be discussing only the most interesting aspects of the code in the following sections. As always, you can see the full executable code on the book's official web page as

well as on the resources page at *https://recluze.net/keras-book* in the notebook **12-01-diffusion-scratch.**

12.3.1 Environment and Data Prep

Let's very briefly cover the setup of our environment and data before we discuss the model in detail. Because Stable Diffusion models are computationally intensive, we need to prepare for the reality of training on free cloud resources. Google Colab's free tier has session time limits that will interrupt our training, so we'll set up persistent storage to save our progress. The simple setup code shown in Listing 12.1 connects your Google Drive to the Colab environment, providing a *persistent* location to store model checkpoints. When Colab's 24-hour limit resets your session, you can reload your saved model and continue training from where you left off. If you're using a paid Colab tier, this step becomes less critical, but it's still good practice for any long-running training job.

```
from google.colab import drive
drive.mount('/content/drive')
```

Listing 12.1 Connect to Google Drive for Checkpoint Persistence

Our preprocessing function in Listing 12.2 handles the essential data preparation steps for training. The function first creates square crops by finding the minimum dimension and cropping from the center. This approach is no longer optional. This is because diffusion models expect consistent input dimensions, and stretching rectangular images to squares would distort important visual features that the model needs to learn. *Center cropping* preserves the most visually important content while giving us the uniform square format required for efficient batch processing. The resize operation standardizes all images to our target size with antialiasing to maintain visual quality during the scaling process. Finally, pixel normalization converts the standard 0–255 range to 0–1 values that work better with our neural network's gradient calculations.

```
def preprocess_img(data):
    height = ops.shape(data["image"])[0]
    width = ops.shape(data["image"])[1]
    crop_size = ops.minimum(height, width)
    image = tf.image.crop_to_bounding_box(
        data["image"],
        (height - crop_size) // 2,
        (width - crop_size) // 2,
        crop_size,
        crop_size,
    )
```

```
        image = tf.image.resize(
            image, size=[img_size, img_size], antialias=True
        )
        return ops.clip(image / 255.0, 0.0, 1.0)
```
Listing 12.2 Preprocessing Data for Stable Diffusion

The dataset preparation pipeline in Listing 12.3 builds an optimized data flow for training. The parallel preprocessing with AUTOTUNE ensures image processing doesn't become a bottleneck during training. Caching stores processed images in memory to avoid redundant preprocessing on repeated epochs. The repeat operation multiplies our dataset size, which is essential because diffusion models require extensive training data to learn the complex noise-to-image mapping. Shuffling with a large buffer prevents the model from learning spurious patterns based on data ordering. As you'll see in a little while, all of these steps are required to make Stable Diffusion work. The computation requirements for this large model are such that, if you skip a single optimization, you can significantly increase the training time.

```
def prepare_dataset(split):
    return (
        tfds.load(dataset_name, split=split, shuffle_files=True)
        .map(preprocess_img, num_parallel_calls=tf.data.AUTOTUNE)
        .cache()
        .repeat(ds_repetitions)
        .shuffle(10 * batch_size)
        .batch(batch_size, drop_remainder=True)
        .prefetch(buffer_size=tf.data.AUTOTUNE)
    )

train_dataset = prepare_dataset(
    "train[:80%]+test[:80%]"
)
val_dataset = prepare_dataset(
    "train[80%:]+test[80%:]"
)
```
Listing 12.3 Dataset Preparation and Splits

12.3.2 Setting Up an Evaluation Measure

When we're training a diffusion model to generate images, we face a unique challenge: How do we measure whether our generated images are actually good? Unlike classification, where we can simply count correct predictions, evaluating generated images requires more sophisticated thinking. Consider an art teacher grading student paintings. There's no direct way to just measure technical accuracy because the teacher needs to assess whether the paintings look realistic, appear diverse, and capture the essence of what they're supposed to represent. This is exactly the challenge we face with generated images. The solution is the *Kernel Inception Distance* (*KID*). This metric is going to be our art critic for generated images. KID works by using a pretrained neural network that has already learned to understand visual features from millions of images. This network acts like an experienced art expert who can immediately recognize the difference between amateur and professional work.

KID approaches the evaluation by taking both real images and our generated images, running them through this expert visual network, and extracting high-level feature representations. These features capture the essential visual qualities that make images look natural and realistic. The clever part is how KID compares these feature sets. Rather than simply measuring average differences, it uses a mathematical technique called *kernel methods* to compare the overall distributions of features. This approach captures whether our generated images have the same kind of visual richness and variety as real images.

What makes KID particularly valuable is its sensitivity to both *quality* and *diversity*. A model that generates only one perfect image over and over would score poorly, as would a model that generates diverse but unrealistic images. KID rewards models that can produce both high-quality and varied outputs, exactly what we want from our diffusion model. Let's see the implementation of this in Listing 12.4. The KID class extends Keras's *metric* system, giving us a tool that can evaluate our generated images throughout training.

The encoder pipeline creates our visual system using *InceptionV3*, a powerful network trained on ImageNet's millions of images. We're essentially borrowing the visual expertise that Google's researchers spent enormous computational resources developing. The preprocessing steps ensure our images match exactly what InceptionV3 expects — proper scaling, sizing, and input format. Notice how we remove the top classification layers with `include_top=False`. We don't want InceptionV3 to classify our flowers as, say, daisy or rose — we want to use its deeper understanding of visual features. These are output as the `GlobalAveragePooling2D` layer that gives us a compact feature representation that captures the essential visual qualities of each image.

> **A Special Case of Transfer Learning**
>
> The use of Inception V3 inside KID is a special case of *transfer learning* where we employ a pretrained model to perform tasks it wasn't originally designed for. Rather than using the model for its intended purpose, we use its learned representations for entirely different applications. In Chapter 8, Section 8.4, we explored the detailed mechanics of using transfer learning for feature extraction. You might want to revisit that section for a refresher on the underlying principles and implementation strategies.

The `polynomial_kernel` function implements the mathematical comparison we described earlier. Rather than simply measuring distances between individual feature vectors `feats_1` and `feats_2`, this kernel function evaluates how similar the overall patterns are between our real and generated image features. The cubic polynomial amplifies meaningful differences while being robust to noise. Instead of checking if each painting matches exactly, we're asking the following: "Do these two portfolios show the same level of artistic sophistication and variety?" This captures both quality and diversity in a single mathematical framework.

```python
@keras.saving.register_keras_serializable()
class KID(keras.metrics.Metric):
    def __init__(self, name, **kwargs):
        super().__init__(name=name, **kwargs)
        self.kid_tracker = keras.metrics.Mean(name="kid_tracker")
        self.encoder = keras.Sequential(
            [
                keras.Input(shape=(img_size, img_size, 3)),
                layers.Rescaling(255.0),
                layers.Resizing(
                    height=kid_img_size, width=kid_img_size
                ),
                layers.Lambda(
                    keras.applications.inception_v3.
                                            preprocess_input
                ),
                keras.applications.InceptionV3(
                    include_top=False,
                    input_shape=(kid_img_size, kid_img_size, 3),
                    weights="imagenet",
                ),
                layers.GlobalAveragePooling2D(),
            ],
            name="inception_encoder",
        )

    def polynomial_kernel(self, feats_1, feats_2):
        feat_dims = ops.cast(ops.shape(feats_1)[1],
                        dtype="float32")
        return (feats_1 @ ops.transpose(feats_2) /
                feat_dims + 1.0) ** 3.0
```

Listing 12.4 The Kernel Inception Distance Measure: Pretrained Model

Continuing in the KID class, the update_state method in Listing 12.5 performs the actual quality assessment. First, both real and generated images get processed through our InceptionV3 encoder, extracting the visual features that capture what makes each image distinctive. The kernel calculations create three important comparisons. We compute the following:

- How similar real images are to each other
- How similar generated images are to each other
- How similar real images are to generated images, which is the most crucial comparison

The diagonal masking with (1.0 - ops.eye(batch_sz)) prevents images from being compared to themselves, which would artificially inflate similarity scores. The final KID calculation follows a specific mathematical pattern: We add the *internal similarities* of real and generated images, and then subtract twice the *cross-similarity*. This formula captures a key insight—if our generated images truly match the real distribution, the cross-similarity should be high enough to make the KID score approach 0. A low KID score means our diffusion model has learned to create images that are statistically indistinguishable from the real dataset.

```
def update_state(self, real_imgs, gen_imgs, sample_weight=None):
    real_feats = self.encoder(real_imgs, training=False)
    gen_feats = self.encoder(gen_imgs, training=False)

    kernel_real = self.polynomial_kernel(real_feats, real_feats)
    kernel_gen = self.polynomial_kernel(gen_feats, gen_feats)
    kernel_cross = self.polynomial_kernel(real_feats, gen_feats)

    batch_sz = real_feats.shape[0]
    batch_sz_f = ops.cast(batch_sz, dtype="float32")
    mean_kernel_real = ops.sum(
        kernel_real * (1.0 - ops.eye(batch_sz))
    ) / (batch_sz_f * (batch_sz_f - 1.0))
    mean_kernel_gen = ops.sum(
        kernel_gen * (1.0 - ops.eye(batch_sz))
    ) / (batch_sz_f * (batch_sz_f - 1.0))
    mean_kernel_cross = ops.mean(kernel_cross)
    kid_val = (
        mean_kernel_real + mean_kernel_gen -
            2.0 * mean_kernel_cross
    )
    self.kid_tracker.update_state(kid_val)
```

```
def result(self):
    return self.kid_tracker.result()

def reset_state(self):
    self.kid_tracker.reset_state()
```

Listing 12.5 Updating States in the KID

The `@keras.saving.register_keras_serializable()` decorator ensures our custom KID metric can be properly saved and restored with our model. Because we established that diffusion training runs for extended periods with inevitable session interruptions, this decorator becomes essential for our checkpoint strategy. Without it, Keras wouldn't know how to reconstruct our custom evaluation component when loading a saved model, breaking our training pipeline at restart.

> **Functionality Enhancement Through Decorators**
>
> *Decorators* are an advanced Python feature that modify the functionality of existing functions without changing their original code. They use the `@decorator_name` syntax and wrap the original function to add new behavior such as logging, timing, or authentication. This allows you to enhance functions cleanly while keeping the original implementation intact. Here, we're using a specialized decorator from Keras to enable proper compatibility with the rest of Keras model handling.

12.3.3 Model Description and Time-Step Encodings

With the evaluation mechanism out of the way, we turn our attention to another unique communication challenge. Our model needs to tell every part of the network that "we're at time step 247 out of 1000" in a language that neural networks understand. The *sinusoidal embedding* function in Listing 12.6 enables this crucial element. Previously in Chapter 9, we discussed position encodings in transformers. This follows the same elegant principle, but now we're encoding time instead of position. The function creates a rich, multidimensional representation of each time step using mathematical waves at different frequencies.

The *logarithmic spacing* of frequencies creates embeddings that capture both fine-grained timing (high frequencies) and broader *temporal patterns* (low frequencies). The sine and cosine pairs ensure that nearby time steps have similar embeddings while maintaining uniqueness across the entire range. This embedding gets broadcast across the entire image, giving every pixel location access to precise temporal information. Our U-Net can then use this time step context to adjust its denoising strategy—being more aggressive early in the process when there's lots of noise and more careful near the end when fine details matter most.

```
def sinusoidal_embedding(x):
    embed_min_freq = 1.0
    frequencies = ops.exp(
        ops.linspace(
            ops.log(embed_min_freq),
            ops.log(embed_max_freq),
            embed_dims // 2,
        )
    )
    angular_speeds = ops.cast(2.0 * math.pi * frequencies,
                              "float32")
    embeddings = ops.concatenate(
        [ops.sin(angular_speeds * x),
         ops.cos(angular_speeds * x)],
        axis=3,
    )
    return embeddings
```

Listing 12.6 Time Step Function

The residual block in Listing 12.7 forms the backbone of our U-Net's processing power. What makes this particularly interesting for diffusion models is the choice of *Swish* activation over the traditional Rectified Linear Unit (ReLU). Swish provides smoother gradients during the delicate denoising process, helping our model make more nuanced decisions about which details to preserve or remove. The *batch normalization* here uses center=False and scale=False. These are options for customizing batch normalizing and stripping away the learned scaling parameters. This choice prevents the normalization from interfering with the careful noise predictions that diffusion models rely on. Finally, the width adjustment at the beginning handles channel dimension mismatches as we move through different scales of the U-Net architecture. This flexible approach allows our architecture to seamlessly transition between the varying feature representations needed for effective denoising at multiple resolutions.

```
def residual_block(width):
    def apply(x):
        input_width = x.shape[3]
        if input_width == width:
            residual = x
        else:
            residual = layers.Conv2D(width, kernel_size=1)(x)
        x = layers.BatchNormalization(center=False,
                                      scale=False)(x)
        x = layers.Conv2D(
            width, kernel_size=3, padding="same",
```

```
            activation="swish"
        )(x)
        x = layers.Conv2D(width, kernel_size=3,
                          padding="same")(x)
        x = layers.Add()([x, residual])
        return x
    return apply
```

Listing 12.7 Residual Block for the U-Net Model

This residual block will be incorporated in the down_block and up_block in Listing 12.8 to create the distinctive U-shaped information flow that gives U-Net its name. Information travels down to gather broad context and then travels back up to restore fine details. We've covered U-Net in detail in Chapter 7, so we'll only go through the model briefly here. The down block processes features at each resolution while storing copies in the skips list before moving to a coarser scale. The up block reverses this operation with an important addition. As it scales back up using bilinear interpolation, it retrieves those stored features from the skip connections. This creates information fusion where broad context from the bottom of the U combines with preserved details from the corresponding down level.

To enable flexibility in code, each block depth parameter controls how many residual layers process the information at each scale, allowing us to adjust the network's capacity for different levels of detail.

```
def down_block(width, b_depth):
    def apply(x):
        x, skips = x
        for _ in range(b_depth):
            x = residual_block(width)(x)
            skips.append(x)
        x = layers.AveragePooling2D(pool_size=2)(x)
        return x
    return apply

def up_block(width, b_depth):
    def apply(x):
        x, skips = x
        x = layers.UpSampling2D(size=2, interpolation="bilinear")(x)
        for _ in range(b_depth):
            x = layers.Concatenate()([x, skips.pop()])
            x = residual_block(width)(x)
        return x
    return apply
```

Listing 12.8 Downscaling and Upscaling Blocks for U-Net

12.3 Implementing Stable Diffusion Models

The network assembly function in Listing 12.9 brings together all of our building blocks into a complete denoising system. Our network has two key inputs: noisy images and noise variance values. The noise variance tells our network how much noise it's dealing with at any given time step. We transform this single number into rich spatial information by creating sinusoidal embeddings and broadcasting them across the entire image dimensions using nearest-neighbor *upsampling*.

The network begins by projecting the input image into our feature space with a simple convolution, and then it immediately fuses the input with the time step information through concatenation. This early fusion ensures that every subsequent operation knows exactly what level of noise it's working with. The U-Net structure unfolds as we descend through progressively wider feature representations, capturing context at multiple scales. At the bottom of the U, we apply additional residual processing to the deepest, most abstract features. Then, we ascend back up, combining deep understanding with preserved spatial details through our skip connections.

The final output layer uses *zero initialization*, which helps training stability by ensuring our network starts by predicting no change. This gives the training process a neutral starting point rather than random predictions that could disrupt early learning.

```python
def get_network(img_size, widths, b_depth):
    noisy_imgs = keras.Input(shape=(img_size, img_size, 3))
    noise_vars = keras.Input(shape=(1, 1, 1))

    e = layers.Lambda(sinusoidal_embedding,
                      output_shape=(1, 1, 32))(noise_vars
    )
    e = layers.UpSampling2D(size=img_size,
                            interpolation="nearest")(e)

    x = layers.Conv2D(widths[0], kernel_size=1)(noisy_imgs)
    x = layers.Concatenate()([x, e])

    skips = []
    for width in widths[:-1]:
        x = down_block(width, b_depth)([x, skips])

    for _ in range(b_depth):
        x = residual_block(widths[-1])(x)

    for width in reversed(widths[:-1]):
        x = up_block(width, b_depth)([x, skips])

    x = layers.Conv2D(3, kernel_size=1,
                      kernel_initializer="zeros")(x)
```

```
        return keras.Model([noisy_imgs, noise_vars],
                           x,
                           name="residual_unet")
```

Listing 12.9 Assembling the U-Net Model from Parts

12.3.4 Diffusion and Reverse Diffusion

With the U-Net denoiser and KID evaluation metric prepared, the `DiffusionModel` class in Listing 12.10 brings together everything we've built into a unified system. This class extends Keras's `Model` class, giving us all the standard training machinery while adding the specialized components that make diffusion work. Notice that the initialization creates two identical networks. Our main `network` learns aggressively during training, updating its weights with each batch. The `ema_network` takes a more conservative approach, slowly incorporating changes through *exponential moving averaging* (EMA) The main network experiments and learns quickly, while the EMA network maintains stability and provides reliable guidance.

The `normalizer` handles the statistical properties of our training data. We've seen this pattern before in other neural networks—keeping our data in a consistent range helps the model learn more effectively. The network will learn these statistics automatically during training. Our `compile` method sets up the metrics we'll track during training. The `noise_loss` measures how well we predict the random noise that was added to images. The `image_loss` tracks how accurately we can reconstruct clean images from noisy inputs. Combined with our `KID` metric, we get a comprehensive view of both the learning process and the final generation quality.

The `denormalize` method serves as our bridge back to the real world. After all of our internal processing with normalized values, this function converts our network's outputs back into actual pixel values that we can display and evaluate.

```
@keras.saving.register_keras_serializable()
class DiffusionModel(keras.Model):
    def __init__(self, img_size, widths, b_depth):
        super().__init__()
        self.normalizer = layers.Normalization()
        self.network = get_network(img_size, widths, b_depth)
        self.ema_network = keras.models.clone_model(self.network)

    def compile(self, **kwargs):
        super().compile(**kwargs)
        self.noise_loss_tracker = keras.metrics.
                                  Mean(name="n_loss")
        self.image_loss_tracker = keras.metrics.
                                  Mean(name="i_loss")
```

```
        self.kid = KID(name="kid")

    @property
    def metrics(self):
        return [self.noise_loss_tracker,
                self.image_loss_tracker,
                self.kid]

    def denormalize(self, images):
        images = self.normalizer.mean + images *
                self.normalizer.variance**0.5
        return ops.clip(images, 0.0, 1.0)
```

Listing 12.10 Merging Components in the DiffusionModel Class

The *diffusion schedule*—an important aspect of the diffusion process—in Listing 12.11 implements the cosine noise schedule we discussed in Section 12.1.2 earlier. The function maps diffusion time steps to angles using arccos, and then applies trigonometric functions to compute signal and noise rates. This approach ensures smooth transitions and maintains the mathematical constraint that signal and noise rates are complementary. The *cosine schedule* provides better training stability compared to linear schedules. Early time steps have high signal rates and low noise rates, while later time steps reverse this relationship. The smooth progression prevents abrupt changes that could destabilize training.

The denoise method on the other hand, performs the core noise prediction and image reconstruction. During training, it uses the actively updating network, while inference relies on the more stable EMA network. The network receives noisy images and noise variance as inputs, outputting predicted noise.

The reconstruction formula implements the mathematical relationship we established: Given a noisy image as a mixture of clean image and noise, we subtract the predicted noise component and rescale by the signal rate to recover the original image. This calculation directly follows from our diffusion process formulation.

```
def diff_schedule(self, diff_times):
    start_angle = ops.cast(ops.arccos(max_sig_rate), "float32")
    end_angle = ops.cast(ops.arccos(min_sig_rate), "float32")
    diff_angles = start_angle + diff_times * (
        end_angle - start_angle
    )
    sig_rates = ops.cos(diff_angles)
    noise_rates = ops.sin(diff_angles)
    return noise_rates, sig_rates
```

```
def denoise(self, noisy_imgs, noise_rates, sig_rates, training):
    if training:
        network = self.network
    else:
        network = self.ema_network
    pred_noises = network(
        [noisy_imgs, noise_rates**2], training=training
    )
    pred_imgs = (noisy_imgs - noise_rates * pred_noises) / 
            sig_rates
    return pred_noises, pred_imgs
```

Listing 12.11 The Diffusion Process and Denoising

12.3.5 The Generation Engine

The `reverse_diffusion` method in Listing 12.12 represents the ultimate goal of our entire training process—this is where random noise is transformed into meaningful images after the training has been completed. Once our model has learned to predict noise accurately during training, this function becomes our primary generation tool. The process begins with pure random noise and systematically removes noise over a specified number of steps. Each iteration calculates the current time step and applies our diffusion schedule to determine the appropriate noise and signal rates for that moment in the reverse process. The core operation happens in the denoise call studied earlier.

After prediction, the function calculates the next state by mixing the predicted clean image with the predicted noise according to the rates for the next time step. This creates a *controlled progression* from high noise to low noise, with each step bringing us closer to a final clean image. This gradual approach leverages the network's training on similar small denoising steps.

When training completes successfully, this function becomes our image generator. We feed it random noise, and it produces novel images that match the statistical properties of our training dataset, creating new flower images that never existed before but look convincingly real. We'll show a few of the samples of our model's generation in a little while.

```
def reverse_diffusion(self, initial_noise, diff_steps):
    num_imgs = initial_noise.shape[0]
    step_size = 1.0 / diff_steps
    next_noisy_imgs = initial_noise
    for step in range(diff_steps):
        noisy_imgs = next_noisy_imgs
        diff_times = ops.ones((num_imgs, 1, 1, 1)) -
                    step * step_size
```

```
        noise_rates, sig_rates = self.diff_schedule(diff_times)
        pred_noises, pred_imgs = self.denoise(
            noisy_imgs, noise_rates, sig_rates, training=False
        )
        next_diff_times = diff_times - step_size
        next_noise_rates, next_sig_rates = self.diff_schedule(
            next_diff_times
        )
        next_noisy_imgs = (
            next_sig_rates * pred_imgs +
            next_noise_rates * pred_noises
        )
    return pred_imgs
```

Listing 12.12 Applying Denoising in Reverse Diffusion

The generate method in Listing 12.13 provides our clean interface to the entire diffusion system. This function encapsulates all the complexity we've built into a straightforward three-step process:

1. We create random noise tensors with the exact dimensions needed for our model. The noise follows a standard normal distribution, providing the chaotic starting point that our trained network will systematically transform. Each tensor contains completely random values, giving us the raw material for generation.

2. The reverse_diffusion call performs the actual transformation work. This step uses all of our trained parameters and learned denoising capabilities to convert random noise into structured image data. The number of diffusion steps determines the quality-speed trade-off—more steps generally produce better results but require more computation time.

3. Finally, denormalize converts our internal representation back to standard image format. Our network operates on normalized data for training stability, but we need pixel values in the 0–1 range for display and evaluation purposes.

```
def generate(self, num_imgs, diff_steps):
    initial_noise = keras.random.normal(
        shape=(num_imgs, img_size, img_size, 3)
    )
    gen_imgs = self.reverse_diffusion(initial_noise, diff_steps)
    gen_imgs = self.denormalize(gen_imgs)
    return gen_imgs
```

Listing 12.13 Wrapper Method for Image Generation

> **Designing for Larger Ecosystems**
>
> The generate function represents the user-facing interface to our entire diffusion system. After training completes, researchers and practitioners will primarily interact with this simple API to generate new images, abstracting away the mathematical complexity of the underlying diffusion process. It's always a good idea to provide these interfaces because, ultimately, the whole point of creating and training these models is to use them in the engineering of other, larger ecosystems.

The train_step method in Listing 12.14 implements the forward diffusion process (discussed earlier in our theory sections), but now in reverse. We're teaching our network to undo what the forward process creates. This function represents the heart of how our model's trained, and it's called automatically by Keras as part of the training process. Each training step begins by normalizing our real images and generating random noise tensors. We then sample random time steps uniformly across our diffusion schedule. This random sampling ensures our network learns to handle denoising at every level of the process, from barely noisy images to pure chaos. Using our diffusion schedule, we calculate the noise and *signal rates* for the sampled time steps, then create the training examples by mixing clean images with noise according to these rates. This step recreates the exact same noisy images that our forward diffusion process would generate at those specific time steps.

The gradient tape captures our network's predictions as it attempts to denoise these artificially corrupted images. We calculate two losses—one measuring how well we predict the added noise and another measuring how accurately we can reconstruct the original clean image. These dual objectives ensure our network learns both the noise patterns and the underlying image structure. The gradient computation and application follows standard neural network training, but with an important addition—the EMA weight updates. After each training step, we slowly incorporate the current network's learned parameters into our more stable EMA network. This EMA creates a secondary network that changes more gradually, providing stability during generation. The amount of incorporation of new information is decided by the ema_weight hyperparameter.

```
def train_step(self, images):
    images = self.normalizer(images, training=True)
    noises = keras.random.normal(
        shape=(batch_size, img_size, img_size, 3)
    )
    diff_times = keras.random.uniform(
        shape=(batch_size, 1, 1, 1), minval=0.0, maxval=1.0
    )
    noise_rates, sig_rates = self.diff_schedule(diff_times)
```

```
    noisy_imgs = sig_rates * images + noise_rates * noises

    with tf.GradientTape() as tape:
        pred_noises, pred_imgs = self.denoise(
            noisy_imgs, noise_rates, sig_rates, training=True
        )
        noise_loss = self.loss(noises, pred_noises)
        image_loss = self.loss(images, pred_imgs)

    gradients = tape.gradient(
        noise_loss, self.network.trainable_weights
    )
    self.optimizer.apply_gradients(
        zip(gradients, self.network.trainable_weights)
    )
    self.noise_loss_tracker.update_state(noise_loss)
    self.image_loss_tracker.update_state(image_loss)

    for weight, ema_weight
        in zip(self.network.weights, self.ema_network.weights):
        ema_weight.assign(ema * ema_weight + (1 - ema) * weight)
    return {m.name: m.result() for m in self.metrics[:-1]}
```

Listing 12.14 Step for the Training Loop

The model compilation in Listing 12.15 instantiates the diffusion system by specifying the optimization algorithm. The *AdamW* optimizer is an upgrade over standard Adam. The *W* stands for *weight decay*, which addresses a critical challenge in training large neural networks. During training, neural networks can develop a tendency to rely too heavily on specific weights, making them brittle and prone to overfitting. Weight decay acts as a regularization mechanism, continuously applying a small penalty that encourages weights to stay reasonable in magnitude. For diffusion models specifically, weight decay becomes essential because these networks must learn incredibly complex mappings from noise to data across thousands of different time steps. Without proper regularization, the model might memorize specific noise patterns rather than learning the underlying generative process. Weight decay keeps the learning process focused on generalizable patterns rather than dataset-specific artifacts.

As always, the learning rate parameter controls how aggressively our model updates its weights during training. For diffusion models, this balance becomes particularly delicate—too high, and the model might overshoot optimal solutions during the intricate denoising process; too low, and training becomes prohibitively slow for the already demanding computationally intensive models.

```
model = DiffusionModel(img_size, widths, block_depth)
model.compile(
    optimizer=keras.optimizers.AdamW(
        learning_rate=learn_rate, weight_decay=weight_decay
    ),
    loss=keras.losses.mean_absolute_error,
)
```

Listing 12.15 Instantiating and Compiling the Model

The checkpoint configuration in Listing 12.16 addresses the practicality issues we discussed earlier. Diffusion models require extensive training time that often exceeds Colab's session limits. This callback system provides our insurance against losing valuable training progress. The `checkpoint_path` directs our model weights to persistent Google Drive storage, ensuring they survive session resets. We save only the weights rather than the complete model structure, which reduces file size. The `save_best_only=True` parameter ensures we maintain only the highest-quality version of our model. As training progresses and our KID scores fluctuate, this mechanism automatically preserves the iteration that produced the most realistic images according to our evaluation metric. This means that you can start the training process during the night, get a good eight hours sleep, and come back to a model that is actually generating flowers. Not only will the model be trained, but you'll also have the trained weights saved persistently in your Google Drive.

The loading code demonstrates the restoration process for continuing training after session interruptions. The `model.build()` call initializes the model architecture with proper input shapes. This is necessary for enabling Keras to load weights into a model that has custom components and hasn't yet been initialized (as will be the case when you're starting from scratch later). This checkpoint strategy is essential for diffusion training, where meaningful progress might require dozens of hours. Without this safeguard, session timeouts would force us to restart training from scratch, wasting enormous computational resources and making practical experimentation impossible on free cloud platforms.

```
checkpoint_path = "/content/drive/MyDrive/colab-data/
                    keras-book/diffusion_model.weights.h5"
checkpoint_callback = keras.callbacks.ModelCheckpoint(
    filepath=checkpoint_path,
    save_weights_only=True,
    monitor="val_kid",
    mode="min",
    save_best_only=True,
)
```

```
# we can load it later using the following
model.build(input_shape=(None, img_size, img_size, 3))
model.load_weights(checkpoint_path)
```

Listing 12.16 Saving and Loading Model Weights for Continuing Training

An important aspect of diffusion models is understanding the statistical landscape of the target dataset. This is achieved using the `normalizer` component shown in Listing 12.17. Our model needs to understand the statistical landscape of our training data—what's typical, what's unusual, and how the data is distributed across different ranges. When we call `model.normalizer.adapt(train_dataset)`, we're giving our model a chance to examine the entire training dataset and learn its statistical fingerprint. The normalizer calculates the mean and variance of pixel values across all of our flower images, building an internal understanding of what "normal" looks like in our dataset.

```
# Calculate mean and variance for normalization
model.normalizer.adapt(train_dataset)
```

Listing 12.17 Understanding the Statistical Landscape of Data

12.3.6 Following Progress in the Training Process

We're now ready to actually start the training. The execution in Listing 12.18 launches our model into its training loop, where thousands of flower images will be turned noisy. As always, our training setup includes both training and validation datasets, allowing our model to learn from one set of images while testing its progress on another. The callbacks serve as our training assistants, each with a specific mission. The `LambdaCallback` triggers our `plot_images` function at the end of each epoch, giving us visual feedback on our model's creative progress. We've omitted this function from code listings for the sake of saving space. You can see the full code in the book resources. For now, it's enough to know that `plot_images` does as the name implies—it generates a few images after each epoch and plots them so that we can monitor progress of our generation visually. Watching these generated images evolve from pure noise to recognizable flowers provides immediate insight into how well our diffusion process is learning.

Our checkpoint callback stands guard throughout training, automatically saving our best-performing model weights based on validation KID scores.

> **Start with Only a Few Epochs**
> When experimenting with code of such large models, it's generally a good idea to start with a couple of epochs to make sure that the model is progressing in the right direction. While the code provided with this book will work fine, you might want to change the hyperparameters and the model architecture to fully understand the workings of this

> complex beast. When you do this, it's best to start with two to three epochs to make sure the model is working fine before you start the full training loop.

When this training completes, we'll have hopefully transformed our random initialization into a sophisticated system capable of generating entirely new flower images that capture the essence of our training data. (During our runs, it took about 65 epochs and three days of loading weights from checkpoints to get to acceptable results. We'll get to this discussion in a while.)

```
model.fit(
    train_dataset,
    epochs=30,
    validation_data=val_dataset,
    callbacks=[
        keras.callbacks.LambdaCallback(on_epoch_end=model.plot_images),
        checkpoint_callback, # Added checkpoint callback
    ],
)
```

Listing 12.18 Training the Model

These initial training results after the first epoch show exactly what we'd expect from our diffusion model in its earliest stages—pure, colorful chaos. Take a look at Figure 12.3.

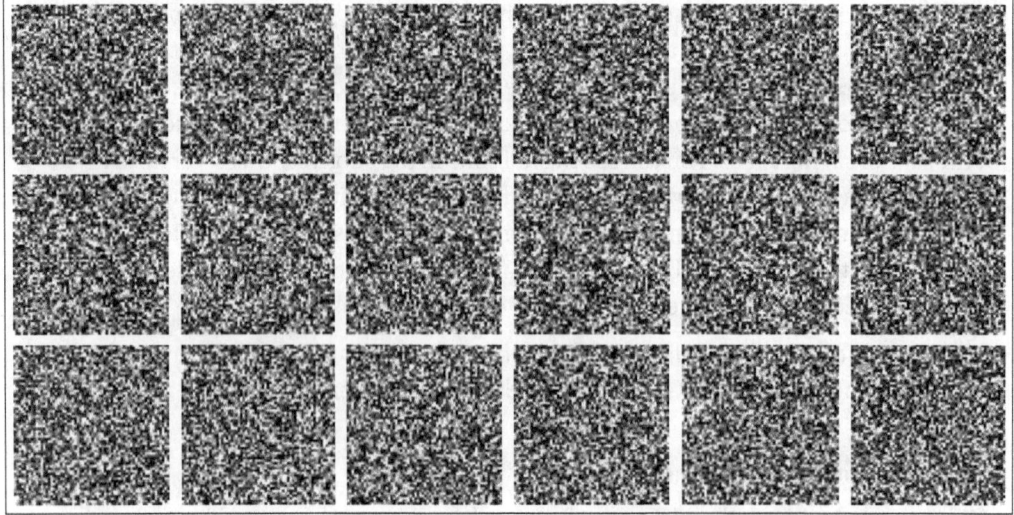

Figure 12.3 Initial Generation After a Single Epoch of Training

Each image represents our network's first attempts at generation. If you look carefully, you'll notice that these aren't purely uniform random pixels. Even in this apparent chaos, there are subtle variations in density and intensity across different regions of each image. Some areas appear slightly more saturated, others show faint clustering of similar colors. These tiny variations become the *seeds* from which our model will learn to recognize and amplify meaningful patterns.

Our network is starting with absolutely no understanding of what a flower should look like. It has no concept of petals, stems, or natural color combinations. Right now, it's simply learning the statistical fingerprint of our training data—understanding that certain color combinations appear more frequently than others, that natural images have specific noise characteristics, and that there are underlying structures worth preserving.

This chaotic starting point is actually perfect—it demonstrates that our model hasn't memorized any specific images and is truly learning to generate from scratch. The shift from this noise to beautiful flower images will showcase the incredible power of the diffusion learning process.

Figure 12.4 shows the images generated after just 10 epochs. Our diffusion model has made remarkable progress from random noise to meaningful imagery.

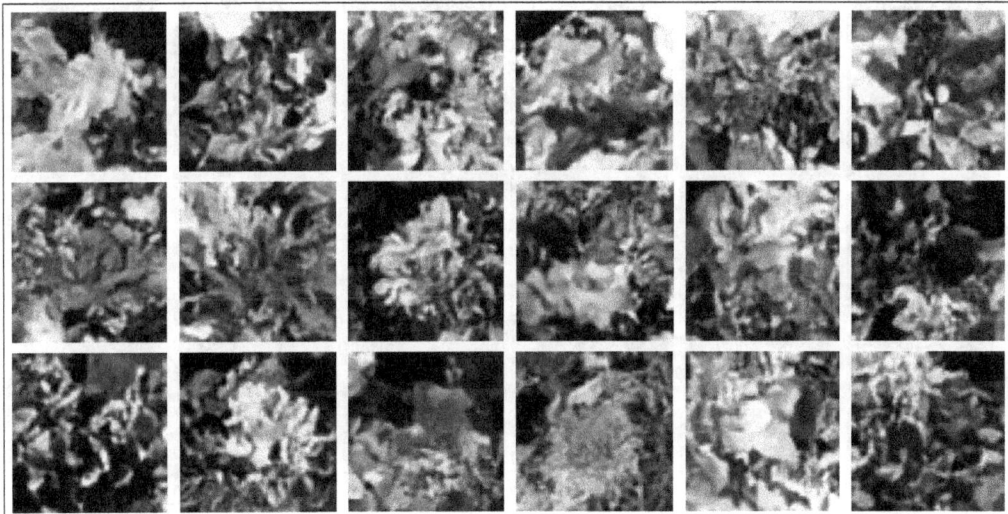

Figure 12.4 Image Generation Results After 10 Epochs

The first time, we can squint our eyes and say, "Yes, I can see some flower-like objects beginning to emerge." Look closely at these results, and you'll notice something interesting. While these images don't yet have the crisp clarity of professional flower photography, they capture something far more important—the essence of what makes a flower look like a flower. The color patterns are right, with natural transitions from darker backgrounds to brighter petals. The spatial arrangements suggest the organic clustering we'd expect from floral structures. The blurriness and abstract quality we're seeing isn't a failure—it's exactly what we'd expect (and want) at this stage. Our model is still learning to balance the complex relationship between removing noise and preserving meaningful details.

This progression from pure chaos to recognizable patterns demonstrates the incredible power of the diffusion learning process. Our model is developing an internal understanding of what being a flower means, building up this knowledge through millions of small denoising decisions across thousands of training examples.

After 65 epochs and multiple checkpointing sessions due to Colab's time limits, our diffusion model has achieved remarkable results, as shown in Figure 12.5. Looking at these final generated images, we've crossed a crucial threshold—these are genuinely convincing flowers that could fool a casual observer. The transformation from our earlier abstract attempts to these crisp, detailed blooms represents thousands of hours of computational learning condensed into visible progress. Each image now displays the hallmarks of photographic realism: individual petals with natural curves, realistic lighting, and shadows, proper depth of field effects, and color gradients that feel organic rather than artificial.

Figure 12.5 Images Generated After 65 Epochs

What makes this achievement even more remarkable is that we've built this entire system from scratch. Apart from the pretrained InceptionV3 weights we borrowed for our KID evaluation metric, every single parameter in our diffusion model learned its knowledge through our training process. We didn't start with a massive foundation model or *transfer learning* from existing image generators. Instead, we watched our network develop its understanding of flowers, noise patterns, and visual composition entirely through its own experience with our training data. While we can still spot occasional inconsistencies if we look closely—perhaps a petal that doesn't quite follow natural curves or lighting that seems slightly off—these generated images represent a technological achievement that would have been impossible just a few years ago. We're watching AI create photorealistic content from pure mathematical noise, with no human guidance beyond the training examples it learned from.

Finally, to drive this point home, we can see our reverse diffusion process captured in a single sequence. Starting with pure random noise, our trained model transforms it step-by-step into a recognizable image. Look at Figure 12.6 for this progression. The process happens gradually rather than all at once. In the first few steps, the changes are subtle—slight shifts in color distribution and hints of structure beginning to form. The chaotic rainbow of pixels starts organizing itself into warmer, more natural tones as the model applies its learned knowledge about which patterns typically appear together in flower images.

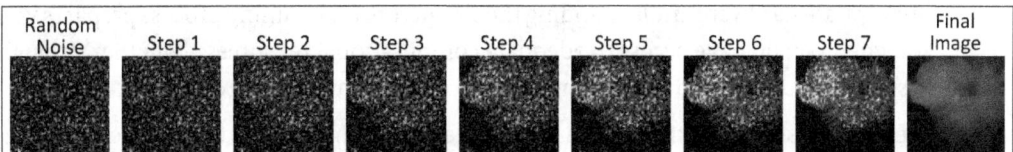

Figure 12.6 Progression from Noise to a Realistic Image in Stable Diffusion

By the middle steps, you can see the random pixels forming into regions that suggest depth and organic shapes. The model is essentially using everything it learned about color relationships and spatial patterns during training to guide the denoising process.

The final image shows the complete transformation from mathematical randomness into a convincing flower with natural lighting and realistic petals. Each step represents a decision about which details to preserve and which noise to remove, guided by the statistical patterns the model absorbed from thousands of training examples.

This step-by-step process shows how diffusion models generate images through a controlled, deliberate approach rather than creating them all at once. In the next section, we'll explore how this same process can be enhanced with text conditioning, allowing us to guide the generation with specific descriptions and prompts.

12.4 Case Study: Image Generation

Before we explore prompt-based image generation, we need to address a significant hurdle in our workflow. The model we just trained, while impressive in its results, contained 3,907,014 parameters and took three full days to complete training. When we scale this up to understand the computational requirements for state-of-the-art models with billions of parameters, the math becomes sobering—our current setup would require approximately 8.5 years to train a 3-billion parameter model. The challenges extend beyond just training time. Models with billions of parameters demand enormous amounts of RAM simply to load into memory, often requiring tens of gigabytes that far exceed what's available on standard hardware. The training data itself becomes another bottleneck, as these larger models typically require massive datasets that can overwhelm storage and memory systems.

For those working with Google Colab's free tier, there's currently no practical workaround for these computational limitations. The hardware constraints are simply too restrictive for training or even running large-scale diffusion models that power modern text-to-image generation systems.

However, we'll still walk through the code architecture for prompt-based generation, providing you with the theoretical foundation and implementation knowledge you'll need. While you may not be able to run these models immediately without access to appropriate hardware, understanding their structure and training process prepares you for when you do have access to adequate computational resources—whether through research funding, cloud computing credits, or more powerful hardware.

We'll also discuss practical alternatives for experimenting with these concepts using existing platforms that provide pretrained models, though these won't offer the same flexibility as writing and training your own implementations. The goal is ensuring you understand the underlying mechanics, so when the opportunity arises to work with sufficient compute resources, you can transition from theory to practice seamlessly.

12.4.1 Loading Pretrained Models from Keras Hub

To work with larger, pretrained diffusion models, we'll use Keras Hub—a powerful ecosystem that shouldn't be confused with Kaggle Hub, which we explored in Chapter 9. While Kaggle Hub focuses on datasets and competitions, Keras Hub serves as a comprehensive repository for state-of-the-art machine learning models.

Because Google Colab doesn't include Keras Hub by default, we need to install it first using the magic command !pip install keras-hub. You can view the full code listing on the book resources page at *https://recluze.net/keras-book* in the notebook **12-02-prompt-based-images**.

Keras Hub provides access to pretrained models across various domains—from language models and computer vision to audio processing and multimodal systems. For our diffusion work, it offers several advantages over building everything from scratch. The platform includes optimized implementations of complex architectures, pretrained weights from models trained on massive datasets, and standardized APIs that make it easy to load, fine-tune, and deploy sophisticated models. What makes Keras Hub particularly valuable for diffusion models is its collection of *text-to-image generators* that have already undergone the computationally intensive training process we can't replicate on free hardware. These models come with their text encoders, diffusion networks, and VAE components already trained and ready for inference or fine-tuning.

The hub also provides preprocessing pipelines, tokenizers, and other essential components that ensure compatibility between different parts of complex multimodal systems. Rather than implementing these from scratch and dealing with the intricate details of model architecture and weight loading, we can focus on understanding how these systems work and experimenting with their capabilities. After having worked through this book's code, you already understand these components. Keras Hub will enable you to now work with them efficiently without sacrificing the deeper understanding.

12.4.2 Loading Models Through Keras Hub

With Keras Hub available, we can load a pretrained text-to-image model using Keras Hub's streamlined interface, as shown in Listing 12.19. The `from_preset()` method loads the complete Stable Diffusion 3 model with pretrained weights. The `stable_diffusion_3_medium` preset specifies which model variant to use—this medium version balances quality and computational requirements. The `dtype="float16"` argument reduces memory usage by using half-precision floating point numbers, essential for running large models on limited hardware.

Interestingly, Keras Hub offers several other model presets beyond Stable Diffusion, including various language models such as Gemma (that we saw in Chapter 9) and BERT, vision models such as ResNet and EfficientNet, and other generative models. For text-to-image specifically, you'll find different Stable Diffusion variants and other diffusion architectures, each optimized for different use cases and computational constraints.

The elegance of this approach lies in its simplicity—a single line loads a multibillion parameter model that would have taken us years to train from scratch. This model includes the text encoder that processes our prompts, the diffusion U-Net that performs the denoising, and the VAE that converts between latent and pixel spaces. Instead of using this `from_preset` method, we can also exercise more flexibility by specifying other options as we'll see in a minute.

```
text_to_image = keras_hub.models.
                StableDiffusion3TextToImage.from_preset(
                    "stable_diffusion_3_medium",
                    dtype="float16",
                    verbose=True)
```

Listing 12.19 Loading a Model Through Keras Hub

Before generating images, we'll set up a helper function for displaying our results. The function in Listing 12.20 handles different image formats that our model might return. For single images (3D arrays), it displays them directly. For multiple images (4D arrays or lists), it concatenates them horizontally for side-by-side comparison. The function is flexible enough to work with various output formats from different models. You can modify this helper function as needed—perhaps to create image grids instead of horizontal strips, add titles, or save images to files rather than just displaying them.

```
def display_generated_images(images):
    display_image = None
    if isinstance(images, np.ndarray):
        if images.ndim == 3:
            display_image = Image.fromarray(images)
        elif images.ndim == 4:
            concated_images = np.concatenate(list(images), axis=1)
            display_image = Image.fromarray(concated_images)
    elif isinstance(images, list):
        concated_images = np.concatenate(images, axis=1)
        display_image = Image.fromarray(concated_images)

    if display_image is None:
        raise ValueError("Unsupported input format.")

    plt.figure(figsize=(10, 10))
    plt.axis("off")
    plt.imshow(display_image)
    plt.show()
    plt.close()
```

Listing 12.20 Helper Function to Display Generated Images

12.4.3 Using Stable Diffusion Models

As mentioned before, Keras Hub is quite flexible. While we can use a single line to load a model, we can also assemble our text-to-image system by constructing its three essential components separately, as shown in Listing 12.21. Recall the backbone concept from Chapter 9. We learned there that the backbone contains the essential model weights and layers, while other components handle specific tasks such as preprocessing. Here, it represents the core neural network architecture that does the heavy lifting—U-Net and its associated components actually perform the diffusion process. We specify the output image dimensions (512x512 pixels with three color channels) and use float16 precision to manage memory efficiently.

The preprocessor handles the text side of our pipeline. It takes our written prompts and transforms them into the numerical representations that our model understands, managing tokenization and encoding behind the scenes. These three components work together seamlessly. When you provide a text prompt, the preprocessor converts your words into mathematical form. The backbone then uses this encoded text to guide its diffusion process, conditioning each denoising step based on your description. Finally, the complete `text_to_image` model orchestrates this entire workflow, taking care of the complex coordination between text understanding and image generation.

This modular approach gives us flexibility—we could swap out different preprocessors for various text formats or modify the backbone for different image resolutions, all while maintaining the same overall architecture.

```
backbone = keras_hub.models.
                StableDiffusion3Backbone.from_preset(
                    "stable_diffusion_3_medium",
                    image_shape=(512, 512, 3),
                    dtype="float16"
)
preprocessor = keras_hub.models.
                StableDiffusion3TextToImagePreprocessor
                .from_preset(
                    "stable_diffusion_3_medium"
)
text_to_image = keras_hub.models.
                StableDiffusion3TextToImage(
                    backbone,
                    preprocessor
)
```

Listing 12.21 Loading Different Components of a Model Separately

We can now directly run the model to make predictions (i.e., generate images) based on the prompts. Take a look at the code in Listing 12.22. Remember that the backbone is already trained and the only input it needs from us is a text string that will be encoded by the transformer. Let's follow our first example prompt, "a beautiful landscape with lots of green grass and a flowing river, wide view 4k," as it transforms from words into a visual image using the three-component architecture we discussed earlier.

Recall from our theory section how Stable Diffusion combines three essential components—text, noise schedule, and time steps. Here, we see that architecture in action with the latter two components built-in to the model we just loaded. The process begins with our preprocessor, which handles the text conditioning we explored in detail. Just as we discussed, it takes our human language and converts it into the numerical format that our neural network can work with. The words "beautiful landscape" get transformed into the dense numerical vectors that capture their meaning—exactly the process we illustrated earlier in this chapter and in Chapter 9.

These encoded words then travel to our backbone, implementing the U-Net architecture. The backbone is very similar to the model we created from scratch in the previous section. It receives two inputs: our encoded text description and random noise. This is where the cross-attention mechanisms we discussed come into play—the backbone knows how to shape that noise, guided by the text instructions, to create the image we described. During this process, the text conditioning works through the attention mechanisms. At each step of the denoising process, the backbone refers back to our encoded prompt through those cross-attention layers, ensuring that the emerging image matches what we described. This is exactly the text-conditioning process we illustrated when we discussed how different components of Stable Diffusion work together.

The backbone performs this transformation through the reverse diffusion process. Each step reduces noise slightly, guided by both the learned denoising patterns and our text conditioning. The model applies the same mathematical principles—using the learned relationships between text descriptions and visual features to guide each denoising decision.

This transformation works because our model embodies all the learning relationships we discussed—the connections between language tokens and visual features, the attention mechanisms that let text guide image generation, and the diffusion process that gradually sculpts noise into meaningful imagery.

```
prompt = """a beautiful landscape with lots of green grass
          and a flowing river, wide view 4k"""

generated_image = text_to_image.generate(prompt)
display_generated_images(generated_image)
```

Listing 12.22 Text-Guided Image Generation

The resulting image in Figure 12.7 seems quite impressive and to the layperson looking at the image without context, there would be little reason to doubt that this image was created as a result of removing noise from completely random pixel values.

> **[!] Working Around Hardware Limitations**
>
> If you're working with Colab's free tier, you'll likely encounter "out of memory" errors when trying to run these large models. The hardware simply isn't equipped to handle the memory demands of billion-parameter diffusion systems.
>
> For those with access to more powerful resources, you have several options. Colab Pro offers upgraded runtimes with more RAM and better GPUs that can handle these models. Cloud platforms such as Microsoft Azure, Google Cloud, or services such as RunPod provide high-end GPU instances specifically designed for machine learning workloads.
>
> In the meantime, you can experiment with the same Stable Diffusion 3 model we've been discussing through the Hugging Face **Spaces** page at *http://s-prs.co/v62404*. This lets you test different prompts and see the results without needing any local computational resources.

Figure 12.7 Result of Text-to-Image Generation

12.4.4 Beyond Image Generation to More Complex Workflows

With our text-to-image foundation in place, we can explore more sophisticated workflows using the same underlying model. One particularly powerful technique is *inpainting*—a process that lets us selectively modify specific regions of existing images while preserving everything else. Think of inpainting as digital restoration work, similar to how art conservators carefully repair damaged paintings. Instead of creating entirely new images from scratch, we're taking existing photos and making precise, targeted changes. You might want to remove an unwanted object from a family photo, replace a cloudy sky with a sunset, or add elements that weren't in the original scene.

Setting aside the ethical considerations of image manipulation for now, let's focus on the technical mechanics of how this process actually works. The beauty of inpainting lies in how cleverly it leverages our diffusion model's existing capabilities. The process begins with three key ingredients:

- Your original image
- A *mask* (black-and-white image) that defines which areas you want to change (areas with white pixels)
- A text prompt describing what should appear in those masked regions

Our diffusion model approaches this challenge with remarkable elegance. Instead of treating the entire image as a blank canvas, it uses the unmasked portions as anchor points—areas that must remain absolutely unchanged. The model then focuses its creative energy exclusively on the masked regions, generating new content that seamlessly blends with the existing image. During each step of the reverse diffusion process, the model performs a careful balancing act. It applies the denoising process only to the masked areas while preserving the original pixel values everywhere else. This creates a natural boundary where new content must harmoniously connect with existing elements—matching lighting conditions, perspective, and style.

The *text conditioning* guides what appears in the masked regions, just like in our earlier text-to-image generation. But now, the model has additional context from the surrounding image content, helping it make more informed decisions about color palettes, lighting direction, and spatial relationships. The result is new content that doesn't just match your text description but also feels like it naturally belongs in the original scene. Let's see this inpainting in action through code that brings together all the concepts we've discussed. The process shown in Listing 12.23 demonstrates how we can selectively modify parts of an existing image while preserving everything else.

We start by creating our inpainting model `StableDiffusion3Inpaint` using the same backbone and preprocessor components we assembled earlier. This gives us access to all that learned knowledge about text-to-image generation, but now with the special ability to work with existing images and masks. The image preparation involves loading our base photograph and resizing it to the standard 512×512 dimensions our model expects.

The mask represents the heart of our inpainting operation. Think of it as a stencil that tells our model exactly where to focus its creative efforts. We load the mask as a grayscale image where white pixels indicate "paint here" and black pixels mean "leave this alone." Converting to Boolean format gives us the precise on/off switches our model needs to know which areas are fair game for modification.

Finally, our prompt guides what should appear in the masked regions, just like our earlier text-to-image generation. Here we're asking for "a photorealistic little tree with green leaves and a few berries" to appear wherever the mask indicates.

The generate function receives all three pieces of information: the original image providing context and unchangeable areas, the mask defining the modification zone, and the prompt describing what should appear there. Our model then works its magic, creating new content that seamlessly blends with the existing image.

```
inpaint = keras_hub.models.
            StableDiffusion3Inpaint(
                backbone,
                preprocessor
        )

image = Image.open("landscape.jpg").convert("RGB")
image = image.resize((512, 512))
image_array = rescale(np.array(image))

# Note that the mask values are of Boolean dtype.
mask = Image.open("landscape-mask.jpg").convert("L")
mask = mask.resize((512, 512))
mask_array = np.array(mask).astype("bool")

prompt = """photorealistic little tree with green
        leaves and a few berries, 4k"""

generated_image = inpaint.generate(
    {
        "images": image_array,
        "masks": mask_array,
        "prompts": prompt,
    }
)
display_generated_images(
    [
        np.array(image),
        np.array(mask.convert("RGB")),
        generated_image,
    ]
)
```

Listing 12.23 Complete Inpainting Workflow with Image, Mask, and Prompt Inputs

The resulting image shown in Figure 12.8 highlights the impressive precision of our inpainting system. In the left image, we see the original river landscape that we generated earlier. The middle image shows our mask, the white shape that instructed our model where to add new content.

The final image demonstrates how effectively our model understood the task. Where the mask indicated, a tall conifer tree has been added, maintaining the natural perspective and lighting of the scene. Notice how the tree appears seamlessly integrated into the landscape; it casts appropriate shadows, aligns with the depth of the terrain, and blends perfectly with the existing moss-covered rocks and flowing water. What's particularly impressive is how the model preserved every detail outside the masked area while creating content that feels like it belonged there all along. The boundaries between original and generated content are nearly invisible, and the new tree adds to the scene's natural composition rather than disrupting it.

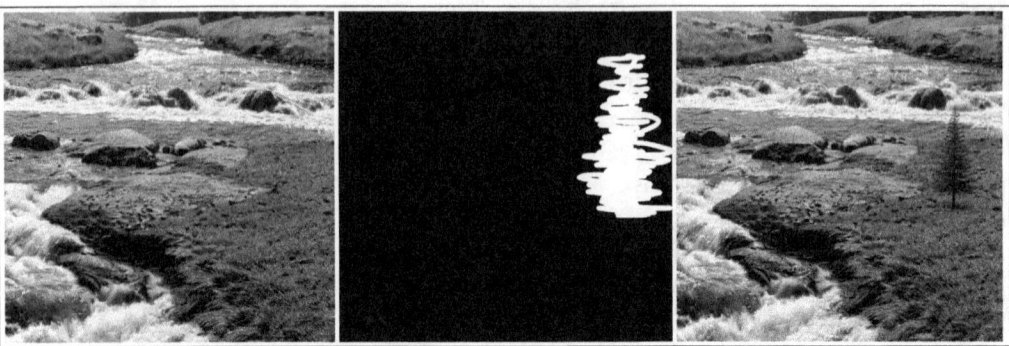

Figure 12.8 Results of Inpainting with Stable Diffusion

This demonstrates the power of combining learned visual understanding with precise spatial control—our model drew upon its training to understand how trees should appear in natural settings, and then applied that knowledge specifically to our defined area.

> [»] **Inpainting on Hugging Face**
>
> As before, if you're facing computational power restrictions, you may use the Hugging Face **Spaces** page to experiment with inpainting for free: *http://s-prs.co/v613205*.

> [+] **Exploring Keras Hub**
>
> Keras Hub offers a complete suite of Stable Diffusion capabilities beyond what we've explored here. You can experiment with *image-to-image* generation, where existing photos serve as starting points for artistic transformations. There's also *outpainting*, which extends images beyond their original boundaries—consider taking a portrait and seamlessly expanding it to show more of the surrounding environment. For those interested in fine-tuning, Keras Hub provides tools to adapt these models to specific styles or domains using your own datasets. Each of these techniques builds on the same foundational concepts we've covered but opens up entirely new creative possibilities for visual generation and manipulation. Take a look at the full pretrained models list on Keras Hub here: *https://keras.io/keras_hub/presets/*.

12.5 Summary

In this chapter, we've gone from mathematical theory to practical implementation, exploring how abstract concepts can become systems that create visual art from text descriptions. We started by understanding the mathematics behind diffusion models—how they learn to reverse the process of adding noise to images, teaching computers to gradually transform chaos into meaningful images. We built our own diffusion model from scratch, watching it progress from generating random pixels to creating recognizable flower images through millions of small denoising decisions. We then explored how text conditioning transforms these image-generation systems into tools that respond to written descriptions, using pretrained models through Keras Hub to generate landscapes from prompts. Finally, we discovered inpainting, where spatial control meets image generation, allowing us to modify existing images with precision.

In the next chapter, we'll provide you with practical guidance and insights for moving forward with these powerful technologies in your own projects and research.

Chapter 13
Recap of Key Concepts

This chapter serves as both a celebration of your learning journey and a compass for the road ahead. We'll explore where deep learning is heading, discover practical strategies for staying current with rapid developments in Keras and AI research, and develop skills for following cutting-edge research. Most importantly, you'll gain the mindset and tools needed to transform from a student of machine learning into a practitioner who can adapt and grow alongside this quickly evolving field.

This recap marks the end of an extensive learning effort, where you've transitioned from a foundational understanding of machine learning and Keras to possessing the practical expertise required to construct a wide array of AI systems. This includes everything from straightforward classifiers to highly advanced generative AI (GenAI) models capable of creating detailed images from text descriptions. This profound transformation signifies more than simply acquiring technical skills. It represents the development of a unique problem-solving mindset. This mindset seamlessly integrates the rigorous logic of mathematics with practical, real-world intuition. You now grasp the intricacies of gradient descent and its role in optimizing complex functions, comprehend how neural networks discern intricate patterns within data, and understand the revolutionary impact of modern architectures such as transformers on the field of AI.

The landscape of deep learning is in a state of continuous and rapid evolution. New architectural paradigms are constantly emerging, novel research techniques are proposed on a weekly basis through countless papers, and the perceived boundaries of what's achievable with AI are constantly expanding. This concluding chapter not only celebrates the significant knowledge you've acquired but also provides a strategic road map for your continued growth and engagement within this exceptionally dynamic field. Our aim is to illuminate the future trajectory of deep learning, equip you with strategies to remain current with rapid advancements, and, most critically, help you sustain the powerful learning momentum you've cultivated throughout this book. The overarching goal is to ensure that the knowledge gained here serves as a robust foundation for lifelong learning (an area of research in its own right), rather than merely a final destination in your educational pursuits.

13.1 Future Trends in Deep Learning

The core mathematical foundations explored within this book remain the bedrock for the most state-of-the-art advancements in AI. Gradient descent, initially introduced as a fundamental optimization technique, has undergone significant evolution, giving rise to highly sophisticated variants that now power the world's largest language models and most advanced GenAI systems. Modern optimizers, such as AdamW, and advanced techniques such as gradient clipping, are direct descendants of these fundamental principles that you mastered in the earlier stages of your learning journey. These evolutions allow for more stable and efficient training of increasingly complex neural networks. If you're interested in this core machine learning research area, one important aspect to explore further is the concept of *momentum*, which forms the basis of most advanced algorithms.

From a more high-level perspective, *Neural Architecture Search* (NAS) stands out as one of the most exciting and transformative developments in this domain. Historically, designing effective neural network architectures (e.g., the number of layers and the interconnections between them) required extensive human expertise and iterative experimentation. With NAS, we can now leverage machine learning itself to automatically discover optimal network designs. These sophisticated systems apply the very same optimization principles you've learned, but at a higher, meta-level, where the optimization target shifts from merely adjusting network weights to optimizing the entire structure and connectivity of the networks. This automates a previously labor-intensive and often intuitive design process. The inherent flexibility of the Keras Functional API has become even more indispensable as network architectures continue to grow in complexity. Contemporary AI systems frequently combine multiple distinct types of networks, are designed to process diverse data *modalities* simultaneously (e.g., text and images), and often employ *dynamic routing* mechanisms between different components. The composable and modular nature of Keras makes it an exceptionally ideal platform for the rapid prototyping and iterative development of these increasingly sophisticated and multifaceted designs.

On the more practical side, transfer learning has advanced significantly beyond its earlier, simpler forms, which often involved merely taking a pretrained convolutional neural network (CNN) and attaching new output layers. Modern *foundation models* represent a paradigm shift. These models are trained on colossal and extraordinarily diverse datasets, after which they are meticulously fine-tuned for a vast array of specific tasks. This fundamentally alters our approach to many machine learning problems. Instead of embarking on the resource-intensive process of training models from scratch for each new task, we're increasingly building upon the rich, learned representations and generalized knowledge embedded within these massive, pretrained systems. This dramatically reduces training time and data requirements for new applications.

This *democratization* of advanced AI capabilities is accelerating at an unprecedented pace. Techniques that just a few years ago demanded substantial computational resources and specialized expertise are now readily accessible through user-friendly, high-level APIs and scalable cloud services. This overarching trend amplifies the value of the foundational understanding you've developed throughout this book, as it empowers you not only to effectively use these powerful tools but also to critically comprehend their inherent strengths and limitations.

> **Bridging Research and Implementation**
>
> One extremely useful resource for keeping up with the latest and greatest research in this area is the Hugging Face Trending Papers page (*https://huggingface.co/papers/trending*). This site has revolutionized research tracking by combining publications with working implementations, creating a direct bridge between theory and practice. Instead of spending weeks deciphering papers, you can immediately experiment with new techniques and evaluate their relevance to your work. The platform's benchmarks provide clear comparisons between methods, helping you identify genuine advances versus incremental improvements. Take a look at the site if you haven't already.

In the following sections, we'll see how these broad architecture changes are leading to improvements in different areas of deep learning both in terms of the core models and those related to reinforcement learning.

13.1.1 Advanced Architecture Trajectories

Transformers have fundamentally redefined our understanding of the capabilities of neural networks. While we initially explored their application primarily within the realm of natural language processing (NLP), their profound influence has extended far beyond text. *Vision Transformers* (*ViT*), for instance, process images by treating them as sequences of distinct patches, applying the very same attention mechanisms that were so instrumental to the success of transformers in language tasks. This successful cross-pollination of architectural concepts between seemingly different domains represents an extremely important trend in modern AI development, indicating a convergence of ideas across different modalities.

As you might be well aware if you keep track of AI news, the scaling of transformer models continues to unveil surprising and emergent capabilities. As these models grow progressively larger and are trained on increasingly diverse and expansive datasets, they begin to exhibit behaviors that weren't explicitly programmed or anticipated. Large language models (LLMs) can now adeptly perform tasks for which they were never specifically trained, demonstrating a remarkable form of few-shot learning that bears a striking resemblance to human cognitive flexibility and generalization abilities. This emergent intelligence is a hallmark of truly large-scale models. This is becoming increasingly evident as multimodal transformers push the boundaries of this architectural evolution. These sophisticated systems are designed to simultaneously process and learn from various data types, including text, images, audio, and more. Think of these as combining transformers with vision transformers. By integrating these diverse streams, they learn rich, unified representations that capture intricate relationships and dependencies across different modalities. The same attention mechanisms that enabled models to understand the nuances of language now empower them to reason effectively about the complex connections between visual information and textual descriptions.

On the implementation side, significant efficiency improvements in transformer architectures are actively addressing one of the primary limitations of these computationally intensive models. Sparse attention mechanisms, for example, dramatically reduce the computational complexity associated with attention from quadratic to near-linear with respect to sequence length. This critical advancement makes it feasible to process much longer sequences of data while largely maintaining the expressiveness and powerful representational capabilities of full attention mechanisms, thus enabling new applications previously constrained by computational cost.

CNNs, far from being replaced, continue to evolve in tandem with transformers. Modern CNN architectures are now incorporating valuable lessons and inductive biases learned from transformer designs, while conversely, transformers are adopting useful elements from convolutional approaches. This reciprocal cross-fertilization of ideas is leading to the development of powerful *hybrid architectures* that ingeniously combine the complementary strengths of both paradigms, yielding superior performance across a range of tasks. Some of these CNN-based models are powering the development of *3D understanding capabilities*. These extend the scope of computer vision beyond static 2D images. Modern architectures are now proficient at processing dynamic video sequences, understanding complex temporal relationships within them, and even generating 3D content. These advanced capabilities are built upon the same fundamental convolutional and attention mechanisms you've learned, but are applied to higher-dimensional, *spatio-temporal data*, pushing the boundaries of visual AI.

13.1.2 Reinforcement Learning Frontiers

In recent times, reinforcement learning has undergone a significant maturation, transitioning from a predominantly academic pursuit to a robust and practical tool for solving complex real-world problems. The powerful integration of deep neural networks with reinforcement learning algorithms has unlocked a vast array of applications in diverse fields such as robotics, advanced game playing, optimized resource allocation, and highly sophisticated autonomous systems. These contemporary advances build directly upon the foundational reinforcement learning concepts we explored, extending them to effectively handle the immense complexities inherent in sequential decision-making environments. For instance, *multi-agent reinforcement learning* represents a particularly exciting and rapidly developing area. Instead of training individual agents in isolation, these systems learn to coordinate, cooperate, or even compete with multiple intelligent entities within a shared environment. This collaborative or competitive learning approach directly mirrors many real-world scenarios where optimal decisions must account for the dynamic actions and reactions of other independent agents, leading to more realistic and robust solutions.

A slightly less mainstream but very useful alternative direction is *offline reinforcement learning*, which directly addresses one of the major practical limitations of traditional reinforcement learning approaches: extensive and costly interaction with the environment. Unlike online methods, offline reinforcement learning techniques can learn highly effective policies directly from static datasets of previously collected experience, without any further interaction with the environment. This crucial advancement makes reinforcement learning broadly applicable to domains where active exploration is prohibitively expensive, potentially dangerous, or simply impractical, opening up new avenues for reinforcement learning deployment.

Hierarchical reinforcement learning takes an alternative approach and tackles the formidable challenge of learning highly complex and extended behaviors by intelligently decomposing them into a structured set of simpler, more manageable subtasks. These advanced approaches operate at multiple time scales, enabling agents to concurrently develop low-level, precise motor skills, as well as high-level, abstract strategic reasoning, facilitating the learning of intricate and long-horizon tasks. Reinforcement learning is one area in which there is a lot to be done. *Reward designing* remains a most important aspect of any system relying on reinforcement learning. Automating rewards is the holy grail for these systems but remains a very challenging and active area of research. Concepts such as *intrinsic curiosity* aim to come up with solutions to the problem of relying on reward from the environment. These are extremely interesting areas of research you are urged to explore if you're interested in the concept of reinforcement learning.

13.1.3 Generative AI Revolution

While reinforcement learning shows high potential, the public attention seems to be more focused on the domain of GenAI. With the emergence of systems capable of creating remarkably realistic images, generating coherent and contextually relevant text, and even writing functional code, GenAI seems to be the most hyped area in machine learning these days. These extraordinary capabilities are built upon the fundamental autoencoder and Generative Adversarial Network (GAN) architectures that we explored earlier, but are significantly extended and enhanced through more sophisticated architectures and advanced training techniques. *Text-to-image* and more recently *text-to-video* generation represents one of the most visible and impactful applications of modern GenAI. These innovative systems seamlessly combine the advanced language understanding capabilities of transformers with the powerful image-generation abilities of diffusion models. The synergistic result is the creation of systems that can translate simple, descriptive text prompts into detailed, creatively rich, and contextually appropriate images or videos, blurring the lines between language and visual art.

The ambitious extension of generative capabilities to diverse modalities such as video, 3D models, and even functional code showcases the remarkable generality and adaptability of these underlying generative approaches. The same fundamental principles and core architectures that enable image or text generation can be applied across these distinct domains, with appropriate modifications and adaptations to account for the specific characteristics and complexities of each data type.

Controllable generation has become an increasingly critical aspect as GenAI systems transition from experimental curiosities to practical applications. Rather than producing random or unguided outputs, modern generative systems now offer fine-grained control over various attributes such as style, specific content elements, and other semantic properties. It also allows us to modify parts of the whole instead of creating complete artifacts from scratch each time. We saw a sample of this in the inpainting example in the last chapter. This enhanced *controllability* transforms GenAI into a powerful and practical tool for professional creative workflows, moving beyond mere experimental exploration.

One particularly exciting area gaining momentum on this platform involves a fundamental shift in how LLMs generate text. While traditional LLMs produce text one word at a time in sequence—much like writing a sentence from left to right without the ability to go back and edit—researchers are now exploring *diffusion-based text generation*. These experimental systems apply the same principles that made diffusion models so successful for image generation to the challenge of creating coherent text. The potential advantages are compelling. Diffusion approaches can generate text significantly faster than traditional autoregressive methods, and they solve a persistent problem with current LLMs: Once an incorrect word is generated early in a sequence, the model can't revisit and correct that mistake. Diffusion-based text generation allows for a more flexible, iterative refinement process where the entire output can be improved simultaneously rather than being locked into early decisions. Early results suggest this approach could reshape how we think about language generation entirely. These systems can refine and improve text globally, potentially leading to more coherent and higher-quality outputs. This represents exactly the kind of fundamental architectural shift worth monitoring closely as it moves from experimental research to practical implementation.

Keeping up with this fast-paced domain is quite difficult though, so we'll share some tips on how to follow the most important concepts without getting burned out.

13.2 Tips for Staying Updated with Advancements

Maintaining proficiency and staying current in the rapidly evolving field of deep learning requires continuous effort and a strategic approach. We'll provide some tips for this in the following sections.

13.2.1 Technical Skills Maintenance

Keras itself is a framework undergoing rapid evolution, with each new version introducing capabilities that would have been considered unstable or leading-edge just months prior. Staying abreast of these developments requires a systematic approach to ongoing learning and hands-on experimentation. Crucially, the framework's strong commitment to backwards compatibility ensures that your existing knowledge and learned practices remain valuable, while new features seamlessly extend and enhance your capabilities. For instance, the multi-backend architecture of Keras 3 signifies a fundamental paradigm shift in how deep learning frameworks are organized and used. Your Keras models can now execute seamlessly on various backends, including TensorFlow, JAX, or PyTorch, often with minimal or no code modifications. When JAX was introduced, support for this backend was added quite quickly to Keras. It's expected that any future backend that gains popularity will be supported by Keras as well. This remarkable flexibility safeguards your investment in learning Keras while simultaneously granting you access to the latest optimizations, performance benefits, and unique features across different computational ecosystems, providing unparalleled adaptability.

New *layer types* are regularly introduced in Keras releases, frequently implementing the very latest published research findings. These additions mean you can readily experiment with cutting-edge techniques and novel architectures without the difficult task of implementing them from scratch. However, a deep understanding of the underlying theoretical principles and mathematical foundations becomes even more critical for effectively using and configuring these advanced tools and layers.

> **Keeping Up with Keras**
> One of the greatest machine learning resources is the Keras official documentation. For instance, you can keep track of all the latest research in layer types simply by following the changes on the **Keras layers API** page at *https://keras.io/api/layers/*. I keep going back to this page regularly and reading through the different links in the sidebar to keep track of the advancements in core deep learning architectures.

Reproducibility has become an increasingly paramount concern as deep learning experiments grow in complexity and computational expense. Modern machine learning workflows strongly emphasize rigorous version control, not just for code but also for datasets, trained model weights, and specific experimental configurations. Dedicated tools such as *MLflow*, *Weights & Biases*, and *TensorBoard* have evolved significantly to support increasingly sophisticated experiment tracking, comprehensive logging, and systematic comparison across multiple runs.

Lastly, *containerization* of machine learning workflows has dramatically simplified the process of sharing and reproducing experiments across diverse computing environments. A practical understanding of these *Machine Learning Operations (MLOps)* technologies such as *Docker* and similar containerization platforms has become an essential skill for serious practitioners in the field. These tools guarantee that your deep learning experiments can run consistently and predictably, whether on your personal laptop, a cloud instance, or a production deployment server.

13.2.2 Following Tutorials and Keras Codebase

Adding to the earlier suggestion about the official Keras documentation, it's important to note that this documentation has evolved far beyond a simple collection of API references. The examples repository on GitHub (*https://github.com/keras-team/keras/tree/master/examples*) contains meticulously implemented versions of recent research papers, often accompanied by detailed explanations of the underlying concepts and practical implementation nuances. These examples serve as excellent starting points for your own experimental endeavors and provide crucial insights into best practices for implementing advanced deep learning techniques.

While it might seem mundane, release notes for Keras updates often contain valuable insights into the design philosophy and rationale behind new features. See an example at *https://github.com/keras-team/keras/releases*. Understanding why certain design decisions were made helps you use new capabilities more effectively and even anticipate future developments in the framework. The Keras development team typically provides thoughtful and clear explanations of their architectural choices and the specific problems they aim to solve with new releases. While you're on the GitHub repository, exploring the Keras source code provides a level of deep understanding that no amount of documentation alone can convey. The actual implementation details of layers, optimizers, and training loops reveal the practical considerations, engineering trade-offs, and subtle nuances that fundamentally shape how these complex systems function. This intimate knowledge proves invaluable when you need to extend, customize, or modify existing Keras functionality for your highly specific use cases.

An easier route to gain a foothold in the community would be to contribute to Keras documentation and examples. This is an excellent way to solidify your own understanding while simultaneously supporting the broader community. Even small contributions, such as fixing typos, clarifying ambiguous explanations, or proposing minor improvements, directly help other learners and provide you with invaluable insight into the collaborative development process of a major open source project.

The broader tutorial ecosystem of Keras provides a wide array of perspectives on how to apply Keras to various problems. However, the quality, currency, and accuracy of these external resources can vary significantly. Developing strong skills for evaluating the credibility and usefulness of tutorials becomes crucial for efficient and effective learning. Prioritize tutorials that not only explain what to do but also provide clear explanations of why certain approaches are preferred or work better than others. Creating your own tutorials forces you to rigorously organize and articulate your understanding of complex topics. The act of teaching others, even through written explanations, invariably reveals gaps in your own knowledge and helps to consolidate what you've learned. The process of striving for clear and concise explanations often leads to deeper insights and a more profound understanding of the material you're attempting to explain. This was part of the motivation for writing this book in the first place even though it took a lot more time than contributing to the academic research community, which is my usual target audience.

13.2.3 Research Consumption Strategy

Speaking of research, the sheer volume of machine learning research being published can feel overwhelming and daunting. However, developing an effective research reading strategy makes it a manageable and productive endeavor. Not every paper warrants equal attention, and learning to quickly identify the most relevant, impactful, and trustworthy work saves an enormous amount of time and effort. *Foundational papers* retain their immense value even years after their initial publication. A deep understanding of seminal works, such as the original transformer paper, ResNet, or the Variational Autoencoder (VAE) paper, provides essential historical context and a robust framework for comprehending subsequent developments. These foundational papers often contain profound insights and fundamental ideas that might get overlooked or diluted in later works focusing on incremental improvements. The text of this book was designed in a way that after completing this book, you should be able to easily parse the full text of the original research papers.

An easier read is the *survey papers* and *review articles* that offer excellent high-level overviews of entire research areas. These types of papers meticulously synthesize large bodies of existing work, identify key trends, highlight significant breakthroughs, and often point out promising directions for future research. They serve as highly efficient ways to gain a broad understanding of a complex area without the necessity of reading dozens of individual research papers.

Learning to critically evaluate research claims is becoming increasingly important as the field grows more competitive and results can sometimes be overstated. A solid understanding of experimental methodology, statistical significance, and the crucial distinction between correlation and causation helps you effectively discern genuine, robust advances from incremental improvements that might be presented as major breakthroughs.

> **[+] Red Flags in Low Quality Research**
>
> As an example, consider the reporting of appropriate model evaluation metrics. We discussed some important metrics in Chapter 5. If a research paper doesn't report the appropriate metrics as part of their results, it's probably not worth reading. For instance, if a research paper has an unbalanced dataset and they are only reporting their accuracy and not their precision, recall, or F1 score, that should be a big red flag and is much more common than you would think. Review Chapter 5, Section 5.4.4, for a detailed discussion about this point.

The practice of implementing research directly from papers deepens understanding far beyond the act of merely reading. Even partial or simplified implementations often reveal subtle assumptions, implicit details, and practical challenges that aren't always explicitly clear from the written text. This hands-on, practical approach also helps you to realistically evaluate the practical applicability and true utility of proposed techniques. The tip about the Hugging Faces Trending Papers site shared earlier in the chapter might be useful here. You may also find great support from the community of developers if you know where to look.

13.2.4 Community Engagement

The machine learning community extends far beyond academic researchers, encompassing a diverse array of practitioners, engineers, and enthusiastic hobbyists from varied backgrounds. Actively engaging with this broader community provides diverse perspectives on shared technical challenges and often reveals practical considerations and deployment nuances that academic papers might overlook. Professional conferences offer highly concentrated learning opportunities, although they can be expensive and time-consuming. Fortunately, many major conferences now provide virtual attendance options, and recorded talks offer invaluable access to forefront research presentations and discussions. While the vital networking aspects of in-person conferences are harder to replicate virtually, active participation in online communities can partially fill this gap. Contributing to open source projects provides invaluable practical experience with professional software development workflows while simultaneously supporting the very tools and frameworks you frequently use. Even small contributions, such as bug fixes, documentation improvements, or minor feature enhancements, help you understand how large software projects are organized, maintained, and collaboratively developed. This experience proves invaluable whether you're working on research prototypes or robust production systems.

Knowledge sharing through platforms such as blogging, giving presentations, or teaching others is mutually beneficial. It significantly helps the broader community by disseminating valuable information and also reinforces your own learning. Explaining complex concepts to others forces you to understand them more deeply and often reveals new insights about familiar material. The feedback you receive from your audience also helps identify areas where your understanding could be further strengthened.

An important balance to strike is between perspectives from industry and academic research. This provides a holistic understanding of both theoretical advancements and practical constraints. Industry presentations and publications often focus on real-world deployment challenges, scalability issues, and operational complexities that academic papers might not address. Conversely, academic research explores groundbreaking possibilities that might not yet be practically viable but could become profoundly important in the future. A strong foundation such as that achieved by fully understanding the concepts in this book will help you avoid buzz words and dead ends coming from the industry as well as weed out poor quality research.

13.3 Following the Latest Research

Navigating the state-of-the-art of AI research requires a structured approach to filter the constant influx of new information. The established academic conference system provides a crucial framework for structuring the otherwise chaotic flow of research publications. Understanding the submission and presentation systems of major conferences helps you anticipate when significant announcements and breakthroughs are likely to appear. Conferences such as Neural Information Processing Systems (*NeurIPS*), International Conference on Machine Learning (*ICML*), and International Conference on Learning Representations (*ICLR*) have become the primary global venues for unveiling substantial advances in machine learning.

> **Pro Tip for Getting Ideas**
>
> One pro tip I can share from my own experience is to submit your research ideas to these top conferences even if you don't think they deserve an acceptance. I once submitted a paper to a Tier-1 conference and while it was rejected, I got a six-page review from the field thought leaders. This detailed review gave me food for thought and research ideas I explored for the next two years! The moral of the story is that even a rejection from these top venues is a blessing in disguise.

Specialized workshops and smaller conferences often serve as vital platforms for showcasing emerging research directions before they gain widespread mainstream attention. These more focused venues can provide early insights into nascent trends, allow for more in-depth discussions on specific topics, and often feature a more informal atmosphere conducive to frank discussions about limitations and even unsuccessful approaches.

Preprint servers such as *arXiv* have fundamentally transformed how research is disseminated. Papers now appear months before formal peer review and publication, enabling much faster communication of new ideas and findings. However, this speed comes with the trade-off of reduced quality control and peer validation, making highly developed critical evaluation skills even more paramount. More specialized preprint servers such as *BioRXiv* are also a great source for learning about the state-of-the-art in specific areas. Such cross-disciplinary influences are increasingly shaping the direction of machine learning research. Techniques and insights from fields such as physics, biology, neuroscience, and mathematics continue to inspire novel approaches to complex AI problems. Staying aware of significant developments in these related scientific and engineering fields can provide invaluable foresight into future directions for machine learning research. Let's walk through some pointers about specific areas, which can serve as a starting point in research exploration.

13.3.1 Technical Deep Dives

Federated learning addresses the growing imperative of *privacy* and *data locality* in modern machine learning applications. These cutting-edge techniques enable the training of models on distributed datasets without the need to centralize or aggregate sensitive information. The significant mathematical challenges associated with optimization across *heterogeneous data distributions* make this an exceptionally active area of both theoretical and practical research.

The intersections between *quantum computing* and machine learning remain largely theoretical at present, but they hold the potential for profound implications if quantum hardware continues its rapid improvement. Grasping the basic principles of quantum mechanics and quantum computation helps you critically evaluate claims about quantum advantage and identify potential applications within your own work, preparing you for future paradigms.

An orthogonal area is *neuromorphic computing* that draws direct inspiration from the highly efficient structure and function of biological neural networks to create more energy-efficient hardware specifically tailored for AI applications. These biologically inspired approaches could dramatically reduce the energy consumption of large-scale AI systems, making powerful models accessible in resource-constrained environments such as edge devices.

Edge AI focuses on bringing AI capabilities closer to the data sources, directly on devices at the edge of the network. This proximity significantly reduces latency and bandwidth requirements while simultaneously improving data privacy. The inherent constraints of edge devices (limited power, memory, and compute) drive intense research into model compression techniques, highly efficient architectures, and novel training methods that maintain accuracy while drastically reducing computational demands.

Finally, something much closer to my own interest is *interpretability* in AI. Modern AI systems exhibit remarkable capabilities—they can reason about complex problems, write poetry, solve mathematical equations, and engage in nuanced conversations—yet we have surprisingly little insight into the internal mechanisms that produce these behaviors. It's as if we've built incredibly sophisticated minds but can only observe their thoughts from the outside, watching their responses without understanding the *cognitive processes* that generate them. Researchers are now developing techniques to peer inside these digital black boxes, using methods such as *attention visualization*, *neuron activation analysis*, and *causal interventions* to map how information flows through the network's layers. Perhaps most intriguingly, these investigations are revealing *emergent properties* that nobody anticipated during training—capabilities that seem to spontaneously arise from the complex interactions between millions of parameters, much like consciousness might emerge from the intricate dance of biological neurons. Understanding these phenomena is becoming essential for building more reliable, controllable, and innovative AI systems. By learning how current models develop their remarkable abilities, we can design architectures that amplify these strengths while addressing their limitations, potentially leading to entirely new forms of machine intelligence that we haven't yet imagined that can be of practical use in many more areas.

13.3.2 Practical Research Integration

Moving from research prototypes to production systems involves navigating numerous complex practical challenges that research papers rarely explicitly address. Considerations such as *system scalability*, *robustness requirements*, stringent *latency targets*, and *integration constraints* often necessitate significant modifications and adaptations to published research techniques, requiring substantial engineering effort.

A/B testing frameworks for machine learning systems are essential for rigorously evaluating whether research improvements genuinely translate into meaningful business outcomes or tangible user benefits in real-world scenarios. These methodologies demand careful experimental design to account for the inherent complexities of real-world deployments, where multiple interacting factors can simultaneously influence observed outcomes.

> **A/B Testing in the Wild**
>
> Sometimes when you're using a chatbot, it will show you two outputs and ask you which one you prefer as the response. This is a type of A/B testing being performed on you and the model to figure out which one works best. You can apply the same principle in practical scenarios to improve your models' output.

Ethical considerations are becoming increasingly paramount as AI systems permeate more aspects of human life. A deep understanding of issues such as algorithmic bias, fairness, interpretability, and safety necessitates both technical knowledge and a keen awareness of their broader social implications. These critical considerations often influence which research directions are deemed most valuable to pursue and how techniques should be ethically implemented.

Similarly, *risk management* in AI deployment requires not only understanding what potential failures or undesirable behaviors can occur but also developing robust mechanisms for detecting and mitigating problems when they arise. Building resilient monitoring systems, reliable fallback mechanisms, and efficient update pipelines often proves to be more challenging and critical than simply developing the initial AI capability itself.

13.3.3 Parting Words

The extensive knowledge you've gained throughout this book provides a solid foundation for your continued growth and impact in the rapidly evolving field of AI. You now possess a comprehensive understanding of the fundamental mathematical principles that underpin modern AI systems, the practical techniques for implementing these systems using powerful frameworks such as Keras. This powerful combination of theoretical understanding and hands-on practical skills positions you exceptionally well for whatever directions the field may take in the coming years. Your learning doesn't end with the final pages of this book though. Instead, it should transition into a new, more advanced phase. You now have the specialized vocabulary and the essential conceptual framework required to confidently engage with current research, interpret and understand new developments, and actively contribute to the field's ongoing evolution. While the breathtaking pace of change in AI might occasionally seem daunting, your newly acquired foundational knowledge provides a critical sense of stability amidst the constant flux of novel techniques and emerging applications.

As you continue your learning and personal growth, always remember that the most profound and valuable insights often emerge from applying these powerful techniques to problems that genuinely ignite your curiosity and passion. The general principles and methodologies you've learned are remarkably adaptable and can be tailored to countless specific domains and innovative applications. Your curiosity and persistent dedication to working through challenging problems will continue to drive your growth and discovery long after you've completed reading this book.

The very future of AI will be shaped by individuals like you—those who possess a deep understanding of both the immense possibilities and the inherent limitations of these powerful techniques. Use your knowledge wisely and responsibly, continue learning with enthusiastic vigor, and always remember that the most effective and impactful AI systems are built through the collaborative efforts of dedicated teams working together toward shared, ambitious goals.

The Author

Dr. Mohammad Nauman is a seasoned machine learning expert with more than 20 years of teaching experience and a track record of educating more than 40,000 students globally through his online courses on platforms like Udemy and YouTube. He holds a PhD in computer science and was a post-doctoral fellow at the Max Planck Institute for Software Systems in Germany. His groundbreaking post-doc research focused on applying machine learning to advance security and privacy solutions. Dr. Nauman's teaching philosophy—rooted in bridging theory and practice—empowers learners to master tools while building robust foundational skills, whether in academic settings or through his widely accessible digital programs.

Index

1D convolution 380
4D tensor 290

A

Accuracy 166
Accuracy paradox 214
Actions 66
Action-value function 450
Activation function 123, 270
Actor-critic 484
Adagrad 135
Adam 135
Adam optimizer 309
Adversarial 496
Adversarial contest 535
Adversarial relationship 536
Agent 66, 431
Agent architecture 439
AI 33
Aligned 429
AlphaDropout 258
AlphaGo 434
AMD GPU 181
Analogical relationship 383
Anomaly detection 61, 330, 515
Ape-X 485
Architectural pattern 326
Architecture design 205
Association rule learning 64
Asymmetric architecture 504
Attention 375
Attention head 397
Attention matrix 390
Attention mechanism 299, 388, 598
Attention score 388
Audio processing 594
AutoDiff 204
Autoencoder 64, 496
AutoGrad 171, 204
Autonomous vehicle 53
Autoregressive generation 413
Averaging effect 533
Axis parameter 198

B

Backbone 419, 596
Backpropagation 142, 395
Backward pass 92, 281, 333
Bagging 245
Batch gradient descent 101
Batching 207
Batch normalization 358, 485, 579
Bellman equation 451
Bernoulli random variable 530
BERT 419, 595
Beta annealing 524
Bias 85
Bias-variance dilemma 502
Bias-variance trade-off 224, 232
Binary classification 117
Binary cross-entropy 510
Binary IoU 361
Boltzmann exploration 446, 458
Boosting 245
Bottleneck 497
Branching pathway 319
Broadcasting 189

C

Caching 210
Callable 324
Callable layer 288
Callback 168, 170, 216
Capacity 539
Catastrophic forgetting 371, 488
Catastrophic oscillation 483
Categorical cross-entropy 309
Causality 404
Causal language model 418
CausalLM 420
Causal self-attention 404
Center 522
Center cropping 573
Chain rule 204, 267
Channel 208, 280
Channel mixing 285
ChatGPT 375
CHW format 283
CIFAR-10 .... 308, 525
Class-conditional generation 542

621

Class conditioning ... 543
Classification ... 51, 117
   binary ... 57
   multi-class ... 58, 147
   multi-label ... 59, 158
Classification confidence ... 571
Classification label ... 120
Classification metrics ... 205
Classification task ... 119
Classifier-free guidance ... 571
Class selector ... 542
Claude ... 375
Clipping ... 516
Clipping mechanism ... 488
Cluster ... 61
Clustering ... 61
   hierarchical ... 63
Clustering problem ... 63
Co-adaptation ... 243
Co-adapted network ... 244
Co-dependencies ... 243
Colaboratory (colab) ... 74
Compact representation ... 350
Competition ... 496
Compile ... 43
Complementary information ... 330
Composition ... 319
Compressed latent space ... 563
Compressed representation ... 499, 555
Compression ratio ... 508
Computation graph ... 203
Computer vision ... 35, 594
Concatenation ... 338
Confusion matrix ... 212
Constitutional AI ... 489
Content-aware relationship ... 565
Context ... 376, 392
Contextual relationship ... 388
Contour plot ... 102
Contractive autoencoder ... 518
Contractive penalty ... 518
Contrastive loss ... 352
Controlled progression ... 584
Control problem ... 464
Conv1D ... 301
Convergence ... 103
Convex function ... 90
ConvNet (CNN) ... 287
Convolutional neural network ... 287
Convolution layer ... 248
Convolution-normalization-activation ... 359
Convolution operator ... 272

Convolution output ... 273
Co-occurrence ... 383
Correlated experiences ... 482
Cosine schedule ... 583
Cost function ... 87
Covariate shift ... 539
Credit assignment problem ... 65, 432
Cross-attention ... 564
Cross-entropy loss ... 118
Cross-similarity ... 577
Cross-training ... 321
CUDA ... 170
cuDNN ... 170
Curiosity mechanism ... 436
Curriculum learning ... 530
Curse of dimensionality ... 451
Cyclical annealing ... 530

D

Data generation ... 504
Data leakage ... 233
Data point ... 81
Data preparation ... 205
Dataset ... 39, 81
   EuroSAT ... 365
   MNIST ... 35
   Oxford-IIIT Pet dataset ... 354
DDPM ... 559
Decay rate ... 472
Decay schedule ... 417
Decision boundary ... 132
Decision-making ... 439
Decision trees ... 245
Decoder ... 375, 404, 497
Deconvolution ... 301, 358, 528
Deep Convolutional GAN ... 541
Deep learning ... 33, 145
Deeply dense connection ... 322
Deep Q-Network (DQN) ... 434, 449, 473
Deep reinforcement learning ... 434
Delayed feedback ... 432
Denoising ... 554
Denoising autoencoder ... 515
Denoising diffusion probabilistic model ... 559
Dense ... 161, 205
Derivative ... 90, 201
Differentiable ... 128
Diffusion schedule ... 583
Dilution ... 333
Dimensionality reduction ... 64
Dimensions ... 39

Index

Direct Preference Optimization 492
Discounted reward 448
Discount factor 448
Discounting 444
Discrete outcomes 117
Discrimination 300
Discriminative model 495
Discriminator-generator imbalance 551
Discriminator network 535
Distributional DQN 463
Divergence 107, 556
Document clustering 61
Dot product 192, 383
dot product 394
Double DQN 463
Downsampling 289, 334, 338
Dropout 205, 243, 256, 407
Drug discovery 435
Dual-model approach 510
Dueling DQN 463
Dying ReLU 527
Dying ReLU problem 270
Dynamic equilibrium 536
Dynamic programming 433

E

Eager execution 178
EarlyStopping 362
Edge detection 274
EfficientNet 595
Elastic Net 252
Element-wise multiplication 277
ELIZA 36
Embedding network 350
Embedding space 353
Emergent behavior 569
Encoder 375, 497
Encoder-decoder 322
End-to-end learning 142
Energy function 87
Energy management 435
Ensemble 244
Ensemble methods 245, 338
Entropy 559
Environment 66, 431
  continuous 441
  deterministic 440
  discrete 441
  stochastic 440
Environment types 439
Episodic memory 483
Epoch 105

Epsilon decay 470
Epsilon-greedy approach 446
Equilibrium distribution 559
Equilibrium statistical mechanics 559
Error 86
Error propagation 202
Evaluation 44
Evaluation metrics 212
Expansive design 538
Experience 80, 431
Experience replay 462, 478
Exploitation 68, 432, 445
Exploration 68, 432, 445
Exploration-exploitation trade-off 432
Exponential moving averaging 582

F

F1 score 215
Factor parameter 362
False alarm 213
False negative 53, 213
False positive 53, 213
Fashion-MNIST 505
Feature 81
Feature discovery 515
Feature extraction 300
Feature extractor 499
Feature map 280
Feature matching 551
Feature reuse 318
Feature selection 252
Feedback 429
Feedback signal 536
Feed-forward neural network 161
Filter 272
Filter progression 529
Financial markets 435
Fine-tuning 435
Fit 43
Flatten 205
Flattening 307
Forward pass 88, 112, 281, 332
Frame skipping 462
Framing 426
Freezing 368
Frobenius norm 518
FrozenLake 465
Fully connected networks (FCN) 266
Functional API 319, 406
Functional composition 324
Functional programming 319
Function approximators 436

623

G

Gaussian distribution	274
GaussianDropout	259
Gaussian Error Linear Unit (GeLU)	271
Gaussian filter	274
GaussianNoise	259
Gaussian noise	515, 557
Gaussian noise injection	545
Gemini	375
Gemma	423, 595
Generation	330
Generative Adversarial Network (GAN)	533, 535
Generative AI (GenAI)	495
Generative model	495
Generator network	535
Global average pooling	545
Global context	354
Glorot initialization	196
GloVe	383
Google Colab	186
Google Drive	573
GoogLeNet	339
Gradient accumulation	204, 417
Gradient descent	92
step	98
Gradient tape	203, 477
Greedy sampling	420
Grid world	440
Ground truth	81
Gymnasium	464

H

Hardware accelerator	187
HDF5	171, 217
Health care	435
Heterogeneous data	338
Hidden layer	141
Hierarchical feature	335
Hierarchical feature learning	292
High bias	231
High variance	232
Hindsight Experience Replay	483
Hugging Face	602
HWC format	283
Hyperband	262
Hyperparameter	105, 407
Hyperparameter optimization	241
Hyperparameter tuning	504
Hypothesis function	83

I

Identity function	342
Identity shortcut	334
Image denoising	504
Image encoder	331
Image segmentation	354
Image space	534
Image-to-image	602
Image-to-image translation	321
Imbalanced dataset	214
Imminent reward	480
Immutability	319
Implicit regularization	570
Inception module	322, 339
InceptionV3	575
Incompatible shapes	303
Inequality detector	137
Inference	503
Information bottleneck	317
Inpainting	599
Input data	81
Input tensor	303
Instability	531, 556
Intelligence	34
Intermediate representation	64
Internal covariate shift	402, 527
Internal similarity	577
Interpolation	533, 555
Interpretability	253, 458, 515
Interpretable	502
Interpretable machine learning	253
Inverted dropout	247

J

Jacobian matrix	518
JAX	170

K

Kaggle API	424
Kaggle Hub	422, 594
Keras	38, 79
Keras Hub	418, 594
Keras Tuner	171, 260
Kernel	272
Kernel Inception Distance	575
Kernel methods	575
Kernel trick	140
Key	388, 393
KID	575

Index

KL collapse 530
KL divergence 488, 523
Kullback-Leibler 488

L

Labeled 50
Labeled data 240
Labeled example 55
Lambda layer 298
Lasso 252
Latent diffusion 562
Latent space 499
Layer 42, 170
Layer block 322
LayerNormalization 410
Layer normalization 402
Layer sharing 321
Learned encoding 399
Learned loss function 536
Learned noise pattern 526
Learned representation 499
Learned structure 502
Learned transformation 394
Learning experience 51
Learning mechanism 439
Learning rate 109, 453
Learning rate warmup 417
Linearly separable 139
Local details 354
Localized connection 265
Logarithmic spacing 578
Logistic regression 131
Log variance 523
Long-range dependency 388
Lookup table 443
Loss 87
   binary cross-entropy 131
   cross-entropy 118
Loss functions 170
Loss landscape 94
Low-rank adaptation 417

M

Machine learning 33
Machine translation 405
Markov chain 557
Markov Decision Processes 447
Markovian 441
Mask 600
Masking 404

matmul 195
matplotlib 39
Mean squared error 87, 380
Mesh plot 95
Metrics 44, 170
Mini-batch 105
Mini-batch gradient descent 105
Minimax framework 540
Minimax game 536
Misalignment 127
Mixed precision training 417
MMD 530
Mode collapse 551, 553
Mode coverage 537
Model 85
   loading 205
   saving 205
ModelCheckpoint 217, 362
Model parameters 85
Model serialization 168
Model storage 170
Molecular structure 435
Momentum 136
Moving target 482
Multi-agent system 445
Multi-channel tensor 283
Multi-head attention 375, 397, 409
Multi-input fusion 319
Multi-label classification 118
Multimodal learning 315
Multimodal systems 594
Multiple inputs and outputs 318
Multi-scale features 338
Multitask learning 321
Mutually exclusive classes 120

N

Nadam 135
Nearest neighbor interpolation 356
Network traffic analysis 62
Neural network 33, 34
   artificial 70
Neuron 69
Noise estimate 558
Noise schedule 557
Nonlinear topology 322
Nonlinear transformation 139
Normalization 205
NumPy 39
NVIDIA GPU 178

Index

O

Observation 438
Off-policy algorithm 456
One-hot encoding 153, 301, 382
One-hot vector 153
One-vs-all 148
On-policy learning 489
OpenAI Gym 464
Operation recording 204
Optimizer 114
Optimizers 170
Ordering 153
Orthogonal 382
Oscillation 556
Outlier 63
Out-of-distribution 490
Outpainting 602
Output data 81
Output layer 141
Output tensor 303
Overfitting 44, 223, 232, 240

P

Padding 276
    valid 288
Parameterized 87
Partial derivative 96
Patience 362
Pattern completion 503
Pattern-matching 36
Performance measure 80
Piecewise function 129
PIL 275
Policy 66, 442
    deterministic 442
    stochastic 442
Polynomial features 224
Pooling
    average 293
    axis-specific 298
    global 295
    max 293
    min 293
Pool size 294
Positional encoding 375, 393
Positional signature 408
Position encoding 578
Positive predictive value 214
Potential negative outcomes 480
Precision 214

Precision medicine 61
Prediction 44, 82
Prefetching 210, 357
Prenormalization 417
Preprocessor 419
Pretrained GloVe embedding 384
Pretrained model 173, 316
Principal Component Analysis (PCA) 64, 385
Prioritized experience replay 463, 483, 486
Prior knowledge 233
Probability cloud 520
Probability distribution 520
Probability map 361
Prompt engineering 425
Proximal Policy Optimization 488
pyenv 177
Python Imaging Library 275
PyTorch 144, 170

Q

Q-learning 433, 449
Quantitative feedback 443
Query 388, 393
Query image 511
Q-value update 472

R

R2D2 486
Random forests 245
Random guess 54
Random noise 231
Random seed 110
Rank 190
Recall 214
Recommendation system 61, 320
Reconstructed 497
Reconstruction failure 530
Reconstruction loss 501, 523
Rectified Linear Unit (ReLU) 163, 270, 509
Recurrent neural network 388
reduce_max 298
ReduceLROnPlateau 362
Regression 58, 81
    polynomial 225
Regret 447
Regularization 224, 407
Regularization term 514
    L1 252
regularization term 234
    L2 234

Regularizer 297
Reinforcement learning 34, 429
Reinforcement learning from human
    feedback 434, 487
Reinforcement learning loop 438
Reinforcement learning optimization 487
Reinforcement learning with AI feedback 492
Relative positional encoding 417
Reparameterization trick 520
Replay buffer 462, 475
Representation learning 565
Reproducible experiments 111
Residual connection 322, 332, 403
Residual Network (ResNet) 341
ResNet 336
Resource management 436
Reverse mode differentiation 204
Reward 66, 429
    cumulative 439
    intermediate 439
Reward clipping 462
Reward engineering 444
Reward function 448
Reward hacking 444
Reward model training 487
Reward signals 436
Reward Structure 439
Risk-averse 480
RMSprop 135
Robotics 435
Rule 35

S

Sampling 520
Sampling problem 555
Scaled dot-product attention 396
Scatterplot 85
Schedule
    cosine 558
    linear 558
scikit-learn 171
Segmentation 61
Self-attention 389
Self-driving car 33
Self-gating behavior 509
Self-normalizing neural networks 258
SELU 258
Semantic binding 569
Semantic essence 565
Semantic image search 511
Semantic loophole 490

Semantic space 384
Semi-supervised learning 34
Sensitivity 215
Sentiment analysis 330
Sequence data 375
Sequence encoder 331
Sequence modeling 377
Sequential 205
Sequential dependency 377
Sequential model 161
Shallow network 336
Shape 41
Shaped reward 444
Shape mismatch 111, 303, 345
Shared-trunk 546
Sharing weights 265
Shuffling 356
Siamese network 321
Sigmoid 122
Signal rate 586
Similarity 62
Similarity detection 349
Sinusoidal embedding 578
Sinusoidal encoding 399
Sinusoidal positional encoding 561
Skip connection 318, 322, 403, 504
sklearn 39, 226
Sliding window 378
Sobel operator 274
Softmax 42, 309
Softmax function 155
Softmax layer 155
Spam filtering 52
Sparse 555
Sparse and delayed reward 482
Sparse autoencoder 514
Sparse categorical cross-entropy 159
Sparse connectivity 281
Sparse model 252
Sparse reward 438
Sparse reward problem 444
SpatialDropout 259
Spatial locality 292
Spatial relationship 292
Spatial relationships 508
Specificity 384
Spectral normalization 551
Speech recognition 330
Spread 522
Squishing 123
Squishing function 123
Stability 462, 484, 504

Stable Diffusion	554
Stable oscillations	541
State	431
State explosion	433
State space explosion	437
State-value function	450
Stochastic gradient descent	104
Stochasticity	532
Stochastic sampling	570
Stride	280
Strided convolutions	295
Supervised fine-tuning	487
Supervised learning	34
SwiGLU	509
Swish	579
Symbolic computation	168, 200
Symmetric architecture	503

T

Target network	462
Task	80
Task-specific head	420
Temperature parameter	446
Temperature scaling	420
Temporal difference	483
Temporal diversity	571
Temporal gap	437
Tensor	197, 282
TensorFlow	42, 144
constants	202
datasets	355
variables	202
Tensor Processing Unit (TPU)	187
Test set	82, 229, 242
Text conditioning	564, 600
Text encoder	564
Text-to-image	565
Text-to-image generator	594
Theano	169
Threshold	124
Time reversal symmetry	559
Time series	298
Time series data	377
Time-step conditioning	561
Token	419
Tokenizer	419, 595
Top-k sampling	420
Top-p (nucleus) sampling	420
torchaudio	171
torchvision	171
Training collapse	545

Training instability	551
Training set	82, 206, 242
Transfer learning	364, 575, 592
Transformed feature	141
Transformer	279
Transformer layer	248
Transformer model	322
Transition dynamics	451
Transition function	448
Translation invariance	291, 545
Transportation	435
Transpose convolution	528
True negative	213
True positive	213
True positive rate	215
t-SNE	385
Tuple	42
Two-dimensional convolutional layer	279
Type I error	213
Type II error	213

U

Uncertainty	496
Underfitting	223, 231, 240
Unfreezing	368
Uniform distribution	526
Unlabeled	50
Unlabeled data	56
Unsupervised learning	34, 59
Update signal	268
Upper Confidence Bound	446
Upsampling	354, 504, 581

V

VAE	520
Validation set	206, 242
Value	393
Value function	436, 449
Vanilla autoencoder	514
Vanishing gradient	268, 316
Variational autoencoder	520
Vectorization	191

W

Wasserstein	530
Weight	146
Weight decay	571
Weight sharing	351
Word embeddings	382

X

XOR 136

Z

Zero initialization 581
Zero-sum game 540

- Your practical introduction to programming neural networks
- Develop and train simple and multi-layer networks with Python
- Learn about algorithms, activation functions, transformers, and more

Joachim Steinwendner, Roland Schwaiger

Programming Neural Networks with Python

Neural networks are at the heart of AI—so ensure you're on the cutting edge with this guide! For true beginners, get a crash course in Python and the mathematical concepts you'll need to understand and create neural networks. Or jump right into programming your first neural network, from implementing the scikit-learn library to using the perceptron learning algorithm. Learn how to train your neural network, measure errors, make use of transfer learning, implement the CRISP-DM model, and more. Whether you're interested in machine learning, gen AI, LLMs, deep learning, or all of the above, this is the AI book you need!

457 pages, pub. 05/2025
E-Book: $54.99 | **Print:** $59.95 | **Bundle:** $69.99

www.rheinwerk-computing.com/6059

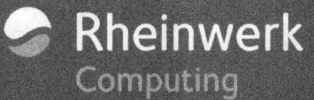

- The complete Python 3 handbook
- Learn basic Python principles and work with functions, methods, data types, and more
- Walk through GUIs, network programming, debugging, optimization, and other advanced topics

Johannes Ernesti, Peter Kaiser

Johannes Ernesti, Peter Kaiser

The Comprehensive Guide

Ready to master Python? Learn to write effective code, whether you're a beginner or a professional programmer. Review core Python concepts, including functions, modularization, and object orientation and walk through the available data types. Then dive into more advanced topics, such as using Django and working with GUIs. With plenty of code examples throughout, this hands-on reference guide has everything you need to become proficient in Python!

1036 pages, pub. 09/2022
E-Book: $54.99 | **Print:** $59.95 | **Bundle:** $69.99

www.rheinwerk-computing.com/5566